D1674265

Edited by
David Farrusseng

Metal-Organic Frameworks

Related Titles

Tremel, W., zur Loye, H. (eds.)

Handbook of Solid State Chemistry

Hardcover
ISBN: 978-3-527-32587-0

Schubert, U., Hüsing, N.

Synthesis of Inorganic Materials

Softcover
ISBN: 978-3-527-32714-0

MacGillivray, L.

Metal-Organic Frameworks

Design and Application

Hardcover
ISBN: 978-0-470-19556-7

Stolten, D. (ed.)

Hydrogen and Fuel Cells

Fundamentals, Technologies and Applications

2010
Hardcover
ISBN: 978-3-527-32711-9

Hirscher, M. (ed.)

Handbook of Hydrogen Storage

New Materials for Future Energy Storage

2010
Hardcover
ISBN: 978-3-527-32273-2

Züttel, A., Borgschulte, A., Schlapbach, L. (eds.)

Hydrogen as a Future Energy Carrier

2008
Hardcover
ISBN: 978-3-527-30817-0

Öchsner, A., Murch, G. E., de Lemos, M. J. S. (eds.)

Cellular and Porous Materials

Thermal Properties Simulation and Prediction

2008
Hardcover
ISBN: 978-3-527-31938-1

Edited by
David Farrusseng

Metal-Organic Frameworks

Applications from Catalysis to Gas Storage

WILEY-VCH

WILEY-VCH Verlag GmbH & Co. KGaA

The Editor

Dr. David Farrusseng
University Lyon 1, CNRS
IRCELYON
2, Av. Albert Einstein
69626 Villeurbanne
France

Cover
The structures on the front covers are based on material supplied by the editor David Farrusseng and images from chapter 1 (authored by Satoshi Horike and Susumu Kitagawa).

All books published by **Wiley-VCH** are carefully produced. Nevertheless, authors, editors, and publisher do not warrant the information contained in these books, including this book, to be free of errors. Readers are advised to keep in mind that statements, data, illustrations, procedural details or other items may inadvertently be inaccurate.

Library of Congress Card No.: applied for

British Library Cataloguing-in-Publication Data
A catalogue record for this book is available from the British Library.

Bibliographic information published by the Deutsche Nationalbibliothek
The Deutsche Nationalbibliothek lists this publication in the Deutsche Nationalbibliografie; detailed bibliographic data are available on the Internet at http://dnb.d-nb.de.

© 2011 Wiley-VCH Verlag & Co. KGaA, Boschstr. 12, 69469 Weinheim, Germany

All rights reserved (including those of translation into other languages). No part of this book may be reproduced in any form – by photoprinting, microfilm, or any other means – nor transmitted or translated into a machine language without written permission from the publishers. Registered names, trademarks, etc. used in this book, even when not specifically marked as such, are not to be considered unprotected by law.

Cover Design   Grafik-Design Schulz, Fußgönheim
Typesetting   Thomson Digital, Noida, India
Printing and Bookbinding Fabulous Printers Pte Ltd, Singapore

Printed in Singapore
Printed on acid-free paper

Print ISBN: 978-3-527-32870-3
ePDF ISBN: 978-3-527-63587-0
ePub ISBN: 978-3-527-63586-3
Mobi ISBN: 978-3-527-63588-7
oBook ISBN: 978-3-527-63585-6

Contents

Preface *XV*
List of Contributors *XIX*

Part One Design of Multifunctional Porous MOFs *1*

1 Design of Porous Coordination Polymers/Metal–Organic Frameworks: Past, Present and Future *3*
Satoshi Horike and Susumu Kitagawa
1.1   Introduction *3*
1.2   Background and Ongoing Chemistry of Porous Coordination Polymers *3*
1.2.1 Frameworks with High Surface Area *5*
1.2.2 Lewis Acidic Frameworks *6*
1.2.3 Soft Porous Crystals *8*
1.3   Multifunctional Frameworks *10*
1.3.1 Porosity and Magnetism *10*
1.3.2 Porosity and Conductivity/Dielectricity *12*
1.3.3 Porous Flexibility and Catalysis *12*
1.4   Preparation of Multifunctional Frameworks *13*
1.4.1 Mixed Ligands and Mixed Metals *13*
1.4.2 Core–Shell *16*
1.4.3 PCPs and Nanoparticles *17*
1.5   Perspectives *18*
       References *19*

2 Design of Functional Metal–Organic Frameworks by Post-Synthetic Modification *23*
David Farrusseng, Jérôme Canivet, and Alessandra Quadrelli
2.1   Building a MOFs Toolbox by Post-Synthetic Modification *23*
2.1.1 Taking Advantage of Immobilization in a Porous Solid *23*
2.1.2 Unique Reactivity of MOFs *24*
2.2   Post-Functionalization of MOFs by Host–Guest Interactions *26*

2.2.1	Guest Absorption 26
2.2.2	Nanoparticle Encapsulation 27
2.3	Post-Functionalization of MOFs Based on Coordination Chemistry 28
2.3.1	Coordination to Unsaturated Metal Centers 28
2.3.2	Coordination to Organic Linkers 30
2.4	Post-Functionalization of MOFs by Covalent Bonds 31
2.4.1	Chemical Modification by Amide Coupling 32
2.4.2	Chemical Modification by Imine Condensation 33
2.4.3	Chemical Modification by Click Chemistry 34
2.4.4	Reactivity of Bridging Hydroxyl Groups 38
2.5	Tandem Post-Modification for the Immobilization of Organometallic Catalysts 39
2.6	Critical Assessment 41
2.6.1	Synthetic Restrictions 43
2.6.2	Balance Between Functionalization Rate and Material Efficacy 43
2.6.3	Characterization of the Functionalized Materials 44
2.7	Conclusion 45
	References 45

Part Two Gas Storage and Separation Applications 49

3 Thermodynamic Methods for Prediction of Gas Separation in Flexible Frameworks 51
François-Xavier Coudert

3.1	Introduction 51
3.1.1	Gas Separation in Metal–Organic Frameworks 51
3.1.2	Dynamic Materials in the MOF Family 52
3.1.3	Possible Applications of Flexible MOFs 54
3.1.4	Need for Theoretical Methods Describing Adsorption and Framework Flexibility 55
3.2	Theoretical Background 56
3.2.1	The Osmotic Ensemble 56
3.2.2	Classical Uses of the Osmotic Ensemble in Molecular Simulation 57
3.3	Molecular Simulation Methods 58
3.3.1	Direct Molecular Simulation of Adsorption in Flexible Porous Solids 58
3.3.2	Use of the Restricted Osmotic Ensemble 60
3.4	Analytical Methods Based on Experimental Data 62
3.4.1	Analytical Methods for Adsorption. Taxonomy of Guest-Induced Flexibility 62
3.4.2	Application to Coadsorption: Selectivity Predictions and Pressure–Composition Phase Diagrams 63

3.5	Outlook 66
	References 67

4	**Separation and Purification of Gases by MOFs** 69
	Elisa Barea, Fabrizio Turra, and Jorge A. Rodriguez Navarro
4.1	Introduction 69
4.2	General Principles of Gas Separation and Purification 72
4.2.1	Some Definitions 72
4.2.2	MOFs: New Opportunities for Separation Processes 73
4.2.3	Mechanisms of Separation and Design of MOFs for Separation Processes 73
4.2.4	Experimental Techniques and Methods to Evaluate/Characterize Porous Adsorbents 77
4.3	MOFs for Separation and Purification Processes 79
4.3.1	MOF Materials as Molecular Sieves 79
4.3.2	Flexible MOFs for Enhanced Adsorption Selectivity 81
4.3.3	MOFs with Coordination Unsaturated Metal Centers for Enhanced Selective Adsorption and Dehydration 86
4.3.4	Hydrocarbon Separation 88
4.3.5	VOC Capture 89
4.3.6	Catalytic Decomposition of Trace Gases 91
4.4	Conclusions and Perspectives 92
	References 92

5	**Opportunities for MOFs in CO_2 Capture from Flue Gases, Natural Gas, and Syngas by Adsorption** 99
	Gerhard D. Pirngruber and Philip L. Llewellyn
5.1	Introduction 99
5.2	General Introduction to Pressure Swing Adsorption 99
5.3	Production of H_2 from Syngas 101
5.3.1	Requirements for CO_2 Adsorbents in H_2-PSAs 103
5.4	CO_2 Removal from Natural Gas 103
5.4.1	Requirements for Adsorbents for CO_2–CH_4 Separation in Natural Gas 104
5.5	Post-combustion CO_2 Capture 105
5.5.1	The State of the Art 105
5.5.2	PSA and VSA Processes in Post-Combustion CO_2 Capture 106
5.5.3	Requirements for Adsorbents for CO_2 Capture in Flue Gases 107
5.6	MOFs 108
5.6.1	Considerations of Large Synthesis and Stability 108
5.6.2	MOFs for H_2-PSA 109
5.6.3	MOFs for CO_2 Removal from Natural Gas 113

5.6.4	MOFs for Post-Combustion CO_2 Capture	113
5.7	Conclusions	116
	References	116

6 Manufacture of MOF Thin Films on Structured Supports for Separation and Catalysis 121
Sonia Aguado and David Farrusseng

6.1	Advantages and Limitations of Membrane Technologies for Gas and Liquid Separation	121
6.2	Mechanism of Mass Transport and Separation	123
6.3	Synthesis of Molecular Sieve Membranes	127
6.3.1	Synthesis of Zeolite Membranes	127
6.3.1.1	Direct Nucleation–Growth on the Support	128
6.3.1.2	Secondary Growth	129
6.3.2	Preparation of MOF Membranes and Films	129
6.3.2.1	Self-Assembled Layers	130
6.3.2.2	Solvothermal Synthesis: Direct and Secondary Growth	131
6.4	Application of MOF Membranes	137
6.4.1	Gas Separation	137
6.4.1.1	Metal Carboxylate-Based Membranes	137
6.4.1.2	Zinc Imidazolate-Based Membranes	138
6.4.2	Shaped Structured Reactors	141
6.4.3	Perspectives for Future Applications	143
6.5	Limitations	143
6.6	Conclusions and Outlook	146
	References	147

7 Research Status of Metal–Organic Frameworks for On-Board Cryo-Adsorptive Hydrogen Storage Applications 151
Anne Dailly

7.1	Introduction – Research Problem and Significance	151
7.1.1	Challenges in Hydrogen Storage Technologies for Hydrogen Fuel Cell Vehicles	151
7.1.2	Current Status of Hydrogen Storage Options and R&D for the Future	152
7.2	MOFs as Adsorptive Hydrogen Storage Options	154
7.3	Experimental Techniques and Methods for Performance and Thermodynamic Assessment of Porous MOFs for Hydrogen Storage	156
7.4	Material Research Results	159
7.4.1	Structure–Hydrogen Storage Properties Correlations	159
7.4.2	Nature of the Adsorbed Hydrogen Phase	162
7.5	From Laboratory-Scale Materials to Engineering	165
7.6	Conclusion	167
	References	168

Part Three Bulk Chemistry Applications   171

8   Separation of Xylene Isomers   173
       Joeri F.M Denayer, Dirk De Vos, and Philibert Leflaive
8.1   Xylene Separation: Industrial Processes, Adsorbents,
        and Separation Principles   173
8.2   Properties of MOFs Versus Zeolites in Xylene Separations   176
8.3   Separation of Xylenes Using MIL-47 and MIL-53   178
8.3.1   Low-Coverage Gas-Phase Adsorption Properties   179
8.3.2   Molecular Packing   180
8.3.3   Separation of Xylene-Mixtures   184
8.4   Conclusions   185
        References   187

9   Metal–Organic Frameworks as Catalysts for Organic Reactions   191
       Lik Hong Wee, Luc Alaerts, Johan A. Martens, and Dirk De Vos
9.1   Introduction   191
9.2   MOFs with Catalytically Active Metal Nodes in the Framework   191
9.2.1   Transition Metal Nodes   192
9.2.2   Coordinatively Unsaturated Metal Nodes   194
9.3   Catalytic Functionalization of Organic Framework Linkers   195
9.3.1   Porphyrin Functional Groups   195
9.3.2   Amine and Amide Functions Incorporated via Grafting   196
9.4   Homochiral MOFs   198
9.4.1   MOFs with Intrinsic Chirality   198
9.4.2   Chiral Organic Catalytic Functions   199
9.4.3   Metalloligands   200
9.5   MOF-Encapsulated Catalytically Active Guests   201
9.5.1   Polyoxometalates (POMs)   201
9.5.2   Metalloporphyrins   203
9.5.3   Metal Nanoparticles   204
9.6   Mesoporous MOFs   206
9.7   Conclusions   209
        References   210

Part Four Medical Applications   213

10   Biomedical Applications of Metal–Organic Frameworks   215
       *Patricia Horcajada, Christian Serre, Alistair C. McKinlay,
       and Russell E. Morris*
10.1   Introduction   215
10.2   MOFs for Bioapplications   216
10.2.1 Choosing the Right Composition   216
10.2.2 The Role of Flexibility   217
10.2.3 The Role of Functionalization   219

10.2.4	Biodegradability and Toxicity of MOFs 219
10.3	Therapeutics 221
10.3.1	Drug Delivery 221
10.3.2	BioMOFs: the Use of Active Linkers 227
10.3.3	Release of Nitric Oxide 228
10.3.4	Activity Tests 231
10.3.4.1	Activity of Drug-Containing MOFs 231
10.3.4.2	Activity of NO-Loaded Samples 233
10.3.4.3	Activity of Silver Coordination Polymers 234
10.4	Diagnostics 235
10.4.1	Magnetic Resonance Imaging 235
10.4.2	Optical Imaging 236
10.5	From Synthesis of Nanoparticles to Surface Modification and Shaping 237
10.5.1	Synthesis of Nanoparticles 237
10.5.2	Surface Engineering 239
10.5.3	Shaping 239
10.6	Discussion and Conclusion 242
	References 244

11 Metal–Organic Frameworks for Biomedical Imaging 251
Joseph Della Rocca and Wenbin Lin

11.1	Introduction 251
11.2	Gadolinium Carboxylate NMOFs 253
11.3	Manganese Carboxylate NMOFs 257
11.4	Iron Carboxylate NMOFs: the MIL Family 258
11.5	Iodinated NMOFs: CT Contrast Agents 260
11.6	Lanthanide Nucleotide NMOFs 262
11.7	Guest Encapsulation within NMOFs 263
11.8	Conclusion 264
	References 264

Part Five Physical Applications 267

12 Luminescent Metal–Organic Frameworks 269
John J. Perry IV, Christina A. Bauer, and Mark D. Allendorf

12.1	Introduction 269
12.2	Luminescence Theory 270
12.2.1	Photoluminescence 270
12.2.2	Fluorescence Quenching 273
12.2.3	Energy Transfer 273
12.3	Ligand-Based Luminescence 274
12.3.1	Solid-State Luminescence of Organic Molecules 274
12.3.2	Ligand-Based Luminescence in MOFs 275
12.3.3	Ligand-to-Metal Charge Transfer in MOFs 280

12.3.4	Metal-to-Ligand Charge Transfer in MOFs	*281*
12.4	Metal-Based Luminescence	*282*
12.4.1	Metal Luminophores	*282*
12.4.2	Lanthanide Luminescence and the Antenna Effect	*282*
12.4.3	Examples of Metal-Based Luminescence	*282*
12.4.3.1	Metal-Centered Luminescence	*282*
12.4.3.2	Metal-to-Metal Charge Transfer (MMCT)	*286*
12.4.4	Lanthanide Luminescence as a Probe of the Metal-Ligand Coordination Sphere	*287*
12.5	Guest-Induced Luminescence	*287*
12.5.1	Encapsulation of Luminophores	*288*
12.5.2	Guest-Induced Charge Transfer: Excimers and Exciplexes	*290*
12.5.3	Encapsulation of Lanthanide Ion Luminophores	*291*
12.6	Applications of Luminescent MOFs	*293*
12.6.1	Chemical Sensors	*293*
12.6.1.1	Small-Molecule and Ion Sensors	*294*
12.6.1.2	Oxygen Sensors	*296*
12.6.1.3	Detection of Explosives	*297*
12.6.2	Radiation Detection	*298*
12.6.3	Solid-State Lighting	*298*
12.6.4	Nonlinear Optics	*300*
12.6.5	Barcode Labeling	*300*
12.7	Conclusion	*301*
	References	*302*

13 Deposition of Thin Films for Sensor Applications *309*
Mark Allendorf, Angélique Bétard, and Roland A. Fischer

13.1	Introduction	*309*
13.2	Literature Survey	*310*
13.3	Signal Transduction Modes	*310*
13.4	Considerations in Selecting MOFs for Sensing Applications	*312*
13.4.1	Pore Dimensions	*312*
13.4.2	Adsorption Thermodynamics	*313*
13.4.3	Film Attachment	*315*
13.4.4	Film Thickness and Morphology	*318*
13.4.5	Response Time	*319*
13.4.6	Mechanical Properties	*320*
13.5	MOF Thin Film Growth: Methods, Mechanisms, and Limitations	*320*
13.5.1	Growth From Aged Solvothermal Mother Solutions	*320*
13.5.2	Assembly of Preformed MOF (Nano-) Particles or Layers	*323*
13.5.3	Electrochemical Deposition	*325*

13.5.4	Liquid-Phase Epitaxy 325
13.5.5	Toward Heteroepitaxial Growth of Multiple MOF Layers 328
13.5.6	Growth of MOF Films in Confined Spaces 329
13.5.7	Comparison of the Different Methods for MOF Thin Film Growth 331
13.6	Conclusions and Perspectives 331
	References 332

Part Six   Large-Scale Synthesis and Shaping of MOFs 337

14   Industrial MOF Synthesis 339
Alexander Czaja, Emi Leung, Natalia Trukhan, and Ulrich Müller

14.1	Introduction 339
14.2	Raw Materials 340
14.2.1	Metal Sources 340
14.2.2	Linkers 340
14.3	Synthesis 343
14.3.1	Hydrothermal Synthesis 344
14.3.2	Electrochemical Synthesis 345
14.4	Shaping 347
14.5	Applications 349
14.5.1	Natural Gas Storage for Automobile Applications 349
14.5.2	Ethylene Adsorption for Food Storage 350
14.6	Conclusion and Outlook 351
	References 352

15   MOF Shaping and Immobilization 353
Bertram Böhringer, Roland Fischer, Martin R. Lohe, Marcus Rose, Stefan Kaskel, and Pia Küsgens

15.1	Introduction 353
15.2	MOF@Fiber Composite Materials 354
15.2.1	MOF-Containing Paper Sheets 354
15.2.2	MOF@Pulp Fibers 355
15.2.3	Electrospinning of MOF@Polymer Composite Fibers 356
15.2.4	MOF Fixation in Textile Structures 359
15.2.4.1	Pretreatment 360
15.2.4.2	Wet Particle Insertion 362
15.2.4.3	Dry Particle Insertion 363
15.3	Requirements of Adsorbents for Individual Protection 367
15.3.1	Relevant Protective Clothing Applications 367
15.3.2	Filter Performance 368
15.3.3	Testing the Chemical Protection Performance of Filters 371
15.3.4	Concepts for Application 373
15.4	MOFs in Monolithic Structures 373

15.4.1	MOF@Polymeric Beads	*374*
15.4.2	Extruded MOF Bodies	*374*
15.4.3	Monolithic MOF Gels	*375*
	References	*379*

Index *383*

Preface

Are porous metal–organic frameworks (MOFs) breakthrough materials, or are they simply an illusion reminiscent of "The Emperor's New Clothes?"

Over the past three decades, the domain of porous solids has been expanded by the discovery of various "cornerstone" materials, such as ALPO molecular sieves (1982), carbon nanotubes (1991), ordered silica mesoporous materials (1992), and CMK (1999), to name just a few. Porous MOFs were first described in Volume 4 of the *Handbook of Porous Solids*, published by Wiley-VCH in 2002. Since that time, this class of materials has become much better known and much more widely studied. The number of publications dealing with MOFs and porous coordination polymers is currently increasing at an exponential rate – with the total doubling every 2 years. In 2009, we could count about 1200 new publications, a rate similar to that observed for ordered mesoporous materials.

Thanks to their hybrid formulation, MOFs bridge the gap between pure inorganic and organic materials, thereby pushing the frontiers of knowledge ever further forward. Initially, MOFs were regarded only as a new type of molecular sieve material with a pore size between those of inorganic zeolites (<1 nm) and ordered mesoporous silica materials (>2 nm). On the other hand, their stimuli-induced flexibility, or more generally their softness, is a common trait with organic enzymes. It is acknowledged that MOFs could mimic enzymes using the concept of molecular recognition, allowing high chemo-, regio-, and enantioselectivity – the ultimate goal in catalysis. With respect to mechanical properties such as hardness and elasticity, the domain corresponding to MOFs can, to some extent, be considered to straddle the borders between purely organic polymers, purely inorganic ceramics, and metallic materials. Some MOFs possess unique features, such as luminescence, that already allow them to surpass benchmark materials.

The ever-increasing demand to develop more complex and integrated processes drives the research and development of advanced "smart" materials, with specific engineering at the molecular level but also at higher scales from the micron to the millimeter. Clearly, MOFs are promising new candidates for addressing current challenges in a number of domains of application. A few MOF solids have recently become commercially available under the trade name Basolite™ – this should greatly accelerate the development of MOF-based processes. Currently, the emphasis of research and development is shifting towards converting the unique properties of MOFs into efficient processes.

This brings us to the purpose of this book – to perform a critical assessment of the properties of MOFs, taking into account the process specifications and performance targets required to allow these solids to be introduced on to the market. It seems that MOF performances are rarely discussed with respect to those of state-of-the-art materials or commercial targets. Furthermore, their shaping and further processing in physical and chemical processes have rarely been reported so far. The ambitious goal here is to measure the gap that exists between the state of the art of MOF and commercial applications in the domains of energy, chemistry, physics, and medicine.

This deliberately application-oriented book is divided into six parts. Each chapter refers to the original literature and can be read independently of the other chapters.

The first part of this book emphasizes the uniqueness of MOFs compared with other porous solids in terms of intrinsic material properties and engineering capabilities. In particular, MOFs are characterized by their softness and by their associated host–guest dynamic properties that make them "smart" materials. The first chapter establishes the mechanisms and provides an outlook on how to proceed in designing multifunctional MOFs, using techniques for addition or modification of physical or chemical features within the frameworks. The second chapter gives a critical review of post-modification methods with emphasis on catalytic applications.

The second part deals with gas storage and separation. The different types of flexibility and the thermodynamic description of breathing are given in Chapter 3, and the associated solids and applications are detailed in Chapter 4. Carbon dioxide capture is treated in detail for PSA/TSA processes in Chapter 5 and for membrane processes in Chapter 6. The topic of hydrogen storage is discussed in Chapter 7.

The third part deals with bulk chemistry. Chapter 8 deals with the separation of xylenes, and Chapter 9 provides a review of MOF applications in catalysis, with particular focus placed on structure–activity relationships.

The fourth part encompasses an overview of medical applications of MOFs (Chapter 10) and imaging (Chapter 11).

In the fifth part, the use of MOFs in the design of small-scale devices and sensors is discussed. Luminescence properties and possible applications are described in Chapter 12. Thin-film preparations for sensor applications are detailed in Chapter 13.

The sixth part discusses the mass production of MOFs, with attention devoted to economic criteria (Chapter 14), and also the shaping of MOFs as large bodies and their immobilization as composite materials with polymer fibers (Chapter 15).

I hope that the information in this book will be of interest both to researchers involved in the development of chemical and physical processes and to scientists focusing on porous solids. I also hope that it will help establish a common ground between different communities by providing a multidisciplinary point of view, including solid-state chemistry, materials science, and process engineering.

The European Community is acknowledged for supporting R&D in this field through the Integrated Projects NanoMOF and Macademia (FP7-NMP).

David Farrusseng

List of Contributors

Sonia Aguado
Université Lyon 1
IRCELYON
CNRS UMR 5256
2 avenue Albert Einstein
69626 Villeurbanne
France

Luc Alaerts
Katholieke Universiteit Leuven
Centre for Surface Chemistry
and Catalysis
Kasteelpark Arenberg 23
3001 Leuven
Belgium

Mark D. Allendorf
Sandia National Laboratories
Department of Energy Nanomaterials
7011 East Avenue
Livermore, CA 94550
USA

Elisa Barea
Universidad de Granada
Facultad de Ciencias
Departamento de Química Inorgánica
Av. Fuentenueva S/N
18071 Granada
Spain

Christina A. Bauer
University of California, Los Angeles
Department of Chemistry and
Biochemistry
607 Charles E. Young Drive East
Los Angeles, CA 90095
USA

Angélique Bétard
Ruhr-Universität Bochum
Anorganische Chemie II –
Organometallics & Materials
Universitätsstrasse 150
44801 Bochum
Germany

Bertram Böhringer
Blücher GmbH
Mettmannerstrasse 25
40699 Erkrath
Germany

Jérôme Canivet
Université Lyon 1
IRCELYON
CNRS UMR 5256
2 avenue Albert Einstein
69626 Villeurbanne
France

List of Contributors

François-Xavier Coudert
Chimie ParisTech
11 rue Pierre et Marie Curie
75005 Paris
France

Alexander Czaja
BASF SE
GCC/PZ – CNSI
570 Westwood Plaza
Los Angeles, CA 90095
USA

Anne Dailly
General Motors Company
R&D Technical Center
Hydrogen Fuel Chemistry and Systems
30500 Mount Road
Warren, MI 48090
USA

Joseph Della Rocca
University of North Carolina
at Chapel Hill
School of Pharmacy
Department of Chemistry
125 South Road
Chapel Hill, NC 27599
USA

Joeri F.M. Denayer
Vrije Universiteit Brussel
Department of Chemical Engineering
Pleinlaan 2
1050 Brussels
Belgium

Dirk De Vos
Katholieke Universiteit Leuven
Centre for Surface
Chemistry and Catalysis
Kasteelpark Arenberg 23
3001 Leuven
Belgium

David Farrusseng
Université Lyon 1
IRCELYON
CNRS UMR 5256
2 avenue Albert Einstein
69626 Villeurbanne
France

Roland Fischer
Norafin GmbH
Gewerbegebiet Nord 3
09456 Mildenau
Germany

Roland A. Fischer
Ruhr-Universität Bochum
Anorganische Chemie II –
Organometallics & Materials
Universitätsstrasse 150
44801 Bochum
Germany

Patricia Horcajada
Université de Versailles St.-Quentin
en Yvelines
Institut Lavoisier
UMR CNRS 8180
45 Avenue des Etats-Unis
78035 Versailles
France

Satoshi Horike
Kyoto University
Graduate School of Engineering
Department of Synthetic Chemistry
and Biological Chemistry
Kyoto Daigaku Katsura
Nishikyo-ku
615-8510 Kyoto
Japan

List of Contributors

Stefan Kaskel
Technische Universität Dresden
Institut für Anorganische Chemie
Bergstrasse 66
01069 Dresden
Germany

Susumu Kitagawa
Kyoto University
Graduate School of Engineering
Department of Synthetic Chemistry
and Biological Chemistry
Kyoto Daigaku Katsura
Nishikyo-ku
615-8510 Kyoto
Japan

Pia Küsgens
Technische Universität Dresden
Institut für Anorganische Chemie
Bergstrasse 66
01069 Dresden
Germany

Philibert Leflaive
IFP-Lyon
Separation Department
Rond-point de l'échangeur de Solaize
69360 Solaize
France

Emi Leung
BASF SE
GCC/PZ – CNSI
570 Westwood Plaza
Los Angeles, CA 90095
USA

Wenbin Lin
University of North Carolina
at Chapel Hill
School of Pharmacy
Department of Chemistry
125 South Road
Chapel Hill, NC 27599
USA

Philip L. Llewellyn
Universités Aix-Marseille I, II,
et III – CNRS
Laboratoire Chimie Provence
(UMR 6264)
Centre de Saint Jérôme
Avenue Escadrille Normandie-Niemen
13397 Marseille
France

Martin R. Lohe
Technische Universität Dresden
Institut für Anorganische Chemie
Bergstrasse 66
01069 Dresden
Germany

Johan A. Martens
Katholieke Universiteit Leuven
Centre for Surface Chemistry
and Catalysis
Kasteelpark Arenberg 23
3001 Leuven
Belgium

Alistair C. McKinlay
University of St. Andrews
EaStChem School of Chemistry
Purdie Building
North Haugh
St. Andrews KY16 9ST
UK

Russell E. Morris
University of St. Andrews
EaStChem School of Chemistry
Purdie Building
North Haugh
St. Andrews KY16 9ST
UK

Ulrich Müller
BASF SE
GCC/PZ – CNSI
570 Westwood Plaza
Los Angeles, CA 90095
USA

John J. Perry IV
Sandia National Laboratories
Department of Energy Nanomaterials
7011 East Avenue
Livermore, CA 94550
USA

Gerhard D. Pirngruber
IFP Energies Nouvelles
Rond-point de l'échangeur de Solaize
69360 Solaize
France

Alessandra Quadrelli
Université de Lyon
ESCPE Lyon
CNRS UMR 9986
43 boulevard du 11 Novembre 1918
69616 Villeurbanne
France

Jorge A. Rodriguez Navarro
Universidad de Granada
Facultad de Ciencias
Departamento de Química Inorgánica
Av. Fuentenueva S/N
18071 Granada
Spain

Marcus Rose
Technische Universität Dresden
Institut für Anorganische Chemie
Bergstrasse 66
01069 Dresden
Germany

Christian Serre
Université de Versailles St.-Quentin en Yvelines
Institut Lavoisier
UMR CNRS 8180
45 Avenue des Etats-Unis
78035 Versailles
France

Natalia Trukhan
BASF SE
GCC/PZ – CNSI
570 Westwood Plaza
Los Angeles, CA 90095
USA

Fabrizio Turra
SIAD SpA
Stabilimento di Osio Sopra (BG)
SS 525 del Brembo no 1
24040 Osio Sopra, BG
Italy

Lik Hong Wee
Katholieke Universiteit Leuven
Centre for Surface Chemistry and Catalysis
Kasteelpark Arenberg 23
3001 Leuven
Belgium

Part One
Design of Multifunctional Porous MOFs

1
Design of Porous Coordination Polymers/Metal–Organic Frameworks: Past, Present and Future

Satoshi Horike and Susumu Kitagawa

1.1
Introduction

At the end of the 1990s, a new porous compound with an inorganic–organic hybrid framework had an impact on the field of porous materials and represented a new family for porous chemistry. Porous coordination polymers (PCPs), also known as metal–organic frameworks (MOFs), have regular pores ranging from micro- to mesopores, resulting in a large pore surface area, and a highly designable framework, pore shape, pore size, and surface functionality. Their structures are based on organic ligands as linkers and metal centers as the connectors. The rich functionality and designability of the organic ligands and the directability and physical properties of the metal ions are fascinating for the design of various functions, not only conventional adsorptive functions such as storage, separation, and catalysis, but also other physical/chemical functions that can be integrated in the frameworks. Whereas the components of PCPs are connected by coordination bonds and other weak interactions or noncovalent bonds (H-bonds, π-electron stacking, or van der Waals interactions), the interactions lead to structural flexibility and dynamics in the crystalline state, which also promotes the unique character of PCPs in the field of porous materials. As synthetic techniques and knowledge have increased in the last decade, we are now ready to design advanced porous functions by making full use of the chemical components and structural topologies. In this chapter, we introduce the background of PCPs/MOFs with some of the main framework designs and describe the unconventional porous properties of multifunctional porous materials based on ligand–metal networks.

1.2
Background and Ongoing Chemistry of Porous Coordination Polymers

Coordination polymers (CPs) are a family of compounds with extended structures formed by metal ions and organic and/or inorganic ligands with coordination bonds.

Metal-Organic Frameworks: Applications from Catalysis to Gas Storage, First Edition. Edited by David Farrusseng.
© 2011 Wiley-VCH Verlag GmbH & Co. KGaA. Published 2011 by Wiley-VCH Verlag GmbH & Co. KGaA.

They can provide various frameworks constructed from one-, two-, and three-dimensional networks. The late transition metal elements (Cu, Ag, Zn, and Cd) tend to provide this type of framework and the chemistry of CPs has been elucidated with the development of single-crystal X-ray crystallography. The term "coordination polymer" was used in a paper in 1916 [1], but there was no means of demonstrating infinite frameworks without single-crystal X-ray crystallography. A three-dimensional coordination framework connected by a CN bridge was realized in 1936 [2], namely the well-known Prussian Blue compounds. Currently, coordination polymers having porous properties are termed PCPs or porous MOFs, and therefore we suggest "coordination framework" as an all-inclusive term because the chemistry of the background is defined as "chemistry of coordination space." To understand the background of this chemistry, there are three important concepts: (1) framework, (2) molecular metal–organic hybrid, and (3) porosity.

1) Concept of Framework

   It is well known that CPs provide us with one-, two-, and three-dimensional motifs. In particular, the structural concept of a framework was demonstrated by Hofmann and Küspert [3], whose compounds are known as the family of Hofmann compounds having a two-dimensional layer-based architecture. The first X-ray crystallographic structure was obtained in 1949 [4]. The complete three-dimensional framework, the so-called Prussian Blue complex, appeared in 1936 and a comprehensive study was performed by Iwamoto et al. in 1967 [2, 5].

2) Molecular Metal–Organic Hybrid

   Hofmann and Prussian Blue compounds have structures bridged by the inorganic ion CN^-, and therefore have a restricted variety of structures. On the other hand, frameworks having organic linkers afford not only designability but also functionality of frameworks. The X-ray crystal structure of the metal–organic coordination framework of $[Cu(adiponitrile)_2]\cdot NO_3$ appeared in 1959 [6]. Since then, many compounds in this category have been synthesized and characterized crystallographically. Yaghi et al. termed these compounds "metal–organic frameworks (MOFs)" in 1995 [7]. $[Cu(adiponitrile)_2]\cdot NO_3$ contains the NO_3^- anion in the voids. Such compounds are regarded as clathrate-type CPs, however, which are not categorized as "porous" compounds. By the late 1990s, many clathrate-type CPs/MOFs had been synthesized.

3) Porosity

   Porosity means "the quality or state of being a porous entity, which has many small holes that allow water, air, and so on, to pass through." The porosity is antithesis to Aristotle's proposition, "Nature abhors a vacuum." Indeed, closely packed solid structures formed by molecules and ions can easily form. Researchers have often misunderstood that the crystallographic structure of MOFs having guest species in their voids is a porous material. In 1997, "porosity" was demonstrated to give a compound that maintains a porous structure without guests in the pores; gas sorption experiments under ambient conditions were carried out for stable apohosts [8, 9]. Reversible gas storage properties were identified and the PCPs have attracted wide attention as new porous materials.

Since that point, the number of reports on PCPs has been increasing rapidly, and many researchers have been developing strategies for the design of porosity, some of which are intrinsically unique to PCP materials.

1.2.1
Frameworks with High Surface Area

One of the great advantages of PCPs/MOFs is their high surface area, attributable to the low density of the porous structure. An MOF composed of Zn_4O clusters connected by benzenedicarboxylate (bdc), $[Zn_4O(bdc)_3]$ (MOF-5), was synthesized in 1999 and possesses a cubic structure with an ordered three-dimensional (3D) porous system (Figure 1.1a) [10]. This compound has a BET surface area of $3800 \, m^2 \, g^{-1}$ [11]. Many porous compounds have been synthesized on the basis of this structural motif, and this approach has been intensively developed to design important porous frameworks. Some related frameworks, $[Zn_4O(btb)_2]$ (MOF-177) and $[Zn_4O(bbc)_2]$ (MOF-200) {btb = 1,3,5-benzenetribenzoate; bbc = 4,4′,4″-[benzene-1,3,5-triyltris(benzene-4,1-diyl)]tribenzoate)} also possess high porosity; the reported BET surface areas for these compounds are 4746 and $6260 \, m^2 \, g^{-1}$, respectively [12, 13]. The self-assembly process of structure growth often faces network interpenetration, which precludes a high surface area, but further

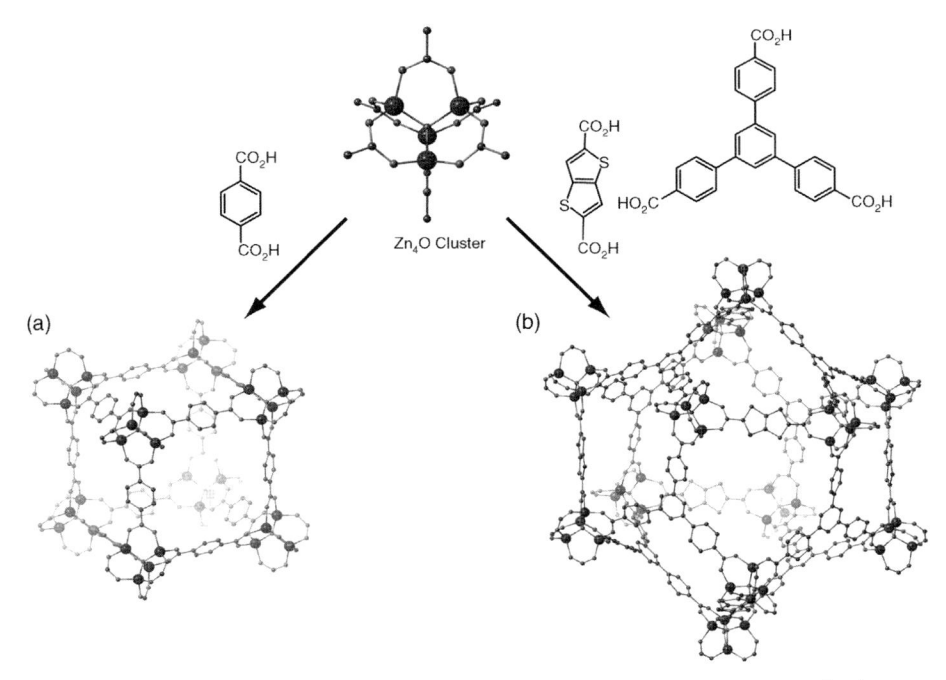

Figure 1.1 Partial crystal structures of (a) $[Zn_4O(bdc)_3]$ (MOF-5, BET surface area = $3800 \, m^2 \, g^{-1}$) and (b) $Zn_4O(t_2dc)(btb)_{4/3}$ (UMCM-2, t_2dc = thieno-3,2-bithiophene-2,5-dicarboxylate, BET surface area = $5200 \, m^2 \, g^{-1}$) constructed from Zn_4O clusters.

improvements in the design of pore network topologies could avoid interpenetration to achieve extremely high surface areas.

Porous frameworks constructed from two or more kinds of ligands are in some cases effective in the design of high surface area compounds. $Zn_4O(t_2dc)(btb)_{4/3}$ (UMCM-2) (t_2dc = thieno-3,2-bithiophene-2,5-dicarboxylate) (Figure 1.1b) is also made up of Zn_4O clusters and two distinct ligands contribute to the construction of the porous framework [14]. There is a narrow distribution of micropores at 1.4–1.6 and 1.6–1.8 nm and a mesopore at 2.4–3.0 nm and the calculated BET surface reaches 5200 $m^2 g^{-1}$.

Another framework, $[Cr_3F(H_2O)O(bdc)_3]$ (MIL-101), is made from the linkage of terephthalate and chromium trimer units that consist of three Cr cations and the μ_3O oxygen anion [15]. The pore space is constructed from two cages with diameters of 2.9 and 3.4 nm which are connected with windows with diameters of 1.2 and 1.45 nm, respectively. The compound has a BET surface area of 4100 $m^2 g^{-1}$ and, compared with the Zn_4O-type metal cluster, the framework is more stable against water and other chemical species and it has also been utilized as a porous matrix for post-synthesis or hybridization with other species such as metal particles [16].

A paddle-wheel-type dimetal cluster is a popular building unit to construct frameworks. Many transition metals can form this type of cluster and it affords square grid extended networks. $[Cu(H_2O)]_3(ntei)$ (PCN-66) is prepared by the combination of 4,4′,4″-nitrilotris(benzene-4,1-diyl)tris(ethyne-2,1-diyl)triisophthalate (ntei) and a Cu^{2+} paddle-wheel cluster and the BET surface area is 4000 $m^2 g^{-1}$ [17]. Isostructures have been made using other hexatopic carboxylate ligands and it is anticipated that even higher surface areas can be designed.

So far, these compounds represent carbon-containing materials with one of the highest surface areas and the feature of complete crystallinity is a significant platform for a high capacity of gas uptake and it also acts as accumulation areas for other materials such as metal particles, functional molecules and polymers, and gases with high density.

1.2.2
Lewis Acidic Frameworks

The design of porous frameworks having guest interaction sites has also been intensively investigated. Especially unsaturated metal sites on the pore interior, which act as Lewis acid sites, have been synthesized because of interest in the storage of gases such as H_2 and CO_2 and for heterogeneous catalysis.

$[Cu_3(btc)_2]$ (HKUST-1), based on Cu_2 paddle-wheel units linked by benzenetricarboxylic acid (btc) is one of the early PCPs with unsaturated metal sites [18]. This compound possesses a 3D channel with a pore size of 1 nm and has high thermal stability and aqueous durability. The axial sites of Cu^{2+} are accessible to guests and gas capture and heterogeneous catalysis have been reported [19, 20]. This motif is available for other metal ions such as W, Fe, and Cr, and $[Cr_3(btc)_2]$ shows O_2 adsorption at 298 K with a Type I isotherm with which adsorption occurs at very low

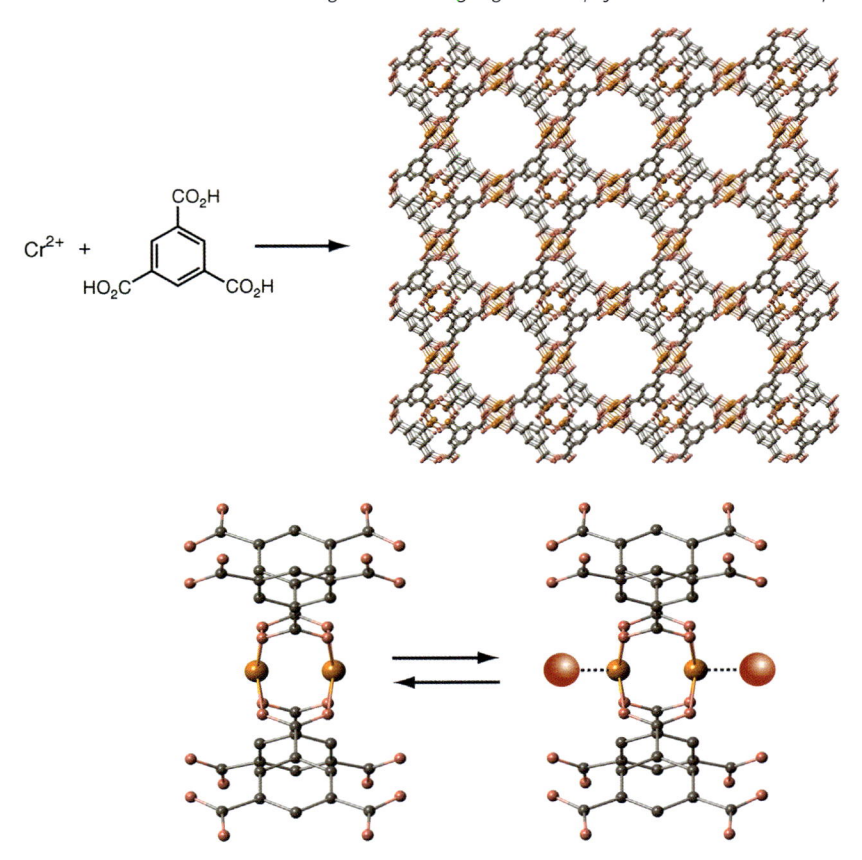

Figure 1.2 3D crystal structure of [Cr$_3$(btc)$_2$] (btc = benzenetricarboxylate) and reversible O$_2$ sorption processes at paddle-wheel Cr dimer unit. Red spheres represent O$_2$ molecules.

pressures [21–23]. In the case, the redox active Cr centers bind O$_2$ molecules (Figure 1.2) to show reversible chemisorption behavior with negligible N$_2$ uptake under the same conditions.

In M$_2$(dhtp) (H$_4$dhtp = 2,5-dihydroxyterephthalic acid; M = Mg, Mn, Co, Ni, Zn), the framework has hexagonal one-dimensional (1D) channels and a high concentration of unsaturated metal sites [24, 25]. One of this series of frameworks, Mg$_2$(dhtp), shows a large CO$_2$ uptake at 298 K and 1 atm (35.2 wt% of CO$_2$) because of the light weight of the framework skeleton and the strong interaction of CO$_2$ and unsaturated metal sites [26]. This framework is relatively stable towards water and has a strong hydrophilic nature. This compound is applicable not only for the capture of CO$_2$ but also other gases such as NO, and it is also promising as a biocompatible material [27].

The large-pore compound [Cr$_3$F(H$_2$O)O(bdc)$_3$] (MIL-101) mentioned above also has unsaturated Cr^{3+} centers in the mesoporous cages and works as a catalyst [28, 29]. The large pores therein are also advantageous for a high rate of substrate diffusion in heterogeneous catalytic reactions.

These compounds produce guest-accessible unsaturated metal centers after proper evacuation (or activation) of coordinating guest molecules and the metal ions simultaneously working as nodes of the frameworks. Further, an unsaturated metal center can also incorporated in the organic linker [30]. The compound [Zn_2(bpdc)$_2$L] {bpdc = 4,4'-biphenyldicarboxylate; L = (R,R)-(2)-1,2-cyclohexanediamino-N,N'-bis[3-*tert*-butyl-5-(4-pyridyl)salicylidene]}, containing chiral salen units, has a grid-type porous framework in which the metal centers of the salen ligand are exposed to the surface of channels [31, 32]. This compound acts as an asymmetric heterogeneous catalyst for alkene epoxidation.

It has been difficult to design very strong Lewis acid sites in the PCP framework via a self-assembly process; however, the high designability of PCPs provides a guideline for multifunctional catalysts. For example, bifunctional-type catalysts such as with acid–base properties for domino reactions and ultra-hydrophobic Lewis acid catalysts are significant targets. On the other hand, incorporation of unsaturated metal centers in porous frameworks can be achieved by applying "post-synthesis" or "grafting" procedures. This is a powerful approach to incorporating functional groups in porous frameworks for catalysis and other functions, and the detailed strategy is described in Chapter 2.

1.2.3
Soft Porous Crystals

In contrast to the robust porous framework, there are studies in which the soft properties of PCPs is evident in terms of structural flexibility and dynamic properties. We can identify compounds showing guest accommodation with reversible nonporous to porous transformations, even if the compound cannot maintain the porous structure in the guest removal process. These compounds are categorized as "soft porous crystals," and are studied owing to their unique properties [33–35]. The adsorption isotherms of this type of compound sometimes cannot be classified according to the conventional IUPAC classification because of the dynamic guest accommodation behavior [36]. For instance, a "gate-type" sorption profile (Figure 1.3) shows no uptake at low concentrations of the guest molecules, and an abrupt increase in adsorption after a threshold concentration. This behavior is associated with a structural transformation from a nonporous to a porous phase and an example is O_2 capture by hemoglobin, which shows gate-opening type sensing and capture of O_2 when the concentration of O_2 reaches a certain level [37]. The softness of the compounds is very sensitive for gas species to be accommodated and often useful for gas separation.

A two-dimensional (2D) layer structure is a typical motif of soft porous crystals showing gate-type sorption behavior because it is readily transformable in response to the guest accommodation [38]. In particular, an interdigitated layer structure is important because we can control interactions between the sheets by modification of the groups to be interdigitated. Several compounds have been reported and each compound shows a characteristic transformation via guest incorporation [39].

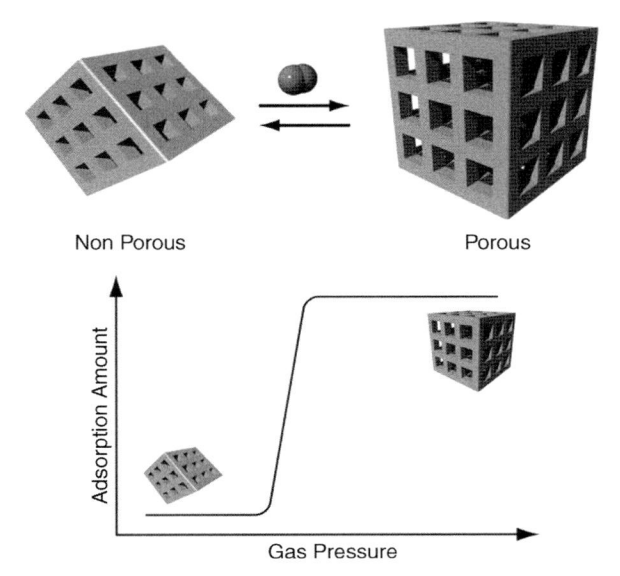

1.2 Background and Ongoing Chemistry of Porous Coordination Polymers | 9

Figure 1.3 Schematic illustration of structure transformation of soft porous crystal. Below is a typical gas adsorption isotherm with gate-opening behavior.

The series of [Zn(dicarboxylate)(bpy)] (CID-1) compounds (CID = coordination polymers with an interdigitated structure; bpy = 4,4′-bipyridyl) provides us with a platform of interdigitated 2D layers [40, 41]. The 2D layer in this series is composed of a V-shaped dicarboxylate, bpy, and dinuclear metal units, and the various dicarboxylate ligands impact on the interaction with the next layers and influences the gate-type sorption behavior. Their flexibility in the layers assists the separation of CO_2 from CH_4–CO_2 gas mixtures and the separately adsorbed CO_2 gas can easily be retrieved with low energy consumption, because we use only structural flexibility to separate the CO_2 without strong interactions. The phenomenon is different from conventional CO_2 separation with strong binding energy.

Even for a 3D porous framework with coordination bonds, some PCPs demonstrates their intrinsic flexibility. [Al(bdc)(OH)] (MIL-53) is a framework containing diamond-shaped 1D channels [42]. One axis is connected by Al–OH–Al chains and the angle of the Al–O–Al bond changes with guest sorption. The softness is derived from the reorientation of the coordination bonds and even with small changes in bond angle/distance, the overall porous structure transforms dramatically. The guest-dependent flexibility in [Al(bdc)(OH)] is applied for the selective sorption of xylene isomers and drug delivery, for example [43, 44].

The flexibility of PCP frameworks is observed both in rearrangements of network topologies and local bond reorientation. Regarding network rearrangement, we have another motif; interpenetration often affords soft properties due to the rearrangement of adjacent networks. [Zn$_2$(bdc)$_2$(bpy)] (MOF-508), which has square grid 3D networks with twofold interpenetration, can have both a closed form and an open form with change in the relative position of the two networks [45, 46].

The transformation from closed to open form depends on the accessing guests and the behavior can separate linear and branched isomers of pentane and hexane like a gas chromatographic separation column.

This type of flexibility in the PCP frameworks is often dramatic and sensitive for identifying the guest molecules. Their binding ability for functional molecules is studied for application of drug delivery systems and gas separation on an industrial scale and conditions. Further, the dynamic behavior is also significant for the design of multifunctional materials in the next step; guest-recognizing flexible catalysts, guest-sensing dielectric materials, and guest-dependent actuating systems are the candidates selected for future design. Some of the ideas are outlined below.

1.3
Multifunctional Frameworks

The high designability and the variety of combinations of organic linkers and metal ions suggest the possibility of the creation of multifunctional porous frameworks, in which two or more physical/chemical properties are integrated in the crystal. For example, some PCPs have permanent porosity and guest-responsive magnetic activity in the framework and these are potentially unique for switching of magnetic properties by use of guest storage/release. In the last 5 years, many researches have focused on the design of multifunctional characteristics in PCP structures, and in this section we introduce some important ideas for such materials.

1.3.1
Porosity and Magnetism

PCP magnets incorporating magnetic properties in the framework are unique multifunctional materials, particularly as chemo-responsive materials, because of the mutual interplay of porous functions and magnetic switching. Many reports on porous magnets using a PCP framework have appeared; however, there are few examples that show a combination of porous and magnetic functions, because most porous magnets undergo a magnetic transformation at very low temperature (critical temperature, T_c), whereas the adsorptive functions occur at ambient temperature. On the other hand, spin crossover (SC), in which electron configurations can be switched between high- and low-spin states in response to external stimuli, producing changes in magnetism, color, dielectric properties, and structure, is often observed at ambient temperature and we could design a real interplay of SC phenomena and adsorption properties.

A representative PCP that shows coupling properties of SC behavior is {Fe(pyrazine)[Pt(CN)$_4$]} [47]. The 2D layers are extended by Pt–CN–Fe linkages and are linked by pyrazine to form a pillared layer-type compound. This compound displays a first-order spin transition at ambient temperature [T_c (up) = 285 K and T_c (down) = 309 K] with 25 K wide hysteresis. The guest-free form adsorbs various guest molecules and the spin state changes depending on the guests (Figure 1.4). The

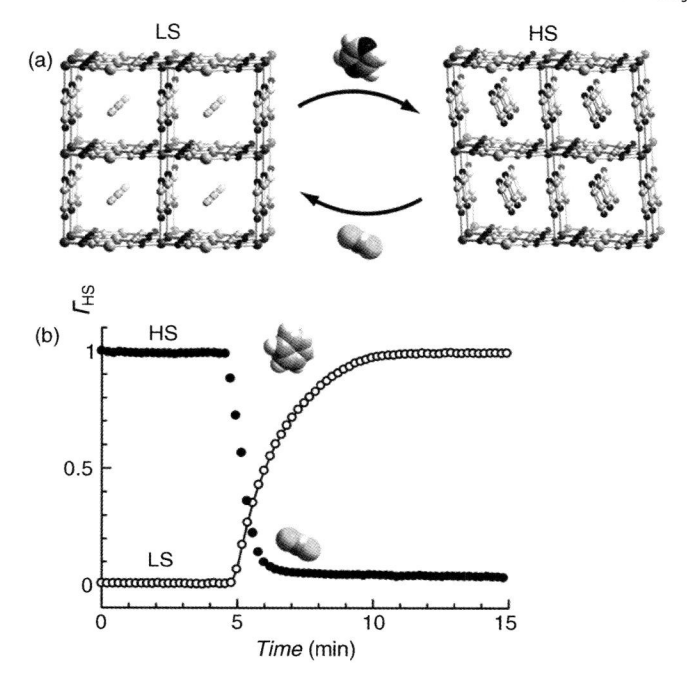

Figure 1.4 (a) Alteration of the porous structure of {Fe(pyrazine)[Pt(CN)$_4$]} by exchange of guest molecules of CS$_2$ and pyrazine. (b) The high-spin (HS) state can be converted to low-spin (LS) state by insertion of CS$_2$ at 298 K. LS to HS conversion can be achieved by accommodation of several guests. Γ_{HS} is the relative existence of the HS state in the system.

reversible conversion of low- and high-spin states is achievable by guest sorption under ambient conditions. The guest-dependent switching of the SC phenomenon is also observed in other compounds [48].

Recently, studies of ion conductivity using a PCP platform by tuning of the pore size and chemical environment have increased. Highly hydrophilic networks with an open structure can store large amounts of H$_2$O molecules resulting from the proton conductivity at high relative humidity [49–52]. Co^{II}[Cr^{III}(CN)$_6$]$_{2/3}$·4.8H$_2$O is a Prussian Blue analog and contains water of crystallization molecules that contribute to the proton conductivity [53]. The temperature–conductivity curve has a flexion point at 313 K that corresponds to the magnetic phase transition temperature. This suggests that there is a coupling effect between magnetostriction and ionic conductivity in the structure and a multifunctional effect between the conductivity and magnetism is expected.

Although there have been many attempts to synthesize frameworks that have both porous and magnetic properties, only a limited number showing "real" interplay have been observed because of the temperature gap issue. One answer is a combination of SCs and we expect that the further design of frameworks, with not only micro-scale but especially also meso-scale integration of porous host and magnetic species (network or molecule), would open new methodologies. Partially related approaches are described below.

1.3.2
Porosity and Conductivity/Dielectricity

Incorporation of electric conductivity in the porous materials is widely regarded as a challenging task. A combination of porosity and electric conductivity is applicable for gas sensors, such as electrodes. Only limited numbers of conductive coordination polymers have been reported and the coexistence of permanent porosity is still rare [54]. The compounds Cu[M(pdt)$_2$] (M = Cu, Ni; pdt = pyrazine-2,3-dithiolate) exhibit electric conductivity [55, 56]. The motif of bisdithiolate complexes is redox-active species, connected by M(pyrazine)$_4$ units to have porous framework with a BET surface area of 385 m^2 g^{-1} (when M = Ni). Even though the framework of Cu[Ni(pdt)$_2$] shows low conductivity (1×10^{-8} S cm^{-1}) at room temperature, it increases to $\sim 1 \times 10^{-4}$ S cm^{-1} on doping with iodine, which act as an oxidant. I$_2$ vapor is introduced into the framework, resulting in [Ni(pdt)$_2$]$^{2-/-}$ oxidation.

Multifunction porosity–dielectric properties have also been studied in recent years. For example, [Mn$_3$(HCOO)$_6$](C$_2$H$_5$OH) having 1D channels occupied by ethanol as guest shows a ferrimagnetic transition at 8.5 K due to the magnetic transition of the Mn^{2+} spin [57]. The dielectric constant of this framework is heavily dependent on the axis of the crystal and the specific axis of the guest ethanol which aligns in parallel has a large constant ($\varepsilon_r = 45$). The temperatures of phase transition and maximum peak dielectric constant are nearly same. Phase transition is attributed to reorientation of the guest molecules and it contributes to the ferroelectricity in the framework. The guest-induced ferroelectricity is unique for porous materials and the interplay between the porosity, magnetism and ferroelectricity, in other words, porous materials with multiferroic behavior, is of great interest for multifunctional porous design [58–61].

1.3.3
Porous Flexibility and Catalysis

As shown above, some PCPs show intrinsic structural flexibility upon guest incorporation and release. This flexibility is one of the unique characteristics among the various porous materials and combination of structural flexibility and catalytic activity is an attractive challenge. As is known, some enzymes show intelligent guest-selective conversion inside the pocket of their structure and the design of such a flexible porous catalyst by use of PCPs is an important aim [62].

There are few examples of PCPs having such catalytic behavior. {[Cd(4-btapa)$_2$(NO$_3$)$_2$]·6H$_2$O·2DMF} 4-btapa = 1,3,5-benzene tricarboxylic acid tris[N-(4-pyridyl)amide] is a flexible PCP having Lewis base catalytic activity [63]. The framework consists of Cd^{2+} and a tridentate pyridyl ligand having amide groups as nodes. The guest-free phase is amorphous and the crystallinity can recover on guest accommodation, especially with small guest molecules such as methanol and ethanol, suggesting their reversible flexibility. The soft compound has amide group-derived Lewis basicity and it promotes Knoevenagel condensation for only small substrates. The compound can activate malononitrile but not ethyl cyanoacetate and cyanoacetic acid *tert*-butyl ester because of the size effect.

1.4
Preparation of Multifunctional Frameworks

Many efforts have been made to create new structures of PCPs with the aid of crystallographic analyses. The number of possible combinations of metal ions and organic ligands is huge and, in principle, the researcher can play the whole field of synthetic chemistry. On the other hand, rather than simple reaction screening of combinations of metal ions and organic ligands, other approaches to designing functional PCPs have been developed in recent years. Some of them are significant for the design of multifunctionality in the framework which is not feasible by simple mixing of metal ions and organic ligands. Especially it has been found that some functions of PCPs originate not only from microscopic structures, but also from more large-scale, so-called meso-domain regions. The functionality of PCPs has been evaluated mainly by the use of bulk solids such as powders and single crystals, and we could expect that the multifunctionality will depend on the method of fabrication. Some of the approaches are introduced below.

1.4.1
Mixed Ligands and Mixed Metals

In principle, PCP frameworks can be constructed from multiple organic ligands and metal ions. However, there are a limited number of reports on frameworks having more than two kinds of ligands or metal ions. Considering that the doping approach has been popular in the area of inorganic materials such as metal oxides and metals, the on-demand doping of metal ions or organic ligands in the PCP frameworks should become another important strategy for multifunctional systems.

Some of the reports about ligand mixing system are outlined here. $[Zn_4O(bdc)_x(abdc)_{3-x}]$ was synthesized, in which the terephthalate linkers are partially substituted by 2-aminobenzene-1,4-dicarboxylate (abdc) [64]. X-ray diffraction revealed the materials to have a random distribution of the two linker molecules according to Vegard's law, as shown schematically in Figure 1.5. The reason for the success of partial ligand doping in $[Zn_4O(bdc)_3]$ is the similar crystal cell parameters of $Zn_4O(bdc)_3$ and $Zn_4O(abdc)_3$, which allows the formation of a grid-type porous framework even with a random distribution of each ligand. The paper also reported the catalytic activity of the mixed-linker compound for formation of propylene carbonate from propylene oxide and CO_2.

The potential of the mixed-ligand approach in the framework of $Zn_4O(L)_3$ (L = terephthalic acid derivatives) was extended by use of a high-throughput technique [65]. In the scaffold of the structure, various terephthalic acid derivatives can be

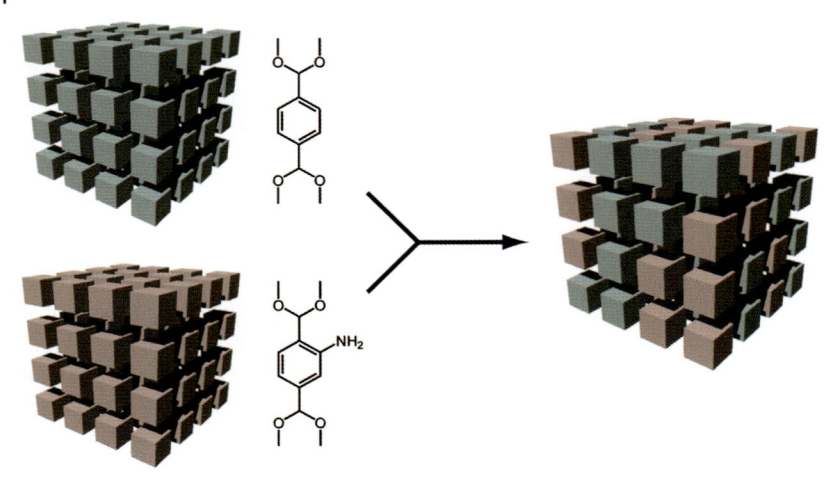

Figure 1.5 Schematic illustration of mixed-ligand structure of [Zn$_4$O(bdc)$_x$(abdc)$_{3-x}$], where bdc = benzenedicarboxylate and abdc = 2-aminobenzenedicarboxylate. Cell parameters are similar with two original structures and each ligand locates randomly in the framework.

incorporated to create a porous structure, and one of them has eight terephthalic acid derivatives in the crystal structure. The distribution of functional groups in the ligands is disordered and the adsorption property of some of the members of this series exhibits up to 400% better selectivity for CO_2 over CO compared with the best same-link counterparts. This also suggests that the matching of the interval of ligands and comparable strengths of coordination bonds are the key to integrating the different ligands in the structure, although each of them has different substituent groups. The reaction conditions should be optimized to synthesize mixed-ligand PCPs and powerful screening with the aid of robots is becoming important.

Not only robust and cubic-type frameworks, but also low-dimensional PCPs with a flexible nature are being developed for mixed-ligand systems [66]. Soft porous coordination polymers with a 2D interdigitated motif [Zn(5-NO$_2$-ip)(bpy)] (CID-5) and [Zn(5-MeO-ip)(bpy)] (CID-6) (ip = isophthalate) are two representatives of soft-type PCP frameworks. CID-5 has high flexibility in the framework and represents a "porous" to "nonporous"-type structure transition, whereas CID-6 has a relatively rigid framework and the crystal structures do not change in desolvated/solvated phases. As synthesized, these compounds have similar cell parameters and they can be mixed with arbitrary ratios of ligand. The series of [Zn(5-NO$_2$-ip)$_{1-x}$(5-MeO-ip)$_x$(bpy)] compounds have their own cell parameters, which means that each compound has a single phase structure. As shown in Figure 1.6, it can be called a ligand-base solid solution and the flexibility of the structures is controllable by tuning each ligand ratio as confirmed by gas adsorption. Optimization of the flexibility in [Zn(5-NO$_2$-ip)$_{1-x}$(5-MeO-ip)$_x$(bpy)] results in better performance in CO_2 separation from CH_4 compared with their original frameworks (CID-5 and CID-6). The control of structural flexibility in the PCP has been a major challenge

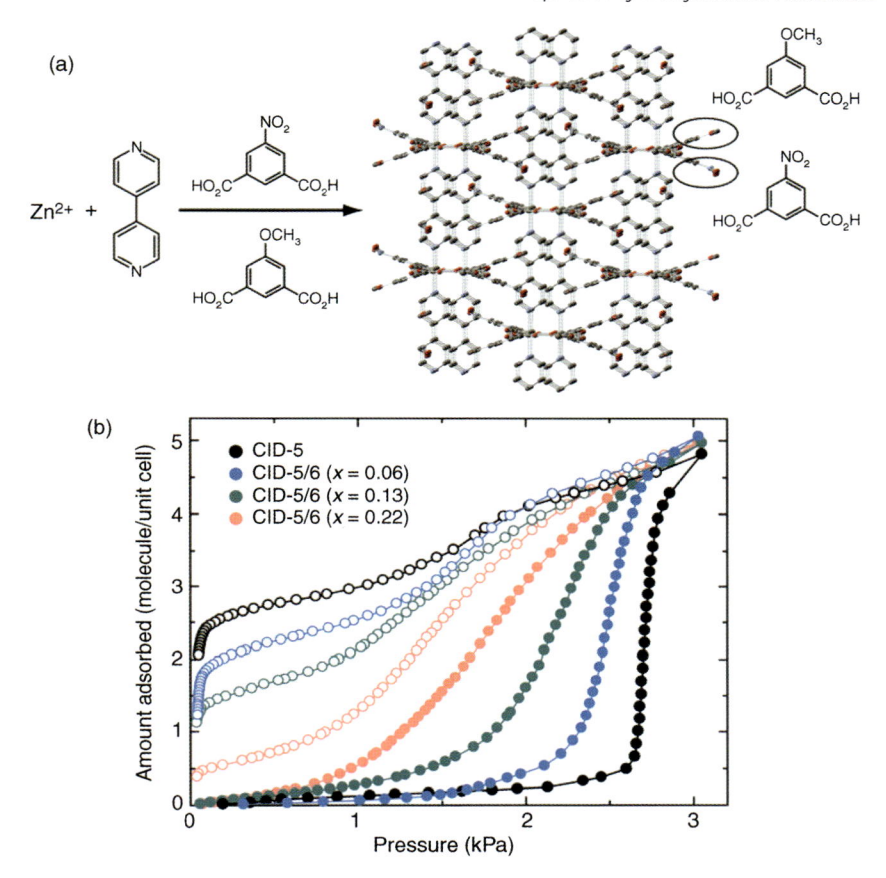

Figure 1.6 (a) Crystal structure of ligand-base solid solution of flexible 2D PCP; [Zn(5-NO$_2$-ip)$_{1-x}$(5-MeO-ip)$_x$(bpy)] (ip = isophthalate), where x is the ratio of 5-MeO-ip ligand in the framework. (b) Adsorption (closed circles) and desorption (open circles) isotherms of water at 298 K for [Zn(5-NO$_2$-ip)$_{1-x}$(5-MeO-ip)$_x$(bpy)] with change in x.

because a small difference in flexibility often contributes to the gas separation. The ligand doping approach for soft-type PCPs is a significant approach to regulate the flexibility.

Not only ligand mixing in the single structure, but also mixing of metal species have been reported. Co^{2+} doping to replace Zn^{2+} ions in the framework of [Zn$_4$O(bdc)$_3$] has been reported [67]. Co is coordinated to six oxygen atoms in the metallic cluster, two of which belong to diethylformamide molecules. The CoZn-MOF-5 materials prepared have higher adsorption capacities for H$_2$, CO$_2$, and CH$_4$ at high pressure than their Co-free homologs. Other well-known PCPs such as [Cu$_3$(btc)$_2$] (HKUST-1) and [Al(bdc)(OH)] (MIL-53) have also been prepared for doping of metal sites and their magnetic behavior and thermal stability have been studied [68, 69].

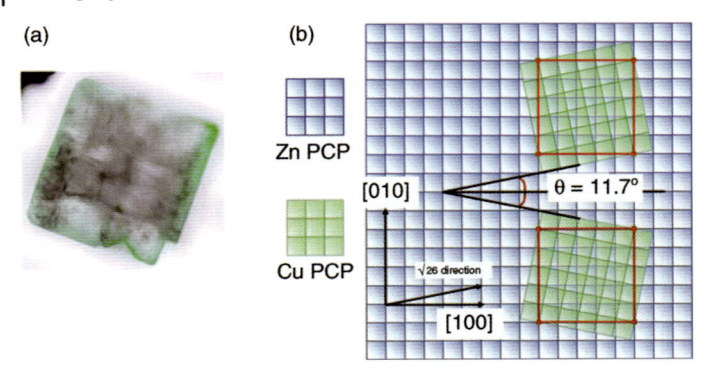

Figure 1.7 (a) Optical microscopic image of the sliced core–shell crystal of [Zn$_2$(ndc)$_2$(dabco)] (core) and [Cu$_2$(ndc)$_2$(dabco)] (shell) (ndc = 1,4-naphthalenedicarboxylate, dabco = 1,4-diazabicyclo[2.2.2]octane). (b) Schematic model of the structural relationship between the core lattice and the shell lattice on the (001) surface. The red lines indicate the commensurate lattice between the core lattice and the shell lattice.

1.4.2
Core–Shell

Incorporation of multiple functions into PCP frameworks has been studied mainly on an atomic scale. The coexistence of different functional groups in the single porous framework described above is an example. In addition to the micrometer-scale integration of different functional moieties, meso-scale combinations of multiple distinct functional PCP frameworks are also of interest for the preparation of multifunctional frameworks.

The core–shell system of PCPs is a candidate for this purpose. Ideally, the epitaxial growth-type core–shell PCP would have multiple functions contributed by the core PCP and shell PCP. Like the core–shell crystal of different metals, some PCPs with an epitaxial core–shell structure have been fabricated. A series of tetragonal frameworks [M$_2$(dicarboxylate)$_2$(N-ligand)], in which dicarboxylate layers link to dimetal clusters to form 2D square lattices, which are connected by dinitrogen pillar ligands to the lattice, have been employed for core–shell fabrication [70]. Step-by-step solvothermal reaction of [Zn$_2$(ndc)$_2$(dabco)] and [Cu$_2$(ndc)$_2$(dabco)] (ndc = 1,4-naphthalenedicarboxylate; dabco = 1,4-diazabicyclo[2.2.2]octane) affords clear core (Zn)–shell (Cu) crystals (Figure 1.7). The 3D configuration was determined by confocal laser scanning microscopy, which showed that the crystals have an anisotoropic configuration; for instance, only four surfaces of the core crystal are covered by the shell crystal. [Zn$_4$O(dicarboxylate)$_3$]-type frameworks are also a good platform to prepare the core–shell-type framework [71, 72]. Two cubic frameworks constructed from the ligands bdc (MOF-5) and 2-NH$_2$-bdc (IRMOF-3) can be crystallized with a core–shell system by a stepwise solvothermal reaction. Although some reports do not discuss cell parameter matching between the core crystal and shell crystal, a multilayered crystal which is interesting for integration of multiple functions in crystals with hetero-junctions has been demonstrated.

Guests such as gases and organic substrates access from the outer surface of PCP crystals and the structural characteristics of the core–shell would control the stepwise adsorptive functions. For instance, integration of a shell crystal having a selective guest sorption ability and a core crystal having catalytic activity make selective conversion of guest molecules possible. Regarding the diffusion of the guest for conversion, the core–shell would have a better diffusion performance for the substrate than a pure selective PCP catalyst and the core–shell system is important for multifunctional PCPs in the sense of time-dependent functions.

1.4.3
PCPs and Nanoparticles

Mutual hybridization of PCP compounds and other solid materials is of interest for complementary multifunctions. The advantages of PCP frameworks again are a high surface area with ordering and there have been several studies on preparation of hybrids of PCPs and nanoparticles which are dispersed inside the porous framework. Introducing metal precursors into the porous framework is mainly executed by chemical vapor deposition (CVD) or immersion methods and additional reduction or hydrogenation of the precursors generates the metal nanoparticles in the PCP framework [73, 74]. The size of the particles depends on the combination of PCPs and method of fabrication of particles, and some studies have reported highly dispersed nanoparticles with retention of the porous structure.

$[Cu_2(BTC)_{4/3}]_6[HnXM_{12}O_{40}]\cdot(C_4H_{12}N)_2$ (X = Si, Ge, P, As; M = W, Mo) is a framework in which the various Keggin polyoxometalates are uniformly incorporated in the nanochannels of HKUST-1 (Figure 1.8) [75]. The size of the polyoxometalate is well

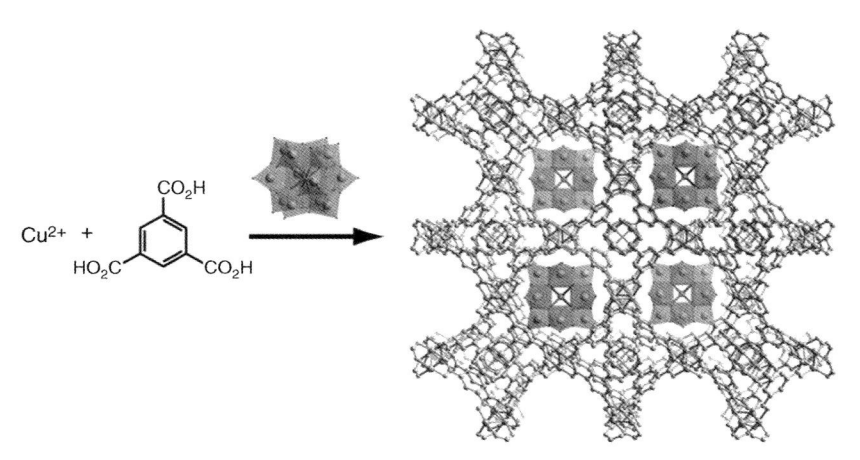

Figure 1.8 Reaction scheme and crystal structure of $[Cu_3(btc)_2]$ (HKUST-1) (btc = benzenetricarboxylate) with Keggin polyoxometalates. Polyoxometalates locate in the pocket of cavities alternately.

fitted to the diameter of the pores and they are distributed with sufficient porosity remaining inside. It has acid catalytic activity for the hydrolysis of esters in excess water and the uniform pores of the hybrid allow only small substrates for the catalytic reaction. This is a notable example of the integration of two different solids, one being the PCP framework.

A further example is the hybridization of a PCP and ammonia borane (AB) (borazane; NH_3BH_3) for the preparation of hydrogen storage materials under moderate conditions [76]. Some of technical challenges with pure AB are a high dehydrogenation temperature and formation of volatile byproducts. [Y(btc)] (JUC-32-Y) was employed to accommodate the AB particles and a simple infusion method gives a uniform hybrid of PCP and AB. The material can release hydrogen even at a low temperature of 85 °C and the AB inside could release 8.0 wt% hydrogen within 10 min. The unsaturated metal Y^{3+} sites of JUC-32-Y interact with AB to prevent the formation of ammonia. The hybridization gives a real synergetic effect of both PCP and AB and the approach would be very suitable for the preparation of other functional particles.

1.5
Perspectives

Since the discovery of the permanent porosity of PCPs/MOFs, confirmed by gas adsorption experiments in the late 1990s, the design of crystal structures has developed extensively. Owing to the huge efforts at synthesis, some conventional functions as adsorbents such as for gas storage and separation have reached a high level even in comparison with other well-known porous materials. However, achieving higher performance and accompanying higher stability and easier handling technique are still needed.

If we take full advantage of the designability of PCPs, we could go further in the design of frameworks as novel functional porous solids. One aspect is multifunctionality in porous structures. The multifunction here ideally is not the independent coexistence of different functions in one framework, but cooperative behavior of multiple chemical/physical properties. For instance, porosity and magnetism or catalysis and flexibility should be closely related each other and the system of interplay in the structure is a significant target. The last decade has mainly focused on the design of frameworks at the atomic scale, in other words, the design of components such as metal clusters and functional organic ligands to create versatile porous frameworks. Together with this approach, the control of integration of the functions of PCPs or with other materials at the meso-scale is also becoming important, and related advanced characterization techniques such as electron microscopy, tomography, and the quartz crystal microbalance [77, 78], and simultaneous *in situ* adsorption measurements to evaluate the porous characteristics, are also significant [79].

We believe that the established multifunctional PCPs would not necessarily be used in large-scale processes in conventional industry. They can be regarded as functional host solids which can open up future application fields such as biosensors, actuators, gas/ion transporters, and continuing efforts to achieve real hybridization of multiple functions and control of material morphology are strongly required.

References

1. Shibata, Y. (1916) *J. Coll. Sci. Imp. Univ. Tokyo*, **37**, 1–31.
2. Keggin, J.F. and Miles, F.D. (1936) *Nature*, **137**, 577–578.
3. Hofmann, K.A. and Küspert, F.A. (1897) *Z. Anorg. Allg. Chem.*, **15**, 204.
4. Powell, H.M. and Rayner, J.H. (1949) *Nature*, **163**, 566–567.
5. Iwamoto, T., Miyoshi, T., Miyamoto, T., Sasaki, Y., and Fujiwara, S. (1967) *Bull. Chem. Soc. Jpn.*, **40**, 1174–1178.
6. Kinoshita, Y., Matsubara, I., Higuchi, T., and Saito, Y. (1959) *Bull. Chem. Soc. Jpn.*, **32**, 1221–1226.
7. Yaghi, O.M., Li, G.M., and Li, H.L. (1995) *Nature*, **378**, 703–706.
8. Kondo, M., Yoshitomi, T., Seki, K., Matsuzaka, H., and Kitagawa, S. (1997) *Angew. Chem. Int. Ed.*, **36**, 1725–1727.
9. Li, H., Eddaoudi, M., Groy, T.L., and Yaghi, O.M. (1998) *J. Am. Chem. Soc.*, **120**, 8571–8572.
10. Li, H., Eddaoudi, M., O'Keeffe, M., and Yaghi, O.M. (1999) *Nature*, **402**, 276–279.
11. Kaye, S.S., Dailly, A., Yaghi, O.M., and Long, J.R. (2007) *J. Am. Chem. Soc.*, **129**, 14176–14177.
12. Chae, H.K., Siberio-Perez, D.Y., Kim, J., Go, Y., Eddaoudi, M., Matzger, A.J., O'Keeffe, M., and Yaghi, O.M. (2004) *Nature*, **427**, 523–527.
13. Furukawa, H., Ko, N., Go, Y.B., Aratani, N., Choi, S.B., Choi, E., Yazaydin, A.O., Snurr, R.Q., O'Keeffe, M., Kim, J., and Yaghi, O.M. (2010) *Science*, **329**, 424–428.
14. Koh, K., Wong-Foy, A.G., and Matzger, A.J. (2009) *J. Am. Chem. Soc.*, **131**, 4184–4185.
15. Férey, G., Mellot-Draznieks, C., Serre, C., Millange, F., Dutour, J., Surble, S., and Margiolaki, I. (2005) *Science*, **309**, 2040–2042.
16. Hong, D.Y., Hwang, Y.K., Serre, C., Férey, G., and Chang, J.S. (2009) *Adv. Funct. Mater.*, **19**, 1537–1552.
17. Zhao, D., Yuan, D.Q., Sun, D.F., and Zhou, H.C. (2009) *J. Am. Chem. Soc.*, **131**, 9186–9187.
18. Chui, S.S.Y., Lo, S.M.F., Charmant, J.P.H., Orpen, A.G., and Williams, I.D. (1999) *Science*, **283**, 1148–1150.
19. Schlichte, K., Kratzke, T., and Kaskel, S. (2004) *Micropor. Mesopor. Mater.*, **73**, 81–88.
20. Alaerts, L., Seguin, E., Poelman, H., Thibault-Starzyk, F., Jacobs, P.A., and De Vos, D.E. (2006) *Chem. Eur. J.*, **12**, 7353–7363.
21. Kramer, M., Ulrich, S.B., and Kaskel, S. (2006) *J. Mater. Chem.*, **16**, 2245–2248.
22. Xie, L.H., Liu, S.X., Gao, C.Y., Cao, R.G., Cao, J.F., Sun, C.Y., and Su, Z.M. (2007) *Inorg. Chem.*, **46**, 7782–7788.
23. Murray, L.J., Dinca, M., Yano, J., Chavan, S., Bordiga, S., Brown, C.M., and Long, J.R. (2010) *J. Am. Chem. Soc.*, **132**, 7856–7857.
24. Dietzel, P.D.C., Morita, Y., Blom, R., and Fjellvag, H. (2005) *Angew. Chem. Int. Ed.*, **44**, 6354–6358.
25. Rowsell, J.L.C. and Yaghi, O.M. (2006) *J. Am. Chem. Soc.*, **128**, 1304–1315.
26. Caskey, S.R., Wong-Foy, A.G., and Matzger, A.J. (2008) *J. Am. Chem. Soc.*, **130**, 10870–10871.
27. McKinlay, A.C., Xiao, B., Wragg, D.S., Wheatley, P.S., Megson, I.L., and Morris, R.E. (2008) *J. Am. Chem. Soc.*, **130**, 10440–10444.
28. Henschel, A., Gedrich, K., Kraehnert, R., and Kaskel, S. (2008) *Chem. Commun.*, 4192–4194.
29. Kim, J., Bhattacharjee, S., Jeong, K.E., Jeong, S.Y., and Ahn, W.S. (2009) *Chem. Commun.*, 3904–3906.
30. Kitaura, R., Onoyama, G., Sakamoto, H., Matsuda, R., Noro, S., and Kitagawa, S. (2004) *Angew. Chem. Int. Ed.*, **43**, 2684–2687.

31 Cho, S.H., Ma, B.Q., Nguyen, S.T., Hupp, J.T., and Albrecht-Schmitt, T.E. (2006) *Chem. Commun.*, 2563–2565.
32 Lee, J., Farha, O.K., Roberts, J., Scheidt, K.A., Nguyen, S.T., and Hupp, J.T. (2009) *Chem. Soc. Rev.*, **38**, 1450–1459.
33 Fletcher, A.J., Thomas, K.M., and Rosseinsky, M.J. (2005) *J. Solid State Chem.*, **178**, 2491–2510.
34 Kitagawa, S. and Uemura, K. (2005) *Chem. Soc. Rev.*, **34**, 109–119.
35 Horike, S., Shimomura, S., and Kitagawa, S. (2009) *Nat. Chem.*, **1**, 695–704.
36 Sing, K.S.W., Everett, D.H., Haul, R.A.W., Moscou, L., Pierotti, R.A., Rouquerol, J., and Siemieniewska, T. (1985) *Pure Appl. Chem.*, **57**, 603–619.
37 Li, G.Q. and Govind, R. (1994) *Ind. Eng. Chem. Res.*, **33**, 755–783.
38 Li, D. and Kaneko, K. (2001) *Chem. Phys. Lett.*, **335**, 50–56.
39 Kitaura, R., Seki, K., Akiyama, G., and Kitagawa, S. (2003) *Angew. Chem. Int. Ed.*, **42**, 428–431.
40 Horike, S., Tanaka, D., Nakagawa, K., and Kitagawa, S. (2007) *Chem. Commun.*, 3395–3397.
41 Nakagawa, K., Tanaka, D., Horike, S., Shimomura, S., Higuchi, M., and Kitagawa, S. (2010) *Chem. Commun.*, **46**, 4258–4260.
42 Loiseau, T., Serre, C., Huguenard, C., Fink, G., Taulelle, F., Henry, M., Bataille, T., and Férey, G. (2004) *Chem. Eur. J.*, **10**, 1373–1382.
43 Horcajada, P., Serre, C., Maurin, G., Ramsahye, N.A., Balas, F., Vallet-Regi, M., Sebban, M., Taulelle, F., and Férey, G. (2008) *J. Am. Chem. Soc.*, **130**, 6774–6780.
44 Finsy, V., Kirschhock, C.E.A., Vedts, G., Maes, M., Alaerts, L., De Vos, D.E., Baron, G.V., and Denayer, J.F.M. (2009) *Chem. Eur. J.*, **15**, 7724–7731.
45 Chen, B.L., Liang, C.D., Yang, J., Contreras, D.S., Clancy, Y.L., Lobkovsky, E.B., Yaghi, O.M., and Dai, S. (2006) *Angew. Chem. Int. Ed.*, **45**, 1390–1393.
46 Barcia, P.S., Zapata, F., Silva, J.A.C., Rodrigues, A.E., and Chen, B.L. (2007) *J. Phys. Chem. B*, **111**, 6101–6103.

47 Ohba, M., Yoneda, K., Agusti, G., Munoz, M.C., Gaspar, A.B., Real, J.A., Yamasaki, M., Ando, H., Nakao, Y., Sakaki, S., and Kitagawa, S. (2009) *Angew. Chem. Int. Ed.*, **48**, 4767–4771.
48 Southon, P.D., Liu, L., Fellows, E.A., Price, D.J., Halder, G.J., Chapman, K.W., Moubaraki, B., Murray, K.S., Letard, J.F., and Kepert, C.J. (2009) *J. Am. Chem. Soc.*, **131**, 10998–11009.
49 Nagao, Y., Fujishima, M., Ikeda, R., Kanda, S., and Kitagawa, H. (2003) *Synth. Met.*, **133**, 431–432.
50 Yamada, T., Sadakiyo, M., and Kitagawa, H. (2009) *J. Am. Chem. Soc.*, **131**, 3144–3145.
51 Bureekaew, S., Horike, S., Higuchi, M., Mizuno, M., Kawamura, T., Tanaka, D., Yanai, N., and Kitagawa, S. (2009) *Nat. Mater.*, **8**, 831–836.
52 Hurd, J.A., Vaidhyanathan, R., Thangadurai, V., Ratcliffe, C.I., Moudrakovski, I.L., and Shimizu, G.K.H. (2009) *Nat. Chem.*, **1**, 705–710.
53 Ohkoshi, S., Nakagawa, K., Tomono, K., Imoto, K., Tsunobuchi, Y., and Tokoro, H. (2010) *J. Am. Chem. Soc.*, **132**, 6620–6621.
54 Fuma, Y., Ebihara, M., Kutsumizu, S., and Kawamura, T. (2004) *J. Am. Chem. Soc.*, **126**, 12238–12239.
55 Takaishi, S., Hosoda, M., Kajiwara, T., Miyasaka, H., Yamashita, M., Nakanishi, Y., Kitagawa, Y., Yamaguchi, K., Kobayashi, A., and Kitagawa, H. (2009) *Inorg. Chem.*, **48**, 9048–9050.
56 Kobayashi, Y., Jacobs, B., Allendorf, M.D., and Long, J.R. (2010) *Chem. Mater.*, **22**, 4120–4122.
57 Cui, H.B., Wang, Z.M., Takahashi, K., Okano, Y., Kobayashi, H., and Kobayashi, A. (2006) *J. Am. Chem. Soc.*, **128**, 15074–15075.
58 Cui, H., Zhou, B., Long, L.S., Okano, Y., Kobayashi, H., and Kobayashi, A. (2008) *Angew. Chem. Int. Ed.*, **47**, 3376–3380.
59 Jain, P., Dalal, N.S., Toby, B.H., Kroto, H.W., and Cheetham, A.K. (2008) *J. Am. Chem. Soc.*, **130**, 10450–10451.
60 Jain, P., Ramachandran, V., Clark, R.J., Zhou, H.D., Toby, B.H., Dalal, N.S., Kroto, H.W., and Cheetham, A.K. (2009) *J. Am. Chem. Soc.*, **131**, 13625–13626.

61 Ye, H.Y., Fu, D.W., Zhang, Y., Zhang, W., Xiong, R.G., and Huang, S.D. (2009) *J. Am. Chem. Soc.*, **131**, 42–43.

62 Farrusseng, D., Aguado, S., and Pinel, C. (2009) *Angew. Chem. Int. Ed.*, **48**, 7502–7513.

63 Hasegawa, S., Horike, S., Matsuda, R., Furukawa, S., Mochizuki, K., Kinoshita, Y., and Kitagawa, S. (2007) *J. Am. Chem. Soc.*, **129**, 2607–2614.

64 Kleist, W., Jutz, F., Maciejewski, M., and Baiker, A. (2009) *Eur. J. Inorg. Chem.*, (24), 3552–3561.

65 Deng, H.X., Doonan, C.J., Furukawa, H., Ferreira, R.B., Towne, J., Knobler, C.B., Wang, B., and Yaghi, O.M. (2010) *Science*, **327**, 846–850.

66 Fukushima, T., Horike, S., Inubushi, Y., Nakagawa, K., Kubota, Y., Takata, M., and Kitagawa, S. (2010) *Angew. Chem. Int. Ed.*, **49**, 4820–4824.

67 Botas, J.A., Calleja, G., Sanchez-Sanchez, M., and Orcajo, M.G. (2010) *Langmuir*, **26**, 5300–5303.

68 Jee, B., Eisinger, K., Gul-E-Noor, F., Bertmer, M., Hartmann, M., Himsl, D., and Poppl, A. (2010) *J. Phys. Chem. C*, **114**, 16630–16639.

69 Marx, S., Kleist, W., Huang, J., Maciejewski, M., and Baiker, A. (2010) *Dalton Trans.*, 3795–3798.

70 Furukawa, S., Hirai, K., Nakagawa, K., Takashima, Y., Matsuda, R., Tsuruoka, T., Kondo, M., Haruki, R., Tanaka, D., Sakamoto, H., Shimomura, S., Sakata, O., and Kitagawa, S. (2009) *Angew. Chem. Int. Ed.*, **48**, 1766–1770.

71 Koh, K., Wong-Foy, A.G., and Matzger, A.J. (2009) *Chem. Commun.*, 6162–6164.

72 Yoo, Y. and Jeong, H.K. (2010) *Cryst. Growth Des.*, **10**, 1283–1288.

73 Hermes, S., Schroter, M.K., Schmid, R., Khodeir, L., Muhler, M., Tissler, A., Fischer, R.W., and Fischer, R.A. (2005) *Angew. Chem. Int. Ed.*, **44**, 6237–6241.

74 Sabo, M., Henschel, A., Froede, H., Klemm, E., and Kaskel, S. (2007) *J. Mater. Chem.*, **17**, 3827–3832.

75 Sun, C.Y., Liu, S.X., Liang, D.D., Shao, K.Z., Ren, Y.H., and Su, Z.M. (2009) *J. Am. Chem. Soc.*, **131**, 1883–1888.

76 Li, Z.Y., Zhu, G.S., Lu, G.Q., Qiu, S.L., and Yao, X.D. (2010) *J. Am. Chem. Soc.*, **132**, 1490–1491.

77 Turner, S., Lebedev, O.I., Schroder, F., Esken, D., Fischer, R.A., and Van Tendeloo, G. (2008) *Chem. Mater.*, **20**, 5622–5627.

78 Biemmi, E., Darga, A., Stock, N., and Bein, T. (2008) *Micropor. Mesopor. Mater.*, **114**, 380–386.

79 Seo, J., Matsuda, R., Sakamoto, H., Bonneau, C., and Kitagawa, S. (2009) *J. Am. Chem. Soc.*, **131**, 12792–12800.

2
Design of Functional Metal–Organic Frameworks by Post-Synthetic Modification

David Farrusseng, Jérôme Canivet, and Alessandra Quadrelli

During the past decade, metal–organic frameworks (MOFs) have become probably the most studied family of porous solids because of their almost infinite variations in structure and composition. However, the use of their full synthetic potential might be further improved, and post-synthetic modification, that is modification of the solid after synthesis, is a powerful tool to achieve that aim.

2.1
Building a MOFs Toolbox by Post-Synthetic Modification

2.1.1
Taking Advantage of Immobilization in a Porous Solid

Zeolites, which belong to the great family of crystalline porous materials, are widely used in gas separation, catalysis (petrochemical cracking) and ion-exchange beds (water purification). However, post-modification of microporous zeolites is limited to just cation exchange or silanation. In addition, zeolites also have a drastic limitation to their pore size. Among other porous materials, mesoporous silicate (MS) materials, such as MCM-41 and SBA-15 [1, 2] are widely used as adsorbents or catalysts. Unlike the highly ordered MOFs, they are amorphous and therefore exhibit relatively disordered hydroxyl groups at the wall surface [3]. In addition, the diversity of MS materials is limited in terms of composition and porous structure, thus narrowing the scope of applications.

Many research groups have already reported the use of post-modified porous solids for adsorption applications. The post-calcination silanation of mesoporous silica such as SBA-15 led to the development of interesting mercury-selective adsorbents, as reported by Jaroniec and co-workers, these functionalized solids being able to remove contaminant mercury from waste oils [4, 5]. These materials were obtained by silanol capping of thiourea derivatives on the silica pore surface. Similar materials have been prepared using amine-terminated organolisilanes, but coverage of the silica surface was complicated by the presence of the basic N atoms and their

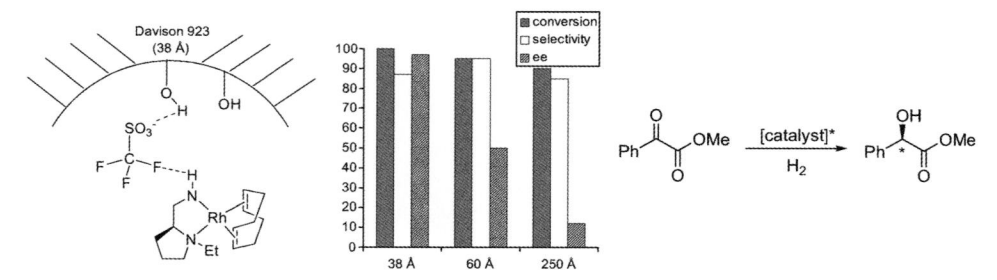

Figure 2.1 Confinement of immobilized, single-site chiral catalyst enhances enantioselectivity.

interactions with the surface silanols and/or the remaining hydroxyl groups [6]. Post-modification involving amine functionalization was also successfully applied to zeolites and mesoporous silicates for the adsorption of carbon dioxide [7].

In addition, the use of porous solids for the immobilization of a catalyst, beyond the formation of a heterogeneous species, has been highlighted, showing their ability to enhance the catalytic properties. In the late 1990s, Corma and co-workers described Rh and Ni complexes anchored to USY-zeolite through a silane bond for hydrogenation reactions, leading to higher catalytic activities compared with the homogeneous catalyst [8, 9]. This was attributed to a cooperative effect of the zeolite which increases the reactant concentration inside the pores. In a similar way, the immobilization of rhodium or nickel complexes of L-proline also on USY-zeolite led to improved enantioselectivity for the hydrogenation of dehydrophenylalanine derivatives [10–13]. More recently, Thomas and co-workers reported the covalent immobilization of $Pd(dppf)Cl_2$ [dppf = 1,1'-bis(diphenylphosphino)ferrocene] in MCM-41 for the allylic amination of cinnamyl acetate, resulting in a change in regioselectivity and enhanced enantioselectivity due to the confinement in the mesopores [14–17]. They also studied the influence of the pore diameter on the efficacy of a silica-anchored chiral catalyst using a nonconvalent approach (Figure 2.1). As a result, they showed that the cavity size determines the face differentiation of the substrate during the asymmetric hydrogenation of methyl benzoylformate: the smaller is the cage, the higher are the constraints and the better is the enantioselectivity.

Inspired by theses pioneering examples, MOFs offer new opportunities as supports for catalytic species with the additional advantages of a possible discrete variation of the pore size and also well-defined single sites allowing complete characterization of the catalytic species such as for molecular catalysts.

2.1.2
Unique Reactivity of MOFs

MOFs are unique two-dimensional (2D) or three-dimensional (3D) hybrid solids formed by metal clusters at nodes connected by organic linkers. They are crystalline materials whose micropore size can be adjusted by varying the ligand length to reach mesoporous size. The attraction of MOFs lies in their well-characterized crystalline

2.1 Building a MOFs Toolbox by Post-Synthetic Modification | 25

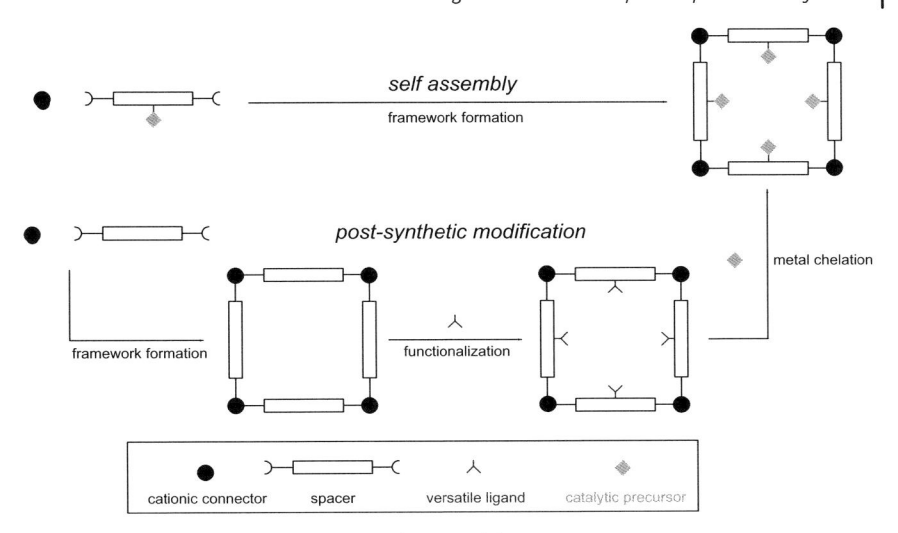

Figure 2.2 Self-assembly versus post-synthetic modification.

architectures combined with a high surface area [18–22]. They also have the great advantage of being finely tunable thanks to their versatile coordination chemistry [23].

MOFs have already found potential applications as adsorbents and catalysts [23–29]. However, the key to multiplying the possible uses of the MOFs is to integrate functionalities in order to obtain materials showing higher sophistication and complexity.

A strategy to access advanced MOFs consists in the use of prefunctionalized moieties such as metalloligands introduced prior to or during the MOF synthesis [30, 31]. Metallic complexes incorporated into the MOF structure have already been prepared by a direct self-assembly approach (Figure 2.2). Here, the spacers possess additional complexing functions, usually donating groups such as hydroxyl, bipyridyl or metalloporphyrin [32–36]. However, this methodology has some limitations and might not be easily generalized to other systems because either the groups could interfere with the formation of the desired material or the properties (thermal stability, solubility) of the functionalized linker might not be compatible with the synthetic conditions. The more complex are the functions to be introduced, the more difficult is the synthesis by self-assembly because of their reactivity towards the metallic precursors.

The post-synthetic modification of MOFs is an appealing alternative to obtain sophisticated structures while avoiding these restrictions. It consists in chemical modification of the solid after formation of the crystalline structure, assuming that the MOFs are sufficiently porous and robust. Post-synthetic modification can provide a wide range of isotopological structures from a single MOF by reacting it as a substrate with a variety of organic reagents. The insertion of new pendant groups on

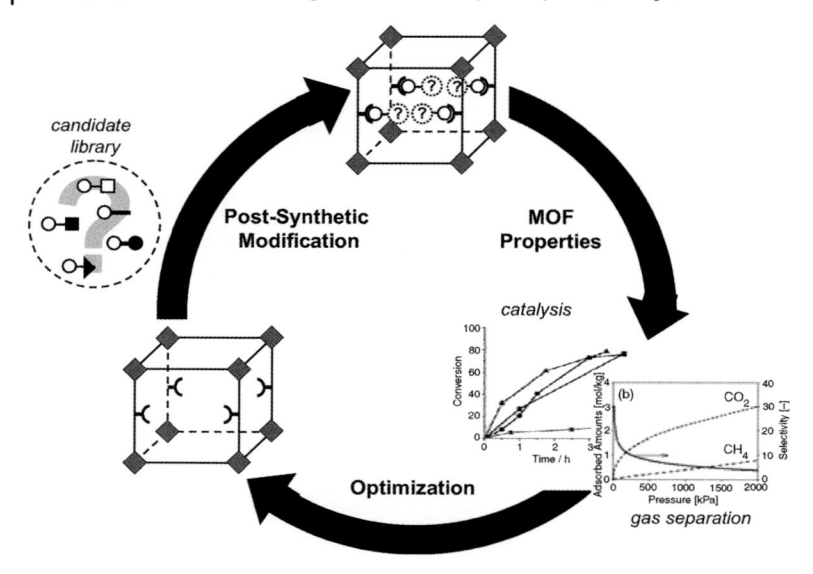

Figure 2.3 Iterative approach for optimization of MOF properties by post-synthetic modification.

or into the MOF allows the modification of its properties while retaining its crystallinity. Thanks to their highly ordered structures, both the position and the post-functionalization rate can be controlled, enabling the MOF properties to be fine-tuned and optimized through combinatorial synthesis (Figure 2.3).

The main difficulty with post-synthetic modification is to not distort the structure of the starting material during the process. Different routes have been developed to access post-functionalized MOFs based on different chemical interactions while keeping the same the native structure. These transformations may involve non-covalent interactions, coordination, or covalent bonds [37].

2.2
Post-Functionalization of MOFs by Host–Guest Interactions

Noncovalent modification of MOFs commonly implies guest exchange or removal from their cavity. The unique host–guest chemistry of MOFs can be used to modify or generate new properties by loading the cavities with functional molecules or metal nanoparticles without compromising the structural integrity.

2.2.1
Guest Absorption

Removing guest molecules from the MOF cavity can lead to a slight change of the structure, potentially reversible, involving modification of properties. Lee and co-workers showed that a 2D MOF based on silver and 1,3,5-tris(3-ethynylbenzonitrile)

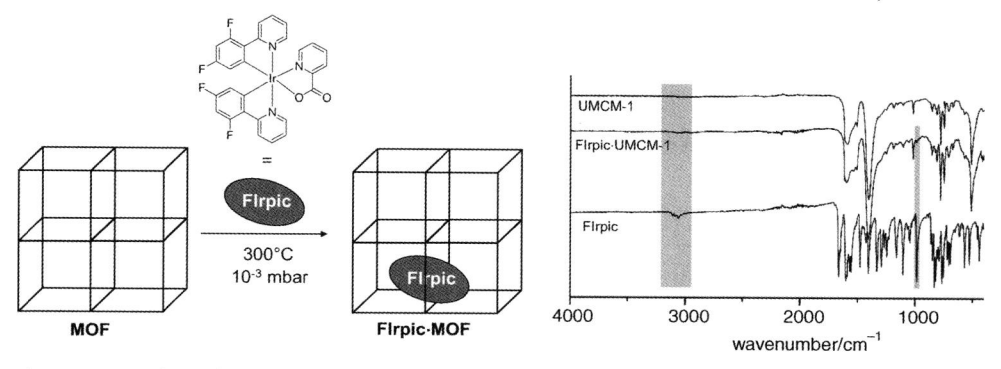

Figure 2.4 Docking of a highly phosphorescent organometallic complex into a 3D MOF without distortion of the initial structure.

benzene ligand could undergo a reversible single-crystal to single-crystal transformation by partially releasing benzene guest molecules upon heating [38]. Similar behavior was observed by Ye and co-workers [39] on two 2D isomorphous complexes, M(obpt)$_2$·0.6H$_2$O [M = Co or Ni; hobpt = 4,6-bis(4-pyridyl)-1,3,5-triazin-2-ol], and by Dietzel et al. [40] on M$_2$(dhtp)(H$_2$O)$_2$·8H$_2$O (M = Ni, Co, Mg, Zn; dhtp = dihydroxyterephthalate) by thermal treatment involving dehydration and rehydration processes. The uncoordinated pyridine rings can link water molecules through hydrogen bonding into 3D threefold interpenetrated networks or water can occupy (and be removed from) a coordination site of the SBU (secondary building unit) metal.

Recently, Fisher and co-workers reported the gas-phase loading of 3D (MOF-5, MOF-177, and UMCM-1) and one-dimensional (1D) [MIL-53(Al)] MOFs with the highly emissive perylene derivative *N,N*-bis(2,6-dimethylphenyl)perylene-3,4,9,10-tetracarboxylic diimide (DXP), and an iridium complex, (2-carboxypyridyl)bis[3,5-difluoro-2-(2-pyridyl)phenyl]iridium (FIrpic), as described in Figure 2.4 [41].

Different spectroscopic analyses showed that neither the structure of the host MOF nor that of the guest was modified during the process; however, the emissive properties of the guest molecule–complex were slightly shifted due to the host–guest interactions The resulting composites showed strong luminescence and exhibited variable stabilities towards guest displacement by solvent molecules.

2.2.2
Nanoparticle Encapsulation

Nanosized metallic particles dispersed in sharply microporous MOFs are of great interest in the field of catalysis. The position of the metallic clusters within the MOF and their size distribution can readily be investigated by electron microscopy, leading to a good knowledge of the potential active sites.

Chemical vapor infiltration of organometallic precursors was shown to be an appropriate method for very high loading of nanoparticles based on Pt, Au,

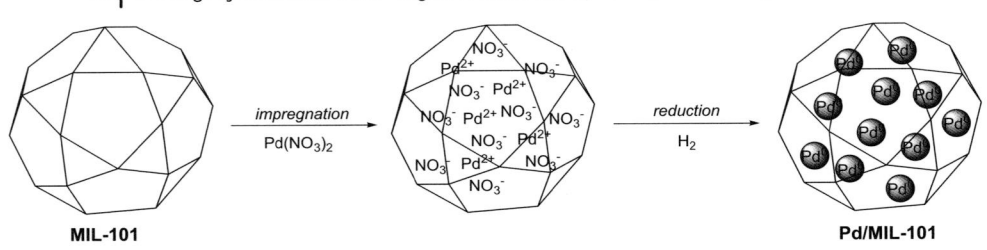

Figure 2.5 MIL-101-supported palladium nanoparticles made by impregnation/reduction.

Pd [42, 43], and Ru [44] into the pores of MOF-5 and MOF-177 [45]. The Pd-based catalyst was found to be active for CO oxidation and alkene hydrogenation.

More recently, chemical-based methods have been developed using wetness impregnation. Kaskel and co-workers reported a composite of Pd-supported MOF-5 prepared by impregnation showing catalytic activity superior to that found for Pd/C for the hydrogenation of various alkenes and esters in three-phase reactions [46]. Haruta and co-workers described nanosized Au hosted in various supports, such as CPL-1 and -2, HKUST-1, MIL-53, and MOF-5, by deposition of the $Me_2Au(acac)$ precursor followed by mild reduction under a low hydrogen flow at 120 °C [47]. A very sharp distribution of Au clusters centered at 1 nm was found for 1 wt% Au/MIL-53(Al). All the Au-supported catalysts were active for the aerobic oxidation of benzyl alcohol in methanol, with a cooperative effect of the MOF support allowing the activation of the alcohol in the absence of a base. MOF-supported Pd nanoparticles were recently reported by Jiang and co-workers (Figure 2.5) [48]. The composite was prepared by impregnation of the large 3D MOF MIL-101 with $Pd(NO_3)_2$, then reduced under a hydrogen stream, resulting on a Pd loading of 1 wt% without loss of crystallinity.

Pd/MIL-101 was found to catalyze the water-mediated Suzuki–Miyaura and Ullmann couplings of aryl chlorides. The MOF-supported palladium nanoparticles were stable over several catalytic cycles with negligible metal leaching, demonstrating that MOFs as catalyst supports could bring new opportunities to heterogeneous catalysis.

2.3
Post-Functionalization of MOFs Based on Coordination Chemistry

Metal–ligand interactions can be involved in two different approaches for the post-synthetic modification of MOFs. One approach targets the unsaturated metal sites, whereas the other exploits the coordination chemistry of the organic linkers, both showing high versatility in incorporating a variety of functionalities into MOFs.

2.3.1
Coordination to Unsaturated Metal Centers

At the end of the 1990s, Williams and co-workers described a highly porous 3D MOF known as HKUST-1, also named CuBTC, formed by paddle-wheel copper clusters linked by 1,3,5-benzenetricarboxylate ligands [49]. They studied the lability of the axial aqua ligands on the metal clusters and their replacement by other molecules. The dehydration of HKUST-1 and its treatment with pyridine led to a framework with new axial ligands but with no change to the 3D lattice, the pyridine-decorated MOF not being directly obtained by addition of pyridine to the reaction mixture.

Hupp and co-workers described the synthesis of a new a 3D non-catenated Zn paddle-wheel MOF, $[Zn_2(L)(dmf)_2]_n(dmf)_m$ (L = 4,4′,4″,4‴-benzene-1,2,4,5-tetraylte-trabenzoic carboxylate; dmf = N,N-dimethylformamide) [50]. After removal of the solvent molecules on the zinc atoms in the axial positions, they were able to introduce various pyridine ligands by immersion of the DMF-free solid into a dichloromethane solution containing the pyridine. As a result, they showed that the post-synthetic modification of a MOF by replacing coordinated solvent molecules with highly polar ligands led to enhancement of the CO_2/N_2 selectivity.

Other investigations were carried out by Férey's group on the chromium-based 3D MOFs MIL-100 [51] and MIL-101 [20]. Both of these materials present potential open metal sites at the Cr^{III} clusters which are occupied by water molecules in their as-synthesized form. The preparation of alcohol-decorated MIL-100 by reaction of the dehydrated MOF with methanol at room temperature was supported by infrared studies, showing the stability of the post-functionalized material even after being evacuated overnight at 373 K [52]. A similar approach was used on MIL-101 to insert organic multifunctional amines, such as ethylenediamine, diethylenetriamine, and 3-aminopropyltriethoxysilane [53]. These newly obtained materials were found to be active as catalysts for the Knoevenagel condensation.

More recently, Banerjee et al. reported the post-modification of MIL-101 by treating it with chiral L-proline derivatives (Figure 2.6) [54]. The functionalization was

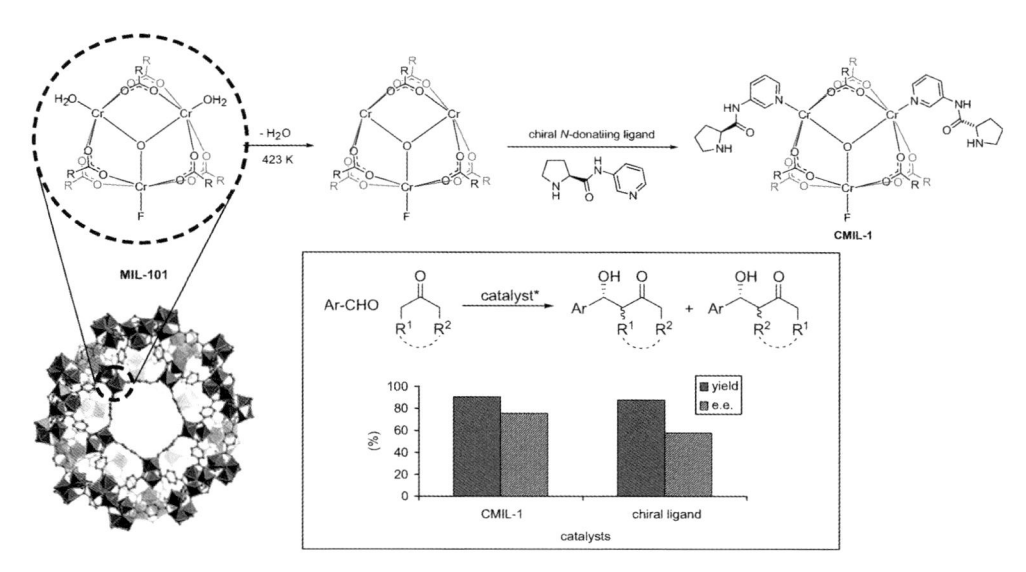

Figure 2.6 Catalytically active homochiral MOF CMIL-1 made by ligand coordination on open metal sites.

achieved by dehydration of the chromium trimers followed by the coordination of the pyridyl moiety of the ligand. These so-called CMIL materials were found to be active catalysts for asymmetric aldols reactions.

2.3.2
Coordination to Organic Linkers

In coordination chemistry, the organic linkers of the MOF structure can be envisaged as ligands thanks to the presence of functional groups. As mentioned above, ligands with potential coordination sites could interfere during the synthesis of the desired MOF. Nevertheless, some reports have already dealt with the direct coordination of a metal to the organic ligands of a MOF structure. This approach can be related to the well-known "surface organometallic chemistry," pioneered by Basset et al. [55–58]. It relies on the bonding of organometallic species to oxides/hydroxides used as supports in order to achieve a single-site heterogeneous catalyst. The role of the support is similar to that of a rigid ligand in the corresponding molecular analogous complex. A key advantage of this "heterogenized" catalyst over molecular species is to access novel chemistry that has no precedent in solution or surface science such as dinitrogen splitting on an isolated silica-grafted tantalum atom [59].

In a similar manner, Lin and co-workers designed a 3D homochiral MOF based on cadmium and on a BINOL derivative (BINOL = 1,1′-bi-2-naphthol) which shows axial chirality [60]. The BINOL ligand presents both pyridyl and orthogonal hydroxyl groups; the OH groups, not being involved in the coordination of the Cd^{III} atoms, are available for further chemical modifications. The MOF obtained was treated with $Ti(O^iPr)_4$, a coordination complex known to form Lewis acid catalysts by coordination to BINOL derivatives (Figure 2.7). Even though the resulting Ti@MOF material was not fully characterized, it showed interesting catalytic activities for the addition of $ZnEt_2$ to a variety of aromatic aldehydes, the enantiomeric excess found being high but in the range of those found for the equivalent homogeneous catalyst. It also

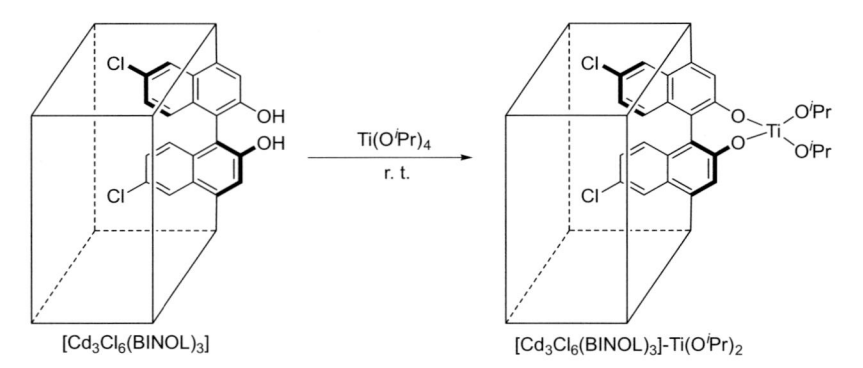

Figure 2.7 MOF-supported chiral titanium–BINOL complex by coordination to the walls functional groups.

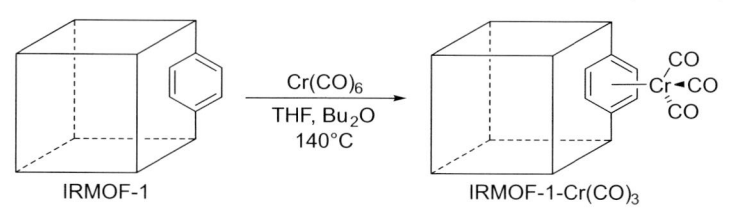

Figure 2.8 "Piano stool" arene–chromium complex supported on IRMOF-1 by coordination to the phenyl walls.

showed remarkable size selectivity, the substrate conversion decreasing as the size of the aldehyde increased.

In the same manner as functionalized linkers, a simple phenyl moiety could be used as an arene ligand in organometallic complexes. This unique approach was adopted by Kaye and Long to functionalize the isoreticular metal–organic framework-1 (IRMOF-1, also known as MOF-5) formed by zinc clusters linked by linear benzenedicarboxylate (bdc) groups [61]. Following the typical synthesis of an arene–chromium tricarbonyl complex in refluxing dibutyl ether and tetrahydrofuran starting from $Cr(CO)_6$, they used IRMOF-1 as a benzene ligand, leading to the formation of the MOF-supported chromium tricarbonyl complex (Figure 2.8).

Spectroscopic analyses supported the attachment of the $Cr(CO)_3$ species to the bdc ring when X-ray analysis and gas sorption confirmed that the crystallinity and the microporosity remained after functionalization. A similar study was performed recently by Lillerud and co-workers on UiO-66 to generate stable arene–chromium moieties [62], more than five different spectroscopic studies being performed in order to determine the chromium coordination [63, 64].

2.4
Post-Functionalization of MOFs by Covalent Bonds

Comparing the above-mentioned methods for post-modifying a MOF structure, the interactions involved are generally weaker than those existing in covalent bonds among main group elements. The creation of covalent bonds through synthetic organic chemistry is a powerful and almost unlimited tool. Although challenging, the application of classical organic chemistry to MOFs, having on their walls reactive groups such as amino in the cases of IRMOF-3 [18], DMOF-1-NH_2 [65], UMCM-1-NH_2 [21], MOF-LIC-1 [66], MIL-53-NH_2, MIL-88-NH_2, and MIL-101-NH_2 [67] and formyl in the case of ZIF-90 [68], has been envisaged by many research teams. They succeeded in creating covalent bonds in MOFs following different approaches based on the reactivity of amino- or formyl-containing MOFs to form amides or imines; some examples even reported sophisticated "click chemistry" applied to MOFs, and others, rather than the modification of the organic linkers, dealt with the reactivity of bridging hydroxyl groups.

2.4.1
Chemical Modification by Amide Coupling

Among simple organic reactions involving an amino substituent, the formation of an amide by reaction with a carboxylic acid or an anhydride would be one of the first to be considered. However, applying a homogeneous chemical mechanism on a solid system is not trivial due to reactivity/stability issues and purification methods. In order to address this challenge, Cohen and co-workers initiated a systematic investigation on the reactivity of porous MOFs based on 2-aminoterepththalate (formally named 2-aminobenzenendicarboxylate, NH_2-bdc) towards a variety of symmetric anhydrides (Figure 2.9) [65, 69–72]. IRMOF-3, which is the amino-substituted equivalent of IRMOF-1, was first chosen in that study owing to its high porosity and crystallinity and the presence of reactive amino groups on the NH_2-bdc linker. The target reaction was the acetylation of the amino group with acetic anhydride solution in chloroform [69]. ^1H NMR analysis of the MOF after digestion, that is, dissolution in an acidic solution, showed that acetylation occurred quantitatively under optimized conditions, giving the IRMOF-3–AM1 product. X-ray powder diffraction (XRPD) and thermogravimetric analysis (TGA) revealed that the integrity of the IRMOF-3 crystalline structure remained after reaction, without loss of thermal stability. Moreover, single-crystal X-ray analysis of the functionalized MOF provided unambiguous evidence of the presence of the amide group.

This strategy was subsequently extended to other anhydrides. A series of 10 straight-chain symmetric alkyl anhydrides having the general formula $O[CO(CH_2)_nCH_3]_2$ ($n = 1$–18) were used as acetylating agents [71]. The reactions yielded to IRMOF-3–AM($n + 1$) solids with a degree of modification essentially ranging from quantitative for short to medium chain length ($n \leq 5$) to less than 10% for the longest ($n = 18$). It was also shown that the surface area of the functionalized MOF correlated inversely with the number of atoms added per unit volume (combining yield and chain length) and that the reactivity trend could be rationalized by a sterically controlled heterogeneous mechanism where the size of the reagents is critical.

Following a similar approach, Cohen and co-workers used cyclic anhydrides to form IRMOF-3 derivatives decorated with free carboxylic groups and a chiral

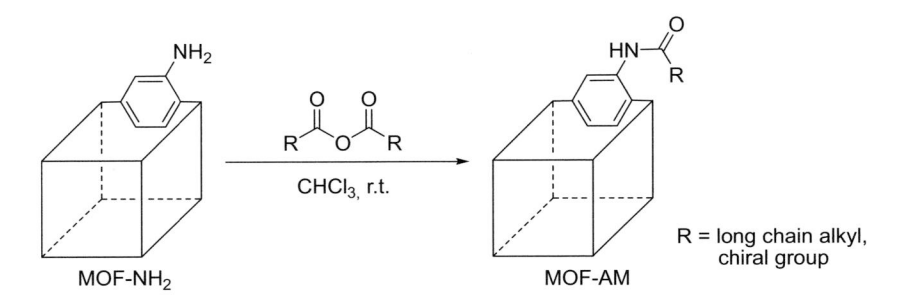

Figure 2.9 Post-modification of NH_2-bdc-based MOF through amide coupling to give amido-functionalized MOF-AM.

anhydride to convert an achiral MOF to a chiral material [73]. More recently, the same group demonstrated the generality of that post-synthetic approach by performing amide coupling using various symmetric anhydrides on other MOFs having morphologies differing from that of IRMOF-3, namely DMOF-1-NH_2, UMCM-1-NH_2 and MIL-53(Al)-NH_2 [65]. Thanks to this post-synthetic modification method, they reported the enhancement of both the hydrogen uptake capacity and heat of adsorption of different MOFs [74]. They also successfully applied this strategy to create superhydrophobic MOFs [75], offering new opportunities for practical applications of these materials.

2.4.2
Chemical Modification by Imine Condensation

Another approach based on the reactivity of amines is their condensation with an aldehyde to form the corresponding imine. Rosseinsky and co-workers choose this strategy and reported the modification of IRMOF-3 by reaction with salicylaldehyde (2-hydroxybenzaldehyde) [76]. Several spectroscopic analyses showed partial modification of the MOF with around 13% conversion. The reaction produced a functionalized MOF having both free imine and hydroxyl moieties in its cavity, making it a good candidate as a ligand for metal complexation. Yaghi and co-workers followed a similar strategy to modify UMCM-1-NH_2, built from both NH_2-bdc and btb (4,4′,4″-benzene-1,3,5-triyltribenzoate) ligands, using 2-pyridinecarboxaldehyde to achieve the formation of a MOF containing both imine and pyridine in its cavity in 87% yield, the crystallinity and the porosity remaining after modification (Figure 2.10) [77]. More surprisingly, the authors reported exactly the same surface area for both the starting and modified MOFs.

In a reverse way, imine condensation can also occur between a formyl group on the MOF and an amine added as reagent. Yaghi and co-workers used also that approach to modify ZIF-90 containing imidazolate-2-carboxaldehyde as a linker [68]. The presence of the free aldehyde functionality inside the framework allows the covalent

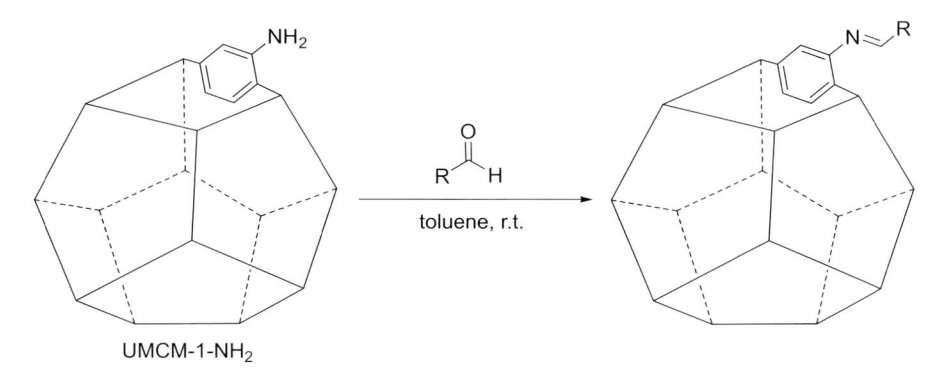

Figure 2.10 Condensation of an aldehyde to the amino moiety of UMCM-1-NH_2.

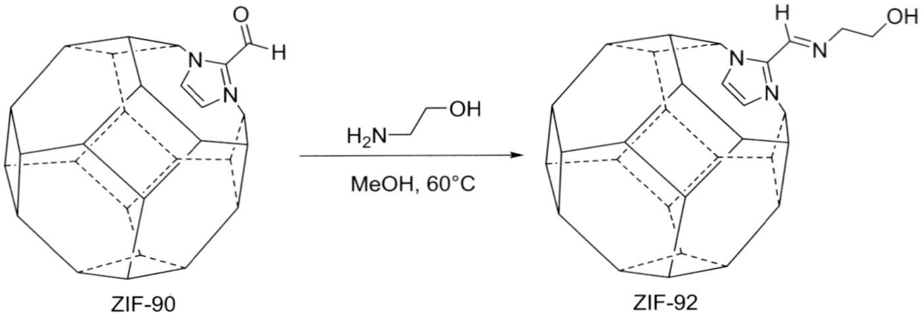

Figure 2.11 Condensation of an amine to the formyl moiety of ZIF-90.

modification of ZIF-90 with ethanolamine to form the corresponding imine (Figure 2.11). The reported conversion was quantitative within 3 h as verified by NMR and FTIR spectroscopy. The high crystallinity of the imine-functionalized ZIF-92 was maintained, as evidenced by the XRPD pattern.

As a result of the high stability of ZIF-90, it was also possible to convert the formyl moiety into the corresponding alcohol by reaction of the MOF with sodium borohydride in methanol at 60 °C. It is worth noting that the porosity of the modified MOF, ZIF-91, was well maintained, with only a slight decrease in surface area when 77% of the formyl groups were reduced to alcohol. This kind of post-modification is anecdotal, however, owing to the harsh reaction conditions involved.

Following a similar strategy, Farrusseng and co-workers applied the imine condensation reaction to a new MOF, SIM-1 (substituted imidazolate material-1), which can grow as a film on alumina supports [78, 79]. SIM-1 is a robust material, isostructural with ZIF-8 and ZIF-90. It consists of ZnN_4 tetrahedra linked by carboxylimidazolates. The aldehyde moiety present on the walls of the structure allows organic modifications in the solid state, such as imine synthesis by condensation with primary amines to give the corresponding imino-functionalized SIM-2 (Figure 2.12) [80]. ^1H NMR analysis of a digested sample of SIM-2(C_{12}) showed 22% functionalization. The crystallinity was found to remain after the according to XRPD, whereas the BET surface area strongly decreased.

As the new SIM-2(C_{12}) was designed to be hydrophobic, the water adsorption isotherms show that SIM-1 adsorbs water at low pressure whereas SIM-2(C_{12}) shows adsorption close to the condensation point only. Moreover, the new water-repellent MOF shows the highest catalytic activity ever found with a MOF for the Knoevenagel condensation.

2.4.3
Chemical Modification by Click Chemistry

"Click chemistry" was introduced by Sharpless in 2001 and consists on the azide–alkyne Huisgen cycloaddition, that is, the reaction of an alkyne with an azido-functionalized compound catalyzed by copper(I) species to form a triazole

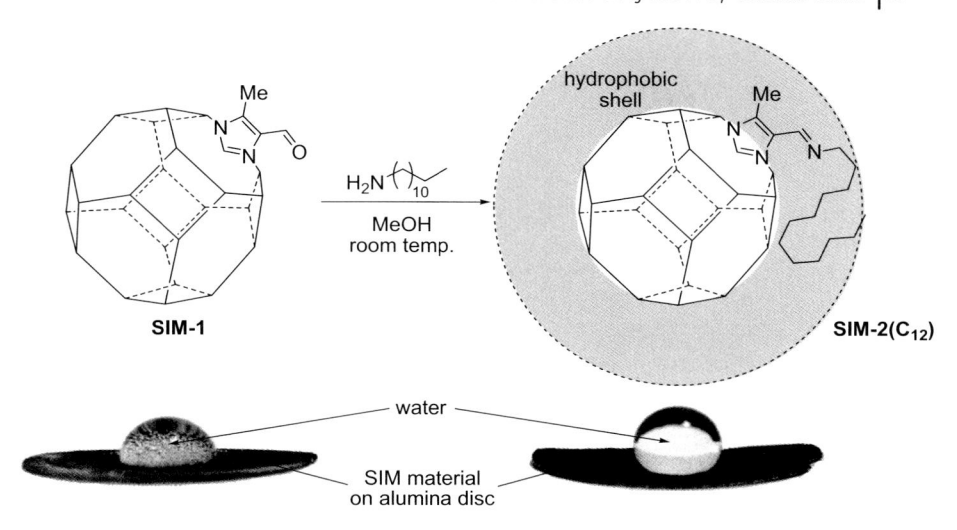

Figure 2.12 Hydrophobization of SIM-1 by imine condensation with C_{12}-alkylamine.

heterocycle [81, 82]. It proceeds efficiently even at micromolar concentrations of reactants with high yields and high specificity in the presence of various functional groups. Such key features prompt its application for the modification of well-defined MOFs.

Hupp and co-workers reported the synthesis of a functionalized MOF composed of dimeric Zn^{II} secondary building units, 2,6-naphthalenedicarboxylate, and a bipyridyl ligand, namely 3-[(trimethylsilyl)ethynyl]-4-[2-(4-pyridinyl)ethenyl]pyridine. The bipyridyl ligand has an acetylene moiety protected by a terminal trimethylsilyl group, which provides a suitable platform for post-synthetic modification (Figure 2.13) [83].

The MOF was initially treated with tetrabutylammonium fluoride in order to desilylate the surface. Ethidium bromide monoazide was chosen as the azide to perform the copper sulfate-catalyzed cycloaddition with the terminal alkynes because its fluorescence can be used for direct visualization of the modification. By comparing the UV–visible absorption spectra of the azide solution before and after the "click" reaction, they determined that <0.8% of the dipyridyl ligands had been "clicked" while the crystallinity and the porosity of the MOF remained. Microscopy imaging confirmed that the modification only occurred at the surface of the material. They successfully extended their approach to a poly(ethylene glycol) azido derivative, namely O-(2-aminoethyl)-O'-(2-azidoethyl)nonaethylene glycol, increasing the hydrophilicity of the material.

In a reverse methodology, Sada and co-workers prepared the N3-MOF-16, derived from its parent IRMOF-16, using a bis(azidomethyl)-functionalized p-terphenyl-4,4''-dicarboxylate derivative as ligand combined with zinc nitrate in N,N-diethylformamide (DEF) [84]. The N3-MOF-16 was allowed to react with terminal alkynes under copper catalysis to form the corresponding MOF decorated with triazole moieties (Figure 2.14). Although no yield was reported, the ^1H NMR spectrum of the

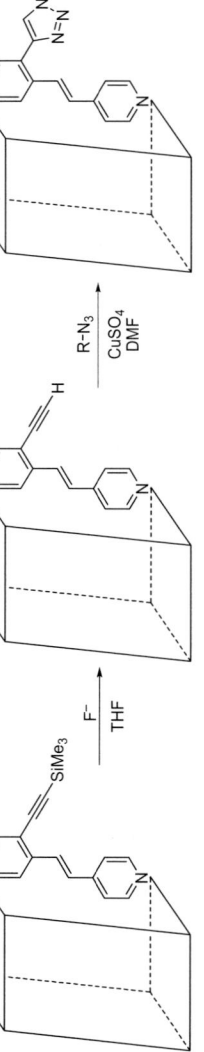

Figure 2.13 From trimethylsilyl-protected alkyne containing MOF to triazole-anchored structure via Huisgen cycloaddition.

Figure 2.14 Reactivity of the "clickable" azido-functionalized N3-MOF-16 towards terminal alkynes under copper(I) catalysis.

"digested" MOF showed that the reaction proceeded almost quantitatively without decomposition of the original MOF network, as confirmed by XRD analysis.

Following a similar strategy, Farrusseng and co-workers used the known DMOF-1-NH_2 as the starting platform for the "click" functionalization. In a first step, the amino group was converted into an azide (Figure 2.15) [85]. As the usual route for preparing azide compounds from the corresponding amines via their diazonium salts was not applicable here because DMOF-1-NH_2 dissolves under acidic conditions, they investigated another route that uses milder conditions and involves stable, nonexplosive compounds.

The azido-functionalized MOF obtained was then reacted with phenylacetylene in the copper-catalyzed Huisgen cycloaddition reaction to form the corresponding triazole-functionalized framework. They also successfully applied their approach to MIL-68(In)-NH_2 [86]. Evidence for the azide formation and the subsequent (3 + 2) cycloaddition were obtained by IR spectroscopy supported by ^{1}H NMR analyses of the different materials after digestion in an acidic deuterated solution. Thanks to the efficiency of the azide formation, the grafting rate can be controlled by adding phenylacetylene in default with respect to –NH_2 functions. For a grafting rate of 50% on MIL-68(In)-NH_2, the surface area decreased by only 55% ($S_{BET} = 571 \, m^2 \, g^{-1}$), which is in line with other methods. Finally, they successfully generalized this

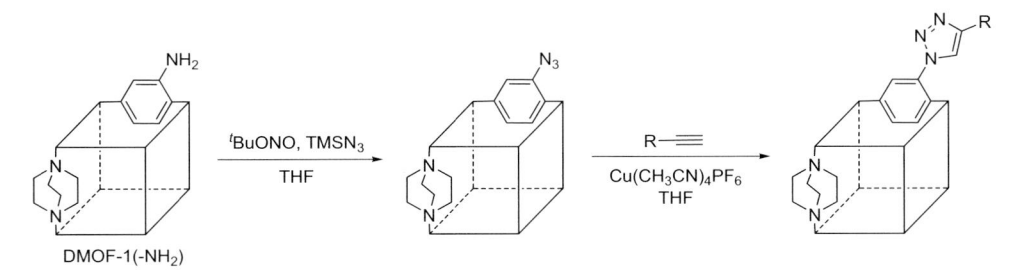

Figure 2.15 Two-step triazoyl-anchored MOF formation from amino-functionalized DMOF-1-NH_2.

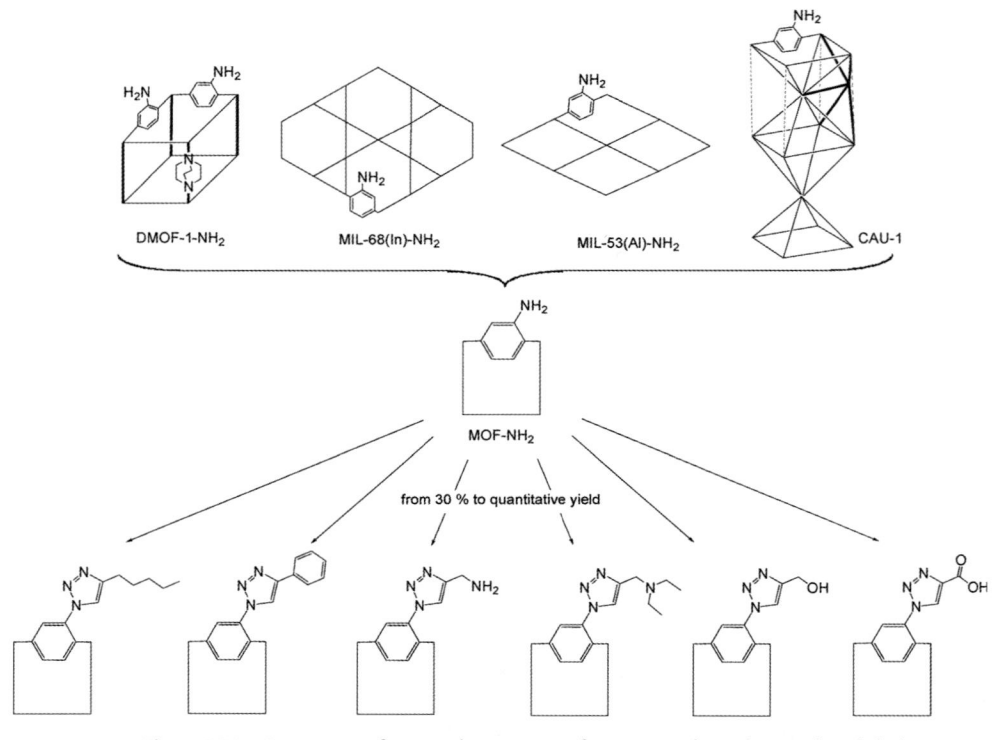

Figure 2.16 Generic post-functionalization route from amino-derived MOFs by "click chemistry."

functionalization methodology to four different MOFs, having different 2D or 3D morphologies, using a wide range of applicable terminal alkynes (Figure 2.16).

As starting platform, in addition to DMOF-1-NH$_2$ and MIL-68(In)-NH$_2$, they used MIL-53(Al)-NH$_2$ [87] and CAU-1 [88], all of them having 2-aminoterephthalate as spacer. After the azido-MOF formation, they applied a range of six different terminal alkynes presenting functional groups such as primary or tertiary amine, hydroxyl, or carboxylic acid with yields ranging from 30% to quantitative, the lowest being found for reactions with MIL-68(In)-NH$_2$. In contrast to the anhydride condensation method reported by Tanabe and Cohen, which has a limited grafting yield (30–50%) [89], they showed that this approach allows complete functionalization even for a bulky group. These findings are also in line with molecular modeling studies which showed a weak steric demand.

2.4.4
Reactivity of Bridging Hydroxyl Groups

Among the few MOFs possessing bridging OH groups, the best known is the 2-dimensional MIL-53 formulated as $M^{III}(OH)(bdc)$, where M^{III} can be Al, Cr [90],

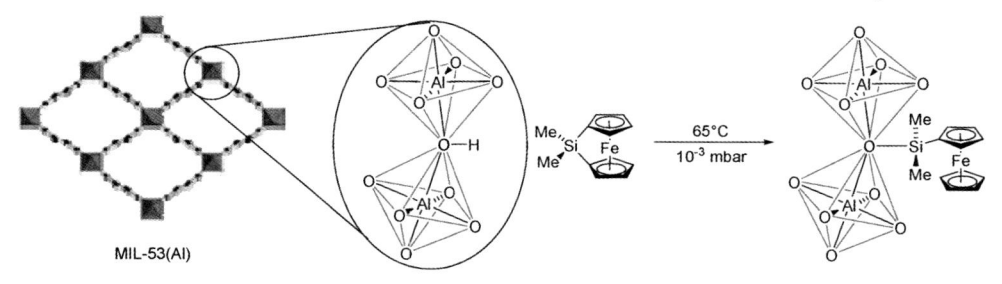

Figure 2.17 Silylation of OH bridging groups in MIL-53(Al).

Ga [91], or Fe [92]. The post-synthetic modification of this inorganic linker, in contrast with the functionalization of organic ligands, is much less studied but must be highlighted as an alternative method, related to the OH reactivity in mesoporous silicates.

Starting from MIL-53(Al), Fischer and co-workers used the highly reactive 1,1′-ferrocenediyldimethylsilane to perform the silylation of the OH group, bridging AlO_6 octahedra, under solvent-free gas-phase conditions in order to form the oxydimethyl (ferrocenyl)silane analog (Figure 2.17) [93].

According to ^2H MAS NMR experiments, 25% of the hydroxyl groups were modified following the protocol. Intensity variations were found in the XRPD data, which were attributed to the pore filling, but TGA showed no weight loss up to the decomposition temperature of the network itself, indicating strong binding of the ferrocenylsilane to the framework. In addition, Fischer and co-workers applies their new ferrocenyl-functionalized MIL-53 as catalyst for liquid-phase benzene oxidation. Unfortunately, the MOF was found to have decomposed slightly after catalysis, the yield also being fairly low.

2.5
Tandem Post-Modification for the Immobilization of Organometallic Catalysts

Concerning MOF applications, mainly gas storage/separation and catalysis promoted by the MOFs *as synthesized* are reported. The tandem post-synthetic modification methodologies could provide access to sophisticated catalytic species, especially MOF-containing organometallics. They consist in the combination of a covalent modification, in order to introduce coordination sites, and coordination chemistry to obtain finally organometallic complexes immobilized in the MOF cavities.

Rosseinsky and co-workers reported the post-modification of IRMOF-3 with salicylaldehyde by imine condensation leading to the formation of the MOF-supported Schiff base in 13% yield. They used the salicylidene-functionalized IRMOF-3 obtained as an N–O ligand for a vanadium oxide complex, characterized by liquid-state NMR and XRPD analysis [76]. The MOF-supported catalyst obtained was found to be active for cyclohexene oxidation with ^{t}BuOOH, although both the conversion

Figure 2.18 IRMOF-3 containing an Au^{III} Schiff base complex by tandem post-functionalization.

and turnover frequency were relatively low, with a possible problem involving framework collapse. Starting with the same approach, Corma and co-workers *et al.* later reported a dichlorogold(III) complex supported in IRMOF-3 through coordination to the salicylidene group (Figure 2.18) [94]. The state of the gold species and also its stability were confirmed by UV–visible spectroscopy, XRPD and transmission electron microscopy, excluding the formation of metallic gold particles. The Au@IRMOF-3 material catalyzed domino coupling and cyclization reactions in the liquid phase with higher activities than the homogeneous and gold-supported catalysts reported earlier. Moreover, the Au^{III} species remained after reaction and the catalyst was fully recyclable.

Tanabe and Cohen applied the post-synthetic modification principle to the synthesis of MOF-supported Cu/Fe catalysts [89]. As shown in Figure 2.19, by reacting the amino-functionalized UMCM-1-NH$_2$ with two different anhydrides under mild conditions, the corresponding amides were obtained in moderate yields

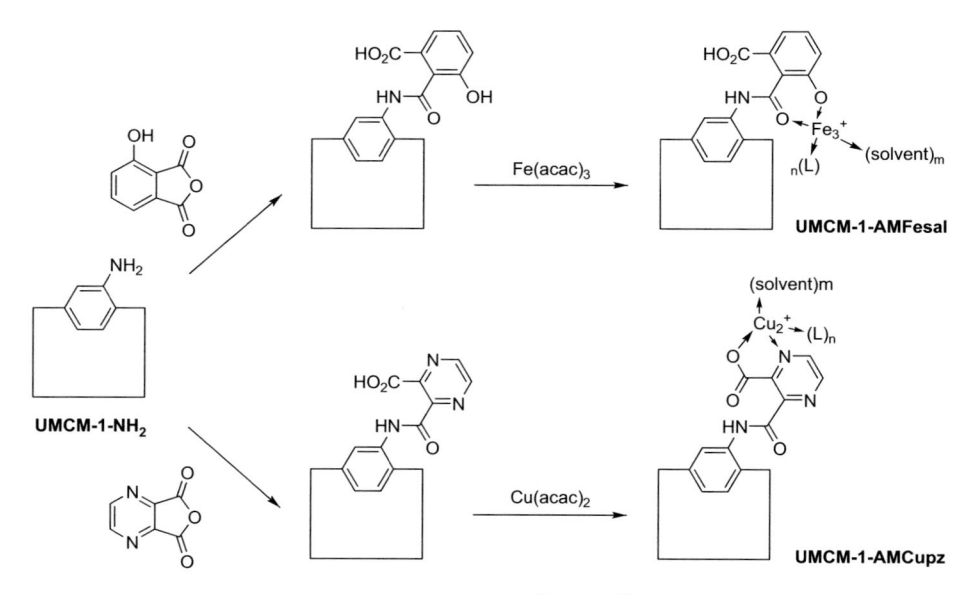

Figure 2.19 Synthesis of UMCM-1-supported Fe^{III} and Cu^{II} complexes.

(35–50%) while the structural integrity of the framework was maintained. Then, addition of an iron(III) or copper(II) salt led to the formation of the supported complexes, UMCM-1-AMFesal and UMCM-1-AMCupz, respectively. Thereby they found new evidence for the great versatility of the post-functionalization technique by achieving two different MOF-supported metal species, containing different ligands and metals, starting from the same MOF carrier. Following the described method, 50% of the potential chelator sites are occupied by a metal. It is noteworthy that the metalated MOFs remain almost as porous as the stating material, with a specific surface area of around $3600 \, m^2 \, g^{-1}$. The resulting material was also characterized by XRPD and TGA. Diffuse reflectance electron spectroscopy supported the formation of Cu^{2+} pyrazine carboxylate and Fe^{3+} salicylate compounds. The latter were tested in the Mukaiyama aldol reaction, which is an extensively studied C−C bond formation reaction involving a Lewis-type catalyst. UMCM-1–AMFesal was found to catalyze the reaction at room temperature with moderate activity but robust, retaining full activity over three catalytic cycles and remaining crystalline.

Yaghi and co-workers described the preparation of a supported palladium–diimino species using the post-modification method starting from the same MOF [77]. By reacting 2-pyridinecarboxaldehyde with UMCM-1-NH_2, they obtained the corresponding N,N-ligand able to chelate palladium dichloride (Figure 2.20).

The Pd-containing MOF was obtained with an overall yield of 74% while the crystallinity remained, according to PRXD data, and the BET surface area decreased from $3200 \, m^2 \, g^{-1}$ for the raw MOF to $1700 \, m^2 \, g^{-1}$ for Pd-containing UMCM-1. The coordination of the palladium atom was supported by EXAFS analysis, indicating the presence of two Pd−Cl and two Pd−N bonds. However, although this promising heterogeneized palladium catalyst is at hand, its catalytic application has not been reported so far.

2.6
Critical Assessment

MOFs display such a great variety of features and physicochemical properties that the number of post-synthetic modification methodologies is almost infinite. These functionalizations are mainly application dependent and often aim to improve adsorption or catalytic properties of the material. Post-synthetic modification was first evidenced as guest exchange processes, with encapsulated solvents. According to the wide range of possibilities offered by these innovative porous materials, the methodology was then extended to coordination chemistry, quickly followed by the use of MOFs as substrates for organic reactions. Whatever the method used, the authors emphasized the great versatility and efficiency of their modification and also the improvement of the material. However, because of the keen interest in MOFs and post-synthetic modification approaches, we need to look critically at the restrictions of the methods in order to develop future applications of these functionalized materials.

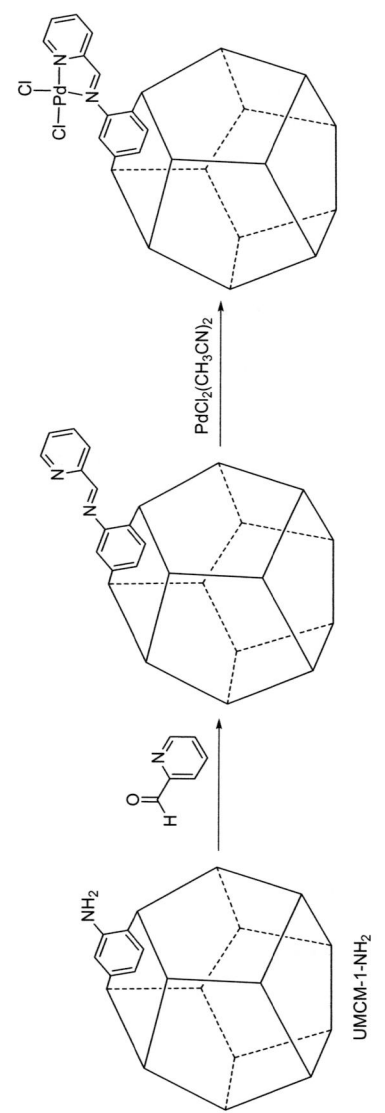

Figure 2.20 Synthesis of UMCM-1-supported Pd^{II} complexes.

2.6.1
Synthetic Restrictions

Regarding all the described post-functionalization approaches to modify a MOF, there are some restrictions concerning the reaction conditions and also the reactants themselves. The greatest issue is the known poor stability of some functionalizable MOFs, depending on the reaction conditions but following the general trend IRMOF-3 < DMOF-1-NH_2 < MIL-68(In)-NH_2 < MIL-53(Al)-NH_2 < ZIF-90.

In the case of noncovalent post-modifications, the size of the guest is crucial as it has to fit perfectly the window (aperture) of the MOF, which narrows the range of applicable MOFs. In addition, the weak interactions involved in such modifications can lead to leaching of the guest species. Concerning the coordination chemistry, this method is restricted to robust MOFs presenting free coordination sites on the ligands or on the metal nodes.

The covalent approach seems very appealing thanks to the strong binding implied and the very high generalization highlighted by Cohen and co-workers, who combined 18 anhydrides with four different MOFs, their stability remaining even in the presence of carboxylic acids formed during the process [65, 71]. However, the generation of water molecules during imine condensation requires moisture-resistant frameworks. On the other hand, the "click" reaction on a MOF needs a copper catalyst, which could remain blocked inside the pores and could interfere during processes involving the functionalized MOF, especially catalysis.

In a more general manner, the size of the MOFs pores, which is critical for accessibility of the reactants inside the cavities, and the stability of the framework towards temperature and acids/bases, required by chemical reactions and also in catalytic applications, narrow the number of MOF candidates.

In the case of substituted heterocycles, such as pyridine derivatives, used as reagents, the competition between the coordination of the heteroatom to the metal nodes and the covalent modification involving the functional group attached to the heterocycle can lead to serious setbacks and to structures too complex for a full characterization.

2.6.2
Balance Between Functionalization Rate and Material Efficacy

Another critical point in the post-functionalization methodologies is the control of the modification yield. By modifying the time, temperature, and/or ratio between reagents, is it possible to fine-tune a MOF in a rational and predictable way? In addition, the "too" high rate of functionalization of a framework with a fairly large molecule could lead to distortion of the structure, a possible loss of crystallinity, and slow destruction of the framework. The correlation between the post-functionalization yield and the application of the modified MOF is thus decisive.

In the case of gas storage, the post-modification rate must take into account the strength of interaction between the gaseous molecules and the MOF container, expected to increase with the functionalization, and the accessible volume in the pores, reduced by the new functional groups introduced. In addition, a high number of more or less bulky groups per volume unit can also hinder the diffusion of gas molecules. In a similar manner for catalytic applications, a high functionalization rate leads to a high number of active sites but also to their low accessibility, slow molecular diffusion, and finally to a decrease in efficiency.

2.6.3
Characterization of the Functionalized Materials

A wide range of characterization techniques are available for materials and most of them are applicable to functionalized MOFs in order to evidence the functionalization and to determine the post-modification yield and the retention of the intrinsic properties of the solids.

Among the characterization analyses, the primary one is XRPD, which is essential to determine whether the material retained the crystal structure through the post-modification steps. Nevertheless, this technique suffers from an inability to detect amorphous by-products, if any, and by the sometimes weak response to the presence of crystalline impurities and/or structural distortions [95]. The measurement of the BET surface area also provides evidence that species, linked or not, are trapped inside the pores, reducing the accessible volume. In addition, elemental analysis of the modified MOFs could be helpful in determining the exact formula of the material. However, the framework often contains an undetermined number of solvent molecules, which could lead to data being hardly interpretable.

A more powerful tool to determine interaction and bonding types is liquid-state NMR spectroscopy, whatever the nucleus observed, but this method remains difficult because of the insolubility of the materials formed. However, characterizations and determinations of modification yields are often achieved by liquid NMR spectroscopy of digested solids. This method can provide useful chemical information, on the basis that the new function is stable under "digestion" conditions, but there is a critical loss of structural data, especially concerning spatial arrangement, function distribution, and interactions. In routine solid-state NMR spectroscopy, resolution problems may occur, especially with paramagnetic metallic nuclei or, to a lesser extent, with materials that are slightly functionalized, for which the peaks observed could be difficult to attribute. Mainly liquid-state NMR spectroscopy after "digestion" is performed to determine the post-functionalization yield, but it may lead to mistakes due to the possible reactions in the NMR tube, lower yields being found due to bond cleavage or decoordination under the drastic acidic or basic conditions required.

Hence these techniques could be too restrictive for sophisticated MOFs, especially supporting organometallic species. In the specific case of organometallics, a lack of unambiguous characterization remains, since numerous converging multi-technique approaches are required. Furthermore, various factors must be taken into account, such as the possible decoordination of the complexes under "digestion" conditions, the possible coordination to unmodified amino moieties, which represent a large percentage of the functions remaining in the cavities in most cases, the nontrivial interpretation of UV–visible diffuse reflectance spectra, and the fact that although EXAFS measurements are reliable, it is noteworthy that only a few examples of unequivocal single-crystal X-ray analysis have been reported [69].

2.7 Conclusion

The chemical modifications in/on the isolated solid offer the possibility of using MOFs as a platform for multifunctional and biocompatible materials. The post-synthetic modification approach provides access to normally inaccessible classes of materials. The ability to develop heterogeneous catalysts, either organocatalytic or transition metal-based systems, is one the area where post-functionalization methods can make crucial contributions. Following the pioneering examples of Corma, Thomas, and co-workers, the additional value of the use of MOFs compared with silica or polymers supports comes from the great versatility of this class of porous materials, their ability to be sophistically functionalized, and the strong bonding of the catalytic species [96]. However, evidence is still required on the effect of the MOF cavity on the catalytic activity and selectivity. Nevertheless, the results from the reported studies highlight the opportunities provided by the covalent post-functionalization approach to produce materials that may be useful as solid-state catalysts by bridging the gap between MOFs and molecular catalysis. Another modification being envisaged is post-synthetic deprotection [97–100]. This method would allow ligands, possessing protected active groups, first to form frameworks by self-assembly then to be deprotected to give the functionalized solids. As a perspective, there is no doubt that enantioselective catalysts supported on MOFs will emerge in the near future by using post-synthetic modification with chiral organometallics.

References

1 Sayari, A. (1996) *Chem. Mater.*, **8**, 1840.
2 Zhao, D.Y., Feng, J.L., Huo, Q.S., Melosh, N., Fredrickson, G.H., Chmelka, B.F., and Stucky, G.D. (1998) *Science*, **279**, 548.
3 Bonelli, B., Onida, B., Chen, J.D., Galarneau, A., Di Renzo, F., Fajula, F., and Garrone, E. (2004) *Micropor. Mesopor. Mater.*, **67**, 95.
4 Olkhovyk, O., Antochshuk, V., and Jaroniec, M. (2004) *Colloid Surf. A: Physicochem. Eng. Aspects*, **236**, 69.
5 Antochshuk, V., Olkhovyk, O., Jaroniec, M., Park, I.S., and Ryoo, R. (2003) *Langmuir*, **19**, 3031.
6 Chong, A.S.M. and Zhao, X.S. (2003) *J. Phys. Chem. B*, **107**, 12650.
7 Su, F.S., Lu, C.Y., Kuo, S.C., and Zeng, W.T. (2010) *Energy Fuels*, **24**, 1441.
8 Corma, A., de Dios, M.I., Iglesias, M., and Sanchez, F. (1997) *Stud. Surf. Sci. Catal.*, **108**, 501.
9 Corma, A., Iglesias, M., and Sanchez, F. (1995) *Catal. Lett.*, **32**, 313.
10 Carmona, A., Corma, A., Iglesias, M., Sanjose, A., and Sanchez, F. (1995) *J. Organomet. Chem.*, **492**, 11.
11 Corma, A., Iglesias, M., Delpino, C., and Sanchez, F. (1993) *Stud. Surf. Sci. Catal.*, **75**, 2293.

12. Corma, A., Iglesias, M., Delpino, C., and Sanchez, F. (1992) *J. Organomet. Chem.*, **431**, 233.
13. Corma, A., Iglesias, M., Delpino, C., and Sanchez, F. (1991) *J. Chem. Soc., Chem. Commun.*, 1253.
14. Thomas, S.J.M. (2010) *ChemCatChem*, **2**, 127.
15. Thomas, J.M. and Raja, R. (2008) *Acc. Chem. Res.*, **41**, 708.
16. Jones, M.D., Raja, R., Thomas, J.M., Johnson, B.F.G., Lewis, D.W., Rouzaud, J., and Harris, K.D.M. (2003) *Angew. Chem. Int. Ed.*, **42**, 4326.
17. Thomas, J.M., Maschmeyer, T., Johnson, B.F.G., and Shephard, D.S. (1999) *J. Mol. Catal. A: Chem.*, **141**, 139.
18. Eddaoudi, M., Kim, J., Rosi, N., Vodak, D., Wachter, J., O'Keeffe, M., and Yaghi, O.M. (2002) *Science*, **295**, 469.
19. Chae, H.K., Siberio-Perez, D.Y., Kim, J., Go, Y., Eddaoudi, M., Matzger, A.J., O'Keeffe, M., and Yaghi, O.M. (2004) *Nature*, **427**, 523.
20. Férey, G., Mellot-Draznieks, C., Serre, C., Millange, F., Dutour, J., Surblé, S., and Margiolaki, I. (2005) *Science*, **309**, 2040.
21. Koh, K., Wong-Foy, A.G., and Matzger, A.J. (2008) *Angew. Chem. Int. Ed.*, **47**, 677.
22. Férey, G. (2008) *Chem. Soc. Rev.*, **37**, 191.
23. Farrusseng, D., Aguado, S., and Pinel, C. (2009) *Angew. Chem. Int. Ed.*, **48**, 7502.
24. Kitagawa, S., Kitaura, R., and Noro, S. (2004) *Angew. Chem. Int. Ed.*, **43**, 2334.
25. James, S.L. (2003) *Chem. Soc. Rev.*, **32**, 276.
26. Janiak, C. (2003) *Dalton Trans.*, 2781.
27. Li, J.R., Kuppler, R.J., and Zhou, H.C. (2009) *Chem. Soc. Rev.*, **38**, 1477.
28. Lee, J., Farha, O.K., Roberts, J., Scheidt, K.A., Nguyen, S.T., and Hupp, J.T. (2009) *Chem. Soc. Rev.*, **38**, 1450.
29. Wang, Z., Chen, G., and Ding, K.L. (2009) *Chem. Rev.*, **109**, 322.
30. Halper, S.R., Do, L., Stork, J.R., and Cohen, S.M. (2006) *J. Am. Chem. Soc.*, **128**, 15255.
31. Kitaura, R., Onoyama, G., Sakamoto, H., Matsuda, R., Noro, S., and Kitagawa, S. (2004) *Angew. Chem. Int. Ed.*, **43**, 2684.
32. Alkordi, M.H., Liu, Y.L., Larsen, R.W., Eubank, J.F., and Eddaoudi, M. (2008) *J. Am. Chem. Soc.*, **130**, 12639.
33. Kitagawa, S., Noro, S., and Nakamura, T. (2006) *Chem. Commun.*, 701.
34. Kosal, M.E., Chou, J.H., Wilson, S.R., and Suslick, K.S. (2002) *Nat. Mater.*, **1**, 118.
35. Shultz, A.M., Farha, O.K., Hupp, J.T., and Nguyen, S.T. (2009) *J. Am. Chem. Soc.*, **131**, 4204.
36. Szeto, K.C., Prestipino, C., Lamberti, C., Zecchina, A., Bordiga, S., Bjorgen, M., Tilset, M., and Lillerud, K.P. (2007) *Chem. Mater.*, **19**, 211.
37. Wang, Z.Q., and Cohen, S.M. (2009) *Chem. Soc. Rev.*, **38**, 1315.
38. Venkataraman, D., Gardner, G.B., Lee, S., and Moore, J.S. (1995) *J. Am. Chem. Soc.*, **117**, 11600.
39. Cao, M.L., Mo, H.J., Liang, J.J., and Ye, B.H. (2009) *CrystEngComm*, **11**, 784.
40. Dietzel, P.D.C., Panella, B., Hirscher, M., Blom, R., and Fjellvag, H. (2006) *Chem. Commun.*, 959.
41. Muller, M., Devaux, A., Yang, C.H., De Cola, L., and Fischer, R.A. (2010) *Photochem. Photobiol. Sci.*, **9**, 846.
42. Hermes, S., Schroder, F., Amirjalayer, S., Schmid, R., and Fischer, R.A. (2006) *J. Mater. Chem.*, **16**, 2464.
43. Hermes, S., Schroter, M.K., Schmid, R., Khodeir, L., Muhler, M., Tissler, A., Fischer, R.W., and Fischer, R.A. (2005) *Angew. Chem. Int. Ed.*, **44**, 6237.
44. Schroeder, F., Esken, D., Cokoja, M., van den Berg, M.W.E., Lebedev, O.I., van Tendeloo, G., Walaszek, B., Buntkowsky, G., Limbach, H.H., Chaudret, B., and Fischer, R.A. (2008) *J. Am. Chem. Soc.*, **130**, 6119.
45. Muller, M., Lebedev, O.I., and Fischer, R.A. (2008) *J. Mater. Chem.*, **18**, 5274.
46. Sabo, M., Henschel, A., Froede, H., Klemm, E., and Kaskel, S. (2007) *J. Mater. Chem.*, **17**, 3827.
47. Ishida, T., Nagaoka, M., Akita, T., and Haruta, M. (2008) *Chem. Eur. J.*, **14**, 8456.

References

48 Yuan, B., Pan, Y., Li, Y., Yin, B., and Jiang, H. (2010) *Angew. Chem. Int. Ed.*, **49**, 4054.

49 Chui, S.S.Y., Lo, S.M.F., Charmant, J.P.H., Orpen, A.G., and Williams, I.D. (1999) *Science*, **283**, 1148.

50 Bae, Y.S., Farha, O.K., Hupp, J.T., and Snurr, R.Q. (2009) *J. Mater. Chem.*, **19**, 2131.

51 Férey, G., Serre, C., Mellot-Draznieks, C., Millange, F., Surblé, S., Dutour, J., and Margiolaki, I. (2004) *Angew. Chem. Int. Ed.*, **43**, 6296.

52 Vimont, A., Goupil, J.M., Lavalley, J.C., Daturi, M., Surblé, S., Serre, C., Millange, F., Férey, G., and Audebrand, N. (2006) *J. Am. Chem. Soc.*, **128**, 3218.

53 Hwang, Y.K., Hong, D.Y., Chang, J.S., Jhung, S.H., Seo, Y.K., Kim, J., Vimont, A., Daturi, M., Serre, C., and Férey, G. (2008) *Angew. Chem. Int. Ed.*, **47**, 4144.

54 Banerjee, M., Das, S., Yoon, M., Choi, H.J., Hyun, M.H., Park, S.M., Seo, G., and Kim, K. (2009) *J. Am. Chem. Soc.*, **131**, 7524.

55 Basset, J.M. (1994) *J. Mol. Catal.*, **86**, 1.

56 Coperet, C., Chabanas, M., Saint-Arroman, R.P., and Basset, J.M. (2003) *Angew. Chem. Int. Ed.*, **42**, 156.

57 Nedez, C., Choplin, A., Corker, J., Basset, J.M., Joly, J.F., and Benazzi, E. (1994) *J. Mol. Catal.*, **92**, L239.

58 Vidal, V., Theolier, A., ThivolleCazat, J., and Basset, J.M. (1997) *Science*, **276**, 99.

59 Avenier, P., Taoufik, M., Lesage, A., Solans-Monfort, X., Baudouin, A., de Mallmann, A., Veyre, L., Basset, J.M., Eisenstein, O., Emsley, L., and Quadrelli, E.A. (2007) *Science*, **317**, 1056.

60 Wu, C.D., Hu, A., Zhang, L., and Lin, W.B. (2005) *J. Am. Chem. Soc.*, **127**, 8940.

61 Kaye, S.S., and Long, J.R. (2008) *J. Am. Chem. Soc.*, **130**, 806.

62 Chavan, S., Vitillo, J.G., Uddin, M.J., Bonino, F., Lamberti, C., Groppo, E., Lillerud, K.P., and Bordiga, S. (2010) *Chem. Mat.*, **22**, 4602.

63 Gianolio, D., Groppo, E., Vitillo, J.G., Damin, A., Bordiga, S., Zecchina, A., and Lamberti, C. (2010) *Chem. Commun.*, **46**, 976.

64 Estephane, J., Groppo, E., Damin, A., Vitillo, J.G., Gianolio, D., Lamberti, C., Bordiga, S., Prestipino, C., Nikitenko, S., Quadrelli, E.A., Taoufik, M., Basset, J.M., and Zecchina, A. (2009) *J. Phys. Chem. C*, **113**, 7305.

65 Wang, Z.Q., Tanabe, K.K., and Cohen, S.M. (2009) *Inorg. Chem.*, **48**, 296.

66 Costa, J.S., Gamez, P., Black, C.A., Roubeau, O., Teat, S.J., and Reedijk, J. (2008) *Eur. J. Inorg. Chem.*, 1551.

67 Bauer, S., Serre, C., Devic, T., Horcajada, P., Marrot, J., Férey, G., and Stock, N. (2008) *Inorg. Chem.*, **47**, 7568.

68 Morris, W., Doonan, C.J., Furukawa, H., Banerjee, R., and Yaghi, O.M. (2008) *J. Am. Chem. Soc.*, **130**, 12626.

69 Wang, Z.Q. and Cohen, S.M. (2007) *J. Am. Chem. Soc.*, **129**, 12368.

70 Dugan, E., Wang, Z.Q., Okamura, M., Medina, A., and Cohen, S.M. (2008) *Chem. Commun.*, 3366.

71 Tanabe, K.K., Wang, Z.Q., and Cohen, S.M. (2008) *J. Am. Chem. Soc.*, **130**, 8508.

72 Wang, Z.Q. and Cohen, S.M. (2008) *Angew. Chem. Int. Ed.*, **47**, 4699.

73 Garibay, S.J., Wang, Z.Q., Tanabe, K.K., and Cohen, S.M. (2009) *Inorg. Chem.*, **48**, 7341.

74 Wang, Z.Q., Tanabe, K.K., and Cohen, S.M. (2010) *Chem. Eur. J.*, **16**, 212.

75 Nguyen, J.G. and Cohen, S.M. (2010) *J. Am. Chem. Soc.*, **132**, 4560.

76 Ingleson, M.J., Barrio, J.P., Guilbaud, J.B., Khimyak, Y.Z., and Rosseinsky, M.J. (2008) *Chem. Commun.*, 2680.

77 Doonan, C.J., Morris, W., Furukawa, H., and Yaghi, O.M. (2009) *J. Am. Chem. Soc.*, **131**, 9492.

78 Aguado, S., Canivet, J., and Farrusseng, D. (2010) *Chem. Commun.*, **46**, 7999.

79 Aguado, S., Nicolas, C.H., Moizan-Baslé, V., Nieto, C., Amrouche, H., Bats, N., Audebrand, N., and Farrusseng, D. (2011) *New. J. Chem.*, **35**, 41.

80 Canivet, J., Aguado, S., Daniel, C., and Farrusseng, D. (2011) *ChemCatChem*, **3**, 675.

81 Sharpless, W.D., Wu, P., Hansen, T.V., and Lindberg, J.G. (2005) *J. Chem. Educ.*, **82**, 1833.

82 Finn, M.G., Kolb, H.C., Fokin, V.V., and Sharpless, K.B. (2008) *Prog. Chem.*, **20**, 1.

83 Gadzikwa, T., Lu, G., Stern, C.L., Wilson, S.R., Hupp, J.T., and Nguyen, S.T. (2008) *Chem. Commun.*, 5493.

84 Goto, Y., Sato, H., Shinkai, S., and Sada, K. (2008) *J. Am. Chem. Soc.*, **130**, 14354.

85 Savonnet, M., Bazer-Bachi, D., Bats, N., Perez-Pellitero, J., Jeanneau, E., Lecocq, V., Pinel, C., and Farrusseng, D. (2010) *J. Am. Chem. Soc.*, **132**, 4518.

86 Savonnet, M. and Farrusseng, D. (2009) FR Patent Application 09/05.101.

87 Ahnfeldt, T., Gunzelmann, D., Loiseau, T., Hirsemann, D., Senker, J., Férey, G., and Stock, N. (2009) *Inorg. Chem.*, **48**, 3057.

88 Ahnfeldt, T., Guillou, N., Gunzelmann, D., Margiolaki, I., Loiseau, T., Férey, G., Senker, J., and Stock, N. (2009) *Angew. Chem. Int. Ed.*, **48**, 5163.

89 Tanabe, K.K. and Cohen, S.M. (2009) *Angew. Chem. Int. Ed.*, **48**, 7424.

90 Férey, G., Latroche, M., Serre, C., Millange, F., Loiseau, T., and Percheron-Guégan, A. (2003) *Chem. Commun.*, 2976.

91 Volkringer, C., Loiseau, T., Guillou, N., Férey, G., Elkaim, E., and Vimont, A. (2009) *Dalton Trans.*, 2241.

92 Whitfield, T.R., Wang, X.Q., Liu, L.M., and Jacobson, A.J. (2005) *Solid State Sci.*, **7**, 1096.

93 Meilikhov, M., Yusenko, K., and Fischer, R.A. (2009) *J. Am. Chem. Soc.*, **131**, 9644.

94 Zhang, X., Llabres, F., and Corma, A. (2009) *J. Catal.*, **265**, 155.

95 Hafizovic, J., Bjorgen, M., Olsbye, U., Dietzel, P.D.C., Bordiga, S., Prestipino, C., Lamberti, C., and Lillerud, K.P. (2007) *J. Am. Chem. Soc.*, **129**, 3612.

96 Corma, A., Garcia, H., and Xamena, F.X.L. (2010) *Chem. Rev.*, **110**, 4606.

97 Yamada, T. and Kitagawa, H. (2009) *J. Am. Chem. Soc.*, **131**, 6312.

98 Deshpande, R.K., Minnaar, J.L., and Telfer, S.G. (2010) *Angew. Chem. Int. Ed.*, **49**, 4598.

99 Gadzikwa, T., Farha, O.K., Malliakas, C.D., Kanatzidis, M.G., Hupp, J.T., and Nguyen, S.T. (2009) *J. Am. Chem. Soc.*, **131**, 13613.

100 Tanabe, K.K., Allen, C.A., and Cohen, S.M. (2010) *Angew. Chem. Int. Ed.*, **49**, 9730.

Part Two
Gas Storage and Separation Applications

3
Thermodynamic Methods for Prediction of Gas Separation in Flexible Frameworks

François-Xavier Coudert

3.1
Introduction

3.1.1
Gas Separation in Metal–Organic Frameworks

During the past decade, the number of metal–organic frameworks (MOFs) synthesized, characterized, and studied for their physicochemical properties has been increasing steadily and at an ever-increasing pace. The main reason behind this enthusiasm and large research effort is the versatility of the porous frameworks belonging to this family, which combines both the richness of metal–organic coordination chemistry and a huge latitude in the choice of the organic linkers and their functionalization. The former offers to materials in this family a very broad array of possibilities in terms of structure geometry, connectivity, and topology. It also influences properties such as thermal and chemical stability, and dynamic properties. The latter allows a wide range of chemistry of the internal surface of the pores, that is, a wide range of sorbate–solid interactions. It also gives these materials an important tunability of pore sizes, by choosing the length of linkers appropriately. In particular, MOFs can be created that feature exceptionally large pore sizes, while still being crystalline solids, and thus regular and well defined.

All these features (tunable pore size, geometry, topology, and internal pore chemistry) give MOFs an edge for gas adsorption and separation applications over other crystalline porous solids such as zeolitic materials, or even mesoporous materials and carbons. For this reason, fluid adsorption and separation have been the object of much research effort. However, this field is still fairly young and the number of materials tested for selective adsorption is much smaller than the total number of synthesized materials. Of these, it is clear that the process of adsorption of small gas molecules in most materials follows a fairly simple phenomenology, and a standard type I isotherm, not dissimilar to many other nanoporous materials. As such, the fluid separation properties of MOFs come from the same physical principles that are in play in zeolites and carbons: competitive adsorption–surface interactions and/or correlation of molecular size of shape with the nanopores. Some members of the MOF family, with pore size at the smaller end of the overall range, may have impressive separation properties, owing to the high tunability of both of these factors, which is hardly possible, or much more difficult, in most other porous solids. For more details about the separation properties of MOFs, readers are referred to Chapters 4 and 8.

3.1.2
Dynamic Materials in the MOF Family

As with any complex molecular structure, all nanoporous frameworks exhibit some degree of flexibility, depending on their chemical nature, structure, and topology. As a consequence, the dynamics of these frameworks can in principle interact with other physicochemical phenomena, including the adsorption of guest molecules, from either the gas or liquid phase. The extent of the influence of structural flexibility on adsorption depends primarily on the degree and type of dynamics displayed by the nanoporous solids. Materials such as zeolites, being built with strong, rigid metal–oxygen bonds, typically display limited structural flexibility. As a consequence, the adsorption of guest molecules can only induce significant deformation in the material at high pressure or at high temperatures. In more standard pressure and temperature ranges, adsorption induces only very limited changes in lattice parameters and pore diameters. Even this limited flexibility, however, is known to have consequences on some physicochemical properties of the materials. One example of that is the phenomenon of *negative thermal expansion*, which occurs for a large number of zeolite structures. Another example of the influence of small structural deformations on the properties of confined fluids in porous materials is the influence of the framework dynamics on guest diffusion and transport properties. Finally, although it is typically noted that the effect of limited flexibility on the thermodynamics of adsorption of guest molecules is in general fairly small, it can play an important role in specific cases. For example, it can lead to some cases of accommodation of larger molecules than is geometrically possible based on to the empty host structures. There are also examples of adsorption-induced guest phase transition between different host structures, as in the case of silicalite-1 [1].

In contrast to zeolites, hybrid framework materials involve significantly weaker bonds (coordinative bonds, π–π stacking, hydrogen bonds, etc.) that are responsible for their intrinsic structural flexibility. Unlike in purely inorganic frameworks, the organic–inorganic connections therefore allow less constrained structural linkages. These are responsible for mechanical properties that are fundamentally different from those of inorganic crystalline materials. As a consequence, physical or chemical stimuli can therefore induce structural transformations inside the material, relying on the allowed distortion of the framework, via stretching, bending, or twisting motions. The range of stimuli-driven phenomena observed in dynamic (or compliant) nanoporous solids is fairly wide, and a tentative list of different categories of stimuli-driven framework dynamics in MOFs are displayed in Figure 3.1. Each category of framework dynamics is accompanied by an example of a known MOF structure displaying this phenomenon.

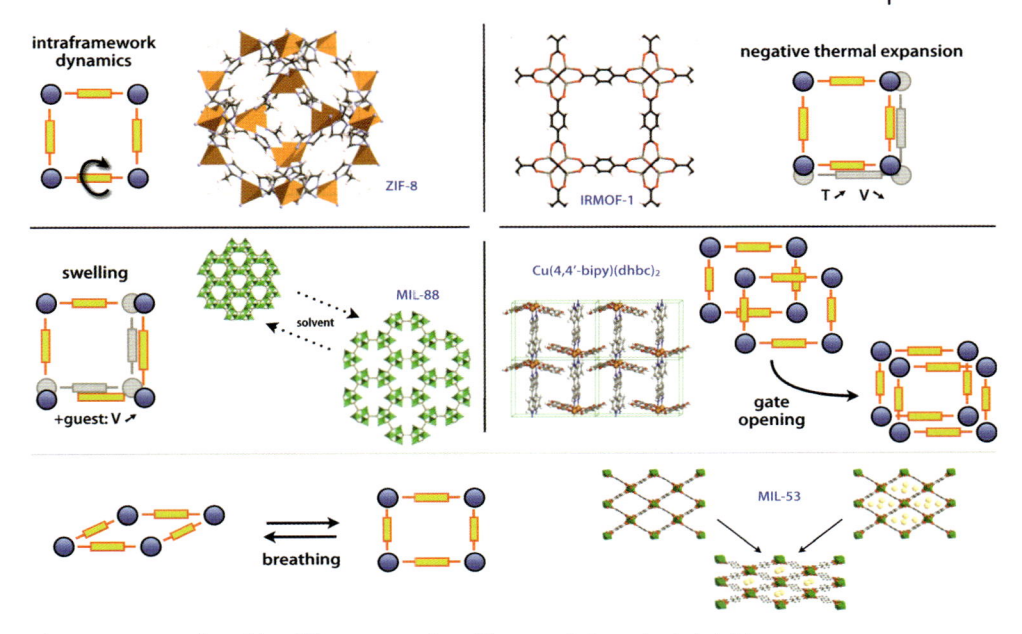

Figure 3.1 Examples of the different categories of framework dynamics in hybrid organic–inorganic materials.

- Most MOFs spontaneously display local deformations of their structure, due to thermal agitation. In particular, in frameworks with flexible organic linkers (or rigid linkers with side chains), these motions can be of large amplitude, yielding important intraframework dynamics, which can also be coupled with the presence of guest molecules inside the nanopores.
- Another phenomenon linked to framework dynamics is that of *negative thermal expansion*, that is, a decrease in the unit cell parameters (or unit cell volume) upon increase in temperature. Although this is an unusual feature for condensed phases and solids, it is found in a large number of zeolitic structures and MOFs. The range of unit cell volume variation, however, is limited, with thermal expansion coefficients in the range 10^{-5}–$10^{-4}\,\text{K}^{-1}$.
- Some other materials display large-amplitude guest-induced *swelling*, as in the case of the MIL-88 family of solids in the presence of pyridine. Swelling denotes a gradual, reversible evolution of unit cell volume (or cell parameters) upon inclusion of a guest molecule (e.g., pyridine in the case of MIL-88) inside the nanopores of the host.
- Finally, a number of MOFs display bistable or multistable structures. For such materials, two or more crystalline structures can be observed under different physicochemical conditions (temperature, mechanical pressure, host inclusion). These different host structures are metastable phases of the same material, and external stimuli can induce transitions between them, by affecting their relative stability. Such stimuli are known to include temperature, mechanical pressure,

and guest adsorption. As a consequence, these materials present a dramatic interplay between fluid adsorption and structural flexibility, with adsorption-induced transitions between structures, a phenomenon which has been called *breathing*.
- One particular family of bistable materials are the porous coordination polymers displaying *gate opening*, which is a transition from an nonporous phase at low vapor pressure to a microporous phase at higher vapor pressure. This family typically includes some twofold interpenetrated frameworks, and also pillared-layer frameworks.

The materials exhibiting large-amplitude, reversible guest-responsive behavior, displaying large modifications of their frameworks upon external stimuli of weak intensity, are called *soft porous crystals* (SPCs). An important, and still rapidly growing, number of these dynamic (or "soft") materials have been reported; for recent reviews, see [2–4]. Research on the design of these materials includes rationalization of what chemical, structural, and topological features of the frameworks allow for the existence or absence of breathing [4], and also classifications of the dynamic porous coordination polymers into six classes, according to the dimensionality of the material's framework and of its organic and inorganic subnetworks [3]. Recent work by Wang and Cohen has shown promising first results on the possibility of tuning the flexibility of a given dynamic material by postsynthetic functionalization, opening the way to nanoporous solids with tailored dynamic behavior [5].

3.1.3
Possible Applications of Flexible MOFs

As the number of SPCs reported in the literature has grown, and a large number of studies describing their physicochemical properties have been published, many potential industrial applications have been proposed for these materials, even though real-life use in industrial processes is still scarce. In addition to the applications of hybrid organic–inorganic frameworks in general, which are abundantly detailed in this book, specific applications of SPCs strive to exploit their structural transitions of large amplitude. Below we present a short overview of some of the specific properties of the SPCs that make them good candidates for some applications, but readers are referred to [4] for a more complete review of this field.

As an example, materials of the MIL-53 family display a pore-shrinking structural transition in the presence of a very low vapor pressure of various organic molecules [6]. Moreover, this exact pressure at which this transitions occurs is highly guest dependent. This sensitivity led to the proposal of applications of these materials in the domain of sensing, for detection of traces of organic molecules, or as molecular actuators. Moreover, these materials also have potential applications in gas separation at higher pressures, as was demonstrated in the case of CO_2–CH_4 mixtures, where the structure with the narrower pores presents a much higher selectivity for CO_2 than the more open framework. Functionalization of the organic linker (e.g., with amine groups) leads to an even better separation of the mixture.

Gate-opening materials, which most often exhibit a wide hysteresis loop in their adsorption–desorption isotherms, and have no microporosity in the closed structure, are considered prospective materials for storing fluids at high pressure, keeping them adsorbed at much lower pressures (which is good for safety reasons), and releasing them completely at low pressures (i.e., complete recovery).

In addition to these adsorption and separation properties, SPCs were investigated for use in drug delivery. Recent work indicates that, compared with rigid frameworks, some soft materials possess a slower release, with kinetics close to zero order. This distinct feature, desirable for long-release, unique-injection therapies, has been attributed to the flexibility of the framework providing an optimal host for the adsorbed molecule, whose unbinding is thus slowed. Finally, it is suspected that flexible MOFs have great potential in catalysis, in part because of the similarities they bear to enzymatic pockets: in both cases, a catalytic site is exposed inside a nanosized cavity, which presents both hydrophobic and hydrophilic regions. Moreover, it is well established that the activity and selectivity of enzymes find their roots in their structural flexibility and dynamic behavior. As a consequence, there is a high hope that carefully designed MOFs with flexible frameworks could be very efficient catalysts [7, 8].

3.1.4
Need for Theoretical Methods Describing Adsorption and Framework Flexibility

As discussed so far, adsorptive gas capture and separation are two of the most highly discussed applications of SPCs. As a consequence, it is desirable to understand what theoretical tools, models, and concepts are used today in this area, in which a large research effort has been concentrated in recent decades. Current industrial processes in these key areas typically use adsorbents such as zeolites, other zeolitic materials, and activated carbons. The technical design of these adsorption processes relies heavily on information about the adsorption equilibria of multicomponent systems in a large number of different thermodynamic conditions. This information is used by process simulation software that integrates this equilibrium information together with kinetic information and structural properties of the simulated design, in order to help choose the best working conditions and parameters for the process (flow rate, column length, regeneration time, etc.).

From the experimental point of view, determination of equilibrium coadsorption data is both expensive and time consuming, considering the very large dimensionality of the parameter space for the problem, that is, the large number of parameters that can vary (temperature, pressure, mixture composition, etc.), in addition to the technical difficulties of the measurement itself. This has led to the development and extensive use of many theoretical methods addressing these issues. For example, the industrial success of adsorptive separation processes, which rely on finding optimal conditions for gas separation in a given adsorbent, is linked to a great extent to the existence of methods that predict multicomponent equilibrium properties based on pure component adsorption data (which are typically much easier to obtain in large quantities). The simplest of these methods is the ideal adsorbed solution theory

(IAST) [9], but many more elaborate methods are used to take into account the nonideality of fluid mixtures. At another level, molecular simulation of coadsorption in rigid nanoporous materials is now part of the standard toolbox in the field and is routinely used to understand and predict the properties of known materials in untested conditions, and also to gain a better understanding of the relation between microscopic and macroscopic properties of confined molecular fluids in nanopores and help design better adsorbents and molecular sieves.

However, the theoretical tools developed to study fluid adsorption in porous materials almost always consider the host matrix as a completely rigid framework. This is, in particular, the case for the IAST method, and for molecular simulations studies of adsorption with the widely-used grand canonical Monte Carlo (GCMC) method. Some studies have included local or continuous deformation of the materials (swelling), but overall, little has been done to understand the thermodynamics at play in the complete host–guest system where structural transitions are induced by adsorption. For example, the well-known coadsorption models (such as IAST) are not applicable to these materials, as they fail to take into account the guest-induced changes in the structure upon adsorption. Finally, even as the number of SPCs synthesized and reported grows, there is a severe lack of experimental data on gas coadsorption and separation in these materials, compared with the data available for pure component adsorption. As a consequence, there is a strong need for theoretical models to rationalize and predict the structural transitions and adsorption properties of mixtures in SPCs. We detail in this chapter the models and methods that have been put forth to answer these questions.

3.2
Theoretical Background

3.2.1
The Osmotic Ensemble

In the following, we describe how theoretical methods allow us to understand and predict the coadsorption of fluid mixtures in flexible porous coordination polymers, either by atomistic molecular simulation or by analytical means, based on experimental data. In order to do so, we first introduce concisely the statistical mechanical basis for all these methods: the osmotic thermodynamic ensemble.

From a theoretical point of view, the adsorption of fluids in flexible porous solids is a thermodynamic equilibrium between guest molecules confined inside the crystalline matrix and an external, infinite reservoir of fluid mixture (whose components we name A, B, C, ...). The parameters that describe each thermodynamic state of this system are the following:

- the temperature T
- the external, mechanical pressure exerted on the solid, σ
- the quantity of matter of the solid, N_{host}
- the chemical potential of each guest species in the reservoir phase, μ_A, μ_B, μ_C, ...

In this ensemble, the volume of the system V and the number of particles of each type in the adsorbed phase $N_{\text{ads},i}$ are not fixed but allowed to fluctuate. This ensemble, which is a mix of the grand canonical ensemble (μ, V, T) and the isothermal–isobaric ensemble (N, σ, T), is also sometimes referred to as the "semi-grand ensemble."

Its configuration integral can be written as

$$Z_{\text{os}}(N_{\text{host}}, \mu_i, \sigma, T) = \sum_V \sum_{N_{\text{ads},i}} \sum_{\mathbf{q}} \exp\left[-\beta U(\mathbf{q}) - \beta \sigma V + \sum_i \beta \mu_i N_{\text{ads},i}\right] \quad (3.1)$$

and its thermodynamic potential as

$$\Omega_{\text{os}}(N_{\text{host}}, \mu_i, \sigma, T) = -kT \log Z_{\text{os}} = U - TS - \sum_i \mu_i N_{\text{ads},i} + \sigma V \quad (3.2)$$

3.2.2
Classical Uses of the Osmotic Ensemble in Molecular Simulation

Before describing how molecular simulation of adsorption processes can be performed in the osmotic ensemble, we give here a brief overview of other systems that can be studied with it. The osmotic ensemble was first proposed by Mehta and Kofke [10] in 1994, and initially implemented for the simulation of coexistence phase diagrams of liquid mixtures. It was presented together with the semi-grand ensemble, and the results and convergence properties of both approaches were compared. They were later theorized, and formulated as instances of the more general concept of pseudoensembles, by Escobedo [11]. Their principal use was the prediction of multicomponent phase equilibria in dense systems, where they allowed one to work around the convergence issues of standard techniques for dense fluids, such as grand canonical or Gibbs ensemble simulations. In contrast to regular ensembles, they were termed "pseudoensembles," due to the limitation that, for mixtures of simple molecular species, the (N_1, μ_2, P, T) ensemble is not actually well defined as the volume is not bounded, as both N_2 and V can grow together to infinity. In this formulation, the use of the osmotic ensemble can be seen as a methodological trick for accelerating convergence, by sampling phase space in a region larger than (but close to) the standard ensembles, rather than a full-bodied statistical mechanics ensemble.

In a second period, starting in 1996 with a study by Theodorou and co-workers [12], a number of research groups proposed various closely related Monte Carlo simulation schemes based on the use of the osmotic ensemble, mainly in the domain of gas solubility in polymers. This particular research area is a good fit with the osmotic ensemble: there, it is both necessary and well defined. First, it is necessary, because swelling of the polymer network has a very large influence on the solubility of gases in polymer melts, which grand canonical simulations would neglect. Second, it is well defined, because the polymer chains in a melt are strongly interlaced and entangled, giving the system a finite extensibility and bounding the sum over phase space, yielding a convergent partition function. As a consequence, Monte Carlo simulation techniques based on the osmotic ensemble are widely used in the domain of phase mixtures containing long-chain molecules.

3.3
Molecular Simulation Methods

There is a lot of literature on the topic of the prediction of adsorption of fluids and fluid mixtures in microporous and mesoporous rigid materials, and on the application of both theoretical models and molecular simulation techniques in understanding the adsorption process at the molecular level. In particular, GCMC simulation methods are now very widely used, and part of the standard molecular simulation toolbox for calculating thermodynamic adsorption and coadsorption properties in rigid porous materials. As space limitations prevent us from covering such wide topics here, we refer the reader to existing material, where a good introduction to this method can be found [13, 14]. Below, we discuss how more complex molecular simulation methods can be constructed from these standard tools, in order to study adsorption in flexible porous solids.

3.3.1
Direct Molecular Simulation of Adsorption in Flexible Porous Solids

In standard GCMC simulations, a rigid framework is assumed for the adsorbent, that is, the host matrix is considered "frozen". This approximation was repeatedly shown to be fairly good for a large number of nanoporous solids (zeolitic materials in particular), for which framework flexibility is assumed to play a role in transport properties, but not much use for the thermodynamics of adsorption. The huge advantage of this "rigid host" assumption is that such simulations only require description of the host–guest and guest–guest interactions. Because the solid is considered rigid, the molecular simulation algorithm does not necessitate any information about the energetics of the host framework. Classical molecular simulation relies on empirical approximations of the intermolecular and intramolecular energies of the simulated species (called *forcefields* or *interaction potentials*). These forcefields need to be parameterized for each species considered in the system and, as a consequence, neglecting framework dynamics in the simulation leads to much less parameterization work.

In contrast with that simple "rigid host" case, two approaches exist for the molecular simulation of adsorption in a flexible porous solid. The first is to consider that, while the adsorption of fluid may influence the structure of the host framework, the deformation of the solid will be local and, in a reasonable range of temperature and pressure, the overall contraction or swelling of the solid is negligible. Such simulations keep the solid unit cell parameters fixed while allowing the individual atoms of the solid to move. They still take place in the grand canonical ensemble. This approximation is particularly suited to describe fluid adsorption in dynamic interpenetrated frameworks, where two or more sublattices move with respect to one another. Many examples of this behavior can be found in the recent literature [15], and theoretical efforts to understand these phenomena have been made using "jungle-gym" structures as an ideal representation of a twofold interpenetrated framework. The fixed unit cell approximation has also been used to study the influence of local

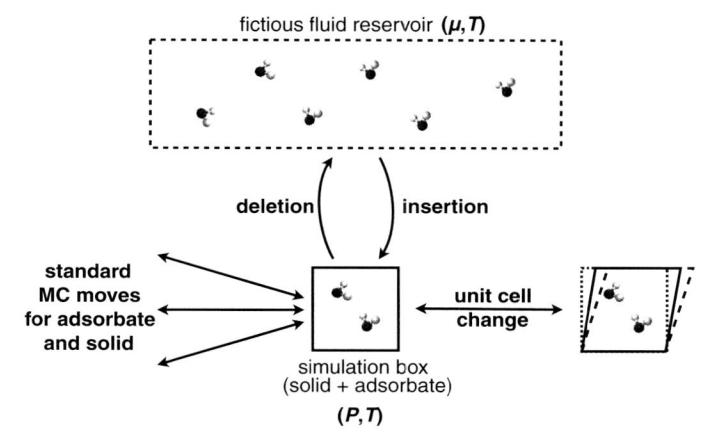

Figure 3.2 Schematic representation of the Monte Carlo moves during a molecular simulation of adsorption performed in the osmotic ensemble.

framework dynamics, for example, in studies of noble gas adsorption in IRMOF-1, where flexibility was demonstrated to have a negligible effect on sorption properties [16].

The second approach, which we now discuss, is to take fully into account the possible deformations of the host framework. This can achieved by performing molecular simulations in the osmotic ensemble, allowing the unit cell parameters of the solid to vary along the simulation, under a given external mechanical constraint (the mechanical pressure). The typical Monte Carlo moves considered during such a simulation are thus (schematized in Figure 3.2): molecular moves (translation, rotation, internal conformation changes, etc.) of the adsorbate, conformation changes of the host framework, molecule insertion or deletion, and changes of unit cell parameters. The nature of the ensemble and the Monte Carlo moves involved places very stringent requirements on the simulation. The first is that a classical atomistic forcefield describing the solid is required, in order to evaluate the energy for each sampled configuration of the framework. Due to the nature of the interactions that need to be reproduced (both bonding and nonbonding), and to the possibly complex molecular structure of the building blocks of the solid, the parameterization of such forcefields is a very complex and time-consuming task. The large number of parameters involved in the analytical description of the interatomic interactions need to be fitted in order to reproduce a number of target properties, which can be gathered from experiments and/or quantum chemistry calculations.

This work can be done in a rather systematic way in flexible solids whose motion is limited to vibrations around an equilibrium configuration. There, the intramolecular potential terms can be linked to vibration modes and frequencies and thus chosen in a reasonably easy way. However, for nanoporous solids that display a large-amplitude swelling, and for bistable materials such as the breathing and gate-opening MOFs that can oscillate between two metastable framework structures, both the functional form of the potential and the optimization procedures are considerably less straightforward. Studies therefore often resort to combining existing forcefields (e.g., CVFF or UFF for organic molecules) and adjusting them on a few selected properties (e.g., the energy difference between the metastable structures). Unfortunately, this results in *ad hoc* forcefields that can hardly be used outside the configurations used for optimization or for calculating radically different properties.

Another (and more fundamental) source of difficulty for molecular simulations in the osmotic ensemble is the typically very low value of the acceptance ratio for unit cell changes. The issue arises from the fact that in a volume change move, the adsorbed species need to be displaced in addition to the unit cell. The way in which this is done is that the unit cell is rescaled by keeping fixed the reduced coordinates of the adsorbed species. This move involves a large number of molecules in a condensed state and, as a consequence, it is often energetically unfavorable, and volume changes therefore have a low acceptance probability. To work around this issue, Monte Carlo simulations in the osmotic ensemble are typically performed following a hybrid Monte Carlo (HMC) scheme, in which short molecular dynamics simulations in the (N,P,T) ensemble are considered as single Monte Carlo steps. The intrinsically collective nature of the motions during the molecular dynamics yields a much higher acceptance probability for the volume changes, and hence a more efficient convergence towards thermodynamic equilibrium. One problem that it does not solve, however, is the slow convergence encountered when the porous solid can oscillate between several metastable structures, being bistable like the MIL-53 family, or oscillating between three close structures like silicalite-1. In this instance, more than in the case of swelling, the barriers present along the free energy landscape are difficult to surmount in a finite simulation time. This was observed, for example, in a recent HMC simulation of CO_2 adsorption in MIL-53(Al), where it was evidenced that only one of the two breathing transitions of the host material is ever observed, even though the total size of the system is relatively modest by today's standards [17].

As a conclusion, direct molecular simulation of multistable soft porous crystals in the osmotic ensemble is certainly not a simple matter. The next section shows how one can, in some cases, avoid the direct simulation of the material's flexibility.

3.3.2
Use of the Restricted Osmotic Ensemble

As a consequence of the issues outlined above, atomistic molecular simulations in the osmotic ensemble are seldom used for adsorption in flexible porous materials. Moreover, most of the studies that have been published concern the phenomenon of adsorption-induced swelling rather than first-order structural transitions and multistable materials. In order to gain thermodynamic insight into the adsorption of fluid inside multistable materials, Jeffroy *et al.* proposed a simulation scheme deriving from the osmotic ensemble in which the number of degrees of freedom of the host material is limited to a set of rigid structures [1]. That is, the porous solid is only allowed to assume a fixed number of conformations, corresponding to the metastable structures of its framework, rather than sampling throughout its entire configuration space. Further, the sampling of this restricted subset of the osmotic ensemble is performed by independent GCMC simulations of each structure. In this scheme, separate simulations for each structure of the porous solid allow the calculation of so-called "rigid host" hypothetical isotherms, $N_k(\mu)$ (where k runs over all metastable host structures). Then, from the fluid adsorption isotherms calculated in each different rigid host structure, one can calculate the corresponding grand canonical potential as

$$\Omega_k^{GC} = -\int N_k(\mu)d\mu \qquad (3.3)$$

As a consequence, if the relative free energies of the host structures, F_k^{host}, are known, their relative stabilities at each vapor pressure can be determined.

This more indirect approach has the advantage of being as simple as a series of GCMC simulations, while the effect of structural transitions and flexibility is accounted for *a posteriori*. The only requirement added to the GCMC simulations is an estimation of the relative energies of host structures, which can be determined experimentally (e.g., by calorimetry), by quantum chemistry calculations, or indirectly from experimental adsorption isotherms, as will be described in the next section. As an example of this method, Figure 3.3 displays the adsorption isotherm calculated for adsorption of tetrachloroethene in the original publication of Jeffroy et al. where this scheme was proposed [1]. The step on the adsorption isotherm, indicative of the structural transition, is clearly reproduced. In addition, the method allows the determination of "probabilities of occurrence" for each structure, related to

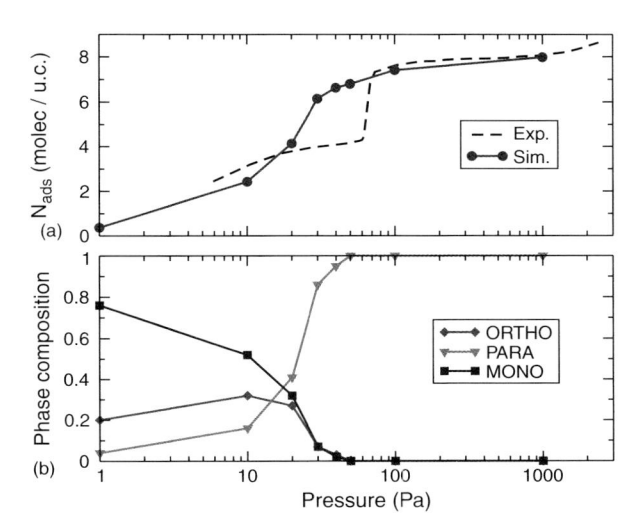

Figure 3.3 (a) C_2H_4 adsorption isotherm calculated in the restricted osmotic ensemble using the three silicalite-1 known structures (MONO, ORTHO and PARA) at 300 K (solid line) compared with experiments (dashed line). (b) The probability of occurrence of each structure [MONO (squares), ORTHO (diamonds) and PARA (inverted triangles)]. Both experimental data and simulation results were extracted from [1].

the grand canonical potentials Ω_k^{GC}. It is also visible on these curves that the ORTHO form of the material, although never the predominant structure, is present in significant amounts according to the simulation results.

3.4
Analytical Methods Based on Experimental Data

As described in the previous section, the molecular simulation tools for studying adsorption in highly flexible materials, for example, those possessing multistable frameworks, are very challenging to implement and use in real-life scenarios where the adsorption strain can be large. Moreover, there is a definite need for theoretical methods and models to help understand experimental results, independently of numerical simulation methods. The recent rapid growth in the number of publications on adsorption properties of SPCs has demonstrated the wide variety of behaviors that they can exhibit. Although most theoretical studies have been focused exclusively on structural features, and the link between flexibility and guest adsorption has mainly been approached from energetic considerations in selected structures, a series of theoretical methods were recently proposed to help develop an understanding of the thermodynamics of adsorption in SPCs, relying on the thermodynamic equations of the osmotic ensemble. We give here an introduction to these methods, stressing how they can help interpret, post-process, and predict experimental adsorption results.

3.4.1
Analytical Methods for Adsorption. Taxonomy of Guest-Induced Flexibility

The analytical methods used to describe adsorption in flexible frameworks are based on the analysis of experimental adsorption isotherms, using a general thermodynamic framework laid out recently [18]. Indeed, the adsorption and desorption isotherms are the most accessible experimental observables in adsorption thermodynamics, as evidenced by the fact that they are routinely reported in the characterization of novel porous materials, in particular those that have possible applications in gas separation. In the case of gas adsorption in SPCs, a substantial number of experimental adsorption data exhibit S-shaped adsorption isotherms. These stepwise isotherms, which most frequently feature hysteresis loops, are typically linked to a structural transition from one host phase to another (pore opening or closing) upon adsorption, although in some cases they can come from structural rearrangements of the adsorbed phase. Some light can be shed on these guest-induced structural transitions by following the "restricted osmotic ensemble" approach, initially proposed as a molecular simulation scheme (see the previous section), to experimental data. Starting from an experimental stepped adsorption isotherm, one has some knowledge of the adsorption isotherms and the transition pressures. Thus, the free energies F_k^{host} can be determined from the experimental data if the full "rigid host" isotherms can be extrapolated from the stepped isotherm.

Hence this method allows one to untangle the thermodynamics of the fluid adsorption and those of the structural deformation of the framework. It provides information on the relative free energies of the metastable structures of the SPC, which are difficult to access experimentally or by quantum chemistry calculations. One approximation is central to the method, however: hypothetical "rigid host" isotherms need to be extrapolated from a single experimental stepped isotherms. This may seem like a fairly large approximation. Nevertheless, application of this method to a wide variety of adsorbate–host pairs has shown that fitting parts of the stepped isotherms by Langmuir equations is both simple and successful. This is due to the fact that adsorption of small gas molecules in common MOFs tends to follow very smooth type I isotherms, which in turn are well described by equations such as the Langmuir equation. Moreover, in cases where a more complex functional form is necessary, other widely used descriptions (such as Langmuir–Freundlich) can be, and have been, used within this method [5].

Moreover, because the mathematical equations of Langmuir-type adsorption in the osmotic ensemble are simple, their behavior can be studied analytically. It was shown that, for a bistable framework where adsorption in each phase follows a Langmuir equation, the existence of adsorption-induced transitions boils down to five key parameters: the two Henry constants, $K_{H,1}$ and $K_{H,2}$, the two saturation uptakes, $N_{max,1}$ and $N_{max,2}$, and free energy difference ΔF_{host}. Depending on the value of these parameters, the taxonomy of guest-induced transitions includes three cases: no, one, or two transitions. Gate opening belongs to the second case, whereas the breathing materials (as the MIL-53 family) belong to the last case. Furthermore, the same material can belong to different cases depending on the adsorption properties, which vary with the nature of the guest or the temperature of the system. For example, the existence or absence of breathing in MIL-53(Al) for different hosts (xenon, CH_4, CO_2, etc.) at room temperature can be rationalized in terms of relative guest–adsorbent affinities. Indeed, if the affinity for the narrow-pore form of the material is sufficiently higher than that of the large-pore form, the former is stabilized sufficiently upon adsorption to be the more stable for a certain range of pressure: the materials "breathes" upon adsorption. Otherwise, no structural transition will be observed.

3.4.2
Application to Coadsorption: Selectivity Predictions and Pressure–Composition Phase Diagrams

As stated earlier, the use of SPCs in adsorption separation processes has been repeatedly proposed, and features high among their possible applications. As a consequence, predictive analytical methods, which proved crucial in separation science and have been widely used for rigid nanoporous solids, need to be extended to flexible host solids. Few direct experimental data on gas separation are available so far, and coadsorption measurements are both complex and time consuming. The OFAST (osmotic framework adsorbed solution theory) method, which we describe in this section, was proposed to predict coadsorption properties from experimental pure-component adsorption data [19, 20].

The OFAST method couples the thermodynamic equations of the osmotic ensemble with the ideal adsorbed solution theory (IAST) [9]. The latter, a coadsorption prediction method introduced by Myers and Prausnitz in 1965, is widely used in the field of adsorption technology. Recent work comparing molecular simulations with the predictions of IAST has shown it to be applicable to adsorption of small gas molecules inside the pores of rigid MOFs. By extending it to the osmotic ensemble, the OFAST method allows the prediction of coadsorption properties using pure-component adsorption data as the only input. It permits calculations such as total adsorbed quantities and adsorption selectivities and, of particular relevance to flexible systems, for a given mixture, to answer the questions of whether structural transitions occur and, if so, at what pressure.

In order to illustrate the capabilities of OFAST, we present two examples of predictions of coadsorption made using this method. Figure 3.4 displays the predicted coadsorption properties of a coordination polymer, $Cu(4,4'\text{-bipy})(dhbc)_2$, synthesized and characterized by Kitagawa's group [21]. This solid, pictured in Figure 3.4a, displays gate opening behavior upon adsorption of various gases, with a gate-opening pressure that depends on the adsorbed gas. From the experimental pure-component adsorption and desorption isotherms for CO_2, O_2, CH_4, and N_2 (Figure 3.4b), the OFAST method can predict the evolution of gate-opening pressure with respect to mixture composition, for binary (Figure 3.4c) and ternary (Figure 3.4d) mixtures. In these plots, the selectivity ϱ_B is defined as

$$\varrho_B = \frac{y_B/y_A}{x_B/x_A} \tag{3.4}$$

where x_i is the external molar fraction of component i and y_i is the corresponding adsorbed molar fraction. Finally, it was further shown that, for this class of gate-opening materials, the gate-opening pressure can be approximately deduced from the pure-component gating pressures by a weighted harmonic mean:

$$1/P_{\text{mix}}(x_i) = \sum x_i/P_i \tag{3.5}$$

As a second example of the application of OFAST, Figure 3.5 displays the coadsorption predictions for a mixture of CO_2 and CH_4 in MIL-53(Al) at room temperature. Figure 3.5a is a pressure–composition phase diagram, showing which structure of the material is stable depending on the thermodynamic conditions, total gas pressure, and mixture composition. Because pure CO_2 induces breathing of the structure whereas pure CH_4 does not (at 300 K), there is a limiting composition (here $x_{CO_2} \approx 0.12$) below which no breathing occurs. Furthermore, the evolution of one of the breathing pressures is not monotonic. Although a complete series of selectivity measurements are lacking for quantitative comparison, the phase diagram

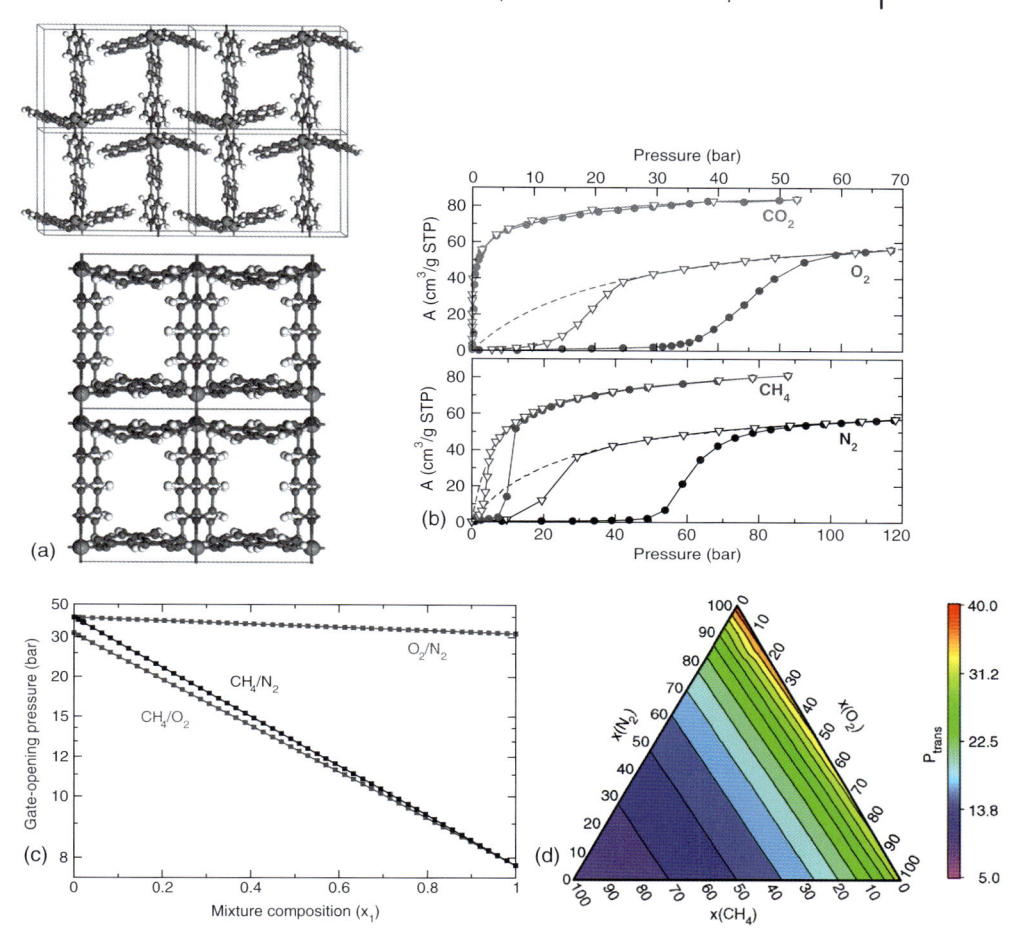

Figure 3.4 (a) Representation of coordination polymer Cu(4,4′-bipy)(dhbc)$_2$. (b) Experimental adsorption and desorption isotherms of CO$_2$, O$_2$, CH$_4$, and N$_2$ in Cu(4,4′-bipy)(dhbc)$_2$. (c) Gate-opening pressure for adsorption of binary mixtures, as predicted by OFAST. (d) Predicted gate-opening pressure for adsorption of a CH$_4$–O$_2$–N$_2$ ternary mixture.

obtained qualitatively agrees with the experimental evidence reported so far [22]. Finally, the prediction of selectivities as a function of pressure and composition (Figure 3.5b) confirms that the narrow-pore form (np phase) has higher CO$_2$ selectivity.

Although the quantitative details of the coadsorption in the two systems highlighted here is still to be fully tested, the picture offered by this method for coadsorption prediction can help in the design processes and applications based on these materials, by offering a continuous picture of their properties in

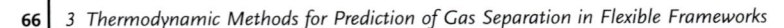

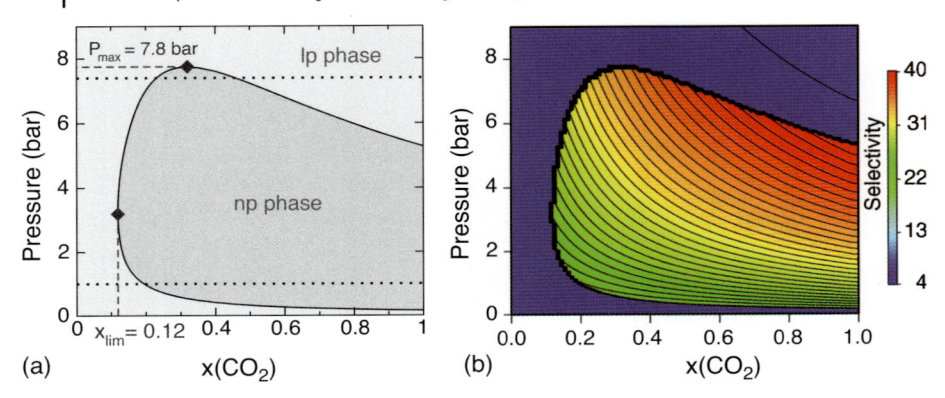

Figure 3.5 Predictions of coadsorption of a mixture of CO_2 and CH_4 in MIL-53(Al). (a) Pressure–composition phase diagram of the material; (b) CO_2 selectivity as a function of total pressure and mixture composition.

the space of thermodynamic parameters, where experimental determinations are difficult and expensive. Such studies can even be further improved by considering the influence of temperature on the coadsorption properties, following the development of recent work in this direction for pure-component adsorption [23].

3.5
Outlook

From the large and ever-increasing number of flexible MOFs being reported and tested for applications in the field of fluid adsorption and separation, it is clear that concepts and methods have to be developed that can help rationalize the many different behaviors observed, and even predict the adsorption of fluids and fluid mixtures in the nanopores of SPCs. We have shown how these issues can be addressed, and what theoretical models and methods, both analytical and numerical, are being developed. This field is, however, still fairly young. The existing methods have to be applied to many of the new systems in order for their robustness and usefulness to be further assessed, and a lot more work will certainly be poured into this area of research in the coming decade.

Acknowledgments

Grateful thanks are due to Alain Fuchs and Anne Boutin for insightful discussions and very fruitful work together on this topic.

References

1 Jeffroy, M., Fuchs, A.H., and Boutin, A. (2008) Structural changes in nanoporous solids due to fluid adsorption: thermodynamic analysis and Monte Carlo simulations. *Chem. Commun.*, 3275–3277.
2 Bradshaw, D., Claridge, J.B., Cussen, E.J., Prior, T.J., and Rosseinsky, M.J. (2005) Design, chirality, and flexibility in nanoporous molecule-based materials. *Acc. Chem. Res.*, **38**, 273–282.
3 Horike, S., Shimomura, S., and Kitagawa, S. (2009) Soft porous crystals. *Nat. Chem.*, **1**, 695–704.
4 Férey, G. and Serre, C. (2009) Large breathing effects in three-dimensional porous hybrid matter: facts, analyses, rules and consequences. *Chem. Soc. Rev.*, **38**, 1380–1399.
5 Wang, Z. and Cohen, S.M. (2009) Modulating metal–organic frameworks to breathe: a postsynthetic covalent modification approach. *J. Am. Chem. Soc.*, **131**, 16675–16677.
6 Bourrelly, S., Llewellyn, P.L., Serre, C., Millange, F., Loiseau, T., and Férey, G. (2005) Different adsorption behaviors of methane and carbon dioxide in the isotypic nanoporous metal terephthalates MIL-53 and MIL-47. *J. Am. Chem. Soc.*, **127**, 13519–13521.
7 Farrusseng, D., Aguado, S., and Pinel, C. (2009) Metal–organic frameworks: opportunities for catalysis. *Angew. Chem. Int. Ed.*, **48**, 7502–7513.
8 Lillerud, K.P., Olsbye, U., and Tilset, M. (2010) Designing heterogeneous catalysts by incorporating enzyme-like functionalities into MOFs. *Top. Catal.*, **53**, 859–868.
9 Myers, A.L. and Prausnitz, J.M. (1965) Thermodynamics of mixed-gas adsorption. *AIChE J.*, **11**, 121–127.
10 Mehta, M. and Kofke, D.A. (1994) Coexistence diagrams of mixtures by molecular simulation. *Chem. Eng. Sci.*, **49**, 2633–2645.
11 Escobedo, F.A. (1998) Novel pseudoensembles for simulation of multicomponent phase equilibria. *J. Chem. Phys.*, **108**, 8761.
12 Spyriouni, T., Economou, I.G., and Theodorou, D.N. (1998) Phase equilibria of mixtures containing chain molecules predicted through a novel simulation scheme. *Phys. Rev. Lett.*, **80**, 4466–4469.
13 Ungerer, P., Tavitian, B., and Boutin, A. (2005) *Applications of Molecular Simulation in the Oil and Gas Industry – Monte Carlo Methods*, Éditions Technip, Paris.
14 Tylianakis, E. and Froudakis, G.E. (2009) Grand canonical Monte Carlo method for gas adsorption and separation. *J. Comput. Theor. Nanosci.*, **6**, 335–348.
15 Watanabe, S., Sugiyama, H., Adachi, H., Tanaka, H., and Miyahara, M.T. (2009) Free energy analysis for adsorption-induced lattice transition of flexible coordination framework. *J. Chem. Phys.*, **130**, 164707.
16 Greathouse, J.A., Kinnibrugh, T.L., and Allendorf, M.A. (2009) Adsorption and separation of noble gases by IRMOF-1: grand canonical Monte Carlo simulations. *Ind. Eng. Chem. Res.*, **48**, 3425–3431.
17 Ghoufi, A. and Maurin, G. (2010) Hybrid Monte Carlo simulations combined with a phase mixture model to predict the structural transitions of a porous metal–organic framework material upon adsorption of guest molecules. *J. Phys. Chem. C*, **114**, 6496–6502.
18 Coudert, F.X., Jeffroy, M., Fuchs, A.H., Boutin, A., and Mellot-Draznieks, C. (2008) Thermodynamics of guest-induced structural transitions in hybrid organic–inorganic frameworks. *J. Am. Chem. Soc.*, **130**, 14294–14302.
19 Coudert, F.X., Mellot-Draznieks, C., Fuchs, A.H., and Boutin, A. (2009) Prediction of breathing and gate-opening transitions upon binary mixture adsorption in metal–organic frameworks. *J. Am. Chem. Soc.*, **131**, 11329–11331.
20 Coudert, F.X. (2010) The osmotic framework adsorbed solution theory:

predicting mixture coadsorption in flexible nanoporous materials. *Phys. Chem. Chem. Phys.*, **12**, 10904–10913.

21 Kitaura, R., Seki, K., Akiyama, G., and Kitagawa, S. (2003) Porous coordination-polymer crystals with gated channels specific for supercritical gases. *Angew. Chem. Int. Ed.*, **42**, 428–431.

22 Hamon, L., Llewellyn, P.L., Devic, T., Ghoufi, A., Clet, G., Guillerm, V., Pirngruber, G.D., Maurin, G., Serre, C., Driver, G., van Beek, W., Jolimaître, E., Vimont, A., Daturi, M., and Férey, G. (2009) Co-adsorption and separation of CO_2–CH_4 mixtures in the highly flexible MIL-53(Cr) MOF. *J. Am. Chem. Soc.*, **131**, 17490–17499.

23 Boutin, A., Springuel-Huet, M.A., Nossov, A., Gédéon, A., Loiseau, T., Volkringer, C., Férey, G., Coudert, F.X., and Fuchs, A.H. (2009) Breathing transitions in MIL-53(Al) metal–organic framework upon xenon adsorption. *Angew. Chem. Int. Ed.*, **48**, 8314–8317.

4
Separation and Purification of Gases by MOFs
Elisa Barea, Fabrizio Turra, and Jorge A. Rodriguez Navarro

4.1
Introduction

From the perspective of saving resources and energy, there is a great need for highly efficient gas separation processes. Separation and purification processes of gases are of paramount importance for gas production and applications in which the use of gases with enhanced purity is required, namely air purification, electronics, fine chemicals, the food industry, fuel cells, power plants, health, and so on. In this regard, current gas purification processes include the use of bases (organic amines and inorganic hydroxides and oxides) for the removal of hydrogen sulfide and carbon dioxide acidic gases, oxidation processes for hydrogen sulfide removal, thermal and catalytic conversion of gas impurities, gas dehydration and purification by adsorption on molecular sieves, and membrane permeation processes [1].

The separation and purification process will also be highly dependent on the source of the feedstock, the production process itself, and the requirements of the product end user. N_2 and O_2 in gaseous and liquid forms are among the top five of the most highly produced industrial chemicals worldwide [2]. Consequently, one of the most important industrial processes is air separation and purification. The process of air separation consists in the separation of the components from which it is made, that is, nitrogen (78%), oxygen (20.9%), argon (1%), and other gases (0.1%). Current systems are based on the evolution of the Claude Linde process, which is summarized in Scheme 4.1. This process consists of an initial suction of atmospheric air followed by its purification by sequential dust filtration, drying on 4 Å molecular sieves, and adsorption of pollutants (i.e., CO_2, hydrocarbons, NO_x, SO_2) on 13X molecular sieves. Its subsequent compression followed by gas cooling through isenthalpic and isentropic expansion cycles leads to its liquefaction. Then a distillation process on two overlapping distillation columns, operating at different pressures, gives rise to oxygen and nitrogen separation. Argon is produced in a third distillation column, where the distillation process is repeated on an oxygen–

Metal-Organic Frameworks: Applications from Catalysis to Gas Storage, First Edition. Edited by David Farrusseng.
© 2011 Wiley-VCH Verlag GmbH & Co. KGaA. Published 2011 by Wiley-VCH Verlag GmbH & Co. KGaA.

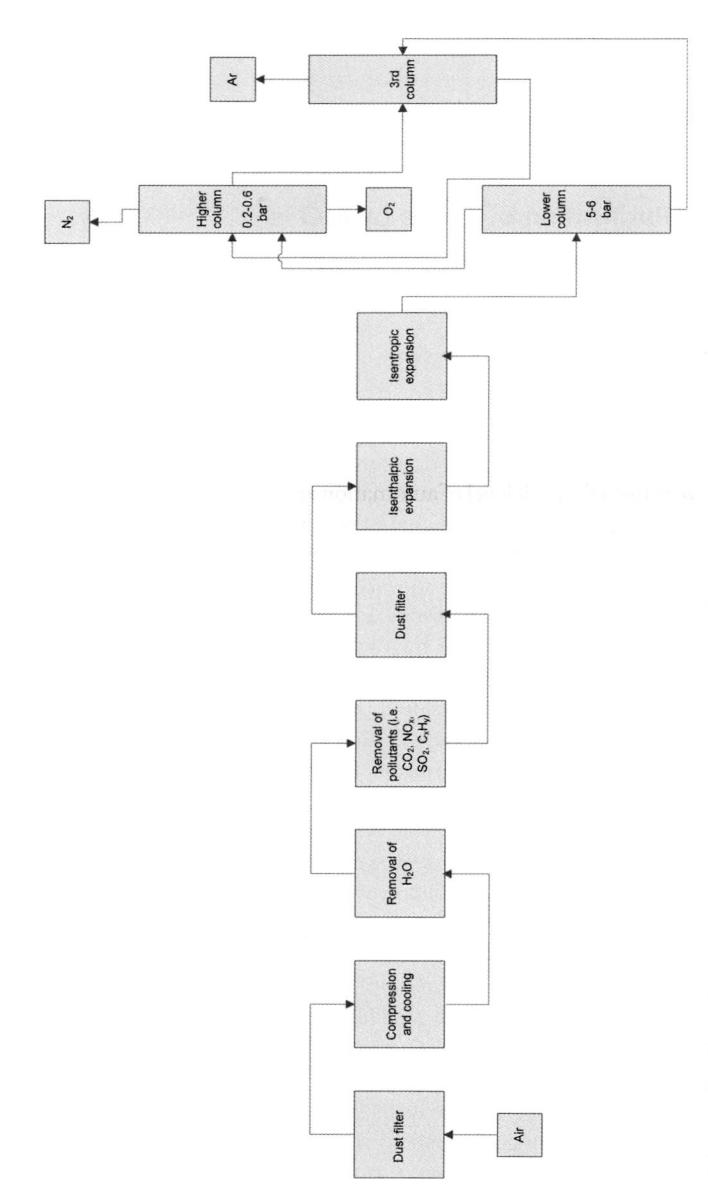

Scheme 4.1 Air purification and separation process on an industrial plant (SIAD).

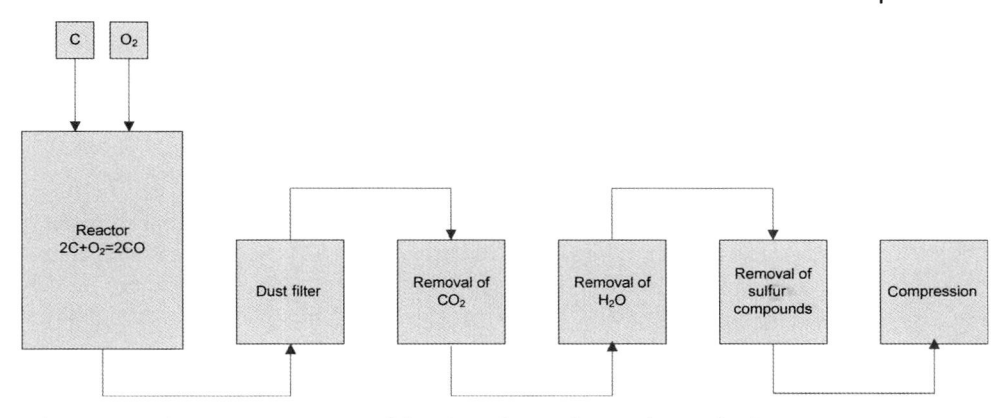

Scheme 4.2 Schematic representation of the CO production from oxidation of carbon, purification, and processing processes of CO at the SIAD production plant of Osio Sopra (Italy).

argon mixture taken from the midpoint of the upper column. The existing air separation units have achieved a high level of automation and their control is effected via a computer control system managing the entire production cycle and the subsequent distribution of the product to storage or use. As a second example, Scheme 4.2 summarizes CO production by partial burning of graphitic carbon and its subsequent purification. In this case, CO_2 by product is removed by the use of inorganic bases and sulfur compounds are eliminated by appropriate adsorbents.

Formerly, distillation (cryogenic separation) and liquid-phase absorption were major processes for gas separation [3]. However, in recent times, adsorption processes have been increasingly applied [4, 5].

Inorganic porous materials include a wide rage of adsorbents, such as silica gels, activated alumina or carbons, molecular sieve carbons, and zeolites. Among them, zeolites are a clear example of microporous systems of increasing technological interest due to their multiple applications in different areas (ion exchange, adsorption, heterogeneous catalysis, etc.), with a world business volume of US$350 billion per year. Nevertheless, these classical materials show some limitations related to the rigidity and anionic nature of the aluminosilicate framework, the difficulty in its functionalization, and the absence of homochirality. The extraordinary utility of zeolites, particularly in the area of molecular sieves, has attracted great attention in the search for alternative materials with improved features. As a result, a new class of synthetic porous materials, metal–organic frameworks (MOFs), also called porous coordination polymers (PCPs), has been developed.

The unique properties of MOFs which permit the fine tuning of the shape, size, and chemical nature of their pores, make them highly promising for carrying out these purification and separation processes in an effective and environmentally friendly way.

In this chapter, we critically review what is expected to be the contribution of the unique features of MOFs in the field of gas separation and purification.

4.2
General Principles of Gas Separation and Purification

4.2.1
Some Definitions

Separation is a process that transform a mixture of substances into two or more products that differ from each other in composition [6, 7]. Adsorptive gas separation processes can be divided into two types according to Keller's definition [8]: bulk separation and purification. The former consists in the adsorption of a significant fraction, more than 10 wt%, from a gas stream, whereas the latter takes place when less than 10 wt% is adsorbed (usually less than 2 wt%).

Separation/purification processes consist in passing a feedstock composed of a mixture of gases throughout a bed containing an appropriate adsorbent so that the resultant mixture is enriched in the more weakly adsorbed components. Adsorbents showing a high adsorption capacity for a specific component are desirable in order to adsorb large amounts of the target substance and to minimize the size of the purification unit. The adsorption capacity will depend on the nature of the adsorbate and on the relative composition of the mixture. On the other hand, the equilibrium selectivity will determine the separation ability of the adsorbent. It is based on differences in affinities of the adsorbent for the different species constituting the fluid phase. In the case of a binary mixture of components N and M, the equilibrium selectivity is generally expressed by using the separation factor $\alpha_{N,M}$: [9]

$$\alpha_{N,M} = (n_N/n_M)(p_M/p_N) \approx K_N/K_M \tag{4.1}$$

where n_N, n_M, p_M, and p_N are values obtained from pure component isotherms. It can be shown that in many cases $\alpha_{N,M}$ can be computed as the ratio of the Henry's constants (K_i) for N and M, and thus often referred as Henry's selectivity. For pressure swing adsorption (PSA) applications, $\alpha_{N,M}$ values between 2 and 104 may be considered acceptable.

The rate at which the adsorption process approaches equilibrium (adsorption kinetics) and the reversibility of the adsorption process (adsorbent regenerability) in a low energy-demanding manner are very important issues [9]. In this sense, it should be noted that fast adsorption kinetics are needed for efficient industrial applications. In addition, adsorbents can be regenerated by a number of methods to be reused in different cycles [4]. Among regeneration methods, temperature swing adsorption (TSA) and PSA are most commonly employed. In TSA, the adsorbent is regenerated by heating using preheated purged gas. As each heating–cooling cycle takes from a few hours to over a day, TSA is mainly used for purification, in which small quantities of adsorptive gases are processed. PSA is a low energy-demanding process operating at near ambient temperatures. In PSA, rapid pressurization–depressurization cycles are possible, which makes it useful for processing large amounts of gases. Adsorbent regeneration is accomplished by lowering the partial pressure of the adsorbed components in the gas phase either by reducing the total pressure or by flowing a portion of the product gas over the adsorbent.

4.2.2
MOFs: New Opportunities for Separation Processes

As mentioned before, in adsorptive separation processes, gas separation is possible because of the different strengths of interaction among the adsorbates and a given adsorbent. The physicochemical properties of the adsorbate molecules and of the adsorbent will determine the performance of the latter towards the different components of a gas mixture. High adsorption capacity, good selectivity, favorable adsorption kinetics, regenerability, and good thermal stability are the properties desirable for a promising adsorbent for gas separation and purification. To satisfy these requirements, porous materials with high surface areas in which their pore properties (pore size, pore volume, and the chemical functionality of the pore walls) can be systematically tuned are needed. In this context, silica gels, activated alumina or carbons, molecular sieve carbons, and zeolites have been traditionally employed as gas adsorbents for industrial applications (Table 4.1) [10, 11]. In recent years, the search for better adsorbents, the attributes of which meet the needs of each specific application, has evolved enormously. In this regard, MOFs are now being explored as adsorbents in gas separation/purification processes, showing promising results (Figure 4.1) [11]. Porous MOF materials exhibit two major differences with respect to classical inorganic adsorbents, which made them more versatile in terms of selectivity. First, they exhibit framework flexibility, in contrast to "rigid" carbons and zeolites. As a consequence, adsorption on MOFs may evolve dynamically depending on the nature and the quantity of host molecules. On the other hand, the fine-tuning of pore functionalization by the right choice of the metals or metal clusters and the organic linkers will also permit adsorbate–adsorbent interactions to be controlled and optimized. Thus, the versatile pore features of MOFs, such as redox activity, Lewis acidity/basicity, hydrophobicity, and chirality, result from the precise arrangement of transition metals and organic functional groups and are of key importance in adsorption processes.

4.2.3
Mechanisms of Separation and Design of MOFs for Separation Processes

In porous solids, separation can be achieved by steric, kinetic, or equilibrium effects [4, 7]. The steric effect refers to the molecular sieving property of materials possessing a uniform aperture pore size, such as zeolites. In this case, only properly sized and shaped adsorbates can diffuse into the adsorbent pores and the rest are excluded. Zeolites and molecular sieves are the most representative examples of adsorbents used in steric separations. The two largest applications are the use of 3 Å and 5 Å zeolites for drying and separation of normal alkanes from isoalkanes and cyclic hydrocarbons, respectively [4]. Kinetic selectivity is possible when the components of the gas mixture show a large difference among their adsorption/desorption

Table 4.1 Classical inorganic adsorbents for gas separation and purification (taken from [10]).

Porous material	Structure and pore features	Pore size	Applications
Silica gels	Amorphous materials with micro- and mesopores of different shapes and sizes, and different degrees of surface hydroxylation	Mean pore diameter: 20–30 Å	(1) Gas drying (2) Production of H_2 from reforming off gas (ROG)
Activated alumina	Amorphous materials with micro- and mesopores of different shapes and sizes, and pore surfaces containing both basic and acidic sites	Mean pore diameter: 20–50 Å	(1) Production of O_2- and N_2-enriched air, H_2, CO, and CO_2 from steam–methane reforming (SMR) off-gas, and H_2 from ROG (2) Solvent vapor recovery (3) VOC removal (4) Electronic gas purification
Activated carbons	Amorphous materials containing interconnected micro- and mesopores with various shapes and sizes, and having different volume fractions and pore walls of different surface chemistry giving rise to different degrees of local surface polarities	Distributed pores with diameter 3–100 Å	(1) Production of H_2, CO, and CO_2 from SMR off-gas, H_2 from ROG (2) Solvent vapor recovery (3) Gas desulfurization (4) VOC removal
Molecular sieve carbons	Amorphous microporous frameworks with larger cavities connected by precisely restricted pore windows	Window diameters: 3–5 Å	(1) Production of O_2- and N_2-enriched air (2) Production of CH_4 and CO_2 from landfill gas
Zeolites – A, X, chabizite, mordenite, silicalite clinoptilolite, and their ion-exchanged forms (H^+, Li^+, Na^+, K^+, Ba^{2+}, Ca^{2+}, Mg^{2+}, Ag^+, etc.)	Crystalline microporous frameworks with well-defined and uniform pore structure. One or more types of hydrated or nonhydrated cations and water molecules are located within the cavities	3–10 Å pore openings (3A, 3 Å; 4A, 4 Å; 5A, 4.9 Å; NaX, 7.5 Å; CaX, 10 Å; mordenite, 4 Å; chabizite, 4.9 Å; clinoptilolite, 3.5 Å; silicalite, 5.3 Å; Ca and Ba mordenites, 3.8 Å)	(1) Production of O_2- and N_2-enriched air, O_2 from air for home medical use, H_2 and CO_2 from SMR off-gas, CH_4 and CO_2 from landfill gas (2) Gas drying, desulfurization (3) Electronic gas purification (4) Separation of normal alkanes from isoalkanes and cyclic hydrocarbons

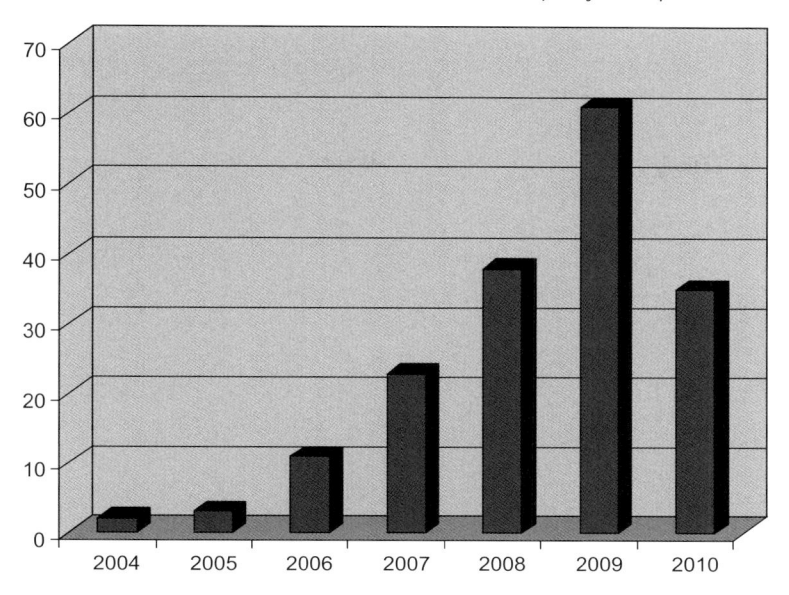

Figure 4.1 Number of publications in 2001–2010 for metal–organic framework or coordination polymers and gas separation or gas purification (from ISI Web of Science up to June 2010).

rates and is used when equilibrium separation is not possible. This is due to sterically hindered diffusion through the pores taking place when the pore window diameter is comparable to the kinetic diameter of the fed species. Air separation by PSA using zeolites and N_2 production from air are performed by this mechanism of adsorption. In the latter case, it should be noted that oxygen diffuses 30 times faster than nitrogen in carbon molecular sieves, although the amounts of both adsorbates at equilibrium are similar. The feasibility of kinetic separation has also been demonstrated in the separation of CH_4–CO_2 mixtures using carbon molecular sieves, propane–propylene separation using $AlPO_4$-14, and the upgrading of natural gas by the removal of N_2 from CH_4 with 4 Å zeolite [7]. Equilibrium separation processes are based on adsorbate–adsorbent interactions. In this case, the adsorbent pore size allows the diffusion of all the components of the mixture. As a consequence, the properties of the targeted molecule, that is to be adsorbed, and the surface of the adsorbent will determine the strength and the nature of such interaction and, as a consequence, the quality of the separation process. In order to select an adsorbent, the physicochemical properties of the targeted adsorbate to be taken into account are polarizability, magnetic susceptibility, acid–base nature, coordinative properties, permanent dipole moment, and quadrupole moment. Table 4.2 gives some physical parameters of selected gas and vapor adsorbates [10–12]. For adsorbates possessing high polarizability, but no polarity, adsorbents with high surface areas are good candidates. For example, in the case of highly polarizable organic molecules, such as benzene, MOFs with large hydrophobic cavities decorated with long π-extended aromatic linkers

Table 4.2 Physicochemical parameters of selected gases and vapors.

Adsorbate	Normal b.p. (K)	T_c (K)	P_c (bar)	Kinetic diameter (Å)	Polarizability $\times 10^{25}$ (cm^3)	$\mu \times 10^{18}$ (esu cm)	Quadrupole moment $\times 10^{26}$ (esu cm^2)	Comments
CO	81.66	132.85	34.94	3.690	19.5	0.1098	2.50	σ-Donor/π-acceptor
CO$_2$	216.55	304.12	73.74	3.3	29.11	0	4.30	Acidic
NO	121.38	180.00	64.80	3.492	17.0	0.15872	—	Redox-active σ-donor/π-acceptor
NO$_2$ (N$_2$O$_4$)	302.22	431.01	101.00	—	30.2	0.316	—	Redox-active σ-donor
N$_2$O	184.67	309.60	72.55	3.828	30.3	0.16083	—	Redox-active
CS$_2$	319.37	552.0	79.03	4.483	87.4–88.6	0	—	Acidic
COS	222.7	378.8	63.49	4.130	52–57.1	0.715189	—	Acidic
SO$_2$	263.13	430.80	78.84	4.112	37.2–42.8	1.63305	—	Acidic σ-donor π-acceptor redox-active
H$_2$S	212.84	373.40	89.63	3.623	37.82–39.5	0.97833	—	Acidic redox-active
(CH$_3$)$_2$S	310.48	503.00	55.30	—	—	1.554	—	σ-Donor
NH$_3$	239.82	405.40	113.53	2.900	21.0–28.1	1.4718	—	Basic σ-donor H-bonding
H$_2$O	373.15	647.14	220.64	2.641	14.5	1.8546	—	σ-Donor H-bonding
CH$_3$OH	337.69	512.64	80.97	3.626	32.3–33.2	1.70	—	H-bonding
C$_2$H$_4$O	283.07	469.00	71.90	—	—	1.8144	—	σ-Donor
C$_2$H$_6$OH	351.80	513.92	61.48	4.530	51.1–54.1	1.69	—	H-bonding
(CH$_3$CH$_2$)$_2$O	307.50	466.70	36.40	5.2	89.8	1.15	—	—
Acetone	329.22	508.10	47.00	4.600	64.7	2.88	—	—
CH$_4$	111.66	190.56	45.99	3.758	25.93	0	0	—
CCl$_4$	249.79	556.30	45.57	5.947	105–112	0	—	—
CF$_4$	145.11	227.51	37.45	4.662	38.38	0	0	—
C$_2$H$_6$	184.55	305.32	48.72	4.443	44.3–44.7	0	0.65	—
C$_2$H$_4$	169.42	282.34	50.41	4.163	42.52	0	1.50	σ, π-Donor/π-acceptor
C$_2$H$_2$	188.40	308.30	61.14	3.3	33.3–39.3	0	—	σ, π-Donor/π-acceptor H-bonding
c-C$_6$H$_{12}$	353.93	553.50	40.73	6.0–6.182	110.4	0	—	—
C$_6$H$_6$	353.24	562.05	48.95	5.349–5.85	104.4	0	—	σ, π-donor/π-acceptor π-stacking

are ideal adsorbents as they may interact with the adsorbate through stacking interactions [13]. Sorbents with highly polar surfaces are ideal for molecules with high dipole moments. This is the case with water, which will interact strongly with MOFs possessing hydrophilic pores functionalized with carboxylic or hydroxyl groups [14]. If the targeted molecule has a high quadrupole moment, sorbents with surfaces that have high electric field gradients are desirable, such as acetylene [11]. Moreover, the possibility of the formation of supramolecular interactions or coordination bonds between the adsorbate and the adsorbent also contributes to the efficiency and selectivity of many separation/purification processes. In this regard, the presence of hydroxyl, amino, carboxylic, or amide groups in the organic linkers of MOFs will constitute attractive interaction sites via hydrogen bonding during recognition processes, and also additional coordination sites for metal ions. Adequate functionalized MOFs with basic surface oxygen or nitrogen atoms are, in principle, good candidates for the selective removal of molecules with H-bonding possibilities such us C_2H_2, in contrast to other molecules, such as CO_2, with similar equilibrium sorption parameters, related physicochemical properties, and molecular shape and size [15]. Moreover, such functional groups can act as catalytic interaction sites on the pore walls [16]. On the other hand, the presence of coordinatively unsaturated metal ions – open metal sites (OMSs) – in the MOF framework paves the way for the use of these materials in enhanced selective adsorption, dehydration, and catalytic decomposition of trace gases as the coordinative unsaturation of such centers may lead to stronger and selective chemisorption of gas molecules. In this case, the mechanism of removal or decomposition of a targeted molecule implies its coordination to metals with vacant coordination sites.

4.2.4
Experimental Techniques and Methods to Evaluate/Characterize Porous Adsorbents

Gas adsorption isotherms (or equilibrium isotherms) characterize the adsorption equilibrium and are the basis for the evaluation of adsorptive separation. An adsorption isotherm is a function that relates at constant temperature the amount of substance adsorbed at equilibrium to the pressure (or concentration) of the adsorptive in the gas phase [17]. Experimentally, adsorption isotherms can be measured by volumetric or gravimetric methods. An issue to be taken into account prior to the measurement of the adsorption isotherms is that the cavities within MOFs are filled with solvent molecules in their as-synthesized forms. Successful establishment of permanent porosity has to be accomplished in order to test the stability of the material upon removal of the guest species, which implies that, after the loss of solvent molecules, the framework is stable and does not collapse. The textural characterization of MOFs is usually performed by measuring their dinitrogen adsorption–desorption isotherms at its normal boiling point (77 K) – previously, the material will be activated heating it at the temperature at which the loss of solvent molecules takes place. The majority of MOFs reported to the date are microporous, having cavities less than 2 nm in size, and display type I isotherms (Brunauer classification).

Subsequently, two values will be calculated from these measurements to allow the porosity to be compared: surface area and pore volume. Type I isotherms can often be described by the Langmuir model [18, 19], which assumes that a homogeneous monolayer of the adsorbate is formed on the walls of the adsorbent. The Brunauer–Emmett–Teller (BET) equation, which is an extension to the former model to describe multilayer adsorption, can also be used [20]. In either case, the apparent surface area will be calculated as the product of the estimated value of monolayer uptake and an accepted value for the area occupied by an adsorbate molecule. The micropore volume, usually calculated by the Dubinin–Radushkevich method [21], is a complementary descriptor of the porosity of a material. Good agreement between pore volumes obtained from crystallographic studies and gas adsorption will be necessary to confirm the purity of the material.

Once the permanent porosity has been established, adsorption–desorption isotherms of the gases to be separated will also be studied so as to determine the adsorption capacity of the MOF under study. High adsorption capacities of the targeted molecule to be removed are desirable. In addition, in order to evaluate the energy of the adsorbate–MOF interaction, the isosteric heats of adsorption have to be calculated according to the Clausius–Clapeyron equation by using the gas adsorption isotherms measured at different temperatures [5]. High isosteric adsorption heats will imply a strong interaction between the guest molecules and the host. It should be noted that the strength of the interaction should be optimized so as to reach high adsorption capacities, which ensure impurity removal and also reversible gas uptake in order to regenerate the adsorbent. Alternative methods for the determination of the heat of adsorption include variable-temperature spectroscopic methods [22] and gas chromatographic methods [5].

Many of the published studies on the possible utility of MOFs for gas separation and purification processes rely on monocomponent isotherms [23]. The real situation may, however, give rise to a different behavior as a consequence of blocking or cooperative effects of the other components [13].

In this regard, measurement of breakthrough curves of multicomponent mixtures provides highly valuable information in order to characterize the separation efficiency of a material towards a gas mixture [4]. Experimentally, a gas mixture is flowed through a thermostated bed containing the activated MOF. The adsorbate concentration in the flow at any given point in the bed is a function of time, resulting from the movement of the concentration front through the bed. In order to determine the potential application of a certain MOF in the separation of a mixture of gases, the composition of the resultant stream is monitored using different techniques, such as mass spectrometry and gas chromatography. The variation of the adsorbate concentration in the column effluent over time is correlated with the selectivity of a MOF for one component relative to another. Ideally, the adsorbate to be removed should be strongly adsorbed in the MOF (it will not be detected in the effluent stream) until saturation (from that moment, it will be detected in the effluent stream) whereas the other components of the mixture pass through the bed and are not retained. Once saturation has been reached, the adsorbent is regenerated by the appropriate method (TSA or PSA) or replaced in the case of a "single-use" cartridge.

4.3
MOFs for Separation and Purification Processes

MOFs are based on metal ions connected by organic linkers defining a porous network ready for molecular recognition. These materials are easily obtained by a self-assembly process using conventional or solvothermal methods. Pioneering investigations of some groups, such as those of Yaghi [24], Kitagawa [25], and Férey [26], showed the structural diversity and the rich variety of properties of MOFs. Over the past 10 years, the synthesis of novel MOFs has been pushed by the widening of applications, including gas separation processes [11], gas storage [27], heterogeneous catalysis [28], sensing [29], and biological applications [30]. In contrast to conventional microporous materials, such as zeolites and activated carbon materials, MOFs have characteristic features that include (i) well-ordered crystalline porous structures, (ii) designable channels surface functionalities that are at the origin of the selective catalytic properties of these materials, and (iii) a framework responsible, as in dense solids, for physical properties such as magnetism [31], conductivity [32], and optical features [33]. It should be emphasized that in contrast to zeolites that exhibit rigid skeletons, MOFs can show rigid or flexible frameworks. The former possess stable and robust porous frameworks with permanent porosity, similar to zeolites, whereas the latter show flexible and dynamic behaviors in response to guest molecules or external stimuli, such as pressure and temperature [34]. Moreover, the crystallinity of MOFs is another interesting feature as it allows their structural characterization through diffraction methods, mainly X-ray diffraction of single crystals and microcrystalline powders. The knowledge of the molecular structure of these compounds is at the basis of the comprehension of their physicochemical properties, such as stability, inertness, magnetism, optics, adsorption, and catalysis, and can therefore be successfully employed to tune the mentioned properties. The right choice of the metal-based building blocks and organic linkers allows the control of the structure and physicochemical properties of MOFs. In addition, organic synthesis may be used as a powerful tool to functionalize rationally the organic spacers even in a postsynthetic stage [35]. Taking into account the above-mentioned characteristics of MOFs, these materials can be considered as potential adsorbents for gas separation and purification. However, current investigations in this field are at an early stage and are mainly focused on gas adsorption studies based on adsorption–desorption isotherm measurements of single gas components. In order to establish the practical application of MOFs in gas separation/purification processes, many other issues should be considered, such as scale-up, shaping, and robustness of MOFs, the evaluation of the selective adsorption of mixtures of gases, and optimization of separation.

4.3.1
MOF Materials as Molecular Sieves

As in the case of molecular sieves, the size of the pores or the pore openings plays a critical role in the separation selectivity by MOFs. For gas separations of small molecules, the size of the pores may be the dominating factor and, as a consequence, the kinetic diameters of the gases to be separated, such as H_2 (2.89 Å), O_2 (3.46 Å), N_2 (3.64 Å), CO (3.76 Å), CO_2 (3.3 Å), and CH_4 (3.8 Å), are of primary concern. However, in the case of the separation of larger molecules, for example, linear or branched aliphatic hydrocarbons and aromatics, not only the size of the pores but also the shape, the nature (hydrophilic/hydrophobic, aliphatic/aromatic), and even the hierarchy of the porosity can be crucial to the selectivity (see Section 4.3.4). The structures of MOFs are typically rigid but some of them exhibit extraordinary structural flexibility upon adsorption/desorption of specific gases or liquids (see Section 4.3.2). In this section, we focus on selective adsorption based mainly on size/shape exclusion in MOFs with rigid structures. In this regard, it should be emphasized that the selective capture of CO_2, in particular at ambient temperature and pressure, from industrial emission streams that contain other gases such as N_2, CH_4, and H_2O still remains challenging. Selective separation of N_2 and O_2 from air is also very interesting as this process is currently performed on a scale of billions of tons per year. On the other hand, separation of H_2 from CO is demanded for fuel cell applications and also H_2 enrichment of the N_2–H_2 exhaust mixture resulting from ammonia synthesis [4].

In this context, some MOFs could potentially be used for the separation of CO_2–CH_4 and CO_2–N_2 mixtures as they selectively adsorb CO_2 into their pores and show almost no N_2 or CH_4 uptake, indicating a general correlation between the pore size and the kinetic diameter of the adsorbates. N_2 and CH_4 cannot be adsorbed because their kinetic diameters are larger than the effective channel size of these compounds. Some examples are $Mn(HCOO)_2$ [36] and MIL-96 [37], which selectively adsorb CO_2 over CH_4 at 195 and 303 K, respectively. In addition, MIL-96 also discriminates CO_2 from N_2 at 195 K, similarly to other MOFs, such as Sm_4Co_3 $(pyta)_6(H_2O)_x$ [38] (pyta = 2,4,6-pyridinetricarboxylate), $[Fe^{III}(Tp)(CN)_3]_2Co^{II}$ [39] [Tp = hydrotris(pyrazolyl)borate], and Zn(dtp) [40] (H_2dtp = 2,3-di-1H-tetrazol-5-ylpyrazine).

On the other hand, preferential adsorption of H_2 over N_2 at 77 K has been observed for $Mn(HCOO)_2$ [36] (Figure 4.2), $Sm_4Co_3(pyta)_6(H_2O)_x$ [38], $Cu(F-pymo)_2$ [41] (F-Hpymo = 5-fluoro-2-hydroxypyrimidine), $[Fe^{III}(Tp)(CN)_3]_2Co^{II}$ [39], $Yb_4(TATB)_{8/3}$ $(SO_4)_2$ [42] (PCN-17) (4,4′,4″-s-triazine-2,4,6-triyltribenzoate), and $Mg(ndc)_3$ [43] (ndc = 2,6-naphthalenedicarboxylate). It should be noted that selective adsorption of O_2 (77 K) over N_2 (77 K) has only been demonstrated for a few known MOFs including $Mn(HCO_2)_2$, Zn(dtp) [40], PCN-17, and $Mg(ndc)_3$. This behavior is analogous to the molecular sieving effect observed in zeolite 4 Å, which shows O_2 but no N_2 uptake at 123 K [44]. Taking this consideration into account, it can be inferred that the pore openings in such compounds are between the kinetic diameters of O_2 (3.46 Å) and N_2 (3.64 Å). This hypothesis is moreover confirmed in PCN-17 and $Mg(ndc)_3$ because in both cases only a very small amount of CO is adsorbed, consistent with prohibited access to the pores due to its larger kinetic diameter of 3.76 Å.

Finally, it should be noted that, in some cases, the activation temperature affects pore size and, as a consequence, MOF selectivity in gas separation processes. This is the case with $Zn_2(cnc)_2(dpt)\cdot G$ [45] (cnc = 4-carboxycinnamate; dpt = 3,6-di-4-pyridyl-1,2,4,5-tetrazine; G = guest). When it is activated under mild conditions, partial

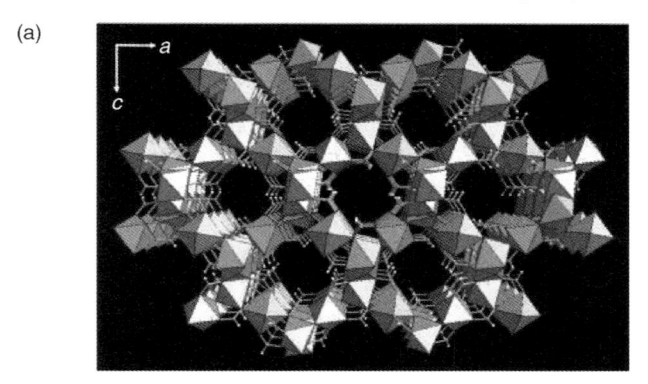

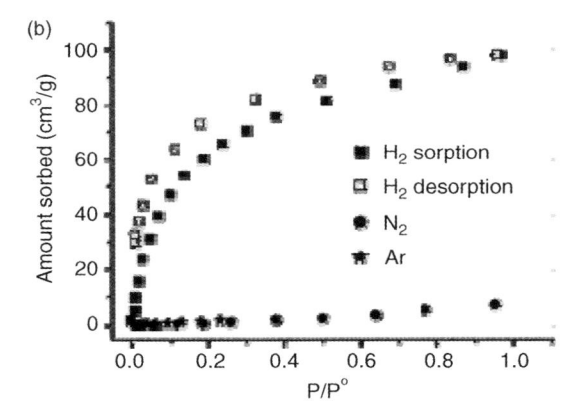

Figure 4.2 (a) X-ray crystal structure of Mn(HCOO)$_2$ showing channels along the b-axis. Guest molecules are omitted for clarity. Mn^{II} ions coordinated by six formate ligands are represented by octahedra. (b) Gas adsorption isotherms at 78 K showing the molecular sieve behavior of H$_2$ over N$_2$.

loss of the guest molecules takes place and the resulting activated phase can take up a fairly large amount of H$_2$ but a negligible amount of N$_2$, underlying its capacity for H$_2$–N$_2$ separation. If this phase is further activated at a higher temperature, all guest molecules are completely released and the selectivity is lost.

4.3.2
Flexible MOFs for Enhanced Adsorption Selectivity

Flexible MOFs respond to external stimuli by reversible structural transformations and, generally, guest molecules may act as the stimuli for such transformations (Figure 4.3) [46]. Much attention has been devoted to these systems because of their particular host–guest chemistry and their potential applications [47]. It should be emphasized that, in the case of flexible MOFs, the combination of flexibility and a functional surface permits an extraordinarily effective selectivity for gas separation

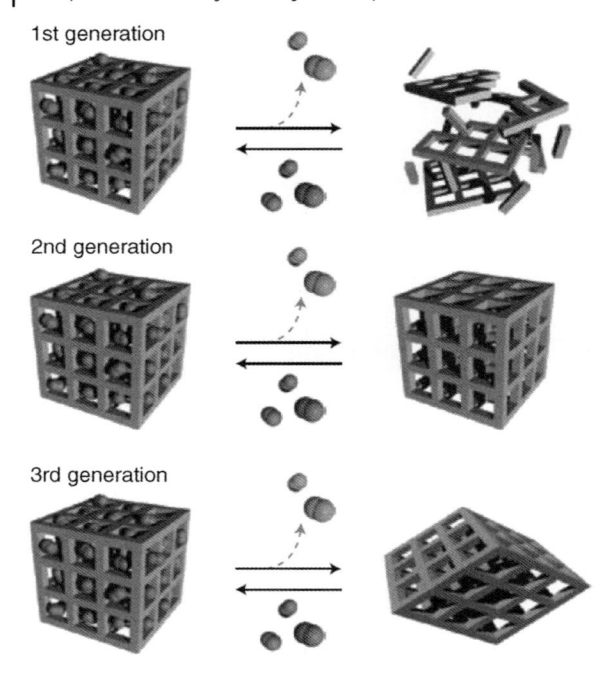

Figure 4.3 Classification of MOFs into three categories according to Kitagawa and Kondo [46]. The first-generation materials collapse on guest removal. The second-generation materials have robust and rigid frameworks, and retain their crystallinity when the guests are not present in the pores. The third-generation materials are transformable accompanied by structural transformation. Reproduced with permission from Kitagawa, S. and Kondo, M. *Bull. Chem. Soc. Jpn.*, 1998, **71**, 1739. Copyright 1998 Chemical Society of Japan.

processes [11, 48]. As mentioned in Section 4.2.4, for the development of a relevant adsorbent, there is a need to control the adsorption isotherms. However, the adsorption isotherms of flexible MOFs sometimes cannot be classified according to the IUPAC scheme [49]. This is related to the dynamic guest accommodation behavior of these materials, which gives rise to a combination of isotherms or to a new isotherm. For instance, a "gate-type" adsorption profile shows no uptake of guest molecules at low pressure, and an abrupt increase in adsorption after a threshold pressure, known as the gate-opening pressure. This behavior is associated with a structural transformation from a nonporous to a porous phase induced by the adsorbate incorporation, with the gate-opening pressure being sensitive to the properties of the gas to be adsorbed (Figure 4.4) [50].

Whereas the role of adsorption for single components on the flexibility is now fairly well established, knowledge is still scarce about the "gate-opening" behavior in the presence of gas mixtures, in particular for those containing a component that induces breathing (see below) and another does not, with eventual cooperative effects [51]. As a consequence, studies in this sense are of great fundamental interest.

(a)

[Cu$_2$(dhbc)$_2$(bpy)]$_n$
Closed form

[Cu$_2$(dhbc)$_2$(bpy)]$_n$
Open form

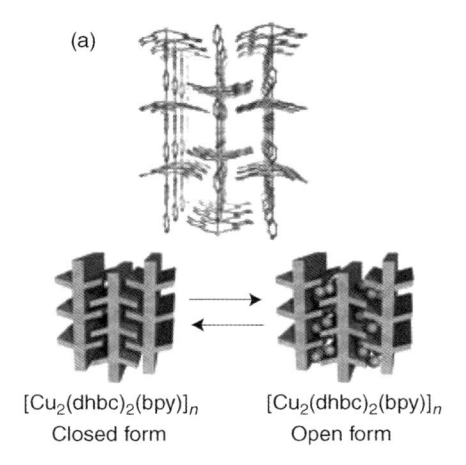

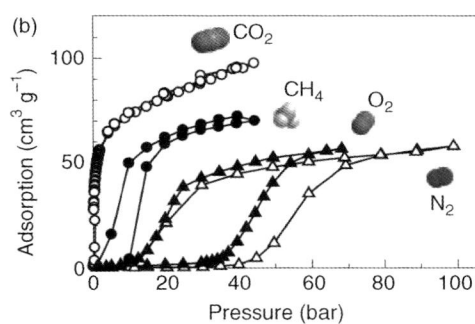

Figure 4.4 (a) 3D crystal structure of interdigitated framework Cu$_2$(dhbc)$_2$(bpy) with illustrations of closed and open structures formed via guest sorption processes. (b) Adsorption isotherms of gases for Cu$_2$(dhbc)$_2$(bpy) at room temperature. Each gas is adsorbed at a different gate-opening pressure. Open circles, CO$_2$; filled circles, CH$_4$; filled triangles, O$_2$; open triangles, N$_2$. Reproduced with permission from Kitaura, R., Seki, K., Akiyama, G., and Kitagawa, S. *Angew. Chem. Int. Ed.*, 2003, **42**, 428. Copyright 2003 Wiley-VCH Verlag GmbH.

One of the most spectacular examples of flexible MOFs is the series of terephthalates MIL-53, M(OH)(O$_2$C–C$_6$H$_4$–CO$_2$) (M = Al^{3+}, Cr^{3+}, Fe^{3+}) [52], which are able to adjust their pore size and shape in response to the adsorption of polar molecules (CO$_2$, H$_2$O) and linear hydrocarbons, except methane [52–54]. Starting from the large-pore form (LP), with rectangular pores (0.85 × 0.85 nm), the structure switches to the narrow-pore (NP) form, with trapezoidal pores (0.26 × 1.36 nm), upon adsorption of the above-mentioned gases or vapors. The reversible transformation of large- to small-pore forms is usually called the "breathing" phenomenon. It should be emphasized that co-adsorption studies of CO$_2$–CH$_4$ mixtures have been performed in the MIL-53(Cr) system in order to understand its breathing mechanism, demonstrating that the closing and opening of the structure is entirely controlled by the partial pressure of CO$_2$ [54]. However, the CO$_2$–CH$_4$ selectivities in the breakthrough experiments are not very high, probably due to a kinetic barrier for the LP to NP transition (the NP form is preferred as it has a stronger affinity for CO$_2$ and virtually excludes CH$_4$), which is not completed during the experiment. Other interesting examples related to the selective adsorption of CO$_2$ from CO$_2$–CH$_4$ mixtures are Zn(Pur)$_2$ (ZIF-20; Pur = purinate) [55], amino-functionalized MIL-53(Al) [56], and our Cu(Hoxonato)(bpy)$_{0.5}$ (H$_3$oxonic = 4,6-dihydroxy-1,3,5-triazine-2-carboxylic acid) complex [57]. In ZIF-20, CO$_2$ uptake at 273 K is five times higher than that of CH$_4$, suggesting a stronger interaction between the framework and the CO$_2$ molecules. Indeed, the breakthrough experiment using a CO$_2$–CH$_4$ (50:50 v/v) gas mixture clearly shows that ZIF-20 can separate CO$_2$ from CH$_4$. The maximum pore aperture (2.8 Å) in ZIF-20, as measured from its crystal structure, is smaller than the kinetic diameters of CO_2 (3.3 Å) and CH_4 (3.8 Å). However, the space inside the structure becomes accessible through a dynamic aperture-widening process wherein the purine swings out of the way to allow gas molecules to pass. The observed gas-separation behavior is probably the result of uncoordinated nitrogen atoms inducing a polar pore wall, and thus favorable CO_2 binding sites, and/or the appropriate pore size that prefers CO_2 rather than CH_4. Moreover, the separation performance of amino-MIL-53(Al) at 303 K using an equimolar CO_2–CH_4 mixture reveals that CH_4 elutes rapidly from the column, whereas CO_2 is strongly retained. Under these conditions, no CH_4 is adsorbed whereas 0.83 mmol of CO_2 is adsorbed per gram of adsorbent. This almost infinite selectivity at 1 bar is a very large improvement relative to its parent MIL-53(Al), which shows a selectivity of ∼7 at 1 bar. It should be noted that, in spite of the lack of a porous structure, Cu(Hoxonato)(bpy)$_{0.5}$ shows the selective retention of CO_2 at 273 K even from CO_2–CH_4 mixtures highly enriched in methane, possessing the typical composition of natural gas (5:95 v/v) (Figure 4.5). The existence of only small, isolated voids in this compound bring us to the conclusion of a dynamic behavior of the flexible network upon its exposure to polar guest molecules (H_2O, CO_2), which is probably related to the typical plasticity of copper(II) coordination polyhedra.

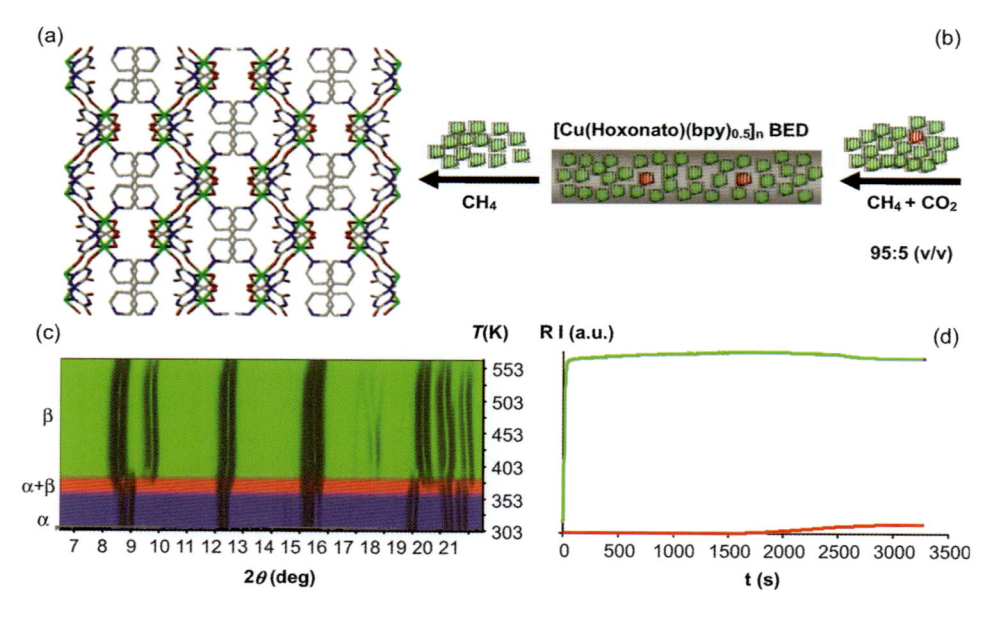

Figure 4.5 (a) Perspective view of Cu(Hoxonato)(bpy)$_{0.5}$ polymer along the 001 direction. (b) Schematic representation of the gas purification process using an activated bed of Cu(Hoxonato)(bpy)$_{0.5}$. (c) Variable-temperature X-ray diffractograms for the as-synthesized hydrated Cu(Hoxonato)(bpy)$_{0.5}$ showing the flexibility of the framework along with the dehydration process. (d) Breakthrough curves of a stream of CH_4–CO_2 (0.95:0.05 v/v) mixture passed through a sample of Cu(Hoxonato)(bpy)$_{0.5}$ showing the retention of CO_2 and passage of CH_4. Reproduced with permission from Barea, E., Tagliabue, G., Wang, W.G., et al. Chem. Eur. J., 2010, **16**, 931. Copyright 2010 Wiley-VCH Verlag GmbH.

On the other hand, selective adsorption of CO_2 over other gases, such as O_2 and N_2, has also been observed in $Cd_2(pzdc)_2L$ [58] [H_2pzdc = 2,3-pyrazinedicarboxylic acid; L = 2,5-bis(2-hydroxyethoxy)-1,4-bis(4-pyridyl)benzene)] and [Ni(4,4′-bpe)$_2$(N(CN)$_2$)](N(CN)$_2$) [59] (4,4′-bpe = *trans*-bis(4-pyridyl)ethylene). In both cases, there is an interesting gate-opening mechanism. $Cd_2(pzdc)_2L$ possesses a rotatable pillar bearing ethylene glycol side chains acting as a molecular gate with locking–unlocking interactions triggered by guest inclusion. As a consequence of this, $Cd_2(pzdc)_2L$ exhibits no uptake of O_2 (kinetic diameter = 3.46 Å) and N_2 (3.64 Å) at 77 K, indicating that there is no porosity corroborating the structure determination. In contrast, CO_2 (3.3 Å), regardless of its comparable size to O_2 and N_2, can be adsorbed. Moreover, the isotherm of CO_2 shows a gate-opening pressure and exhibits a large hysteresis. The inclusion of CO_2 only occurs in the higher vapor pressure region because the molecular gate is strongly locked by the formation of hydrogen bonds between the ethylene glycol side chains, which cannot be broken by CO_2 at low vapor pressure due to its limited ability to form hydrogen bonds. Therefore, the gate only opens at a higher vapor pressure of CO_2 ($p/p° = 0.9$) when the hydrogen bonds between the ethylene glycol side chains are cleaved and the pillars are free to rotate. In addition, in [Ni(bpe)$_2$(N(CN)$_2$)](N(CN)$_2$), despite an adequate effective pore size, no N_2 (16.3 Å^2, 3.64 Å) or O_2 (14.1 Å^2, 3.46 Å) diffusion into the micropores is observed at 77 K, whereas a significant CO_2 uptake is observed at 195 K, despite its similar size to O_2 and N_2. This type of selectivity is unique and can be explained by taking into account that O_2 and N_2 adsorbates at 77 K interact very strongly with the pore windows, blocking other molecules from passing into the pores. In the case of CO_2 sorption (at 195 K), such interactions are overcome by the thermal energy and the host, which contains Ni^{II}, polar groups and π-electron clouds from the bpe ligands inside the pores. This gives rise to an electric field, which is effective in CO_2 sorption. Such dipole–induced dipole interactions, where the quadrupole moment of CO_2 interacts with the electric field gradient, make a further contribution to the potential energy of adsorption for carbon dioxide.

Finally, fewer examples have been published on the separation of H_2 from other gases. In this regard, Co(ncd)(bpy)$_{0.5}$ [60] and Cu(fma)(bpe)$_{0.5}$ [61] (fma = fumarate) selectively adsorb H_2 over N_2 and H_2 over N_2, CO, and Ar at 77 K, respectively. These compounds belong to a family in which the framework is composed of paddle-wheel dinuclear M_2 units bridged through dicarboxylate dianions to form 2D layers, which are further pillared in a 3D network by bipyridine-type ligands {similarly to $Cd_2(pzdc)_2L$ [58]} When adsorbates enter the pores, the interpenetrated frameworks alter their pore sizes, adjusting the inter-framework distances. In the same way, very few MOFs have been shown to favor the uptake of O_2 over N_2 or CO_2. Cu(BDTri)L [62] [H_2BDTri = 1,4-benzenedi(1*H*-1,2,3-triazole; L = dimethylformamide (dmf), diethylformamide (def)] are relevant examples that can accommodate O_2 in their pores, in contrast to N_2. Cu(BDTri)L possesses the same structure type but with different coordinated solvent molecules, L = dmf and def), directed toward the cavity interiors. Cu(BDTri)(dmf) shows two-step isotherms at 77 K for O_2 and N_2 adsorption, which is associated with the permanent porosity of the closed phase and a pore-opening process. With larger def bridging molecules, the analogous Cu(BDTri)(def) exhibits selective pore opening with O_2 but not with N_2 gas. At 200 mbar, the selectivity of this compound of O_2 over N_2 reaches 13.5, representing the highest value yet observed for microporous materials under similar conditions. Zn(TCNQ–TCNQ)(4,4′-bpy) [63] (TCNQ = 7,7,8,8-tetracyano-p-quinodimethane) can accommodated NO and O_2 preferentially over a variety of other gas molecules, such as C_2H_2, Ar, CO_2, N_2, and CO. This unusual selectivity arises from the combination of a charge-transfer interaction between these adsorbates and the TCNQ ligands, and the gate opening and closing of the pores. The fact that the framework does not adsorb CO_2 and C_2H_2 at 193 K excludes the quadrupole moment or polarizability of the guest molecules as factors responsible for the selectivity. Further, the fact that the framework adsorbs NO but not CO (77 K) indicates that the selectivity is also not induced by the electric dipole moment of the guest molecules.

4.3.3
MOFs with Coordination Unsaturated Metal Centers for Enhanced Selective Adsorption and Dehydration

The presence of open metal sites (OMSs) ensures a high interaction with adsorbate molecules exhibiting basic Lewis/coordination characteristics [22]. This feature was initially used with the aim of increasing the gas storage capacity of generally weakly binding molecules, that is, hydrogen [64]. The presence of OMSs is also useful for enhancing the efficiency in the storage of other energetically relevant molecules with coordinative active groups, such as the σ,π-donor/π-acceptor nature of acetylene [65]. The presence of coordinative active binding sites can also become a highly valuable feature for gas purification purposes [66]. In this regard, most of the studies on MOFs with coordinatively unsaturated metal centers have been carried out on the HKUST-1 $Cu_3(btc)_2$ (btc = benzene-1,3,5-tricarboxylate) material and on the CPO-27 or MOF-74 family $M_2(dhtp)$ (M = Mg^{2+}, Mn^{2+}, Co^{2+}, and Zn^{2+}; H_4dhtp = 2,5-dihydroxyterephthalic acid) [65]. Indeed, $Cu_3(btc)_2$ has been shown to be particularly effective for the removal of sulfur-containing molecules, such as tetrahydrothiophene and thiophene, from natural gas [13, 67], gasoline, and diesel oil [68] as a consequence of the formation of Cu–S coordinative bonds. This material has also been shown to be very effective for the capture of other small molecules bearing σ-donor coordinatively active groups, such as ethylene oxide and ammonia [69], and molecules with σ,π-donor/π-acceptor characteristics, which gives rise to the efficient separation of propylene from propane as concluded from breakthrough curve experiments [70]. Another interesting feature of $Cu_3(btc)_2$ is that adsorbate binding to OMSs is responsible for a color change of the material, which is diagnostic for determining the efficiency of the process and the replacement of the cartridge [69]. We have also recently demonstrated that the coupling of OMSs and polarity gradients in the anionic framework $NH_4[Cu_3(\mu_3\text{-}OH)(\mu_3\text{-}4\text{-}$carboxypyrazolato$)_3]$ leads to the resolution of acetylene, methane, and CO_2 complex mixtures, as deduced from variable-temperature reverse pulse gas chromatography [71].

The [M$_2$(dhtp)] CPO-27, also known as MOF-74, series possesses a very high density of active metal sites. The strength of the interaction of this system towards different probe molecules (NO, H$_2$, CH$_4$, CO, and CO$_2$) has been studied by means of variable-temperature IR spectroscopy, giving rise to values as high as 50 kJ mol^{-1} for the interaction of [Ni$_2$(dhtp)] with CO, making this material of interest for H$_2$ purification purposes [72]. Measurements of breakthrough curves also show that [Zn$_2$(dhtp)] is particularly effective for the capture of sulfur dioxide [69]. [Mg$_2$(dhtp)] might eventually be of interest for gas drying purposes as an alternative to molecular sieves, as deduced from reversible hydration–dehydration cycles [73].

The mixed metal Zn$_3$(bdc)$_3$[Cu(Pyen)] (H$_2$bdc = benzene-1,4-dicarboxylic acid; PyenH$_2$ = 5-methyl-4-oxo-1,4-dihydropyridine-3-carbaldehyde) system possesses Cu OMSs, which are responsible for a very high adsorption enthalpy for hydrogen molecules at zero coverage (12.3 kJ mol^{-1}) (Figure 4.6) [74, 75]. Another very significant feature of this system is the kinetic isotope quantum molecular sieving for H$_2$–D$_2$, which is related to a combination of OMSs and ultramicropores.

It has also been shown recently that coupling redox activity with coordinatively unsaturated metal centers in the controlled reduction process of Fe^{III} centers in MIL-100(Fe) to mixed-valence Fe^{II}/Fe^{III} MIL-100(Fe) [{Fe$_3$O(H$_2$O)$_2$F$_{0.81}$(OH)$_{0.19}$}{C$_6$H$_3$(CO$_2$)$_3$}$_2$] gives rise to a pronounced enhancement of the π-back-bonding features of the Fe^{II} metal centers towards the π* antibonding orbitals of CO and

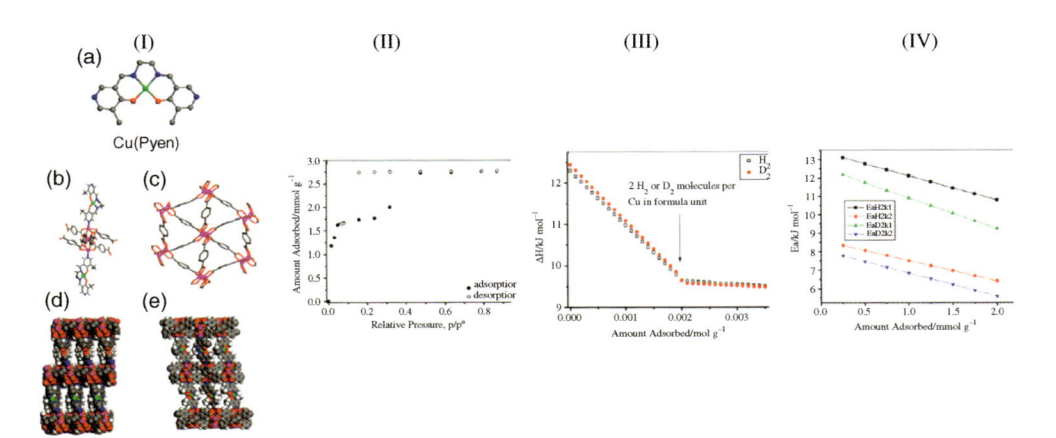

Figure 4.6 (I) X-ray crystal structure of Zn$_3$(bdc)$_3$[Cu(Pyen)] showing (a) [Cu(Pyen)], (b) one trinuclear Zn$_3$(COO)$_6$ secondary building unit, (c) one Zn$_3$(bdc)$_3$ 2D sheet that is pillared by the Cu(Pyen) to form a 3D microporous framework, (d) curved pores of about 5.6 × 12.0 Å along the c-axis, and (e) irregular ultramicropores along the b-axis. Color scheme: Zn (magenta), Cu (green), O (red), N (blue), C (gray), H (white). (II) Adsorption isotherm of benzene at 298 K. (III) Variation of enthalpy of adsorption (kJ mol^{-1}) with amount adsorbed (mol g^{-1}) for H$_2$ and D$_2$ adsorption. (IV) Variation of activation energy (E_a, kJ mol^{-1}) with amount adsorbed (mmol g^{-1}) for H$_2$ and D$_2$ adsorption. Reproduced with permission from Chen, B., Xiang, S., and Qian, G. *Acc. Chem. Res.*, **43**, 1115. Copyright 2010 American Chemical Society.

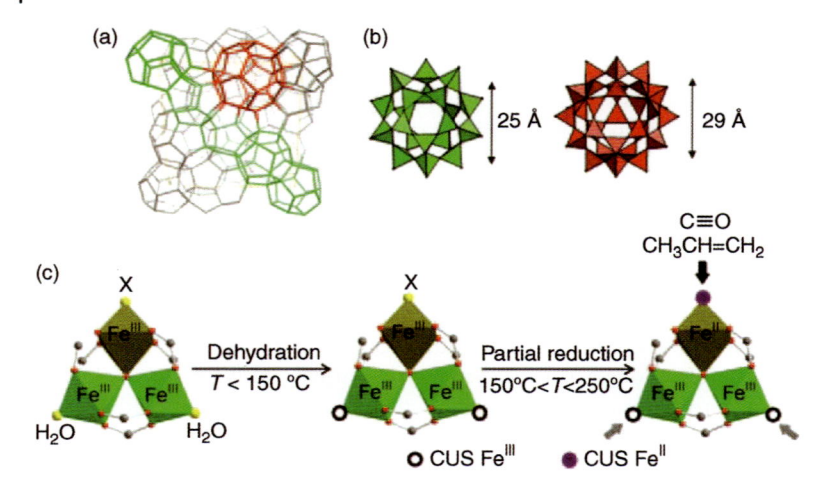

Figure 4.7 Representations of MIL-100(Fe). (a) One unit cell; (b) two types of mesoporous cages shown as polyhedra; (c) formation of Fe^{III} coordination unsaturated metal sites (CUSs) and Fe^{II} CUSs in an octahedral iron trimer of MIL-100(Fe) by dehydration and partial reduction from the departure of anionic ligands ($X^- = F^-$ or OH^-). Reproduced with permission from Yoon, J.W., Seo, Y.K., Hwang, Y.K., et al. Angew. Chem. Int. Ed., 2010, **49**, 5949. Copyright 2010 Wiley-VCH Verlag GmbH.

alkenes, which might be useful for the removal of these gas impurities (Figure 4.7) [76].

4.3.4
Hydrocarbon Separation

Despite the immense amount of research that has been reported regarding MOFs, work concerning the adsorption of volatile organic compounds (VOCs) is still relatively scarce. In this context, the separation of mixed C_8-alkylaromatics, such as xylenes and ethylbenzene, is one of the most challenging issues in the chemical industry because pure isomers are of high value. For instance, p-xylene is oxidized to terephthalic acid for poly(ethylene terephthalate) (PET) manufacture and ethylbenzene is dehydrogenated to styrene for polystyrene production [77]. The separation and detection of individual xylene isomers and ethylbenzene are also of environmental concern and of great practical interest in air monitoring [78] and blood analysis [79]. Distillation is only feasible for the removal of o-xylene (b.p. = 144 °C), but it fails for other isomers, such as p-xylene, m-xylene, and ethylbenzene, because of the similarity of their dimensions and boiling points (138, 138–139 and 136 °C, respectively). Nowadays, Zeolites X and Y, exchanged with Na^+, K^+, and Ba^{2+} cations, are used in industrial adsorptive separations as they discriminate selectively between the different C_8-alkylaromatic isomers [80, 81]. Recently, some MOFs have appeared as novel selective adsorbents for the mentioned separation. For example, the chiral porous 3D MOF [Cd(ClO$_4$)$_2$(L)$_2$]·H$_2$O [82] [L = 4-amino-3,5-bis(4-pyridyl-3-phenyl)-1,2,4-triazole] can effectively recognize and separate methylaromatics. This compound exhibits similar affinity for benzene and toluene, but no affinity for *o*-, *m*-, or *p*-xylene, when exposed to a mixed vapor containing all of these components. In the same manner, this compound shows similar affinity for *o*- and *m*-xylene, but no affinity for *p*-xylene, in the presence of *o*- and *m*-xylene competitors. Moreover, breakthrough experiments on MIL-53 [83] also demonstrated that at pressures much higher than the transition pressure at which the pore system adopts the open form (0.003 bar), a significant difference in breakthrough time between ethylbenzene and *o*-xylene is observed (average separation factor: 6.4) and pure ethylbenzene is eluted from the column. Recently, baseline separation of *p*-xylene, *o*-xylene, *m*-xylene, and ethylbenzene was achieved in a MIL-101-coated capillary column by gas chromatography within 1.6 min without the need for temperature programming [84]. In this case, the excellent selectivity of the MIL-101-coated capillary column originated not only from the host–guest interactions, but also from the Cr coordinatively unsaturated sites and the suitable polarity of MIL-101.

On the other hand, distillation separation of benzene and cyclohexane is a highly demanded and energy-consuming process in the petrochemical industry, because C_6H_{12} is produced by hydrogenation of C_6H_6 in the C_6H_6–C_6H_{12} miscible system and the two substances have very similar boiling points (C_6H_6, 80.1 °C; C_6H_{12}, 80.7 °C). However, the chemical (C_6H_6, aromatic; C_6H_{12}, saturated) and molecular geometry (C_6H_6, 3.3 × 6.6 × 7.3 Å; C_6H_{12}, 5.0 × 6.6 × 7.2 Å) [85] differences between the two molecules can be utilized for their effective separation. In this regard, Cu(etz) [86] (etz = 3,5-diethyl-1,2,4-triazole) can readily adsorb large amounts of C_6H_6 but only exhibits particle surface adsorption for C_6H_{12}. Moreover, we have recently synthesized the series A$_x$(NH$_4$)$_{1-x}$[Cu$_3$(μ$_3$-OH)(μ$_3$-4-carboxypyrazolato)$_3$] [71] (A = Li$^+$, Na$^+$, K$^+$, Ca^{2+}/2, La^{3+}/3), which upon exposure to benzene–cyclohexene (1:1) vapor mixtures produce a significant enrichment of the adsorbate phase in the benzene component. Higher selectivity has been observed for increasing bulk A ions (Figure 4.8).

Another important potential application for selective vapor adsorption is the separation of hydrocarbons. This has been achieved in Cu(hfipbb)(H$_2$hfipbb)$_{0.5}$ [H$_2$hfipbb = 4,4′-(hexafluoroisopropylidene)bis(benzoic acid)], which leads to the selective adsorption of normal C_4 over higher alkanes and alkenes at room and higher temperatures due to both shape and size exclusions. Additionally, ZIF-8 [Zn(2-methylimidazole)$_2$] [87] and their analogs Zn(2-chloroimidazole)$_2$ and Zn(2-bromoimidazole)$_2$ [88] have great potential for the kinetic separation of propane and propene at 303 K based on the remarkable differences in their diffusion rates through the pore system.

4.3.5
VOC Capture

Finally, the effective capture of harmful chemicals is of great importance both for the protection of the environment and for those who are at risk of being exposed to such materials. In this regard, [Cu$_3$(btc)$_2$] [89] has proved to be very effective in removing

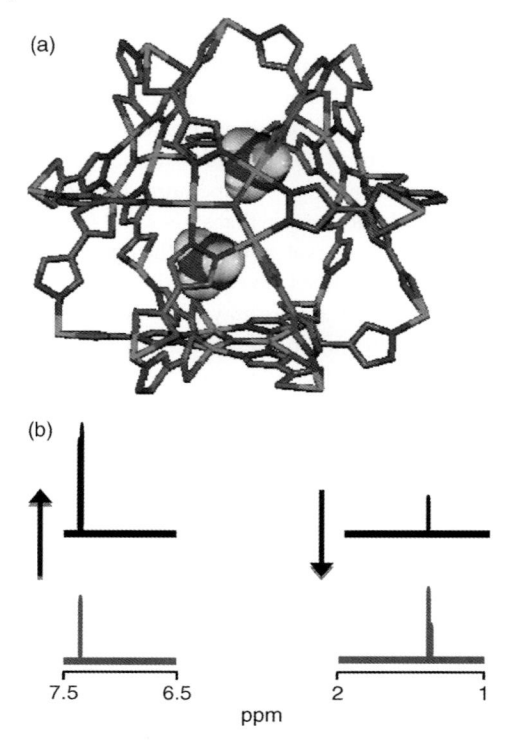

Figure 4.8 (a) View of the tetrahedral cages in the crystal structure of the anionic framework $(NH_4)[Cu_3(\mu_3\text{-}OH)(\mu_3\text{-}4\text{-carboxypyrazolato})_3]$. (b) ^{1}H NMR spectra of the initial 1:1 benzene–cyclohexane mixture (black) and of the benzene-enriched mixture after adsorption in $(NH_4)[Cu_3(\mu_3\text{-}OH)(\mu_3\text{-}4\text{-carboxypyrazolato})_3]$ (gray). Reproduced with permission from Quartapelle Procopio, E., Linares, F., Montoro, C., et al. Angew. Chem. Int. Ed., 2010, **49**, 7308. Copyright 2010 Wiley-VCH Verlag GmbH.

vapors contaminants, such as tetrahydrothiophene, benzene, dichloromethane, and ethylene oxide, from a gas stream, outperforming BPL carbon by an order of magnitude. The presence of coordinatively unsaturated Cu metal sites in $Cu_3(btc)_2$ is responsible for the strong interaction with such molecules. However, $Cu_3(btc)_2$ is ineffective in the presence of humidity because of the irreversible water coordination to the OMSs [13]. As a consequence, this feature is the main drawback to its application if it has to compete with classical hydrophobic adsorbents such as activated carbons. In this context, we have reported a novel flexible MOF, namely Ni(bpb) [13] [H_2bpb = 1,4-(4-bispyrazolyl)benzene], which is efficient for tetrahydrothiophene removal from CH_4–CO_2 mixtures even in the presence of humidity, thus overcoming the still unsolved problems raised by classical carboxylate-based MOFs in real practical applications, such as their incorporation in filters for air and gas purification processes.

4.3.6
Catalytic Decomposition of Trace Gases

Catalytic reactions involving gases are commonly used in gas purification processes. It should be noted that one of the major topics of interest in the field of MOF chemistry is their application as heterogeneous catalysts [90]; however, there are still very few reports in which this feature has been applied for gas purification purposes. Nevertheless, different gases have been found to carry out reactions catalyzed by MOFs and some of them might be of interest for gas purification processes. One of the earliest reports of relevance to NO_x abatement from exhaust gases is the NO decomposition reaction to N_2 and O_2 catalyzed by $[Cu(bpe)_2]NO_3$ [bpe = 1,2-bis(4-pyridyl)ethane] [91]. Cu(5-methylisophthalate) [92] and Ni(imidazole-4,5-dicarboxylato) clusters linked by alkali metal ions [93] are able to catalyze the oxidation of CO to CO_2. This reaction might be useful for hydrogen purification purposes. Three-dimensional $[Yb(OH)(2,6\text{-aqds})(H_2O)]$ (aqds = anthraquinone-2,6-disulfonate) [94] and Yb(succinato)$_{1.5}$ [95] polymeric frameworks act as good catalysts in hydrodesulfurization reactions of thiophene under mild conditions (7 bar H_2 and 70 °C) as a consequence of low coordination for the Yb atom and/or lability of coordinated ligands (Figure 4.9).

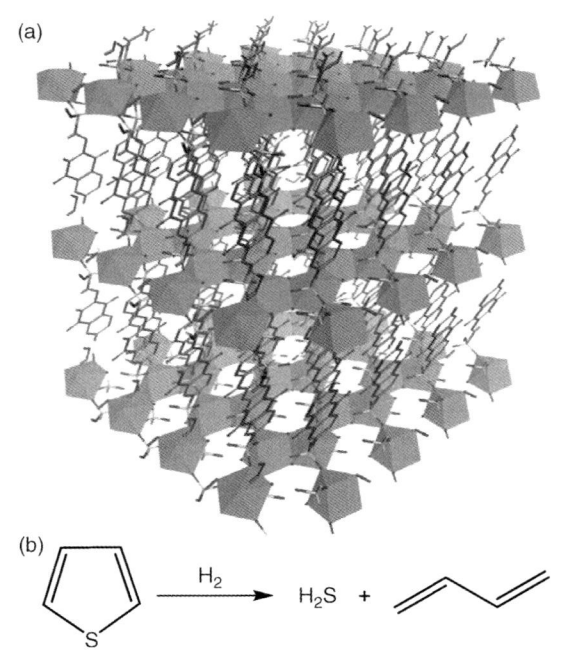

Figure 4.9 (a) Structure of 3D Yb(OH)(2,6-aqds)(H$_2$O) (aqds = anthraquinone-2,6-disulfonate) framework. (b) Hydrodesulfurization reaction of thiophene catalyzed by Yb(OH)(2,6-aqds)(H$_2$O).

4.4
Conclusions and Perspectives

The unique properties of MOFs, which permit the fine-tuning of the shape, size, and chemical nature of their pores, make them highly promising materials for gas separation and purification processes. In this regard, the electronic and fuel cell industries, in which the use of high-purity gases is required, will benefit greatly from MOF research. However, currently, investigations in this field are still in their infancy since most of the effort is mainly focused on gas adsorption studies based on adsorption–desorption isotherm measurements of single gas components. In this regard, experiments on the selective adsorption of gas mixtures and catalytic decomposition of trace gases need to be carried out more extensively in order to establish the real possibilities of MOFs in gas separation–purification processes. On the other hand, the separation of gas mixtures or organic vapors in the presence of water is of great technical and industrial importance as it is desirable that these adsorptive separation/purification processes should be performed under ambient conditions. This will principally hamper the application of MOFs with unsaturated coordination sites in which the separation/purification processes will rely on the selective binding of the targeted adsorbate over water. Hydrophilic/hydrophobic affinity and pore size considerations should be used to engineer porous materials with different binding abilities between water and gases/organic solvents. Other important issues to be considered are MOF scale-up, shaping, and separation optimization under real conditions. Possible directions of research in this field are to exploit more profoundly the framework flexibility features of MOFs, especially the kinetics of phase transformation, and, in addition, to study the effect of multicomponent mixtures on the selectivity of the adsorption processes. The effect of the structural flexibility on molecule diffusion and/or molecular sieving [e.g., Cu(Hoxonato)(bpy)$_{0.5}$, ZIF-20] has not been investigated so far. The incorporation of basic sites in MOFs for the selective capture of gases with acidic characteristics (e.g., SH_2, CO_2, C_2H_2) is also poorly studied and there are still many opportunities for future advancement. Finally, the highly selective catalytic features of MOFs will be a very valuable feature for the decomposition of trace impurities and may permit access to high-purity gases.

References

1 Kohl, A.L. and Nielsen, R. (1997) *Gas Purification*, 5th edn, Gulf Publishing, Houston, TX.
2 Greenwood, N.N. and Earnshaw, A. (1985) *Chemistry of the Elements*, Pergamon Press, Oxford.
3 Kerry, F.G. (2007) *Industrial Gas Handbook: Gas Separation and Purification*, CRC Press, Boca Raton, FL.
4 Yang, R.T. (1987) *Gas Separation by Adsorption Processes*, Imperial College Press, London.
5 Rouquerol, F., Rouquerol, J., and Sing, K. (1999) *Adsorption by Powders and Porous Solids*, Academic Press, London.
6 King, C.J. (1980) *Separation Progress*, 2nd edn, McGraw-Hill, New York.

7 Yang, R.T. (2003) *Adsorbents: Fundamentals and Applications*, John Wiley & Sons, Inc., Hoboken, NJ.

8 Keller, G.E. (1983) Gas adsorption processes: state of the art, in *Industrial Gas Separations* (eds T.E. Whyte Jr., C.M. Yon, and E.H. Wagener), ACS Symposium Series, Vol. 223, American Chemical Society, Washington, DC.

9 Tagliabue, M., Farrusseng, D., Valencia, S., Aguado, S., Ravon, U., Rizzo, C., Corma, A., and Mirodatos, C. (2009) Natural gas treating by selective adsorption: material science and chemical engineering interplay. *Chem. Eng. J.*, **155**, 553.

10 Sircar, S. (2006) Basic research needs for design of adsorptive gas separation processes. *Ind. Eng. Chem. Res.*, **45**, 5435.

11 Li, J.-R., Kuppler, R.J., and Zhou, H.-C. (2009) Selective gas adsorption and separation in metal–organic frameworks. *Chem. Soc. Rev.*, **38**, 1477.

12 Lide, D.R. (ed.) (2010) *CRC Handbook of Chemistry and Physics*, 91th edn, CRC Press, Boca Raton, FL.

13 Galli F S, Masciocchi, N., Colombo, V., Maspero, A., Palmisano, G., López-Garzón, F.J., Domingo-Garcúa, M., Fernández-Morales, I., Barea, E., and Navarro, J.A.R. (2010) Adsorption of harmful organic vapors by flexible hydrophobic bis-pyrazolate based MOFs. *Chem. Mater.*, **22**, 1664.

14 Henninger, S.K., Habib, H.A., and Janiak, C. (2009) MOFs as adsorbents for low temperature heating and cooling applications. *J. Am. Chem. Soc.*, **131**, 2776.

15 Matsuda, R., Kitaura, R., Kitagawa, S., Kubota, Y., Belosludov, R.V., Kobayashi, T.C., Sakamoto, H., Chiba, T., Takata, M., Kawazoe, Y., and Mita, Y. (2005) Highly controlled acetylene accommodation in a metal–organic microporous material. *Nature*, **436**, 238.

16 Uemura, T., Kitaura, R., Ohta, Y., Nagaoka, M., and Kitagawa, S. (2006) Nanochannel-promoted polymerization of substituted acetylenes in porous coordination polymers. *Angew. Chem. Int. Ed.*, **45**, 4112.

17 McNaught, A.D. and Wilkinson, A. (1997). *IUPAC Compendium of Chemical Terminology*, 2nd edn., Blackwell Science, Oxford.

18 Langmuir, I. (1916) The constitution and fundamental properties of solids and liquids. Part I. Solids. *J. Am. Chem. Soc.*, **38**, 2221.

19 Langmuir, I. (1918) The adsorption of gases on plane surfaces of glass, mica and platinum. *J. Am. Chem. Soc.*, **40**, 1361.

20 Brunauer, S., Emmett, P.H., and Teller, E. (1938) Adsorption of gases in multimolecular layers. *J. Am. Chem. Soc.*, **60**, 309.

21 Dubinin, M.M. and Radushkevich, L.V. (1947) Equation of the characteristic curve of activated charcoal. *Proc. Acad. Sci. USSR*, **55**, 331.

22 Chavan, S., Vitillo, J.G., Groppo, E., Bonino, F., Lamberti, C., Dietzel, P.D.C., and Bordiga, S. (2009) CO adsorption on CPO-27-Ni coordination polymer: spectroscopic features and interaction energy. *J. Phys. Chem. C*, **113**, 3292.

23 Bae, Y.-S., Spokoyny, A.M., Farha, O.K., Snurr, R.Q., Hupp, J.T., and Mirkin, C.A. (2010) Separation of gas mixtures using Co(II) carborane-based porous coordination polymers. *Chem. Commun.*, **46**, 3478.

24 Chae, H.K., Siberio-Perez, D.Y., Kim, J., Go, Y., Eddaoudi, M., Matzger, A.J., O'Keeffe, M., and Yaghi, O.M. (2004) A route to high surface area, porosity and inclusion of large molecules in crystals. *Nature*, **427**, 523.

25 Kitagawa, S., Kitaura, R., and Noro, S.I. (2004) Functional porous coordination polymers. *Angew. Chem. Int. Ed.*, **43**, 2334.

26 Férey, G. (2008) Hybrid porous solids: past, present, future. *Chem. Soc. Rev.*, **37**, 191.

27 Morris, R.E. and Wheatley, P.S. (2008) Gas storage in nanoporous materials. *Angew. Chem. Int. Ed.*, **47**, 4966.

28 Mueller, U., Schubert, M.M., and Yaghi, O.M. (2008) Chemistry and applications of porous metal–organic frameworks, in *Handbook of Heterogeneous Catalysis*, 2nd edn (eds G. Ertl, H. Knözinger, F. Schüth, J. Weitkamp) Wiley-VCH Verlag GmbH, Weinheim.

29 Chen, B.L., Wang, L.B., Zapata, F., Qian, G.D., and Lobkovsky, E.B. (2008) A luminescent microporous metal–organic

30. Horcajada, P., Chalati, T., Serre, C., Gillet, B., Sebrie, C., Baati, T., Eubank, J.F., Heurtaux, D., Clayette, P., Kreuz, C., Chang, J.S., Hwang, Y.K., Marsaud, V., Bories, P.N., Cynober, L., Gil, S., Férey, G., Couvreur, P., and Gref, R. (2010) Porous metal–organic-framework nanoscale carriers as a potential platform for drug delivery and imaging. *Nat. Mater.*, **9** 172.

31. Maspoch, D., Ruiz-Molina, D., and Veciana, J. (2004) Magnetic nanoporous coordination polymers. *J. Mater. Chem.*, **14**, 2713.

32. Xu, Z.T. (2006) A selective review on the making of coordination networks with potential semiconductive properties. *Coord. Chem. Rev.*, **250**, 2745.

33. Rieter, W.J., Taylor, K.M.L., and Lin, W.B. (2007) Surface modification and functionalization of nanoscale metal–organic frameworks for controlled release and luminescence sensing. *J. Am. Chem. Soc.*, **129**, 9852.

34. Kitagawa, S. and Uemura, K. (2005) Dynamic porous properties of coordination polymers inspired by hydrogen bonds. *Chem. Soc. Rev.*, **34**, 109.

35. Cohen, S.M. (2010) Modifying MOFs: new chemistry, new materials. *Chem. Sci.*, **1**, 32.

36. Dybtsev, D.N., Chun, H., Yoon, S.H., Kim, D., and Kim, K. (2004) Microporous manganese formate: a simple metal–organic porous material with high framework stability and highly selective gas sorption properties. *J. Am. Chem. Soc.*, **126**, 32.

37. Loiseau F T., Lecroq, L., Volkringe, R.C., Marrot, J., Férey, G., Haouas, M., Taulelle, F., Bourrelly, S., Llewellyn, P.L., and Latroche, M. (2006) MIL-96, a porous aluminum trimesate 3D structure constructed from a hexagonal network of 18-membered rings and mu(3)-oxo-centered trinuclear units. *J. Am. Chem. Soc.*, **128**, 10223.

38. Li, C.J., Lin, Z.J., Peng, M.X., Leng, J.D., Yang, M.M., and Tong, M.L. (2008) Novel three-dimensional 3d–4f microporous magnets exhibiting selective gas adsorption behaviour. *Chem. Commun.*, 6348.

39. Zhang, Y.-J., Liu, T., Kanegawa, S., and Sato, O. (2010) Interconversion between a nonporous nanocluster and a microporous coordination polymer showing selective gas adsorption. *J. Am. Chem. Soc.*, **132**, 912.

40. Li, J.R., Tao, Y., Yu, Q., Bu, X.H., Sakamoto, H., and Kitagawa, S. (2008) Selective gas adsorption and unique structural topology of a highly stable guest-free zeolite-type MOF material with N-rich chiral open channels. *Chem. Eur. J.*, **14**, 2771.

41. Navarro, J.A.R., Barea, E., Rodriguez-Dieguez, A., Salas, J.M., Ania, C.O., Parra, J.B., Masciocchi, N., Galli, S., and Sironi, A. (2008) Guest-induced modification of a magnetically active ultramicroporous, gismondine-like, copper(II) coordination network. *J. Am. Chem. Soc*, **130**, 3978.

42. Ma, S.Q., Wang, X.S., Yuan, D.Q., and Zhou, H.C. (2008) A coordinatively linked Yb metal–organic framework demonstrates high thermal stability and uncommon gas-adsorption selectivity. *Angew. Chem. Int. Ed.*, **47**, 4130.

43. Dinca, M. and Long, J.R. (2005) Strong H_2 binding and selective gas adsorption within the microporous coordination solid $Mg_3(O_2C–C_{10}H_6–CO_2)_3$. *J. Am. Chem. Soc.*, **132**, 9376.

44. Breck, D.W., Eversole, W.G., Milton, R.M., Reed, T.B., and Thomas, T.L. (1956) Crystalline zeolites. I. The properties of a new synthetic zeolite, type-A. *J. Am. Chem. Soc.*, **78**, 5963.

45. Xue, M., Ma, S.Q., Jin, Z., Schaffino, R.M., Zhu, G.S., Lobkovsky, E.B., Qiu, S.L., and Chen, B.L. (2008) Robust metal–organic framework enforced by triple-framework interpenetration exhibiting high H_2 storage density. *Inorg. Chem.*, **47**, 6825.

46. Kitagawa, S. and Kondo, M. (1998) Functional micropore chemistry of crystalline metal complex-assembled compounds. *Bull. Chem. Soc. Jpn.*, **71**, 1739.

47. Horike, S., Shimomura, S., and Kitagawa, S. (2009) Soft porous crystals. *Nat. Chem.*, **1**, 695.

48 Bureekaew, S., Shimomura, S., and Kitagawa, S. (2008) Chemistry and application of flexible porous coordination polymers. *Sci. Technol. Adv. Mater.*, **9**, 1.

49 Sing, K.S., Everett, D.H., Haul, R., Moscou, L., Pierotti, R.A., Rouquerol, J., and Siemieniewska, T.W. (1985) Reporting physisorption data for gas solid systems with special reference to the determination of surface-area and porosity (recommendations 1984). *Pure Appl. Chem.*, **57**, 603.

50 Kitaura, R., Seki, K., Akiyama, G., and Kitagawa, S. (2003) Porous coordination-polymer crystals with gated channels specific for supercritical gases. *Angew. Chem. Int. Ed.*, **42**, 428.

51 Nakagawa, K., Tanaka, D., Horike, S., Shimomura, S., Higuchia, M., and Kitagawa, S. (2010) Enhanced selectivity of CO_2 from a ternary gas mixture in an interdigitated porous framework. *Chem. Commun.*, **46**, 4258.

52 Serre, C., Millange, F., Thouvenot, C., Nogues, M., Marsolier, G., Louer, D., and Férey, G. (2002) Very large breathing effect in the first nanoporous chromium (III)-based solids: MIL-53 or Cr-III(OH)·{$O_2C–C_6H_4–CO_2$}·{$HO_2C–C_6H_4–CO_2H$}$_x$·H_2O_y. *J. Am. Chem. Soc.*, **124**, 13519.

53 Serre, C., Bourrelly, S., Vimont, A., Ramsahye, N.A., Maurin, G., Llewellyn, P.L., Daturi, M., Filinchuk, Y., Leynaud, O., Barnes, P., and Férey, G. (2007) An explanation for the very large breathing effect of a metal–organic framework during CO_2 adsorption. *Adv. Mater.*, **19**, 2246.

54 Hamon, L., Llewellyn, P.L., Devic, T., Ghoufi, A., Clet, G., Guillerm, V., Pirngruber, G.D., Maurin, G., Serre, C., Driver, G., van Beek, W., Jolimaitre, E., Vimont, A., Daturi, M., and Férey, G. (2009) Co-adsorption and separation of CO_2–CH_4 mixtures in the highly flexible MIL-53(Cr) MOF. *J. Am. Chem. Soc.*, **131**, 17490.

55 Hayashi, H., Côté, A.P., Furukawa, H., O'Keeffe, M., and Yaghi, O.M. (2007) Zeolite A imidazolate frameworks. *Nat. Mater.*, **6**, 501.

56 Couck, S., Denayer, J.F.M., Baron, G.V., Rémy, T., Gascon, J., and Kapteijn, F. (2009) An amine-functionalized MIL-53 metal–organic framework with large separation power for CO_2 and CH_4. *J. Am. Chem. Soc.*, **131**, 6326.

57 Barea, E., Tagliabue, G., Wang, W.G., Pérez-Mendoza, M., Mendez-Liñan, L., López-Garzon, F.J., Galli, S., Masciocchi, N., and Navarro, J.A.R. (2010) A flexible pro-porous coordination polymer: non-conventional synthesis and separation properties towards CO_2/CH_4 mixtures. *Chem. Eur. J.*, **16**, 931.

58 Seo, J., Matsuda, R., Sakamoto, H., and Bonneau, C., and Kitagawa, S. (2009) A pillared-layer coordination polymer with a rotatable pillar acting as a molecular gate for guest molecules. *J. Am. Chem. Soc*, **131**, 12792.

59 Maji, T.K., Matsuda, R., and Kitagawa, S. (2007) A flexible interpenetrating coordination framework with a bimodal porous functionality. *Nat. Mater.*, **6**, 142.

60 Chen, B.L., Ma, S.Q., Hurtado, E.J., Lobkovsky, E.B., Liang, C.D., Zhu, H.G., and Dai, S. (2007) Selective gas sorption within a dynamic metal–organic framework. *Inorg. Chem.*, **46**, 8705.

61 Chen, B., Ma, S., Zapata, F., Fronczek, F.R., Lobkovsky, E.B., and Zhou, H.C. (2007) Rationally designed micropores within a metal–organic framework for selective sorption of gas molecules. *Inorg. Chem.*, **46**, 1233.

62 Demessence, A. and Long, J.R. (2010) Selective gas adsorption in the flexible metal–organic frameworks Cu(BDTri)L (L=DMF, DEF). *Chem. Eur. J.*, **16**, 5902.

63 Shimomura, S., Higuchi, M., Matsuda, R., Yoneda, K., Hijikata, Y., Kubota, Y., Mita, Y., Kim, J., Takata, M., and Kitagawa, S. (2010) Selective sorption of oxygen and nitric oxide by an electron-donating flexible porous coordination polymer. *Nat. Chem.*, **2**, 633.

64 Murray, L.J., Dinca, M., and Long, J.R. (2009) Hydrogen storage in metal–organic frameworks. *Chem. Soc. Rev.*, **38**, 1294.

65 Xiang, S.-C., Zhou, W., Zhang, Z., Green, M.A., Liu, Y., and Chen, B. (2010) Open metal sites within isostructural metal–organic frameworks for differential

recognition of acetylene and extraordinarily high acetylene storage capacity at room temperature. *Angew. Chem. Int. Ed.*, **49**, 4615.
66 Dietzel, P., Blom, R., Chavan, S., and Bordiga, S. (2008) Hydrogen purification using Ni$_2$(dhtp) (H$_4$dhtp=2,5-dihydroxyterephthalic acid) metal–organic framework as CO adsorbent. Presented at the 1st International Conference on Metal–Organic Frameworks and Open Framework Compounds, Augsburg, Germany.
67 Mueller, U., Schubert, M., Teich, F., Puetter, H., Schierle-Arndt, K., and Pastre, J. (2006) Metal–organic frameworks. Prospective industrial applications. *J. Mater. Chem.*, **16**, 626.
68 Achmann, S., Hagen, G., Hämmerle, M., Malkowsky, I.M., Kiener, C., and Moos, R. (2010) Sulfur removal from low-sulfur gasoline and diesel fuel by metal–organic frameworks. *Chem. Eng. Technol.*, **33**, 275.
69 Britt, D., Tranchemontagne, D., and Yaghi, O.M. (2008) Metal–organic frameworks with high capacity and selectivity for harmful gases. *Proc. Nat. Acad. Sci. USA*, **105**, 11623.
70 Yoon, J.W., Jang, I.T., Lee, K.-Y., Hwang, Y.K., and Chang, J.-S. (2010) Adsorptive separation of propylene and propane on a porous metal–organic framework, copper trimesate. *Bull. Korean Chem. Soc.*, **31**, 220.
71 Quartapelle Procopio, E., Linares, F., Montoro, C., Colombo, V., Maspero, A., Barea, E., and Navarro, J.A.R. (2010) Cation exchange porosity tuning in anionic metal–organic frameworks for selective separation of gases and vapours and catalysis. *Angew. Chem. Int. Ed.*, **49**, 7308.
72 Chavan, S., Vitillo, J.G., Groppo, E., Bonino, F., Lamberti, C., Dietzel, P.D.C., and Bordiga, S. (2009) CO adsorption on CPO-27-Ni coordination polymer: spectroscopic features and interaction energy. *J. Phys. Chem. C*, **113**, 3292.
73 Dietzel, P.D.C., Blom, R., and Fjellvåg, H. (2008) Base-induced formation of two magnesium metal–organic framework compounds with a bifunctional tetratopic ligand. *Eur. J. Inorg. Chem.*, 3624.
74 Chen, B., Zhao, X., Putkham, A., Hong, K., Lobkovsky, E.B., Hurtado, E.J., Fletcher, A.J., and Thomas, K.M. (2008) Surface and quantum interactions for H$_2$ confined in metal–organic framework pores. *J. Am. Chem. Soc.*, **130**, 6411.
75 Chen, B., Xiang, S., and Qian, G. (2010) Metal–organic frameworks with functional pores for recognition of small molecules. *Acc. Chem. Res.*, **43**, 1115
76 Yoon, J.W., Seo, Y.K., Hwang, Y.K., Chang, J.-S., Leclerc, H., Wuttke, S., Bazin, P., Vimont, A., Daturi, M., Bloch, E., Llewellyn, P.L., Serre, C., Horcajada, P., Grenèche, J.-M., Rodrigues, A.E., and Férey, G. (2010) Controlled reducibility of a metal–organic framework with coordinatively unsaturated sites for preferential gas sorption. *Angew. Chem. Int. Ed.*, **49**, 5949.
77 (2010) *Ullmann's Encyclopedia of Industrial Chemistry*, 7th edn, Wiley-VCH Verlag GmbH, Weinheim, electronic release.
78 Yassaa, N., Brancaleoni, E., Frattoni, M., and Ciccioli, P. (2006) Isomeric analysis of BTEXs in the atmosphere using beta-cyclodextrin capillary chromatography coupled with thermal desorption and mass spectrometry. *Chemosphere*, **63**, 502.
79 Hattori, H., Iwai, M., Kurono, S., Yamada, T., Watanabe-Suxuki, K., Ishii, A., Seno, H., and Suxuki, O. (1998) Sensitive determination of xylenes in whole blood by capillary gas chromatography with cryogenic trapping. *J. Chromatogr. B*, **718**, 285.
80 Mèthivier, A. (2002) in *Zeolites for Cleaner Technologies. Catalytic Science Series*, vol. 3 (eds M. Guisnet and J.P. Gilson), Separation of Paraxylene by Adsorption. Imperial College Press, London.
81 Kulprathipanja, S. and Johnson, J. (2002) in *Handbook of Porous Solids* (eds F. Schüth, K. Sing, and J. Weitkamp) Liquid Separations. Wiley-VCH Verlag GmbH, Weinheim, 2568.
82 Liu, Q.K, Ma, J.P., and Dong, Y.B. (2009) Reversible adsorption and separation of aromatics on Cd^{II}-triazole single crystals. *Chem. Eur. J.*, **15**, 10364.

83 Finsy, V., Kirschhock, C.E.A., Vedts, G., Maes, M., Alaerts, L., De Vos, D.E., Baron, G.V., and Denayer, F.M. (2009) Framework breathing in the vapour-phase adsorption and separation of xylene isomers with the metal–organic framework MIL-53. *Chem. Eur. J.*, **15**, 7724.

84 Gu, Z.Y. and Yan, X.P. (2010) Metal–organic framework MIL-101 for high resolution gas-cromatographic separation of xylene isomers and ethylbenzene. *Angew. Chem. Int. Ed.*, **49**, 1477.

85 Webster, C.E., Drago, R.S., and Zerner, M.C. (1998) Molecular dimensions for adsorptives. *J. Am. Chem. Soc.*, **120**, 5509.

86 Zhang, J.P. and Chen, X.M. (2008) Exceptional framework flexibility and sorption behavior of a multifunctional porous cuprous triazolate framework. *J. Am. Chem. Soc.*, **130**, 6010.

87 Park, K.S., Ni, Z., Cote, A.P., Choi, J.Y., Huang, R.D., Uribe-Romo, F.J., Chae, H.K., O'Keeffe, M., and Yaghi, O.M. (2006) Exceptional chemical and thermal stability of zeolitic imidazolate frameworks. *Proc. Natl. Acad. Sci. USA*, **103**, 10186.

88 Li, K., Olson, D.H., Seidel, J., Emge, T.J., Gong, H., Zeng, H., and Li, J. (2009) Zeolitic imidazolate frameworks for kinetic separation of propane and propene. *J. Am. Chem. Soc.*, **131**, 10368.

89 Britt, D., Tranchemontagne, D., and Yaghi, O.M. (2008) Metal–organic frameworks with high capacity and selectivity for harmful gases. *Proc. Natl. Acad. Sci. USA*, **105**, 11623.

90 Farruseng, D., Aguado, S., and Pinel, C. (2009) Metal–organic frameworks: opportunities for catalysis. *Angew. Chem. Int. Ed.*, **48**, 7502.

91 Parvulescu, A.N., Marin, G., Suwinska, K., Kratsov, V.C., Andruh, M., Parvulescu, V., and Parvulescu, V.I. (2005) A polynuclear complex, $\{[Cu(bpe)_2](NO_3)\}$, with interpenetrated diamondoid networks: synthesis, properties and catalytic behaviour. *J. Mater. Chem.*, **15**, 4234.

92 Zou, R.Q., Sakurai, H., Han F S., Zhong, R.-Q., and Qu, X. (2007) Probing the Lewis acid sites and CO catalytic oxidation activity of the porous metal–organic polymer [Cu(5-methylisophthalate)]. *J. Am. Chem. Soc.*, **129**, 8402.

93 Zou, R.Q., Sakurai, H., Han, S., Zhong, R.-Q., and Qu, X. (2006) Preparation, adsorption properties, and catalytic activity of 3D porous metal–organic frameworks composed of cubic building blocks and alkali-metal ions. *Angew. Chem. Int. Ed.*, **45**, 2542.

94 Gandara, F., Gutierrez-Puebla, E., Iglesias, M., Proserpio, D., Snejko, N., and Monge, M.A. (2009) Controlling the structure of arenedisulfonates toward catalytically active materials. *Chem. Mater.*, **21**, 655.

95 Bernini, M.C., Gandara, F., Iglesias, M., Snejko, N., Gutierrez-Puebla, E., Brusau, E.V., Narda, G.E., and Monge, M.A. (2009) Reversible breaking and forming of metal–ligand coordination bonds: temperature-triggered single-crystal to single-crystal transformation in a metal–organic framework. *Chem. Eur. J.*, **15**, 4896.

5
Opportunities for MOFs in CO_2 Capture from Flue Gases, Natural Gas, and Syngas by Adsorption

Gerhard D. Pirngruber and Philip L. Llewellyn

5.1
Introduction

The high surface area and pore volume of many metal–organic framework (MOF) materials has created great expectations in the scientific community that MOFs could bring about the long-awaited breakthrough in materials for CO_2 capture. A large number of research papers and patents have already dealt with CO_2 adsorption on MOFs and the subject has been reviewed [1–6]. Unfortunately, the CO_2 adsorption and separation properties of MOFs are often discussed only from a scientific point of view, without going into the details of application-related aspects. The demands in terms of adsorption capacity, selectivity, pressure range, stability, and so on differ from one application to another. Hence there is no such thing as a good general CO_2 adsorbent, as we have to identify the application for which the adsorbent is best suited. In this chapter, we therefore do not start by discussing the CO_2 adsorption properties of MOFs, but we set out from three industrial processes in which MOFs may potentially be used as CO_2 capture agents: (1) the purification of synthesis gas to produce hydrogen, (2) the removal of CO_2 from natural gas, and (3) the removal of CO_2 from flue gases produced by combustion processes (post-combustion CO_2 capture). We first describe the state-of-the-art technology used in these processes and discuss what precisely the CO_2 capture agent is expected to do in each of them. Based on the analysis of the requirements of these three processes, we then discuss how and if performance breakthroughs could be achieved by introducing MOF materials as CO_2 adsorbents.

5.2
General Introduction to Pressure Swing Adsorption

Pressure swing adsorption (PSA) is a process by which it is possible to separate gases. The first PSA patents date back to the 1930s and the first commercial systems were developed in the 1950s by Air Liquide and Esso. The development of this process was

Metal-Organic Frameworks: Applications from Catalysis to Gas Storage, First Edition. Edited by David Farrusseng.
© 2011 Wiley-VCH Verlag GmbH & Co. KGaA. Published 2011 by Wiley-VCH Verlag GmbH & Co. KGaA.

accelerated in the 1970s, in the aftermath of the first oil crisis. Several books are devoted to PSA for those who are interested in further reading [7, 8].

The PSA process itself involves two basic steps:

- An *adsorption* step. The feed mixture is introduced into the adsorption column under a given pressure (upper working pressure). The most strongly adsorbed species are retained by the solid (adsorbent), allowing the least strongly adsorbed species to break through at the end of the column. The species recovered in this step is known as the *raffinate*.
- A *desorption* or *regeneration* step. Here the pressure in the bed is reduced (lower working pressure), which allows the recovery of the preferentially adsorbed species previously adsorbed. This is known as the *extract*.

It is possible to recover either a highly purified raffinate or a highly purified extract. Recovery of two pure products is also possible, but is more demanding. The main advantage of PSA over other separation process, such as distillation, is its rapidity and the possibility of having fairly small, inexpensive units. The main restriction on PSA is that the process is limited to gas–adsorbent pairs in which the adsorption is not very strong. PSA is generally an adiabatic process.

Figure 5.1 illustrates the (co-)adsorption isotherms, as a function of total pressure, of two species, A and B, that are to be separated by a PSA process. Species A is more strongly adsorbed than B. The PSA process works between the upper and lower working pressures. The upper pressure is often imposed by the pressure of the upstream process, while the lower pressure is usually around 1 bar. To help the regeneration, it also possible use vacuum conditions for the lower pressure. In that case, one uses the term vacuum swing adsorption (VSA). Regeneration can also be assisted by heating, in which case one uses the term mixed PSA–TSA (temperature swing adsorption). One aims to work in a region where the working capacity (i.e., the

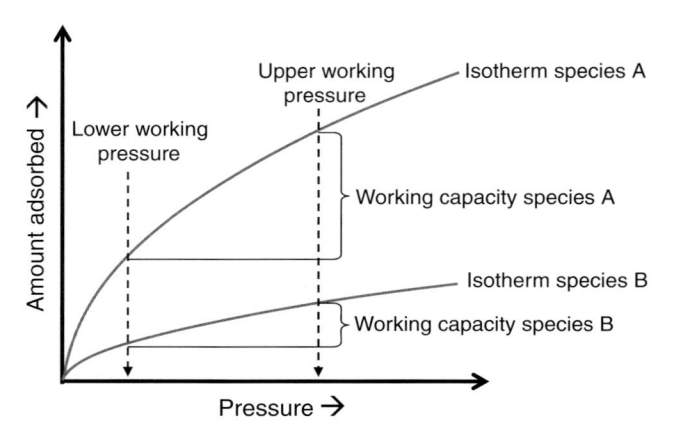

Figure 5.1 Representation of the pure component isotherms and the working capacities in a PSA process.

uptake between the upper and lower working pressures) is the most favorable for species A with respect to B.

The main separation processes in which PSA units can be found are air separation, gas drying, hydrogen purification, carbon dioxide recovery, and natural gas purification. Only some of these will be considered here with respect to the potential use of MOFs.

5.3
Production of H_2 from Syngas

The demand for hydrogen is constantly growing. A large share of the increasing demand comes from refineries, which consume hydrogen for the upgrading of heavy petroleum feedstocks. Depending on the development of hydrogen storage systems and an infrastructure for its distribution, hydrogen may itself become a transportation fuel and energy carrier in the long term. At present, H_2 is mainly (96%) produced from fossil fuels, either by steam reforming of natural gas or liquid hydrocarbons or by gasification of coal and heavy petroleum fractions. Both produce synthesis gas (syngas), a mixture of H_2, CO_2, CO, CH_4, H_2O, and other minor components. The syngas mixture is sent to a water gas shift reactor, which converts a large fraction of the CO to CO_2 and at the same time increases the yield of H_2:

$$CO + H_2O \rightarrow CO_2 + H_2$$

After the water gas shift reactor and condensation of the excess water, the syngas has a composition of roughly 70–80% H_2, 15–25% CO_2, 3–6% CH_4, and 1–3% CO (on a dry basis). The pressure is between 20 and 25 bar. The state-of-the-art technology for producing pure H_2 from this synthesis gas mixture is PSA. A multilayer adsorbent bed successively removes the impurities from the feed gas at high pressure. The first layer, which is either an activated carbon, an alumina, or a silica gel, removes H_2O and other heavy components from the feed, the second layer, an activated carbon, removes CO_2 and part of the CO and CH_4, and the final layer, which is either zeolite 5A (CaNaA) or 13X (NaX), assures the final purification, that is, it removes the last traces of CO, CH_4 and N_2 (if present in the feed). The adsorber is regenerated by depressurization and subsequent purge with part of the purified H_2 (see the paper by Sircar and Golden [9] for more details of the PSA cycle). The idea behind the multilayer adsorption column is to find the best compromise between adsorption capacity and ease of regeneration for each component of the feed. Current PSA units produce up to 200 000 $Nm^3 h^{-1}$ of H_2 of extremely high purity (>99.995%). The recovery of H_2 is between 70 and 90%; the rest is sacrificed for the abovementioned purge steps, which are necessary to clean the bed before a new adsorption cycle starts. The "by-product" of an H_2-PSA unit is a waste gas at low pressure, in which the impurities CO_2, CH_4, and CO are concentrated. The waste gas also contains 20–40% H_2, originating from the purge step. The waste gas is usually valorized as a fuel for the steam reforming furnace (Figure 5.2).

$CO + H_2O \rightarrow CO_2 + H_2$

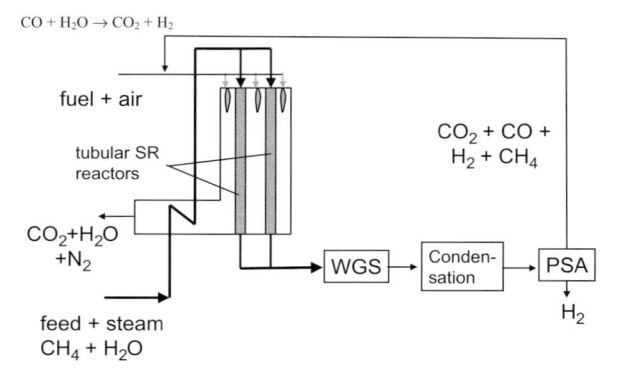

Figure 5.2 Scheme of a steam reforming plant for H_2 production.

Hydrogen is generally regarded as a clean, "carbon-free" fuel. However, the production of H_2 by steam reforming (or gasification) co-produces large amounts of CO_2, which is released to the atmosphere. The concern about the role of CO_2 emissions in global warming (see Section 5.5) provides a strong incentive to capture and sequester the CO_2 co-product. Air Products has developed a modified PSA process [10], called Gemini, which produces two pure product streams, that is, CO_2 and H_2, and a third waste stream containing mainly H_2, CH_4, and CO and a small fraction of CO_2. This is achieved by using two types of bed in series, A-beds and B-beds (Figure 5.3). The A-beds selectively remove CO_2 from the gas mixture.

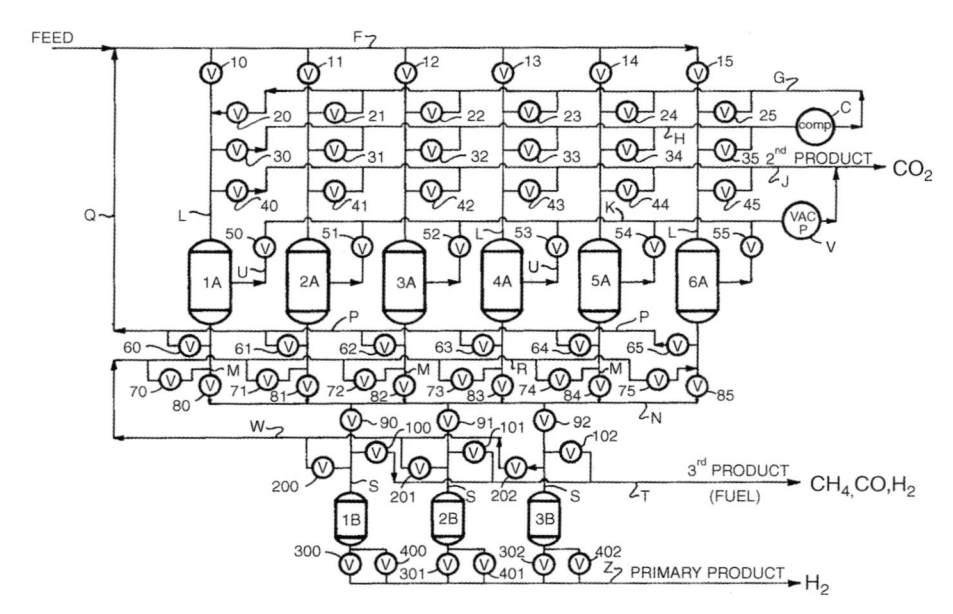

Figure 5.3 Two-bed PSA system for the simultaneous production of pure H_2 and pure CO_2.

The high purity of the CO_2 product is assured by a purge of the column with pure CO_2 before the depressurization step. For this purpose, part of the CO_2 product stream that is produced during depressurization and subsequent evacuation of the bed is recompressed to the feed pressure and recycled to the bed to purge it from all the lighter components. After the purge, the bed is depressurized and evacuated. Pure CO_2 is released during these steps. The B-beds, which are connected in series with the A-beds, work in a similar way as in conventional PSAs. An efficient exchange of depressurization and purge streams between the two types of bed leads to high recoveries of both CO_2 and H_2.

5.3.1
Requirements for CO_2 Adsorbents in H_2-PSAs

In conventional H_2-PSAs, the main technical criterion for the selection of a CO_2 adsorbent is the adsorption capacity. Since the other impurities (CH_4 and CO) have to be adsorbed anyway in the following layers, the selectivity for adsorption of CO_2 versus CH_4 and CO is not an important issue. The higher the CO_2 adsorption capacity, the smaller is the volume of adsorbent layer required. The productivity of the process and the H_2 recovery both increase. What counts, however, is not the absolute capacity of CO_2 adsorption at a given feed partial pressure, but the working capacity (or delta loading), which represents the difference in the adsorbed amount between the feed and the regeneration step (see Figure 5.1). The real working capacity depends on the operating conditions of the process and is best determined by a simulation of the PSA cycle, but for a first approximation, the delta loading can be estimated from the co-adsorption isotherms, using the composition of the feed and the waste gas, respectively, for calculating the adsorbed amounts at the end of the feed and the regeneration step.

For a PSA that co-produces CO_2 and H_2, the CO_2/CH_4 and CO_2/CO selectivities are important selection criteria for the adsorbent in the A-beds. Although a high purity of the CO_2 product stream can always be achieved by means of the CO_2 purge (which recycles part of the product to the column), the amount of purge decreases when the adsorbent is very selective. This reduces the power consumption for recompression of the CO_2 purge stream.

5.4
CO_2 Removal from Natural Gas

The CO_2 content of raw natural gas varies widely from one gas field to another, from 1 to 40%. CO_2 has to be removed from natural gas to increase its heating value and also to meet the specifications for pipeline transport (<3–4%). For the production of liquefied natural gas (LNG), the CO_2 content has to be brought down to as low as 50 ppm. Natural gas also invariably contains some H_2S (the concentration may vary between 0 and 15%). Usually, H_2S and CO_2 are removed simultaneously by a process gas acid gas removal (AGR). The state-of-the-art AGR technology is chemical or physical absorption by solvents. Chemical absorption relies on the acid–base interaction of H_2S and CO_2 with a solution of amines or of K_2CO_3. In physical absorption, CO_2 and H_2S are simply dissolved in the liquid [methanol, poly(ethylene glycol) dimethyl ether, morpholine, etc.]. Some solvents are selective for the absorption of H_2S, others capture both H_2S and CO_2. The CO_2/H_2S-rich solution is regenerated by raising the temperature, producing "acid gas," a mixture of essentially H_2S and CO_2. If a selective solvent is used that preferentially absorbs H_2S, it is possible to generate two acid gas streams, one rich in H_2S, which is further treated in the so-called Claus process, and the other rich in CO_2, which can be used for sequestration of CO_2.

Several gas plants use membrane separation to remove CO_2. The membranes are generally nonporous polymer membranes (cellulose acetate or polyimide) [11, 12], which selectively solubilize CO_2. Membrane processes are advantageously employed when the partial pressure of CO_2 is high and when the purity requirements are not very stringent. A high partial pressure assures a high driving force for permeation. The membrane area depends strongly on the percentage of CO_2 removal [13]. It is therefore difficult to achieve very low CO_2 concentrations with reasonable membrane areas (more selective membranes would be needed for that), but membranes perform well for producing a gas that meets pipeline specifications (CO_2 <3–4%). Compared with solvent-based systems, they also have the advantage of operational simplicity and high reliability, which makes them especially suited for off-shore applications. To avoid fouling of the membranes, however, it is necessary to remove H_2S and heavy hydrocarbons from the gas.

Adsorption processes (PSA) are another alternative technology for CO_2 removal. The characteristic of PSA processes is that they produce a gas of very high purity, that is, they are particularly suited to produce a gas of LNG quality. The most commonly used adsorbents are zeolite 13X (Na-X), 4A (Ca-A), and 5A (Na-A). Zeolite adsorbents simultaneously dehydrate the gas and remove H_2S and mercaptans (thiols). The order of affinity is $H_2O > RSH > H_2S > CO_2$. Hence CO_2 moves fastest through the adsorption column and is gradually replaced by incoming H_2O and sulfur compounds at the inlet of the adsorption column. A simultaneous removal of CO_2, H_2O, H_2S, and mercaptans is therefore possible only if the concentrations of H_2O, H_2S, and RSH are low compared with CO_2.

5.4.1
Requirements for Adsorbents for CO_2–CH_4 Separation in Natural Gas

PSA processes suffer a major handicap in the treatment of natural gas, due to the very high pressure of the feed (60–70 bar). As a consequence of the high pressure, a large amount of gas is contained in the interstitial volume of the adsorber column and this gas is very rich in CH_4. The large quantity of gas-phase CH_4 has to be compensated by a very high capacity for CO_2 in the adsorbed phase, otherwise the losses of CH_4 during the depressurization and purge steps will become unacceptable. The CO_2/CH_4 selectivity is a second very important criterion, in order to assure a high purity and recovery of CH$_4$. Moreover, in the view of what was said above, the adsorbent must be able to cope with small quantities of H$_2$S, mercaptans, and H$_2$O.

5.5
Post-combustion CO$_2$ Capture

There is now a wide consensus that global climate change is, to a large extent, caused by anthropogenic CO$_2$ emissions, which have increased exponentially in the last century. In order to limit the negative effects of climate change, it will therefore be necessary to reduce the CO$_2$ footprint of human activity by a number of measures: (1) a (partial) switch from fossil fuels to renewable fuels, (2) an increase in renewable energies (solar, wind, hydro, etc.) in the overall energy mix, (3) an increase in energy efficiency, and so on. However, mankind will still largely rely on fossil fuels for transportation and energy production for some time. Therefore, carbon capture and storage (CCS) will have to make an important contribution to the overall reduction of anthropogenic CO$_2$ emissions.

CCS can be best applied to large, stationary CO$_2$ emitters, that is, to industrial plants that produce large amounts of CO$_2$ emissions. The main target is coal-fired power plants, which are the world's largest emitters of CO$_2$. Therefore, we will only address CO$_2$ capture in flue gases produced by the combustion of coal (or other fossil fuels).

5.5.1
The State of the Art

The combustion of fossil fuels (coal, petroleum, biomass, etc.) by air produces a gas containing 10–15% CO$_2$, 5–15% H$_2$O, 3–4% O$_2$ (that was not consumed in the combustion process), and traces of NO$_x$ and SO$_x$, the remainder being N$_2$ (originating from the air used for combustion). The flue gas comes at high temperature and low pressure (in essence at atmospheric pressure). The volume of the flue gases is huge: a 600 MW coal-fired power plant produces 1.8×10^6 Nm^3 h^{-1} of flue gases (compared with 0.2×10^6 Nm^3 h^{-1} for the largest existing H$_2$ plant). Since the partial pressure of CO$_2$ in the flue gases is very low, CO$_2$ capture is difficult.

An industrial-scale unit for CO$_2$ capture in post-combustion flue gases does not yet exist. However, among the numerous pilot plants that are in operation, the majority use absorption of CO$_2$ by an aqueous solution of monoethanolamine (MEA) (Figure 5.4). The MEA process can therefore be considered as the state of the art in post-combustion CO$_2$ capture. Before entering into the MEA absorber, NO$_x$ and SO$_x$ have to be removed from the flue gases or they would form irreversible reaction products with MEA. The CO$_2$ in the flue gases is selectively chemisorbed by the amine solution, at 30–50 °C. The heat of absorption is very high (-85 kJ mol^{-1}), which makes it possible to capture CO$_2$ even at very low partial pressures. On the other hand, the price to be paid for the strong affinity between MEA and CO$_2$ is the high energy input necessary to regenerate the MEA solution by heating to 120 °C.

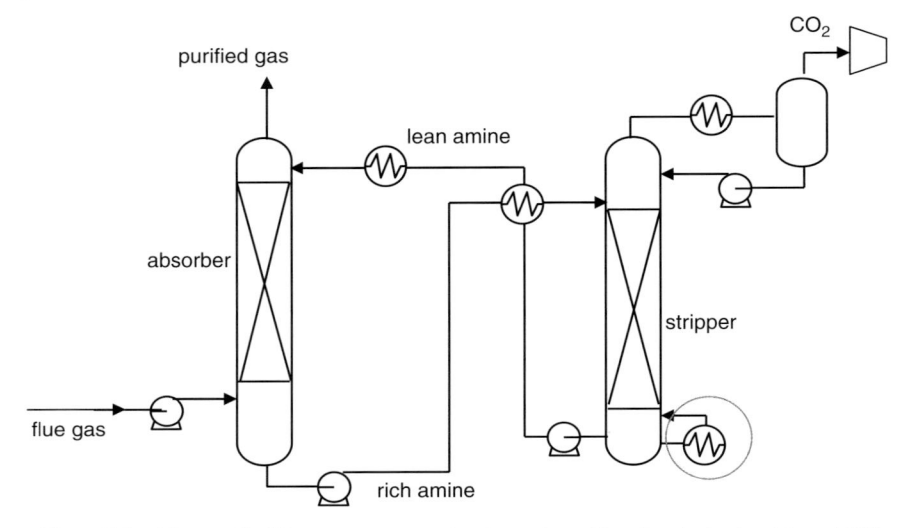

Figure 5.4 Scheme of a CO_2 capture process by absorption with a chemical solvent (e.g., MEA).

In a power plant, this energy is taken from the plant and thereby reduces its net energy output.

Other drawbacks of the MEA technology are the following:

- Degradation of MEA occurs, mainly due to reactions induced by oxygen in the flue gases. Degradation makes it necessary to replace continuously part of the MEA solution and leads to serious corrosion problems in the reactor.
- MEA and volatile degradation products are carried away with the flue gases and make a post-treatment necessary.
- The amines themselves induce corrosion problems in the system which have to be taken into account.

Numerous alternative solutions for post-combustion CO_2 capture are currently under investigation [14]: improved amine solvents, absorption by chilled ammonia, high-temperature carbonation–decarbonation cycles ($CaO-CaCO_3$), polymer or ceramic membranes, PSA, to just name a few. It is beyond the scope of this chapter to provide an exhaustive comparison of all these technologies, and we focus just on PSA and VSA in post-combustion CO_2 capture.

5.5.2
PSA and VSA Processes in Post-Combustion CO_2 Capture

Compared with absorption by MEA, PSA/VSA processes based on physisorption (we exclude immobilized amines from our discussion) potentially have the following advantages: no handling of liquids is necessary (no pumps), there are no corrosion problems with a solid adsorbent, and there are no degradation products. Provided that the adsorbent is fully regenerable and stable, PSA/VSA processes are very robust and can run for many years without requiring a replacement of the adsorbent. A huge handicap compared with absorption columns is, however, the pressure drop, which is several orders of magnitude higher in a fixed-bed adsorber than in a liquid absorption column. This severely limits the maximum height for a fixed-bed adsorber and eventually requires alternative solutions that help to reduce the pressure drop.

Another challenge is the high water content of the flue gases. The partial pressure of water is at least equal to its vapor pressure at the temperature at which the flue gas enters the adsorption process, that is, 0.04 bar at 303 K, 0.07 bar at 313 K, 0.12 bar at 323 K, and so on. Since the flue gas is at atmospheric pressure and its CO_2 concentration is between 10 and 15%, this means that the molar ratio of CO_2 to water is between 3 and 1. This is a large difference from synthesis gas, where the ratio of CO_2 to H_2O (after condensation) is much higher, due to the higher pressure of the feed.

Physisorbents fatally adsorb H_2O more strongly than CO_2. This effect is especially dramatic for polar adsorbents such as zeolites, which can totally lose their CO_2 adsorption capacity in the presence of water in the feed [15–17]. Even supposedly hydrophobic adsorbents such as activated carbons are strongly affected by the presence of water [18]. Therefore, it will usually be necessary to remove a large fraction of the water before contacting the flue gas with the CO_2 adsorbent. A frequently used option is to eliminate water by TSA on zeolite NaX, but the high energy consumption for the regeneration of the adsorbent seems unacceptable for application in CO_2 capture. Gases can also be dried by absorption of water in glycol, but our estimations show that the size of the glycol unit would be huge. The most attractive option may be to integrate the removal of water into the PSA/VSA unit for removal of CO_2, by using a first adsorbent layer that specifically adsorbs H_2O, as practiced in H_2-PSAs. Due to the large amount of water to be removed, this will necessarily have a large impact on the performance of the PSA/VSA process. It is not surprising, therefore, to see that some research groups have recently begun to readdress the problem of water removal by PSA [16, 17].

5.5.3
Requirements for Adsorbents for CO_2 Capture in Flue Gases

The purity requirements for CO_2 destined for transport and sequestration are fairly stringent. The concentration of non-condensable gases in the CO_2 extract must be less than 5%, otherwise the compression of CO_2 to high pressure (110 bar), that is, to the supercritical state in which it is transported to the storage site, becomes difficult. Contamination with H_2O does not pose a problem, because H_2O will condense in the CO_2 compression chain. The 5% limit therefore mainly applies to N_2 (O_2) that is co-captured along with CO_2. By a simple calculation, we can see that the purity requirement imposes a lower limit on the working capacity of the adsorbent. We postulate that the average composition of the recovered gas is given by the amount of CO_2 adsorbed in each adsorption cycle plus the amount contained in the gas phase of the adsorption column at the end of the adsorption step. We further suppose that the adsorbent is perfectly selective, that is, that there is no co-adsorption of N_2. Then, the CO_2 content of the recovered gas is

$$x_{CO_2} = \frac{\Delta q_{CO_2} \varrho (1-\varepsilon) + y_{CO_2} \left(\frac{\varepsilon p_{feed}}{RT}\right)}{\Delta q_{CO_2} \varrho (1-\varepsilon) + \frac{\varepsilon p_{feed}}{RT}} > 0.95$$

where Δq_{CO_2} is the working capacity of the adsorbent (mol kg^{-1}), ϱ is its grain density (kg m^{-3}), ε is the void fraction of the adsorption column, p_{feed} is the pressure of the feed, and T is the temperature. Using $\varepsilon = 0.37$, $p = 1$ bar and $T = 323$ K, we obtain a minimum working capacity of $\Delta q_{CO_2} \varrho = 440$ mol m^{-3}.

The reasoning above is, of course, oversimplified. In practice, not all the N_2 contained in the gas phase of the column will end up in the CO_2 product, because the gas-phase N_2 concentration can be reduced by depressurization and CO_2 purge steps. On the other hand, we made the very optimistic assumption that no N_2 is co-adsorbed. We therefore think that the value of 440 mol m^{-3} is fairly a good estimate of the lower limit working capacity of an adsorbent. The above reasoning also makes it clear that the CO_2/N_2 selectivity of the adsorbent must be very high (almost infinite). Even a small co-adsorption of N_2 will quickly reduce the purity of the CO_2 product and make CO_2 purge steps necessary. This is technically feasible, but reduces the productivity of the process and will further increase the volume of adsorbent required (which is already huge, according to our estimates).

5.6
MOFs

5.6.1
Considerations of Large Synthesis and Stability

A large number of MOFs with very interesting adsorption properties have been reported recently [1–5]. Some of them will probably never make it to a practical application, not for lack of performance, but for other practical reasons. If the synthesis of MOFs is to be brought to industrial scale, the preparation procedure should, if possible, avoid the use of expensive starting materials and of hazardous or toxic solvents. For example, DMF is commonly used as a solvent for the synthesis of MOFs, but it is toxic, in particular for the fetus. Some MOF syntheses require the use of HF, but the use of HF has been prohibited in Europe for safety reasons. Corrosion of the synthesis reactor may also pose a problem for certain synthesis protocols in acidic media. Further, the availability of the starting materials must be considered for production on a large scale. The simple message is that what is all right for gram-scale synthesis in the laboratory may not be all right for the synthesis of tons of adsorbents. The ease of large-scale synthesis will be a crucial criterion for the choice of a MOF adsorbent.

A further point that has to be considered is the stability of MOFs with respect to contaminants that can be found in the various processes. Such contaminants may include SO_x, NO_x, H_2S, and above all water. While it has been shown that the CO_2

uptake on some materials, including Cu-btc, depends on the aging [19], it was Willis's group [20] who developed a systematic evaluation of the stability of MOFs with respect to steaming. Here, small quantities of materials were subjected to steam treatments under various conditions of temperature and steam levels. The samples were then analyzed by X-ray powder diffraction to evaluate the extent of degradation. This experimental approach was compared with a theoretical approach developed by Snurr's group [19] using quantum mechanical calculations on cluster models. Good agreement was found between these two approaches, allowing a steam stability map to be constructed. This approach provides a good starting point as to which MOFs to study with respect to a given application.

Although this approach is fairly thorough, a simpler method for following the stability of a material with respect to water is the use of cycling experiments, following the variation of uptake after each cycle [21]. Concerning other contaminants, several studies have looked at the specific stability with respect to H_2S [22] and SO_x [23].

5.6.2
MOFs for H_2-PSA

The selection criteria for CO_2 adsorbents in H_2-PSAs are the working capacity, that is, the delta loading between adsorption and desorption conditions, and, if CO_2 recovery is desired, the CO_2/CH_4 and CO_2/CO selectivities. Synthesis gas is usually produced at a pressure of 20–25 bar, and a typical composition of the gas entering the PSA unit is 75% H_2, 20% CO_2, 4% CH_4, and 1% CO. The partial pressure of CO_2 is approximately 4 bar. For calculating the working capacity of CO_2, we should calculate the amount of CO_2 adsorbed at 4 bar, if possible in a mixture with CH_4 and CO at the proportions of the feed, and subtract the amount adsorbed at the regeneration pressure, i.e., 1 bar. Since isotherms of CO on MOFs can rarely be found in the literature, it is possible to simplify the problem and calculate the working capacity for pure CO_2, between 4 and 1 bar. The isotherm data (303 K) can be taken either from the literature or from in-house measurements. Some representative CO_2 isotherms of MOFs are shown in Figure 5.5, in comparison with zeolite NaX. We selected two microporous MOFs with open metal sites, that is, Cu-btc and CPO-27(Ni), a large-pore MOF with open metal sites, MIL-100(Cr), a large-pore MOF without open metal sites, MOF-177, and microporous ZIF (ZIF-8). The isotherms are presented per unit adsorbent mass and per unit adsorbent volume. Figure 5.5 shows that the microporous MOFs with open metal sites [in particular CPO-27(Ni)] have fairly steep CO_2 isotherms, similar to NaX. The isotherm of the large-pore MOF MIL-100(Cr) is less concave and that of MOF-177 even becomes convex (type V), because of its very weak interactions with CO_2 (absence of strong adsorption sites). Figure 5.5 further demonstrates that the large-pore MOFs have low adsorption capacities per unit volume in the pressure range in which we are interested for H_2-PSAs, due to their low crystallographic density.

The CO_2/CH_4 and CO_2/CO selectivities are best measured directly by co-adsorption experiments, but such data are difficult to find in the literature. When co-

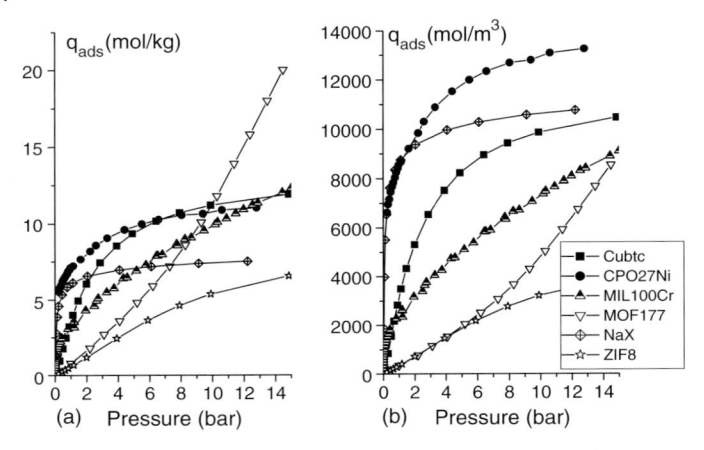

Figure 5.5 Isotherms of a selection of materials at 303 K, (a) in mol CO_2 per unit mass of adsorbent and (b) in mol CO_2 per unit volume of adsorbent.

adsorption data are not available, it is possible to use ideal adsorbed solution theory to calculate the CO_2/CH_4 selectivity from available single-component isotherms. A third important parameter is the heat of adsorption of CO_2. A high heat of adsorption leads to strong temperature excursions in the adsorbent bed (the increase can be more than 50 K), which reduces the adsorption capacity and also the efficiency of the separation. Therefore, the heat of adsorption should be as low as possible to avoid strong thermal effects. On the other hand, a high heat of adsorption usually ensures a high capacity and selectivity of the adsorbent, hence, there is a trade-off to make.

Table 5.1 compiles the working capacity (per unit mass and per unit volume), selectivity, and heat of adsorption of CO_2 for some selected MOFs. Cu-btc shows by far of the highest working capacities, by both unit mass and unit volume. It outperforms activated carbons in terms of selectivity and adsorption capacity and is therefore a very promising adsorbent for H_2-PSAs, with or without recovery of CO_2. Moreover, Cu-btc is fairly easily synthesized in a wide range of conditions including solvothermal (in H_2O and EtOH as solvent), electrochemical [24], ultrasound [25], microwave [26, 27], and mechanochemical [28]. It would seem that its large-scale synthesis should not pose any major problems as BASF are already trading the product through Sigma-Aldrich. The question that remains about this product is its water stability. Indeed, both stability [20] and sorption studies [19] suggest that aging may occur at high humidity levels.

The next best samples with respect to working capacity are the CPO-27 materials. CPO-27 samples are characterized by very strong adsorption of CO_2 and CH_4 on the open metal sites, which leads to isotherms that can be best described by a dual-site Langmuir model: a very strong initial adsorption on the metal sites, followed by a weaker adsorption that fills the remaining pore space. Unfortunately, no quantitative co-adsorption data have yet been published for CPO-27 samples and the dual-site behavior limits the applicability of IAST. The Langmuir co-adsorption model predicts

Table 5.1 Selection criteria for MOF adsorbents for H_2-PSA: working capacity of pure CO_2 between 4 and 1 bar; CO_2/CH_4 selectivity at 4 bar and a CO_2/CH_4 ratio of 4, measured or calculated by IAST; heat of adsorption of CO_2.

MOF adsorbent	Working capacity (mol kg^{-1})	Working capacity (mol m^{-3})	CO_2/CH_4 selectivity	$-\Delta H_{ads}$ CO_2 (kJ mol^{-1})
Cu-btc [29], 303 K	5.15	4520	6.7	28
CPO-27-Ni [30], 298 K	2.45	2940	Very high	37
CPO-27-Mg [30], 298 K	3.22	2970	Very high	42
MIL-53-Cr [31, 32], 303 K	2.1	2695/1995[a]	4.4	25
MIL-100-Cr [33], 303 K	2.96	2072	4.5	62 (42–25[b])
MIL-100-Fe [34], 303 K	3.5	2275	4	34
MIL-101-Cr [33], 303 K	3.8	1672	4	45 (26–23[b])
MOF-177, 298 K	2.77 [35]	1184	4.4 [36][c]	n.d.
ZIF-8 [37], 303 K	2.06	1240	3.2	n.d.
Ni-STA-12 [38], 304 K	1.84	2708	12	32
MOF-5, 298 K	2.6 [35]	1530	5[d]	
NaX [39], 303 K	0.87	1240	90	45 (35[b])
BPL carbon [40], 298 K	2.19	1120	3.4	26 [41]

a) Capacities in the narrow- and large-pore forms of MIL-53.
b) Enthalpies in the 1–4 bar pressure range considered here.
c) Selectivity obtained from isotherms up to 1 bar only, may not be reliable at higher pressure.
d) Estimated from CO_2 data in [35] and CH_4 data in [42]

that the CO_2/CH_4 selectivity should be very high, but this needs to be confirmed by experimental data. Another issue is the co-adsorption of CO, which gives rise to fairly high adsorption enthalpies on the order of -59 kJ mol^{-1} [43]. This is a significant energy which corresponds to the formation of monocarbonyl adducts. Such behavior will probably preclude the use of CPO-27 for syngas separations.

Ni-STA-12 is, like CPO-27 and Cu-btc, a MOF with open metal sites, but it differs in having a phosphonate ligand. Its adsorption capacity per unit mass does not appear impressive but, because of its high density, the working capacity of Ni-STA-12 is comparable to that of the CPO-27s and IAST also predicts a high CO_2/CH_4 selectivity. However, the open metal sites of Ni-STA-12 may irreversibly adsorb CO and thus affect the co-adsorption properties.

Table 5.1 shows that the large-pore MOFs MIL-100(Cr) and MIL-101(Cr) do not seem to be of such great interest as Cu-btc and CPO-27. Nevertheless, as there may be issues with respect to CO poisoning and/or water stability with the CPO-27 family, the MIL-100 and MIL-101 materials could still be of interest as they still seem to outperform the current standard sorbents, NaX and BPL carbon. The initial adsorption heats are fairly high for these solids but they decrease significantly with pressure and at 1 bar (our lower working pressure) they are in the region of -25 kJ mol^{-1}. These initial heats reflect that the coordinatively unsaturated sites (open metal sites) on these large-pore MOFs which equally preferentially adsorb CO. Nevertheless, the CO_2 isotherms are reversible between the two working pressures, suggesting fully regenerable materials. This was confirmed by IR spectroscopy [33]. Interestingly, the smaller pore material (MIL-100) may be of more interest, in terms of amount adsorbed per unit volume, than the larger pore material (MIL-101), even though the final CO_2 uptake of the latter at 60 bar is the current record for any porous material. Finally, one can question the use of potentially toxic chromium in a MOF. However, the MIL-100 sample can be prepared equally well with other less dangerous elements such as aluminum or iron.

The other large-pore, high-capacity MOF that can be found in the literature is MOF-177, which similarly shows exceptionally high CO_2 uptakes. However, in a separation process with the pressure boundaries set here, the material is not of interest owing to the unusual sigmoidal shape of the isotherm with fairly low uptakes in the pressure range considered. Two further considerations are the fairly low stability of this material to humidity [44] and the high price of the linker.

Much interest has been devoted to flexible MOFs with respect to CO_2 capture. Indeed, the idea that an intelligent MOF may be able to accommodate uniquely the desired species via a structural transition is appealing. Several groups have looked at this problem, showing stepped uptakes of CO_2 due to breathing in the MIL-53 family [31, 45] or gate-opening effects, for example for the MOF ELM-11 [46]. The delta loading of ELM-11 in H_2 purification would be very low, since the gate opening pressure for CO_2 on this material is 30 bar, but the delta loading of MIL-53(Cr) is similar to that of MIL-100(Cr). However, the first mixture adsorption experiments carried out with MIL-53(Cr) showed that the CO_2 selectivity of the breathing material is not as good as hoped [32, 48]. Whereas separation factors of around 10 were obtained when the material was in its narrow-pore (NP) form, they decreased significantly with pore opening to the large-pore (LP) form, at which point the CH_4 seemed to be able to enter the porosity simultaneously with the CO_2. Thus, when the CO_2 induces a breathing effect in the MIL-53 material, any other gas present, even if it does not induce breathing, will co-adsorb. The introduction of polar functional groups in the organic leakers can be used to maintain MIL-53 in its highly selective NP form, but this gain in selectivity is achieved at the expense of the adsorption capacity (which is much lower in the NP form) [49].

ZIFs are a very interesting class of materials, because of their supposedly high thermal stability. ZIF-8, with sodalite (SOD) topology, is easy to synthesize (the product is commercially available from Sigma Aldrich) and very stable towards humidity, but its affinity towards CO_2 is fairly low. The adsorption capacity is only ~ 2.5 mol kg^{-1} at 4 bar, hence the working capacity is low. Moreover, ZIF-8 behaves like an apolar adsorbent, and its selectivity is not better than that of an activated carbon. Also, ZIF-69, which is made using 2-nitroimidazole and chlorobenzimidazole ligands and has gmelinite (GME) topology, has a CO_2/CH_4 selectivity similar to that of activated carbon [37, 50]. Banerjee *et al.* showed, however, that it is possible to increase the affinity for CO_2, and thereby the CO_2/CH_4 selectivity, by replacing chlorobenzimidazole by more polar ligands, in particular nitrobenzimidazole (ZIF-78) or 4-cyanoimidazole (ZIF-82) [50]. The selectivity (calculated from the ratio of the Henry constants) increases up to 10-fold. Unfortunately, high-pressure adsorption data were not provided, but an extrapolation of the reported isotherms indicates that the delta loadings of these materials might be lower than those of ZIF-69 and ZIF-8. In conclusion, the concept of improving the CO_2/CH_4 selectivity by increasing the polarity of the ligand is very promising, but the applicability to H_2 purification has yet to be proven by high-pressure adsorption data.

5.6.3
MOFs for CO_2 Removal from Natural Gas

The upgrading of natural gas is certainly the most challenging application for MOFs, because they have to outperform not only the existing adsorbent materials, but also the competing technologies, that is, absorption and membranes. As mentioned before, the very high pressure of the feed (60–70 bar) is generally considered too high for PSA technology. However, for gases that are very rich in CO_2, the high pressure may turn into an advantage because this allows exploitation of the enormous CO_2 adsorption capacity of certain large-pore MOFs, such as MIL-100, MIL-101, and MOF-177, at high partial pressures of CO_2. Large-pore MOFs with open metal sites outperform activated carbons in terms of adsorption capacity and selectivity. Compared with polar zeolites (NaX), the selectivity is lower, but the adsorption capacity at high CO_2 partial pressure is higher. A concern is, however, the stability of these MOFs towards sulfur compounds. It has been shown that several iron-based MOFs [MIL-53(Fe), MIL-100(Fe)] react with H_2S probably to form FeS [22], whereas the aluminum- and chromium-based analogues were stable towards this contaminant. Similar stability problems can be expected for nickel-based MOFs.

Microporous MOFs with very high CO_2/CH_4 selectivity may find niche applications for natural gas at medium pressures and medium to low CO_2 contents. Table 5.1 indicates that CPO-27 materials are highly selective for the separation of CO_2 and CH_4, but the question of potential poisoning by H_2S has, to the best of our knowledge, not yet been addressed.

For natural gas with medium to low CO_2 contents, functionalization of the MOF with "CO_2-phile" groups may also be of interest. As mentioned in the previous section, amine-functionalized MIL-53(Al) shows very high selectivities towards CO_2 with respect to CH_4 [49]. At relatively low CO_2 pressures (up to 10 bar) and/or in the presence of humidity, the MIL-53(NH_2) is in the NP form in which the methane seems to have no access but where the CO_2 specifically interacts with the NH_2 groups.

Other MOFs stable towards H_2S and mercaptans (thiols) could also be of interest here, for example, the recently published UiO-66 structure [51].

5.6.4
MOFs for Post-Combustion CO_2 Capture

For the application of MOFs for post-combustion CO_2 capture, the first selection criterion is a minimum adsorption capacity of 440 mol CO_2 m^{-3} at the CO_2 partial pressure of the feed, that is, 0.15 bar. Many MOFs, including Cu-btc, UMCM-150, ZIF-8, MOF-5 [57] and some materials reported recently by Bae and co-workers [52, 53], do not reach that barrier (Table 5.2). Figure 5.6 shows some selected

Table 5.2 Selection criteria for MOF adsorbents for post-combustion CO_2 capture: adsorption capacity at 0.15 bar CO_2 at 298 K, CO_2/N_2 selectivity (at low loading), and heat of adsorption of CO_2.

MOF adsorbent	q_{ads} CO_2 (mol kg^{-1})	q_{ads} CO_2 (mol m^{-3})	CO_2/N_2 selectivity	$-\Delta H_{ads}$ CO_2 (kJ mol^{-1})
Co^{II} carborane [52] [a]	<0.25		65	n.d.
Zn-MOF [53] [b]	0.2		45	n.d.
Cu-btc	0.48	420	28 [19]	28
CPO-27(Ni) [30]	4.94	5930	High	38
CPO-27(Mg) [30]	6.0	5500	High	42
ZIF-8 [37]	0.09	50	9.5	n.d.
ZIF-78 [50]	0.71	834	33	30
ZIF-82 [50]	0.53	495	24	28
Bio-MOF-11 [54]	1.33	1630	75	32
NaX	4.0	5720	520 [55]	45 [56]

a) Carborane = 1,12-dihydroxycarbonyl-1,12-dicarba-*closo*-dodecacarborane.
b) Zn$_2$(benzene-1,2,4,5-tetrayltetrabenzoic acid)(py-CF$_3$)$_2$.

CO_2 isotherms of MOFs at low pressure. The isotherms of ZIF-8 and Cu-btc are quasi-linear, i.e., still in the Henry regime, which would be good for regeneration, but the slope is not sufficiently high to assure the required minimum adsorption capacity.

The MOFs with the by far the highest adsorption capacity at low pressure are those of the CPO-27 family, especially with Ni and Mg as the metal center [30, 57]. Due to the strong interaction between CO_2 and the unsaturated metal sites [58], the isotherms are very steep until the majority of the metal sites is occupied. This boosts the capacity at low pressure, but also implies that regeneration will be difficult. The capacity per unit mass at low pressure is even higher than that of NaX, and the capacity per unit volume is comparable. To the best of our knowledge, no room temperature N_2 isotherms have yet been published for CPO-27, but we can assume that the CO_2/N_2 selectivity will be very high. The difference in the heats of adsorption of CO_2 and N_2 is -26 kJ mol^{-1} [58].

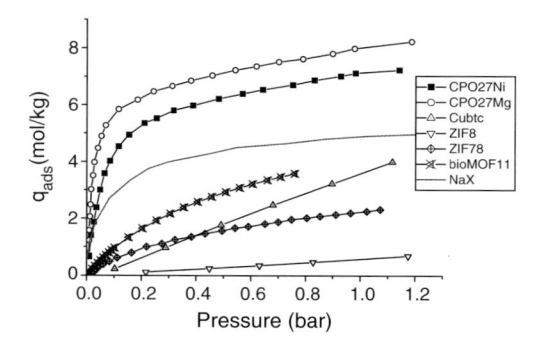

Figure 5.6 CO_2 isotherms between 0 and 1 bar (at 298 K) of some selected MOFs.

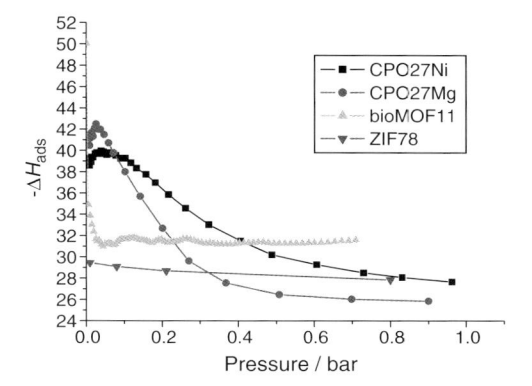

Figure 5.7 Heat of adsorption of CO_2 as a function of pressure for some selected MOFs.

ZIF-78 and ZIF-82, which we already discussed in Section 5.6.2, present isotherms that are slightly curved at low pressure, which is sign of a strong initial interaction with CO_2. The volumetric capacity is above the lower limit of 440 mol m^{-3}, with an acceptable selectivity.

A very interesting sample is bio-MOF-11 [54], which is constructed from Co^{II} with adenine and acetate ligands. It has a significantly higher capacity than the two ZIFs mentioned above and is also very selective.

Figure 5.7 compares the evolution of the heat of adsorption of the CPO-27 materials, ZIF-78, and bio-MOF-11 in the low pressure range. The heat of adsorption of the CPO-27 materials in the range 0.02–0.2 bar is significantly higher, which is undesirable, because this will lead to strong thermal effects in the adsorbent bed and thereby reduce the efficiency of the capture process. It is difficult to judge whether the higher heat of adsorption of the CPOs will entirely cancel their higher isothermal adsorption capacity under adiabatic conditions. Adsorption tests under cyclic conditions and also considerations of stability towards vapor, cost of the material, and ease of synthesis will determine whether one of above-mentioned materials can be used in an industrial process. With respect to the water tolerance of CPO-27(Ni), a recent study showed that even very low partial pressures of H_2O will lead to saturation of the open Ni sites and, as a consequence, drastically reduce the adsorption capacity of the sample [59]. The competition between H_2O and CO_2 is not as strong as for the zeolites NaX and 5A, but it will still be necessary to dry the flue gases upstream of the CPO-27(Ni) adsorption bed.

Finally, in an attempt to render materials more selective to CO_2, functionalization of the organic linker has been carried out [49, 60, 61]. In such cases, the main choice has been functionalization with $-NH_2$ tethers in analogy with the current technology of amine baths for CO_2 recovery under such conditions. However, other electron-donating groups such as $-COOH$, $-NO_2$, and $-SO_3H$ may also be of interest [62, 63]. In the case of the amine-functionalized MIL-53 material [49], an increase in selectivity was observed with a CO_2–CH_4 mixture, and this should also be the case for CO_2–N_2 although it has not been reported so far.

5.7
Conclusions

Enormous challenges are faced by MOFs with respect to CO_2 capture using PSA. The range of operating conditions with respect to various capture processes allow for different types of materials to be adapted to each process.

In materials testing, the CO_2 capacity and selectivity are important factors. In terms of capacity, one should compare uptakes in terms of volume of adsorbent material and not just weight of material. The former representation is the one that will be used in process design even though the latter is often flattering for the lightweight MOFs. One should not forget the stability towards potential contaminants and especially towards humidity. In terms of scale-up, other factors that need to be considered concern the synthesis (cost, environmental issues, etc.) and the shaping of materials (not considered in this chapter).

Nevertheless, there are still enormous opportunities for this fascinating family of novel materials to be used to attack the issue of CO_2 capture, which is one of the major challenges facing society today.

References

1 Ma, S-Q. and Zhou, H.-C. (2010) Gas storage in porous metal-organic frameworks for clean energy applications. *Chem. Commun.*, (46), 44–53.

2 Kuppler, R.J., Timmons, D.J., Fang, Q.-R., Li, J.-R., Makal, T.A., Young, M.D., Yuan, D.-Q., Zhao, D., Zhuang, W.-J., and Zhou, H.-C. (2009) Potential applications of metal–organic frameworks. *Coord. Chem. Rev.*, 253 (23–24), 3042–3066.

3 D'Alessandro, D.M., Smit, B., and Long, J.R. (2010) Carbon dioxide capture: prospects for new materials. *Angew. Chem. Int. Ed.*, 49 (35), 6058–6082.

4 Jones, C.W. and Maginn, E.J. (2010) Materials and processes for carbon capture and sequestration. *ChemSusChem.*, 3 (8), 863–864.

5 Keskin, S., van Heest, T.M., and Sholl, D.S. (2010) Can metal–organic framework materials play a useful role in large-scale carbon dioxide separations? *ChemSusChem.*, 3 (8), 879–891.

6 Hedin, N., Chen, L.J., and Laaksonen, A. (2010) Sorbents for CO_2 capture from flue gases - aspects from materials and theoretical chemistry. *Nanoscale*, 2, 1819–1841.

7 Ruthven, D.M., Farroq, S., and Knaebel, K.S. (1994) *Pressure Swing Adsorption*, John Wiley & Sons, Inc., New York.

8 Yang, R.T. (1997) *Gas Separation by Adsorption Processes*, Imperial College Press, London.

9 Sircar, S. and Golden, T.C. (2000) Purification of hydrogen by pressure swing adsorption. *Sep. Sci. Technol.*, 35 (5), 667–687.

10 Sircar, S. (1979) Separation of multicomponent gas mixtures. US Patent 4,171,206.

11 Bernardo, P., Drioli, E., and Golemme, G. (2009) Membrane gas separation: a review/state of the art. *Ind. Eng. Chem. Res.*, 48 (10), 4638–4663.

12 Baker, R.W. (2002) Future directions of membrane gas separation technology. *Ind. Eng. Chem. Res.*, 41 (6), 1393–1411.

13 Dortmundt, D. and Doshi, K. (1999) *Recent Developments in CO_2 Removal Membrane Technology*, UOP LLC, Des Plaines, IL.

14 Figueroa, J.D., Fout, T., Plasynski, S., McIlvried, H., and Srivastava, R.D. (2008) Advances in CO_2 capture technology – the U.S. Department of Energy's carbon

sequestration program. *Int. J. Greenhouse Gas Control*, **2**, 9–20.

15 Brandani, F. and Ruthven, D.M. (2004) The effect of water on the adsorption of CO_2 and C_3H_8 on type X zeolites. *Ind. Eng. Chem. Res.*, **43** (26), 8339–8344.

16 Li, G., Xiao, P., Webley, P., Zhang, J., Singh, R., and Marshall, M. (2008) Capture of CO_2 from high humidity flue gas by vacuum swing adsorption with zeolite 13X. *Adsorption*, **14** (2–3), 415–422.

17 Li, G., Xiao, P., Webley, P.A., Zhang, J., and Singh, R. (2009) Competition of CO_2/H_2O in adsorption based CO_2 capture. *Energy Procedia*, **1** (1), 1123–1130.

18 Adams, P.M., Katzman, H.A., Rellick, G.S., and Stupian, G.W. (1998) Characterization of high thermal conductivity carbon fibers and a self-reinforced graphite panel. *Carbon*, **36** (3), 233–245.

19 Liang, Z.J., Marshall, M., and Chaffee, A.L. (2009) CO_2 adsorption-based separation by metal organic framework (Cu-BTC) versus zeolite (13X). *Energy Fuels*, **23**, 2785–2789.

20 Low, J.J., Benin, A.I., Jakubczak, P., Abrahamian, J.F., Faheem, S.A., and Willis, R.R. (2009) Virtual high throughput screening confirmed experimentally: porous coordination polymer hydration. *J. Am. Chem. Soc.*, **131** (43), 15834–15842.

21 Wiersum, A.D., Soubeyrand-Lenoir, E., Yang, Q., Moulin, B., Guillerm, V., Ben Yahia, M., Bourrelly, S., Vimont, A., Miller, S., Vagner, C., Daturi, M., Clet, G., Serre, C., Maurin, G., and Llewellyn, P.L. A strategy for the evaluation of MOFs for gas-based applications: case of UiO-66. *Chem. Eur. J.* Phil.

22 Hamon, L., Serre, C., Devic, T., Loiseau, T., Millange, F., Ferey, G., and De Weireld, G. (2009) Comparative study of hydrogen sulfide adsorption in the MIL-53(Al, Cr, Fe), MIL-47(V), MIL-100(Cr), and MIL-101(Cr) metal–organic frameworks at room temperature. *J. Am. Chem. Soc.*, **131** (25), 8775.

23 Dathe, H., Peringer, E., Roberts, V., Jentys, A., and Lercher, J.A. (2005) Metal organic frameworks based on Cu^{2+} and benzene-1,3,5-tricarboxylate as host for SO_2 trapping agents. *C. R. Chim.*, **8** (3–4), 753–763.

24 Mueller, U., Schubert, M., Teich, F., Puetter, H., Schierle-Arndt, K., and Pastre, J. (2006) Metal–organic frameworks – prospective industrial applications. *J. Mater. Chem.*, **16** (7), 626–636.

25 Khan, N.A. and Jhung, S.H. (2009) Facile syntheses of metal–organic framework $Cu_3(BTC)_2(H_2O)_3$ under ultrasound. *Bull. Korean Chem. Soc.*, **30** (12), 2921–2926.

26 Seo, Y.K., Hundal, G., Jang, I.T., Hwang, Y.K., Jun, C.H., and Chang, J.S. (2009) Microwave synthesis of hybrid inorganic-organic materials including porous $Cu_3(BTC)_2$ from Cu(II)-trimesate mixture. *Micropor. Mesopor. Mater.*, **119** (1–3), 331–337.

27 Khan, N.A., Haque, E., and Jhung, S.H. (2010) Rapid syntheses of a metal–organic framework material $Cu_3(BTC)_2(H_2O)_3$ under microwave: a quantitative analysis of accelerated syntheses. *Phys. Chem. Chem. Phys.*, **12** (11), 2625–2631.

28 Klimakow, M., Klobes, P., Thunemann, A.F., Rademann, K., and Emmerling, F. (2010) Mechanochemical synthesis of metal–organic frameworks: a fast and facile approach toward quantitative yields and high specific surface areas. *Chem. Mater.*, **22** (18), 5216–5221.

29 Hamon, L., Jolimaitre, E., and Pirngruber, G.D. (2010) CO_2 and CH_4 separation by adsorption using Cu-BTC metal–organic framework. *Ind. Eng. Chem. Res.*, **49** (16), 7497–7503.

30 Dietzel, P.D.C., Besikiotis, V., and Blom, R. (2009) Application of metal–organic frameworks with coordinatively unsaturated metal sites in storage and separation of methane and carbon dioxide. *J. Mater. Chem.*, **19** (39), 7362–7370.

31 Bourrelly, S., Llewellyn, P.L., Serre, C., Millange, F., Loiseau, T., and Férey, G. (2005) Different adsorption behaviors of methane and carbon dioxide in the isotypic nanoporous metal terephthalates MIL-53 and MIL-47. *J. Am. Chem. Soc.*, **127** (39), 13519–13521.

32 Hamon, L., Llewellyn, P.L., Devic, T., Ghoufi, A., Clet, G., Guillerm, V., Pirngruber, G.D., Maurin, G., Serre, C., Driver, G., Van Beek, W., Jolimaitre, E., Vimont, A., Daturi, M., and Férey, G. (2009) Co-adsorption and separation of CO_2–CH_4 mixtures in the highly flexible MIL-53(Cr) MOF. *J. Am. Chem. Soc.*, **131** (47), 17490–17499.

33 Llewellyn, P.L., Bourrelly, S., Serre, C., Vimont, A., Daturi, M., Hamon, L., De Weireld, G., Chang, J.S., Hong, D.Y., Hwang, Y.K., Jhung, S.H., and Férey, G. (2008) High uptakes of CO_2 and CH_4 in mesoporous metal–organic frameworks MIL-100 and MIL-101. *Langmuir*, **24** (14), 7245–7250.

34 Yoon, J.W., Seo, Y.K., Hwang, Y.K., Chang, J.S., Leclerc, H., Wuttke, S., Bazin, P., Vimont, A., Daturi, M., Bloch, E., Llewellyn, P.L., Serre, C., Horcajada, P., Greneche, J.M., Rodrigues, A.E., and Férey, G. (2010) Controlled reducibility of a metal–organic framework with coordinatively unsaturated sites for preferential gas sorption. *Angew. Chem. Int. Ed.*, **49** (34), 5949–5952.

35 Millward, A.R. and Yaghi, O.M. (2005) Metal–organic frameworks with exceptionally high capacity for storage of carbon dioxide at room temperature. *J. Am. Chem. Soc.*, **127** (51), 17998–17999.

36 Saha, D., Bao, Z.B., Jia, F., and Deng, S.G. (2010) Adsorption of CO_2, CH_4, N_2O, and N_2 on MOF-5, MOF-177, and zeolite 5A. *Environ. Sci. Technol.*, **44** (5), 1820–1826.

37 Perez-Pellitero, J., Amrouche, H., Siperstein, F.R., Pirngruber, G., Nieto-Draghi, C., Chaplais, G., Simon-Masseron, A., Bazer-Bachi, D., Peralta, D., and Bats, N. (2010) Adsorption of CO_2, CH_4, and N_2 on zeolitic imidazolate frameworks: experiments and simulations. *Chem. Eur. J.*, **16** (5), 1560–1571.

38 Miller, S.R., Pearce, G.M., Wright, P.A., Bonino, F., Chavan, S., Bordiga, S., Margiolaki, I., Guillou, N., Férey, G., Bourrelly, S., and Llewellyn, P.L. (2008) Structural transformations and adsorption of fuel-related gases of a structurally responsive nickel phosphonate metal–organic framework, Ni-STA-12. *J. Am. Chem. Soc.*, **130** (47), 15967–15981.

39 Belmabkhout, Y., Pirngruber, G., Jolimaitre, E., and Methivier, A. (2007) A complete experimental approach for synthesis gas separation studies using static gravimetric and column breakthrough experiments. *Adsorption*, **13** (3–4), 341–349.

40 Wilson, R.J. and Danner, R.P. (1983) Adsorption of synthesis gas-mixture components on activated carbon. *J. Chem. Eng. Data.*, **28**, 14–18.

41 Sircar, S. and Golden, T.C. (1995) Isothermal and isobaric desorption of carbon dioxide by purge. *Ind. Eng. Chem. Res.*, **34** (8), 2881–2888.

42 Duren, T., Sarkisov, L., Yaghi, O.M., and Snurr, R.Q. (2004) Design of new materials for methane storage. *Langmuir*, **20** (7), 2683–2689.

43 Chavan, S., Vitillo, J.G., Groppo, E., Bonino, F., Lamberti, C., Dietzel, P.D.C., and Bordiga, S. (2009) CO adsorption on CPO-27-Ni coordination polymer: spectroscopic features and interaction energy. *J. Phys. Chem. C*, **113** (8), 3292–3299.

44 Saha, D. and Deng, S.G. (2010) Structural stability of metal organic framework MOF-177. *J. Phys. Chem. Lett.*, **1** (1), 73–78.

45 Serre, C., Bourrelly, S., Vimont, A., Ramsahye, N.A., Maurin, G., Llewellyn, P.L., Daturi, M., Filinchuk, Y., Leynaud, O., Barnes, P., and Férey, G. (2007) An explanation for the very large breathing effect of a metal–organic framework during CO_2 adsorption. *Adv. Mater.*, **19** (17), 2246.

46 Kitagawa, S., Kitaura, R., and Noro, S. (2004) Functional porous coordination polymers. *Angew. Chem. Int. Ed.*, **43** (18), 2334–2375.

47 Kanoh, H., Kondo, A., Noguchi, H., Kajiro, H., Tohdoh, A., Hattori, Y., Xu, W.C., Moue, M., Sugiura, T., Morita, K., Tanaka, H., Ohba, T., and Kaneko, K. (2009) Elastic layer-structured metal organic frameworks (ELMS). *J. Colloid Interface Sci.*, **334** (1), 1–7.

48 Finsy, V., Ma, L., Alaerts, L., De Vos, D.E., Baron, G.V., and Denayer, J.F.M. (2009)

Separation of CO_2/CH_4 mixtures with the MIL-53(Al) metal–organic framework. *Micropor. Mesopor. Mater.*, **120** (3), 221–227.

49 Couck, S., Denayer, J.F.M., Baron, G.V., Remy, T., Gascon, J., and Kapteijn, F. (2009) An amine-functionalized MIL-53 metal–organic framework with large separation power for CO_2 and CH_4. *J. Am. Chem. Soc.*, **131** (18), 6326.

50 Banerjee, R., Furukawa, H., Britt, D., Knobler, C., O'Keeffe, M., and Yaghi, O.M. (2009) Control of pore size and functionality in isoreticular zeolitic imidazolate frameworks and their carbon dioxide selective capture properties. *J. Am. Chem. Soc.*, **131** (11), 3875–3877.

51 Cavka, J.H., Jakobsen, S., Olsbye, U., Guillou, N., Lamberti, C., Bordiga, S., and Lillerud, K.P. (2008) A new zirconium inorganic building brick forming metal organic frameworks with exceptional stability. *J. Am. Chem. Soc.*, **130** (42), 13850–13851.

52 Bae, Y.S., Farha, O.K., Spokoyny, A.M., Mirkin, C.A., Hupp, J.T., and Snurr, R.Q. (2008) Carborane-based metal–organic frameworks as highly selective sorbents for CO_2 over methane. *Chem. Commun.* (35), 4135–4137.

53 Bae, Y.S., Farha, O.K., Hupp, J.T., and Snurr, R.Q. (2009) Enhancement of CO_2/N_2 selectivity in a metal–organic framework by cavity modification. *J. Mater. Chem.*, **19** (15), 2131–2134.

54 An, J., Geib, S.J., and Rosi, N.L. (2010) High and selective CO_2 uptake in a cobalt adeninate metal–organic framework exhibiting pyrimidine- and amino-decorated pores. *J. Am. Chem. Soc.*, **132** (1), 38–39.

55 Kim, J.-N., Chue, K.-T., Kim, K.-I., and Cho, S.-H. (1994) Non-isothermal adsorption of nitrogen–carbon dioxide mixture in a fixed bed of zeolite X. *J. Chem. Eng. Jpn.*, **27** (1), 45–51.

56 Maurin, G., Llewellyn, P.L., and Bell, R.G. (2005) Adsorption mechanism of carbon dioxide in faujasites: grand canonical Monte Carlo simulations and microcalorimetry measurements. *J. Phys. Chem. B*, **109** (33), 16084–16091.

57 Yazaydin, A.O., Snurr, R.Q., Park, T.H., Koh, K., Liu, J., Levan, M.D., Benin, A.I., Jakubczak, P., Lanuza, M., Galloway, D.B., Low, J.J., and Willis, R.R. (2009) Screening of metal–organic frameworks for carbon dioxide capture from flue gas using a combined experimental and modeling approach. *J. Am. Chem. Soc.*, **131** (51), 18198–18199.

58 Valenzano, L., Civalleri, B., Chavan, S., Palomino, G.T., Arean, C., and Bordiga, S. (2010) Computational and experimental studies on the adsorption of CO, N_2, and CO_2 on Mg-MOF-74. *J. Phys. Chem. C*, **114** (25), 11185–11191.

59 Liu, J., Wang, Y., Benin, A.I., Jakubczak, P., Willis, R.R., and Levan, M.D. (2010) CO_2/H_2O adsorption equilibrium and rates on metal–organic frameworks: HKUST-1 and Ni/DOBDC. *Langmuir*, **26** (17), 14301–14307.

60 Devic, T., Horcajada, P., Serre, C., Salles, F., Maurin, G., Moulin, B., Heurtaux, D., Clet, G., Vimont, A., Greneche, J.M., Le Ouay, B., Moreau, F., Magnier, E., Filinchuk, Y., Marrot, J., Lavalley, J.C., Daturi, M., and Férey, G. (2010) Functionalization in flexible porous solids: effects on the pore opening and the host–guest interactions. *J. Am. Chem. Soc.*, **132** (3), 1127–1136.

61 Deng, H.X., Doonan, C.J., Furukawa, H., Ferreira, R.B., Towne, J., Knobler, C.B., Wang, B., and Yaghi, O.M. (2010) Multiple functional groups of varying ratios in metal–organic frameworks. *Science*, **327** (5967), 846–850.

62 Torrisi, A., Mellot-Draznieks, C., and Bell, R.G. (2009) Impact of ligands on CO_2 adsorption in metal–organic frameworks: first principles study of the interaction of CO_2 with functionalized benzenes. I. Inductive effects on the aromatic ring. *J. Chem. Phys.*, **130** (19), 194703.

63 Torrisi, A., Mellot-Draznieks, C., and Bell, R.G. (2010) Impact of ligands on CO_2 adsorption in metal–organic frameworks: first principles study of the interaction of CO_2 with functionalized benzenes. II. Effect of polar and acidic substituents. *J. Chem. Phys.*, **132** (4), 044705.

6
Manufacture of MOF Thin Films on Structured Supports for Separation and Catalysis

Sonia Aguado and David Farrusseng

6.1
Advantages and Limitations of Membrane Technologies for Gas and Liquid Separation

Membranes have been increasingly used in industrial applications for the last two decades [1, 2]. It is estimated that the annual revenue of the worldwide membrane industry is over US£1 billion, and an annual growth rate of about 10% has been forecast for this industry. Currently, the industry is dominated by polymeric membranes that have been used in a variety of applications ranging from food and beverage processing, desalination of seawater, and gas separations to medical devices. Recently, research directed at the development and application of inorganic membranes has been gaining momentum because of their high demand in new application fields, such as fuel cells [proton exchange membrane fuel cells (PEMFCs), solid oxide fuel cells (SOFCs)], membrane reactors, and other high-temperature separations (water electrolysis).

Membrane separation processes provide several advantages over other conventional adsorption/regeneration-based separation techniques. In contrast to pressure swing adsorption (PSA) systems, the membrane process is a continuous separation process. Since it does not require alternate pressurization–depressurization phases, membrane processes are usually energy-efficient solutions (Figure 6.1). Second, the necessary process equipment is very simple, with no moving parts, and relatively easy to operate and control, and also easy to scale up.

Inorganic membranes can be classified into two categories: porous and dense. The latter, which includes polymer and solid oxide membranes, will not be discussed here. In porous inorganic membranes, a porous, thin top layer is supported on a porous metal or ceramic support, which provides mechanical strength with a minimum mass-transfer resistance. Alumina, carbon, glass, silicon carbide, titania, zeolite, and zirconia membranes are classical porous inorganic thin membranes supported on different porous substrates. Supports can be made of metal, carbon, or ceramic materials, and typically they are asymmetric, formed by layers with different porosity and pore size (Figure 6.2).

Metal-Organic Frameworks: Applications from Catalysis to Gas Storage, First Edition. Edited by David Farrusseng.
© 2011 Wiley-VCH Verlag GmbH & Co. KGaA. Published 2011 by Wiley-VCH Verlag GmbH & Co. KGaA.

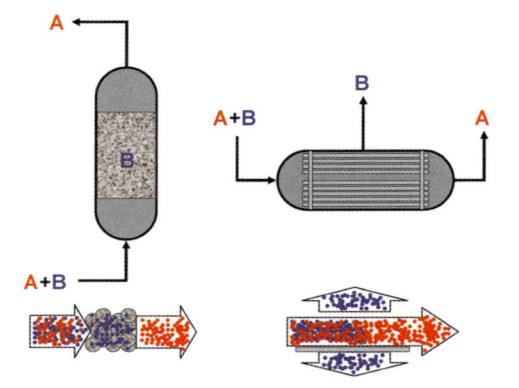

Figure 6.1 Pressure swing adsorption (PSA) and membrane technology.

When looking at the typical pore size of metal–organic frameworks (MOFs) and the application domains of membranes, it becomes clear that MOF membranes could potentially play a role in numerous separation processes, such as gas and liquid separations, reverse osmosis, and nanofiltration.

For gas separation or pervaporation, microporous membranes have been developed, mainly zeolite membranes. However, diffusion through such narrow pores is usually slow [4]. The main issues with nanofiltration membranes are (1) membrane fouling, (2) insufficient separation between solutes, and (3) the low chemical resistance of membranes [5]. Pervaporation is a membrane separation technique in which a liquid feed mixture contacts the feed side of a selective membrane and the other side is typically under vacuum to provide vapor permeate.

Pervaporation has been widely studied as a means of dehydrating solvents, whose recovery in industry is frequently sought but difficult when an azeotrope is involved. Mitsui-BNRI has played a pioneering role in the cost reduction of membrane separations by integrating distillation and membrane separation in the so-called membrane separation and distillation (MDI)). In the MDI process, dehydrated ethanol with less than 0.4 wt% residual water is produced from a liquid containing

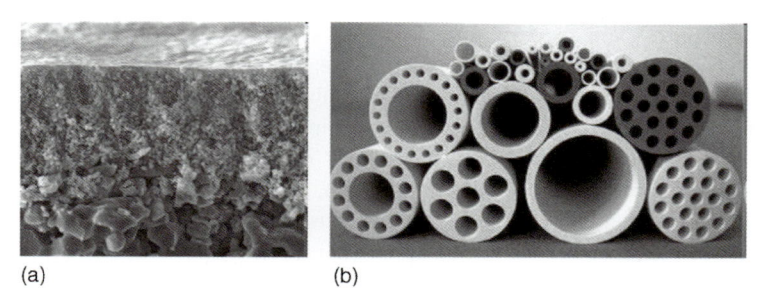

Figure 6.2 (a) SEM image of an Al_2O_3 support and (b) different configurations of tubular support [3].

8 wt% ethanol. The water flux measured in pervaporation operation for 90 wt% ethanol solution at 75 °C is about $7\,\text{kg}\,\text{m}^{-2}\,\text{h}^{-1}$. Ethanol hardly permeates through the membrane, resulting in a separation factor (water/ethanol) of around 10 000. The NaA membranes showed good stability with an operating lifetime of 3–5 years. A few attempts have been reported for a preferred separation of organic compounds from water applying Mitsui X, Y, and ZSM-5 type membranes, achieving high separation factors but lower fluxes.

In this chapter, the synthesis, characterization, and performances of MOF membranes will be presented. There are many common aspects between zeolites and MOFs in terms of membrane synthesis and characterization but also adsorption and diffusion. Therefore, advantages and limitations of MOFs with respect to this benchmark will be discussed.

The main limitations of zeolite membranes are as follows:

- Low permeance, resulting in a large membrane surface that has to be manufactured and thus a high capital cost investment.
- Synthesis costs associated with scale-up and reproducibility issues, especially for hydrothermal synthesis involving the removal of organic templates. Removal of the template to open up the pore structure requires heating in an oxidative atmosphere to decompose and burn out these organics. Differences in thermal expansion between the support and zeolite layer and changes in unit cell size may induce cracks and loss of performance.
- The narrow range of pore sizes of zeolites (ultramicroporous range), although intense efforts are currently being devoted to the discovery of aluminophosphates and zeolites with very large pores. On the other hand, mesoporous silicate materials have pore sizes that are too large to allow gas or vapor separations. Among about 150 zeolite structures, only a few topologies can be prepared in as membranes (about 15 structures have been tried).

MOF membranes can, in principle, address these issues. MOF materials actually bridge the gap between zeolite and mesoporous material types. Indeed, MOFs with pore systems ranging from the ultramicroporous to mesoporous have been reported. The abundant choice of MOF materials with structural controlled pore size opens up great opportunities for separative applications (Figure 6.3). In addition, MOFs can be obtained in very thin layers, which will enhance mass transport. Finally, MOFs are synthesized without the use of templates, making activation no longer an issue.

6.2
Mechanism of Mass Transport and Separation

The permeance denotes the rate per unit membrane surface area and transmembrane pressure at which a gas traverses a membrane. It is the ratio of the flux to the transmembrane pressure. The permeance of a membrane towards gases is a function of membrane properties (texture and structure), the nature of the permeate species (size, shape, and polarity), and the interaction between membrane and permeate

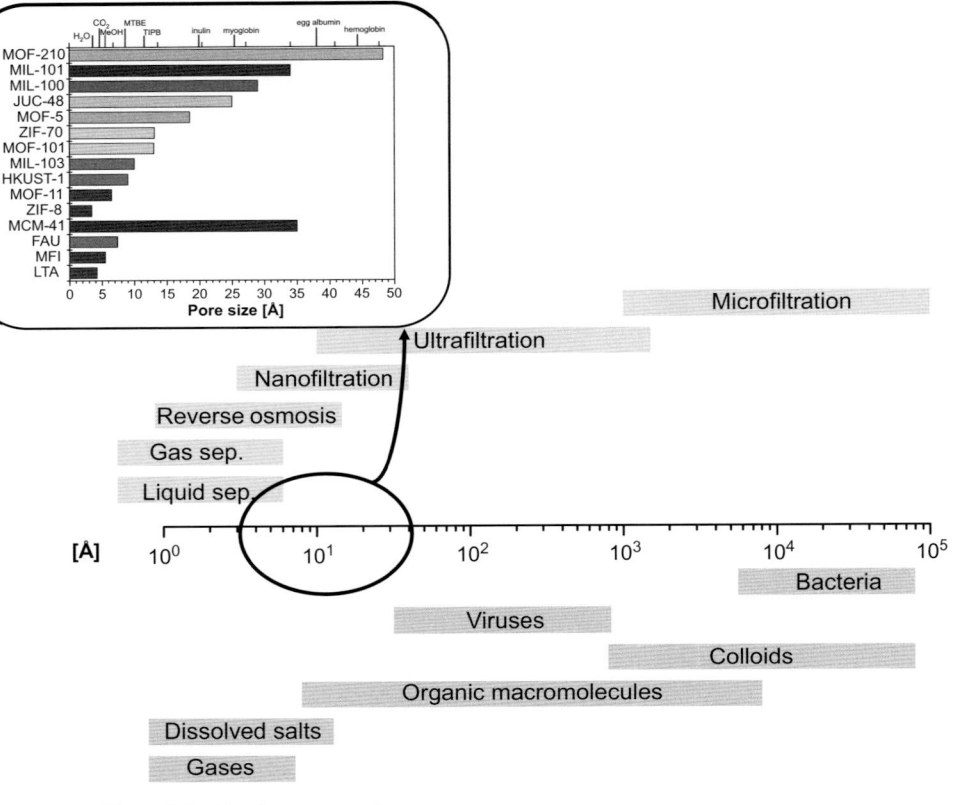

Figure 6.3 Membrane separation processes.

species. Membranes utilized in separations need to possess both high selectivity and high permeation. The selectivity of the membrane to a specific component is subject to the ability of the component to diffuse through the membrane. The permselectivity or ideal separation factor (pure gas permeation), α, is the ratio of two gases, 1 and 2, being separated:

$$\alpha_{1/2} = \frac{P_1}{P_2} \tag{6.1}$$

The selectivity or real separation factor (mixed gas permeation) can be defined as

$$\alpha_{1/2} = \frac{x_1/x_2}{y_1/y_2} \tag{6.2}$$

where y_1 and y_2 are the mole fractions of species 1 and 2, respectively, in the feed, and x_1 and x_2 are the corresponding mole fractions in the permeate.

There are four main transport regimes (Figure 6.4), which mainly depend on the size of the pores with respect to the size of the substrates or, more specifically, the

6.2 Mechanism of Mass Transport and Separation

	Mass transport	Separation mechanism
λ/d	Viscous flux	No separation $\alpha=1$
	Knudsen diffusion	$\alpha_{2/1}=(M_1/M_2)^{1/2}$
	Capillary condensation (high P)	Gas/vapors (absolute separation to the highest ΔH)
	Surface diffusion (low P)	• Molecular sieving (absolute separation) • Adsorption-diffusion (depending on ΔH)

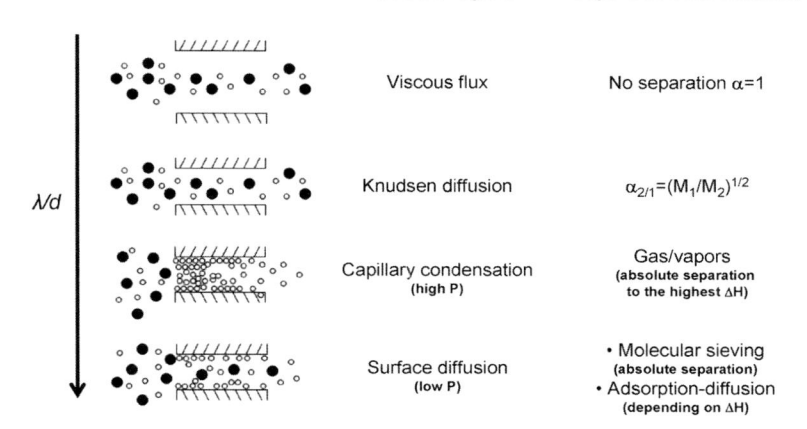

Figure 6.4 Mass transport and separation mechanism as a function of the pore size (mean free path:pore size ratio).

mean free path. The kinetic diameter is employed to determine the accessibility of molecules to zeolite channels and is related to the minimum equilibrium diameter of a molecule, r_{min}, given by a Leonard-Jones potential and assuming that the molecule is effectively spherical [6]. The mean free path (λ) is the average distance covered by a molecule between successive collisions against the adsorbent wall which modify its direction or energy. The separation mechanism depends on the type of mass transport regime [7, 8].

Viscous flow takes place when the pore size is larger than λ. In this regime, the gas molecules collide with the others and interactions with the wall of the solid are negligible so that there are no separation effects.

Knudsen diffusion occurs in the gas phase through the pores in the membrane layer having diameters (d) smaller than λ in the gas mixture [i.e., the Knudsen number (λ/d) is close one or infinite]. As a result, the movement of molecules inside the narrow pore channels takes place through collision of the diffusing molecules with the surface (wall) rather than with each other. Since the driving force for transport is the partial pressure of the gas species, Knudsen transport can occur by pressure gradients (Figure 6.5). The relative permeation rate of each component is inversely proportional to the square root of its molecular weight. As consequence, in this regime separation factors for light gases are relatively small, for example CO_2/N_2 0.79, CO_2/CH_4 0.6, O_2/N_2 0.93, H_2/nC_4 4.38, H_2/CH_4 2.83, H_2/CO_2 4.69 and H_2/CO 3.74. Nevertheless, high separation factors can be obtained by integrating several membrane separation stages. In the 1960s in France, the enrichment of ^{238}U was achieved by multiple membrane separation steps although the Knudsen separation factor of $^{238}UF_6/^{235}UF_6$ is only 1.0043.

In the surface diffusion mechanism, the diffusing species adsorb on the walls of the pore, and are readily transported across the surface in the direction of decreasing surface concentration. In mixture conditions, the competition of adsorption occurs

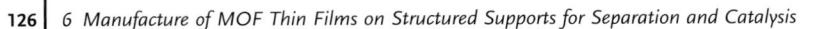

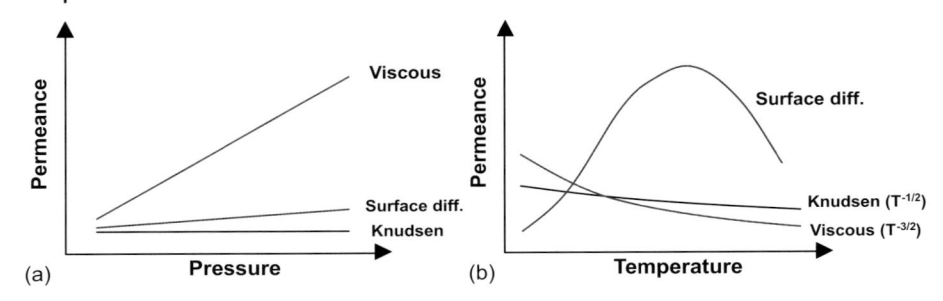

Figure 6.5 Typical profiles of permeation versus (a) pressure and (b) temperature.

between the different adsorbates. The adsorbate, which covers most of the surface, hinders the others from being adsorbed and thus from permeating. Hence, at equivalent partial pressure, the adsorbate with the highest adsorption enthalpy permeates preferentially. This regime usually exhibits opposite selectivity to the Knudsen regime. Generally, in the Knudsen regime the smallest (lightest) substrate permeates preferentially whereas for surface diffusion the largest (heaviest) is favored due to the usual correlation between the size/weight of the substrate and their associated enthalpy of adsorption [9, 10]. As an example, nC_4 can permeate with high selectivity by preferential adsorption from an nC_4–H_2 mixture at low temperature on a ZMS-5 membrane [11]. As the temperature increases, the surface becomes less and less covered by the strongest adsorbate, thus allowing H_2 to permeate (Figure 6.6). At high temperature, there is a change in the separation selectivity, because H_2 permeates preferentially. At 500 °C, where the adsorption is very limited, the permeation is inversely proportional to the square of the molecular weight, corresponding to a Knudsen regime.

In the case, the adsorbates have very similar adsorption enthalpies, and the separation is controlled by diffusion coefficients and/or entropic phenomena. An

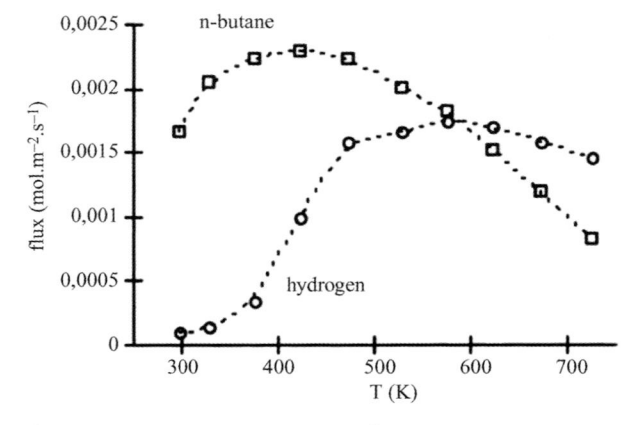

Figure 6.6 Permeation in H_2–n-C_4H_{10} binary mixture [11].

example is the mixtures *n*- and *i*-butanes or *p*- and *o*-xylenes relative to the 0.55 nm pore diameter of a ZSM-5 membrane. This mechanism requires a low or medium pore filling of the zeolite pores, which allows the mixture components to diffuse undisturbed by molecule–molecule interactions. The separation of xylene isomers on ZSM-5 and silicalite-1 membranes is one of the most often studied permeation systems. As the diffusion coefficient of *p*-xylene is larger than that of *o*-xylene by a factor of 1000, *p*-xylene can be separated from a binary mixture with *o*-xylene [12–16].

In theory, the absolute separation factor can be obtained upon size exclusion in a molecular sieve-type membrane. This is the basic mechanism for nanofiltration membranes. However, for light molecules, this is not implemented in practice. Indeed, for ZSM-5 or silicalite-1 membranes, most of the molecules having a kinetic diameter larger than 5.5 Å can easily condensed using physical processes. Note that, molecular sieve membranes cannot be considered as perfect monocrystal layers. Indeed crystalline defects and intergrowth lead to intercrystalline micro- and mesoporosity, which affects the separation performances.

In practice, all these mass transport regimes can coexist in a certain range of pressure and temperatures. In addition, porous membranes always exhibit a pore size distribution, which may originate from intercrystalline defects. Hence it is not always obvious how to identify properly the mass transport regime of a new membrane. Results on ideal selectivity are usually sufficient to characterize transport mechanisms and to quantify the dominant transport regime. To do so, a systematic investigation of permeation on different pure gases and mixtures as a function of the temperature and pressure should be investigated.

6.3
Synthesis of Molecular Sieve Membranes

6.3.1
Synthesis of Zeolite Membranes

Zeolite membranes have typically been prepared from hydrogels or sols composed of a source of SiO_2, Al_2O_3, a mineralizing agent (OH^- or F^-), water, and an organic structure-directing agent (SDA) for zeolites with high silica content. The support is immersed in the mixture and heated under autogenic pressure.

Among the different designs of zeolite membranes, the composite ceramic–zeolite is the most commonly encountered architecture. In this case, the nucleation–growth takes place into the inorganic layers, leading to a composite layer (Figure 6.7) [17]. The thin zeolite top layer can be crystallized hydrothermally in one step (so-called *in situ*) on the top of the support or inside the pores of the support (pore plugging). Seeding-supported crystallization occurs when the zeolite layer is formed in several steps using seed crystals that grow together in a subsequent hydrothermal synthesis.

On the other hand, there are other methods that are less significant in the preparation of zeolite films, probably due to the difficulties with scale-up and

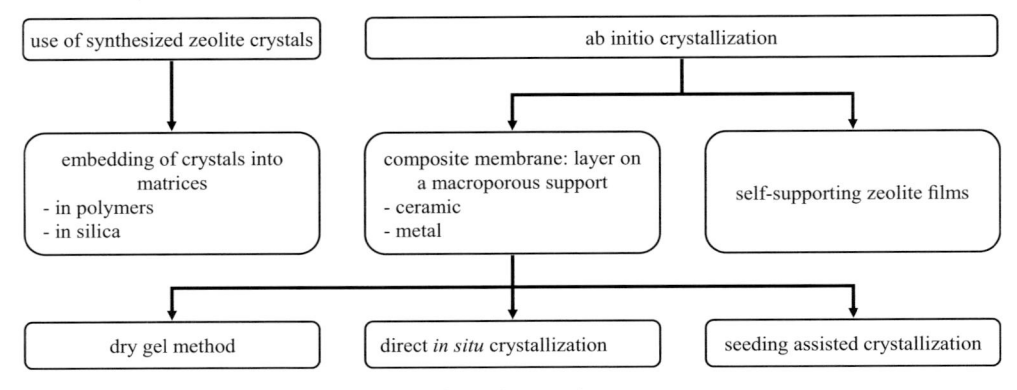

Figure 6.7 Variety of concepts for zeolite membrane preparation [17].

the issues for industrial application. The preparation of self-supporting films essentially consists of hydrothermal synthesis in the presence of various phases or mixtures containing nuclei with nanometer dimensions [18]. Zeolite layers can be synthesized on Teflon surfaces and detached from the substrate after crystallization [19]. Another method of direct synthesis to obtain zeolitic membrane is to bind together polymeric materials with a high content of zeolite crystals. It is possible to distinguish different composite membranes according to the host polymeric matrix [20].

The dry gel method includes two different procedures, vapor-phase transport (VPT) and steam-assisted crystallization (SAC). In both strategies, a layer with the necessary nutrients for the formation of the zeolite (silica, alumina, and the mineralizing agent) is deposited on the support by a sol–gel technique. Subsequently, the support with the dry gel is exposed to the vapors of water (SAC) or water and SDA [21, 22].

In all cases, the synthesis of zeolite membranes of high quality requires intense experimental efforts. They are numbers of parameters that have to be dealt with in order to optimize the procedure: (i) parameters related to the reaction media, such as the nature of the reagents, nature of the support, alkalinity, dilution, and ratio of reagents, and (ii) parameters related to the method of synthesis, such as reaction temperature, crystallization time, aging, mixing of the reagents, and stirring, to cite just a few [23].

6.3.1.1 Direct Nucleation–Growth on the Support

In the case of MFI zeolite, with a high silica content, the crystallization starts at the phase boundary between a liquid phase and a gel layer. Only in the solution are the template ions found and the gel layer on the surface of the support is formed by precipitation of the silica sol particles in certain concentration ranges and at given temperatures. The crystal grows into the gel layer, consuming the gel until the growing crystals reach the support [24, 25].

The gel for the zeolite crystallization can be soaked into the pore system of the support, forming zeolite plugs. The resulting nanocomposite zeolite membranes exhibit improved mechanical stability [26].

6.3.1.2 Secondary Growth

For controlling the preparation of supported zeolite membranes, decoupling of the nucleation and the crystal growth steps is proposed. In the first stage, the seeding step, the surface of the macroporous support is covered with submicron seed crystals.

Since the concentration needed for secondary growth is lower than that required for direct *in situ* synthesis, further nucleation is strongly reduced and almost all the crystal growth takes place over the existing seed crystal [27].

The most popular seeding processes reported in the literature are the following:

- Dip coating: the supports are seeded with nanocrystals from a colloidal suspension.
- Rubbing: based on manual deposition of zeolite crystals as a powder on the surface of the support [28].
- Brushing: this process is similar to the rubbing, but the seeds are rubbed on the inner of tubular supports by using a test-tube brush [29].
- Laser ablation: in this process, a high-intensity excimer laser beam strikes a pressed zeolite pellet that generates a plume comprised of fragments of the target that deposit on a temperature-controlled substrate [30].
- Electrophoresis: charged particles dispersed in a liquid are deposited on a substrate under the force of an applied electric field [7].

Zeolite films prepared by *in situ* synthesis, secondary growth, and pore plugging are shown in Figure 6.8.

6.3.2
Preparation of MOF Membranes and Films

The synthesis of MOF crystals occurs by self-assembly of cationic systems acting as nodes with polytopic organic ligands acting as linkers. As for zeolite frameworks, MOFs are constructed by motives of larger size that are usually called secondary building units (SBUs). It was showed that SBUs exist in solution before they crystallize. Although the synthesis mechanisms of both zeolites and MOFs are far from being completely understood, they have common traits. Hence strategies to prepare MOF membranes are mostly derived from the synthesis of zeolite membranes. There are three different approaches to the fabrication of MOF thin films: (a)

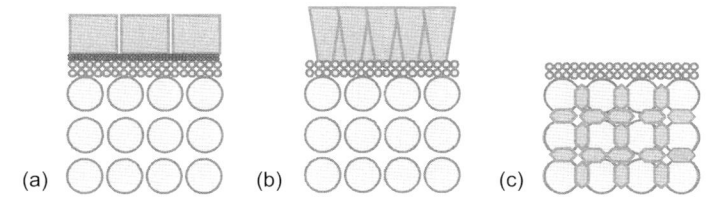

Figure 6.8 Zeolite films prepared by (a) *in situ* synthesis, (b) secondary growth, and (c) pore plugging.

growth of functionalized self-assembly monolayers (layer-by-layer), (b) secondary growth after the assembly of preformed, ideally size- and shape-selected, nanocrystals, and (c) direct nucleation–growth by solvothermal synthesis.

The influence of the surface chemistry (functionality) of the chosen substrate and the use of MOF seeds on the nucleation, orientation, and adhesion of the MOF films are important factors.

6.3.2.1 Self-Assembled Layers

Self-assembled organic monolayers (SAMs) have been used to direct the nucleation, orientation, and structure of the deposited MOFs. The approach is to control the crystallization of hybrid materials by the influence of organic macromolecules, based on the principle of biomineralization. It is the process by which living organisms produce minerals, often to harden or stiffen existing tissues. The concept of biomineralization refers to oriented nucleation, control over crystal morphology, formation of unique composites of proteins and single crystals, and the production of ordered multicrystal arrays. More details can be found in Chapter 13.

Regarding carboxylate-based MOFs, directed heterogeneous nucleation on solid substrates should be possible if solvated or polynuclear Zn^{2+} complexes or even larger MOF nuclei selectively bind to carboxylate-terminated SAMs via a terephthalate bridge connected to a surface-bound Zn cation.

Hermes et al. reported the deposition of MOF-5 crystals on a patterned SAM of 16-mercaptohexadecanoic acid and $1H,1H,2H,2H$-perfluorododecanethiol on Au(111) [31]. After synthesis, a thin film of MOF-5 particles was obtained and anchored selectively at the carboxylate-terminated areas of the SAM. The MOF film remained on the surface even after several washing cycles with ethanol. Similar experiments were carried out with bare Au(111) and CF_3- and COOH-terminated non-patterned SAMs. Only in the latter case were well-adherent MOF-5 films formed (Figure 6.9).

Later, the same group presented the results of a comparison of the growth of MOF-5 crystals at bare alumina and silica surfaces with silica surfaces modified with COOH- or CF_3-terminated SAMs [32]. Self-assembled monolayers of the different organosilanes were prepared by immersing the substrates in a solution (pentane) of 10% of 10-undecenyltrichlorosilane or $1H,1H,2H,2H$-perfluorododecyltrichlorosilane. Thin MOF-5 films are likely to be grown on any type of substrate which can be

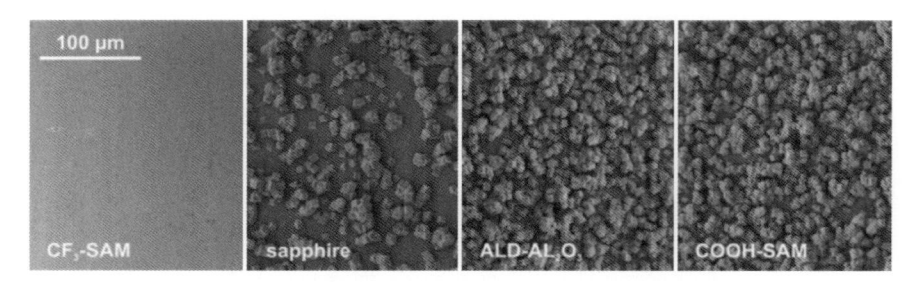

Figure 6.9 SEM images of MOF-5 thin films on different pretreated alumina substrates [32].

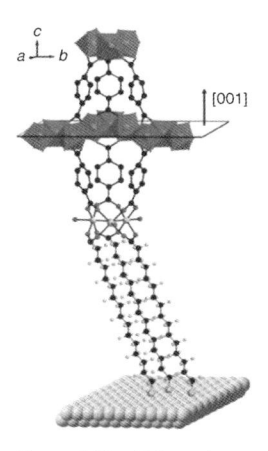

Figure 6.10 Oriented growth of MIL-88B crystals on MHDA SAMs on an Au(111) surface. From [34].

previously modified either by application of a suitably functionalized SAM (e.g., COOH-terminated). The remarkable difference between SiO_2 and Al_2O_3 surfaces is attributed to the isoelectric points of these materials, and the obvious requirement of a basic, that is, electrostatically compatible, surface for anchoring [32].

Following similar procedures, Biemmi et al. studied the deposition of HKUST-1 silane-based SAMs on SiO_2/Si substrates (known to be much more thermally robust than thiol-based SAMs) [33]. They proposed two different termini (–COOH and –OH) to force the oriented crystal growth on the molecular layer on the substrate.

Oriented growth of MIL-88B on mercaptohexadecanoic acid SAMs, where MIL-53 is the product of homogeneous nucleation, has also been reported (Figure 6.10) [34].

As indicated earlier, the polytopic organic linkers are what make these materials unique and extremely versatile, and, at the same time, the organic linkers can be the major obstacle to growing MOF films on substrates. These organic linkers typically do not provide additional linkage groups that can form bonds with linkage groups on the surface of the supports (i.e., hydroxyl groups in case of metal oxides). As a result, the surfaces of the support typically have to be either chemically modified or electrostatically compatible in order to grow MOF films. In applications of this approach to synthesize supported membranes possess, some important issues arise. The first is that the selected supports for self-assembly are nonporous metal and ceramic supports, with a shape that is difficult to scale up and useless for the preparation of membranes. The laborious preparation of the substrates and the lack of binding strength of MOF crystals on the substrates are also of critical importance.

6.3.2.2 Solvothermal Synthesis: Direct and Secondary Growth

6.3.2.2.1 Metal Carboxylate-Based Membranes

The first attempt at the *in situ* crystallization of a MOF on a porous support was made by Arnold et al. [35]. They studied the deposition of $[Mn(HCOO)_2]$ under solvothermal conditions at 115 °C

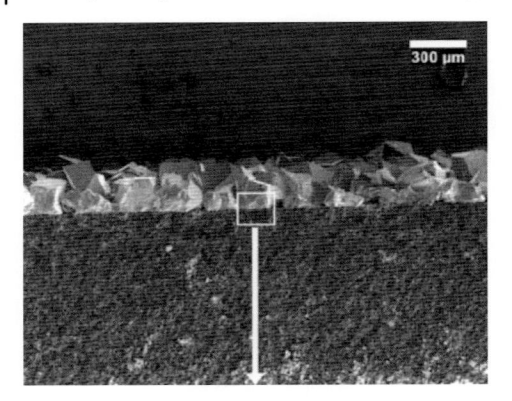

Figure 6.11 Profile of Mn(HCOO)$_2$ crystals grown on a graphite support disc after two *in situ* crystallizations by the "formate" route [35].

[1,4-dioxane and *N,N*-diethylformamide (DEF)] on bare alumina and graphite discs. They observe a low crystal density, with the problem of inhibited nucleation, 10 crystals mm^{-2} for alumina and 80 crystals mm^{-2} for graphite.

By developing the so-called "formate" route (i.e., using sodium formate rather than formic acid) and oxidized graphite substrates, nucleation is enhanced and fairly dense coatings of [Mn(HCOO)$_2$] with crystallite sizes >100 μm are obtained. The problem with this material is its one-dimensional (1D) channel system, so an orientation of the channel structure perpendicular with respect to the porous substrate is essential. Also, only in the case of the "formate" route with functionalized graphite as a substrate were the authors successful in obtaining a coating with a preferred orientation and a tilt angle of about 34° of the 1D channel system (Figure 6.11).

Gascón *et al.* reported dense coatings of HKUST-1 on α-alumina supports by a combination of suitable seeding with low concentration mother liquors [36]. They carried out seeding of the support with a slurry of the 1D coordination polymer copper(II) *catena*-triaqua-*m*-(1,3,5-benzenetricarboxylate) [Cu(Hbtc)(H$_2$O)$_3$], obtained by modification of the original HKUST-1 recipe. Dense coatings of small intergrown octahedral microcrystals (2 μm) were formed by immersion of the seeded supports int a diluted solvothermal solution at 110–120 °C over a period of 12–18 h. The authors claimed that the coatings consisted of phase-pure HKUST-1, without preferential orientation, mentioning that no traces of the characteristic peaks of the 1D seeding MOF were detected in the X-ray diffraction (XRD) pattern of the coating, suggesting fairly complete conversion of the seeding material into the HKUST-1 phase under the conditions of the experiment. However, data on the porosity and transport properties of the deposited MOF coatings were not given.

Yoo and Jeong demonstrated the rapid deposition of MOF-5 on carbon-coated anodic aluminum oxide (AAO) by microwave-induced thermal deposition [37]. They showed that microwave irradiation synthesis is a rapid method for the preparation of nanoporous MOF thin films and patterns on porous alumina substrates.

6.3.2.2.2 Zinc Imidazolate-Based Membranes

Microwave synthesis is the preparation route used in Caro's group to prepare MOF membranes. They focused their research on a subclass of MOFs, the zeolite imidazolate frameworks (ZIFs).

First, they proposed a rapid and cheap synthetic protocol at room temperature that yields pure-phase nanoscale ZIF-8 material with a narrow size distribution, good thermal stability, and large accessible internal surface area [38]. In contrast to reported protocols, where the design to produce large microcrystals is based on mixtures with the zinc salt and the imidazolate ligand in a molar ratio of 1:2, their best results were obtained when employing $Zn(NO_3)_2 \cdot 6H_2O$, 2-methylimidazole, and methanol in a molar ratio of approximately 1:8:700. Therefore, methanol can be removed much more easily from the pore network than N,N-dimethylformamide (DMF) and, importantly for membrane synthesis, the stress to the crystals is strongly reduced. The time of synthesis is substantially reduced to 4 h by using microwave-assisted heating. By applying this improved synthetic protocol in membrane preparation, they were able to obtain a crack-free, dense, polycrystalline layer of ZIF-8 on a porous titania disc support [39].

In later work, they showed that ZIF-7 can grow *ex situ* on an alumina support. Like ZIF-8, ZIF-7 crystallizes in a SOD structure with benzimidazolate (bim) as linker [40]. They prepared ZIF-7 nanoseeds at room temperature, modifying the synthetic protocol reported by Yaghi and co-workers [41] in which the linker (bim) to zinc ratio was increased from 0.74 to 6.5. It is speculated, as for the ZIF-8 nanocrystals, that one can expect that an equilibrium of the cationic (protonated) and neutral forms exists in solution and deprotonation of the linker is driven only by the crystallization of ZIF. The excess of bim allows it to act both as a linker in its deprotonated form and as a growth terminator and stabilizing agent in its neutral form. This line of argument suggests that this method of excess protic linker might be a general one and transferable to other ZIFs. The synthesized ZIF-7 nanoseeds of 30 nm can be dispersed in methanol or DMF to form stable colloidal dispersions. When these colloidal dispersions are used to seed alumina supports, however, the seed layer can easily peel away from the supports. To address this problem, they dispersed the ZIF-7 nanoseeds in a polyethylenimine (PEI) solution. PEI can effectively enhance the linkage between the seeds and the support through H-bonding interactions. A dip-coating technique was used for the surface seeding of the alumina support. Microwave-assisted solvothermal synthesis was carried out to perform the secondary growth. The seeded support was placed vertically in a clear synthesis solution of Zn^{2+}–bim–DMF with a molar composition of 0.75:1:150, and then microwave heated at 100 °C for 3 h. After secondary growth, a 1–2 µm ordered ZIF-7 layer, without any pinholes or cracks, was formed on the alumina support (Figure 6.12).

Liu *et al.* prepared a ZIF-69 membrane on porous α-alumina support by an *in situ* solvothermal synthesis procedure [42]. The ZIF-69 framework consists of zinc nitrate coordinated with 2-nitroimidazole and 5-chlorobenzimidazole, forming a zeolite GME topology that has 12-membered ring (MR) straight channels along the *c*-axis and 8 MR channels along the *a*- and *b*-axes. The membrane obtained is continuous and has a preferred *c*-orientation (Figure 6.13). The membrane shows

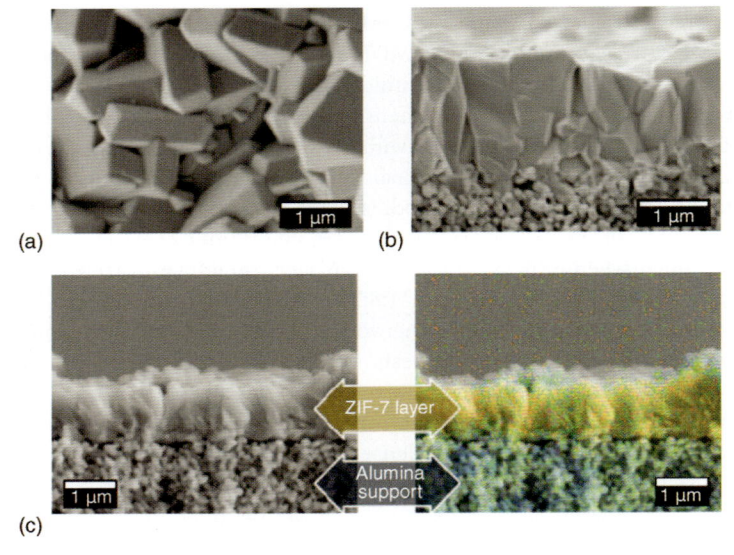

Figure 6.12 (a) Top view and (b) cross-section SEM images of the ZIF-7 membrane; (c) EDXS mapping of the ZIF-7 membrane. Orange, Zn; cyan, Al. From [40].

good thermal stability, and also good chemical stability in boiling methanol and boiling benzene.

A crucial point in the preparation of MOF membranes is the removal of the solvent involved in the synthesis. DMF has a large molecular diameter that can generate problems with its removal, creating cracks in the film. In order to avoid this difficulty, the option is to wash the membrane several times with methanol. Another alternative is to modify the reported synthetic protocols, replacing DMF or aqueous mixtures with pure methanol. Methanol has a much smaller kinetic diameter than DMF. Bux *et al.* demonstrated that in air, methanol readily escaped

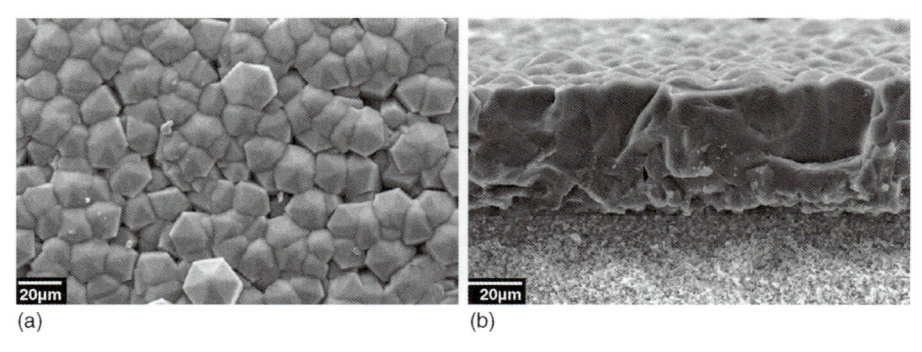

Figure 6.13 ZIF-69 membrane: (a) SEM images of top view and (b) cross-section [42].

Figure 6.14 SEM image of (a) the cross-section (left) and (b) the surface (right) of the SIM-1 membrane [44].

completely from the cavities even at room temperature, yielding a guest-free activated ZIF-8 membrane [39].

Later, Caro and co-workers used an on-stream method to open the pores of a ZIF-7 membrane, monitoring the activation process simultaneously by using a Wicke–Kallenbach permeation cell with a 1:1 mixture of H_2 and N_2 on the feed side. Coinciding with the thermogravimetric analysis, the guest molecules begin to leave the cavities when the cell is heated to 100 °C, at which temperature the ZIF-7 membrane becomes gas permeable. Complete activation is accomplished after the temperature has been maintained at 200 °C for ∼40 h, and the H_2 permeance reaches a plateau value [43].

Recently, Aguado *et al.* prepared a substituted imidazolate-based MOF (SIM-1) membrane *in situ* on a tubular asymmetric alumina support [44]. They reported a very reproducible one-step process operating at atmospheric pressure to prepare a thin MOF, which meets the first criterion enabling scale-up for the preparation of large surface area membranes. The new SIM-1 belongs to the class of ZIF materials and it is isostructural with ZIF-8 and ZIF-7. SIM-1 has no group at position 2 of the imidazolate linkers but it contains a methyl and an aldehyde at positions 4 and 5, the latter conferring polar features due to the dipolar moment of the C=O bond. The synthesis was carried out at 358 K for 72 h, and the resulting membrane was washed with ethanol and dried at room temperature. From the surface view (Figure 6.14), it can be clearly seen that the SIM-1 crystals merge compactly Also, the cross-section view at low magnification proves the absence of defects over a long distance. The thickness of about 25 μm is uniform along the membrane. Excellent attachment of the SIM-1 top layer to the porous support can be observed.

6.3.2.2.3 Chemical Modifications of Substrate Surfaces
Chemical modifications of substrate surfaces have been proposed for directing the nucleation and orientation of the deposited MOF layers, improving the quality of the MOF layer.

Guo et al. used a copper net to provide homogeneous nucleation sites to support continuous HKUST-1 film growth [45]. A copper net (400 mesh) is cut into circular wafers (10 mm in diameter), washed with ethanol, then with water under ultrasound, and placed in an oven at 100 °C for oxidation. The modified copper support is placed vertically and introduced into the synthesis mixture, and the synthesis is carried out at 120 °C for 3 days. By oxidizing the copper net before the hydrothermal synthesis, homogeneous nucleation sites are formed for continuous film growth. The Cu^{2+} ions both on the copper net and in the reaction solution provided a metal source for crystal growth.

Yoo et al. reported the fabrication of continuous MOF-5 membranes with a controllable out-of-plane orientation by a secondary growth method, using graphite-coated alumina discs. Densely packed MOF-5 seed layers on the support are rapidly prepared by microwave-induced thermal deposition. The MOF-5 seed layers are then solvothermally treated to grow into continuous MOF-5 membranes. The presence of a proton-scavenging amine (N-ethyldiisopropylamine) in the precursor solution during secondary growth is critical to prevent seed crystals from dissolving and simultaneously to permit their growth [46].

A seeded growth method is used to prepare preferentially oriented and well-intergrown films of microporous metal–organic frameworks (MMOFs) on modified alumina discs by Ranjan and Tsapatsis [47]. They prepared an MMOF, Cu(hfipbb)$(H_2hfipbb)_{0.5}$ [hfipbb = 4,4'-(hexafluoroisopropylidene)bis(benzoic acid)], having pore channels (0.32 nm) in only one direction, so the orientation of the crystals becomes critical for membrane application (Figure 6.15). The alumina support surface is modified using PEI based on the expectation of enhanced attachment of seeds via H-bonding. The seeding crystals are obtained by solvothermal growth, and later crushed into submicrometer-sized crystals. Seed layers of these particles are deposited by rubbing on the modified supports, following by a secondary growth at 150 °C for 12 h. The preferred orientation is attributed to faster growth along the b-direction during the secondary growth on the randomly oriented seeds. This situation is analogous to the well-established case of c-oriented columnar MFI membranes formed by secondary growth of randomly oriented seeds.

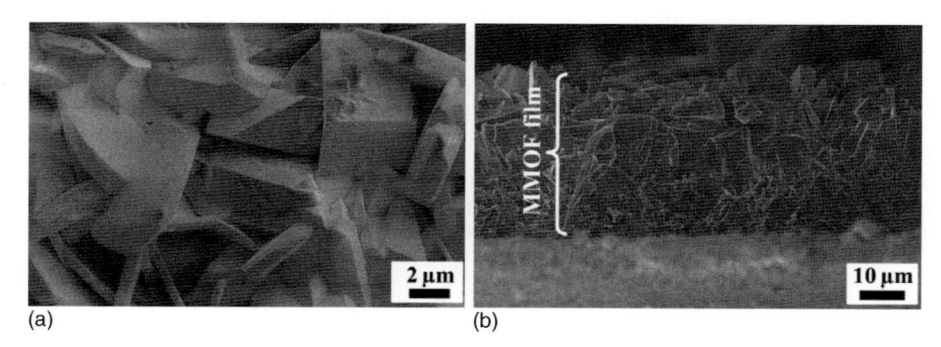

Figure 6.15 SEM image of (a) the surface and (b) the cross-section of the MMOF membrane [47].

6.4
Application of MOF Membranes

6.4.1
Gas Separation

6.4.1.1 Metal Carboxylate-Based Membranes

The study of the ability of MOF membranes to effect separations is still in its early stages. The initial attempts were based on removal of H_2 by molecular sieving and separation of CO_2 by adsorption differences.

The essential points in the separation of H_2 are the pore size of the membrane material, taking into account that the molecular diameter of H_2 is 0.29 nm, and the dimensionality of the channel structure of the MOF.

The copper net-supported HKUST-1 membrane reported by Guo *et al.* exhibits an H_2/N_2 selectivity of 7 (calculated as explained in Section 6.2), being the first MOF membrane to show gas separation performance beyond Knudsen diffusion behavior [45]. The permeation flux of H_2 is much higher than those of other gases, showing that the membrane has a higher size selectivity preference for H_2. It is worth mentioning that the sorption capacities of CO_2, CH_4, and N_2 are much higher than that of H_2. However, the separation factors of H_2-N_2 and H_2-CO_2 gas mixures are higher than the ideal separation factors. This can be explained by the pore size of HKUST-1 (0.9 nm), large enough to allow the passage of H_2, but too large to permit blockage of hydrogen due to the stronger adsorption of CO_2 and CH_4. In order to measure the thermal stability of the $Cu_3(btc)_2$ (btc = benzenetricarboxylate) membrane, the separation temperature is increased from 0 to 70 °C. The membrane maintains its separation factors and permeations over a 24 h period. Furthermore, the membrane was used repeatedly over a long period of time (over 6 months).

The MMOF membrane prepared by Ranjan and Tsapatsis [47] has an ideal H_2/N_2 selectivity of 4 at room temperature and 23 at 190 °C. This higher selectivity, compared with the report of Guo *et al.* [45], might be a result of the smaller effective pore size (0.32 nm), which results in a relatively low H_2 permeance of this MMOF membrane (10^{-8} mol m^{-2} s^{-1} Pa^{-1} at room temperature). The authors attributed this finding to the blockage of the 1D straight-pore channels in the membrane [47].

Takamizawa *et al.* observed selective permeation and anisotropic permeation properties using a single-crystal membrane of $[Cu_2(bza)_4(pyz)]_n$ [48]. They prepared two kinds of membranes with orthogonal crystal orientation, one with exposed (001) crystal surfaces (channel membrane) and the other with exposed (001) crystal surfaces (non-channel membrane) in holes of aluminum plates using epoxy bonding. Permeability values along the channel membrane were larger than with the non-channel membrane for He, H_2, and CO_2, and undetectable for N_2, Ar, CO, O_2, and CH_4. The reason given by the authors was that He and H_2 permeate by the crystal defect, whereas the high adsorption ability of CO_2 can explain its non-zero permeance. The permselectivities of H_2/N_2 and H_2/CH_4 are 10 and 19, respectively. In gas permeation of an H_2-CO_2 gas mixture, the permeance of H_2 decreased in the presence of CO_2 whereas that of CO_2 gas remained unchanged, suggesting that the high adsorption and later diffusion of CO_2 in the channels are blocking the H_2 from passing. Therefore, with regard to H_2 separation, small-pore MOFs having three-dimensional (3D) channel structures are considered to be ideal membrane materials [48].

6.4.1.2 Zinc Imidazolate-Based Membranes

Because of the narrow size of the six-membered ring pores (0.34 nm), it was expected that a ZIF-8 membrane would be able to separate H_2 from larger molecules by a molecular sieving mechanism. Unexpectedly, the single gas permeances of H_2, CO_2, N_2, and CH_4 show no sharp cutoff at 0.34 nm. This indicates than the framework structure of ZIF-8 is more flexible than static in nature. In spite of that, the membrane reaches a relatively high H_2/CH_4 selectivity of 11.2 in a 1:1 mixture at 25 °C and 1 bar [39].

The same group chose ZIF-7, formed by a benzimidazole linker and sodalite topology, as a promising candidate for the development of an H_2-selective membrane [40, 43]. The pore size of ZIF-7 estimated from crystallographic data is about 0.3 nm, which in between those of H_2 (0.29 nm) and CO_2 (0.33 nm). The permeance of H_2 increases quickly with increase in temperature, indicating an activated diffusion process (Figure 6.16). An apparent activation energy (E_{act}) for H_2 diffusion of 11.9 kJ mol^{-1} is obtained by fitting the experimental gas permeance data to an Arrhenius equation. The activation energy for H_2 permeance gives a good correlation with the separation factor and can be used as a measure of membrane quality (experimental data suggest a minimum of 10 kJ mol^{-1}). Finally, they evaluated the influence of the presence of water in the H_2–CO_2 gas feed mixture, exposing the membrane to 3 mol% steam at 220 °C. The ZIF-7 membrane showed very good stability of more than 50 h. Again, the absence of a clear cutoff was attributed to a certain influence of the nonselective mass transport through the grain boundaries of the polycrystalline ZIF-7 layer. Aguado *et al.* recently showed that ZIF-7 exhibits a reversible breathing effect upon changes in temperature or CO_2 partial pressure and

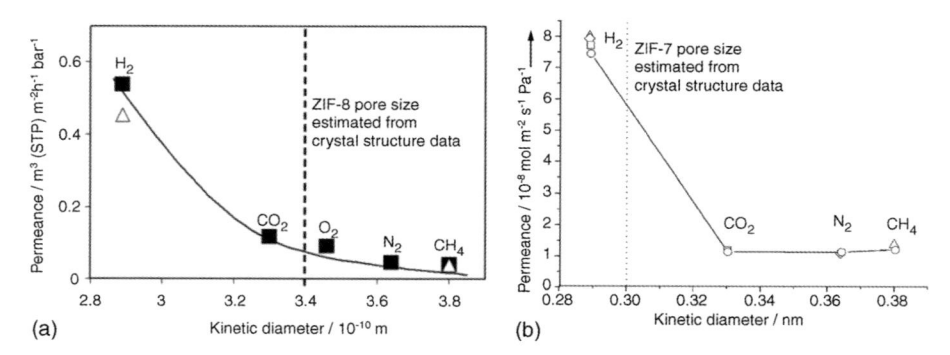

Figure 6.16 Permeances of single gases of the (a) ZIF-8 [39] and (b) ZIF-7 [40] membranes as a function of molecular kinetic diameters.

which is accompanied by reversible phase-to-phase transformation. This phenomenon may also explain the peculiar performances of ZIF-7 membranes [49].

As described in Chapter 5, much attention has been focused on MOFs for CO_2 capture from natural gas land field and flue gases. Screening and understanding of the fundamental structure–function relationships are very important for developing new processes based on MOF membranes. Snurr and co-workers used data on the adsorption of CO_2 on MOFs to validate a generalized strategy for molecular modeling. They found that at low pressures, uptake in different MOFs correlates with the heat of adsorption; at intermediate pressures, uptake correlates with the MOF surface area; and at the highest pressures, uptake correlates with the free volume available within the MOFs. There is a clear correlation between the CO_2 uptake and heat of adsorption at pressures <1 bar [50]. Table 6.1 shows the uptake (experimental and simulated data) and heat of adsorption of CO_2 for several MOFs and zeolites. This gives hints for the selection of appropriate MOFs to be prepared as membranes for CO_2 separation. Whereas the total CO_2 capacity is an important adsorbent criterion for PSA/TSA processes, this is not the case for membrane processes which operate in a continuous fashion. Instead, high surface area and 3D connectivity of the porous networks are the most important factors that will determine the permeability of the MOF membrane.

In spite of the broad diversity of MOFs with high capacity towards CO_2, only a few membranes for CO_2 capture have been reported.

Liu *et al.* characterized the ZIF-69 membrane (pore size of 0.78 nm) by separation of CO_2/CO mixture [42]. Yaghi and co-workers reported ZIF-69 selectivities (based on adsorption data) of 5, 20, and 21 for CO_2 compared with CH_4, N_2, and CO, respectively [51, 57]. They performed the separation of the binary mixture in the

Table 6.1 Uptake of CO_2 (experimental and simulated data) and heat of adsorption at 0.1 bar and 25 °C.

MOF	Q_{st} (kJ mol^{-1})	CO_2 uptake (mg g^{-1})		Ref.
		Experimental	Simulated	
HKUST-1	25.1	27.3	23.7	[50]
MOF-5	15.8	3.5	4.2	[50]
CPO-27(Mg)	26.7	261.6	51.5	[50]
ZIF-8	18.2	5.3	5.5	[50]
MIL-47	20.8	7.8	15.7	[50]
SIM-1	33.0	17.1	15.3	[44]
ZIF-69		9.0		[51]
SAPO-5	22.2	4.2		[52]
LiCHA	34.0	143		[53]
NaX	45.9	170		[54]
NaZSM-5	36.0	39.5		[55]
Silicalite	19.2	15.2		[56]

Wicke–Kallenbach mode, with a feed mixture containing a 1:1 volume ratio of CO_2–CO mixture, at room temperature. The CO_2/CO selectivity was around 3.5, and the permeance of CO_2 was in the region of 3×10^{-8} mol m^{-2} s^{-1} Pa^{-1}. Both the permeance and selectivity are higher for the binary mixture permeation than the single-component mixture. This is due to selective adsorption of CO_2, which enhanced the transport rate of CO_2. Comparing these results with those for zeolite membranes that have similar pore size to zeolite Y, Kusakabe et al. reported a CO_2/CH_4 selectivity of 20 with a permeation of CO_2 of 5×10^{-8} mol m^{-2} s^{-1} Pa^{-1} [58]. This result indicates that the ZIF-69 membrane could potentially be used for CO_2 capture.

The ZIF-8 membranes prepared by Venna and Carreon on alumina tubes displayed unprecedented high CO_2 permeances for an MOF membrane [59] The membrane has a permeance of 2.4×10^{-5} mol m^{-2} s^{-1} Pa^{-1} and CO_2/CH_4 selectivities from 4 to 7 at a 22 °C feed pressure of 1.4 bar. These values are comparable to those for SAPO-34 aluminosilicate membranes, that can effectively separated CO_2–CH_4 mixtures at pressures up to 17 bar [60]. The authors claimed that the explanation for the high flux obtained is a combination of the use of a macroporous support and preferential adsorption sites for CO_2 molecules. Surprisingly, the permeance of CO_2 is of the same order as for the raw material, although the membrane has a layer thickness of 9 μm, comparable to other reported MOF layers.

The SIM-1 membranes were tested in a single gas permeance setup with H_2, CO_2, N_2, and CH_4 using the Wicke–Kallenbach technique. Single permeances calculated from the volumetric flow rates through the SIM-1 membrane are in the region of 10^{-8} mol m^{-2} s^{-1} Pa^{-1}. Ideal selectivity data calculated from single gas permeances at 303 K deviated slightly from the Knudsen value for $H_2/N_2 = 2.5$ (3.7) but reversed for $CO_2/N_2 = 1.1$ (0.78), indicating an adsorption–diffusion-based mechanism. The permeance as a function of temperature was investigated to characterize the transport mechanism further (Figure 6.17). At low temperatures, the permeance

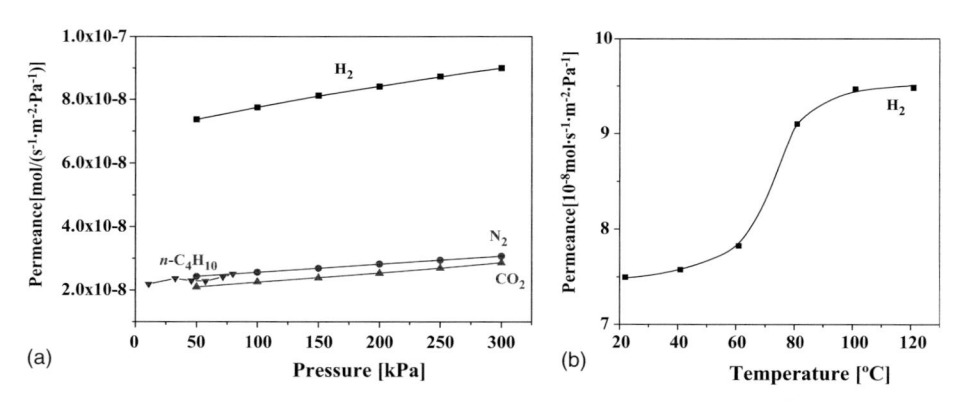

Figure 6.17 (a) Permeances of single gases of the SIM-1 membrane at 303 K as a function of the pressure and (b) permeance of H_2 as function of temperature [44].

profiles as a function of temperature corresponded to an activated transport mechanism such as that found for microporous membranes [44].

For the ternary mixture CO_2–N_2–H_2O (10:87:3 vol.%, 324 K, 4 bar and $\Delta P = 40$ mbar), a CO_2/N_2 separation factor of 4.5 is measured, which is much higher than the corresponding Knudsen separation factor (0.78). Therefore, under mixture conditions, a much larger preferential permeance of CO_2 (versus N_2) is obtained with respect to single permeance results. Similar selectivities have been obtained with MFI zeolite membranes [61]. It should be noted that much higher selectivity values have been reported for CO_2/N_2 separation on zeolite membranes, such as with DDR membranes, but usually in dry conditions [62]. In addition, SAPO-34 membranes are reported to have the highest performance; with selectivity around 150 [63, 64]. Nevertheless, the results observed on the SIM-1 membrane demonstrate that the transport mechanisms are similar to those for zeolite membranes. Clearly, surface transport takes place in the SIM-1 membrane, which allows the separation of gases by preferential adsorption, with the most adsorbed component reducing the diffusion of other molecules.

6.4.2
Shaped Structured Reactors

So far, few studies of catalytic applications of MOF films have been reported. One such study investigated the performance of MIL-101(Cr) immobilized on a monolith structure for the oxidation of tetralin [65].

The authors prepared a MIL-101(Cr) film on an alumina-coated cordierite monolith by secondary growth. Two different procedures were applied to cover the discs with seeds, spin coating and dip-coating, with a suspension of crystals of 150 nm dispersed in ethanol. Once the discs had been seeded, a secondary growth was performed by rotating synthesis at 220 °C for 8 h.

They compared the catalytic performance of crystals in a slurry reactor and the performance of the monolithic stirrer reactor. The MIL-101(Cr)-coated monolith showed a slightly higher activity than the crystals in the slurry. This result indicates good contact between the liquid phase and solid in the monolithic reactor and a lack of transport limitations. Furthermore, the easy recovery of the catalyst allows as many reuses as necessary to be performed, after washing in boiling chlorobenzene together with overnight treatment in air at 250 °C.

Using the same *in situ* procedure as for the SIM-1 membrane, Aguado *et al.* carried out synthesis of SIM-1 films on ceramic beads and used them for the reduction of acetophenone to phenylethanol by transfer hydrogenation in 2-propanol [66].

In the synthesis, α- or γ-alumina beads were immersed in a glass vial containing the synthesis mixture. After heating at 85 °C for 48 h, the resulting supported material was washed with ethanol to remove unreacted precursors and fine unsupported SIM-1 particles. In both types of alumina, the loading of SIM-1 was ∼10 wt% (±1 wt%). In the SIM-1–γ-Al_2O_3 composite, crystals are embedded in cavities of the support and some crystals grow on the surface of the bead, without forming a continuous film (Figure 6.18). The microporous SIM-1 fills about 25% of the

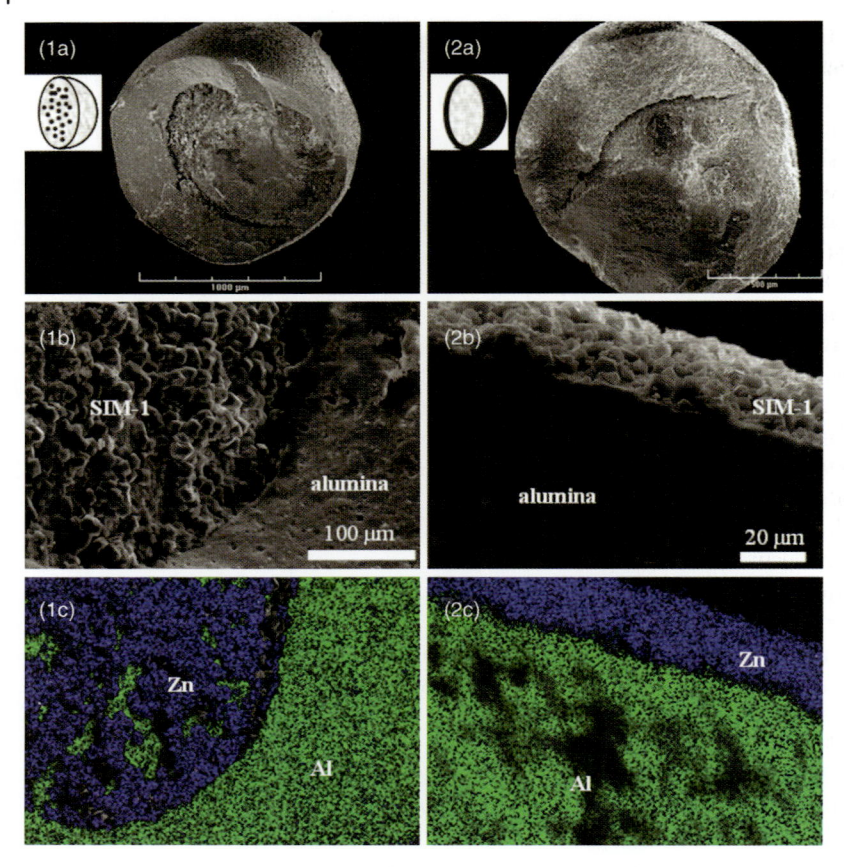

Figure 6.18 SIM-1 supported on (1) γ- and (2) α-alumina beads. (a) SEM image of the bead, view of the cross-section; (b) SEM image; and (c) EDXS mapping of the core (1) or surface (2). Blue, Zn; green, Al [66].

mesoporous volume of the γ-alumina. On the other hand, only small scattered crystals appear inside the α-alumina beads, while the formation of a homogeneous layer of 15 μm takes place on the outer surface.

For the reduction of acetophenone, using 5 mol% of SIM-1 supported in the presence of 25 mol% of potassium hydroxide at 80 °C, more than 90% of the substrate was converted after 10 h, whereas less than 10% conversion was observed when the reaction was performed with alumina beads alone (without SIM-1). Even using magnetic stirring (400 rpm), there was no notable weathering of the alumina beads. Once the reaction was finished, the beads could easily be separated from the reaction mixture by removing the solution. After washing with ethanol and drying, these beads were introduced into a second catalytic run under the same conditions without activity loss, which illustrates that it is possible to reuse the catalyst.

6.4.3
Perspectives for Future Applications

A summary of gas separation performances of MOF membranes is given in Table 6.2. As shown previously, the range of pore size of MOFs makes them suitable for nanofiltration separation processes. In fact, there has already been preliminary work on the use of mixed matrix membranes (MMMs), incorporating MOFs as fillers in polydimethylsiloxane (PDMS) membranes [68]. $Cu_3(btc)_2$, MIL-47, MIL-53(Al), and ZIF-8 were used as dispersed phases and the membranes were applied in solvent-resistant nanofiltration (SRNF), in particular in the separation of Rose Bengal (RB) from 2-propanol. From this preliminary study, we can anticipate that ZIF-type membranes, which are the most advanced so far and which show high water stability, will become the subject of new developments.

Catalytic membrane reactors are an elegant means to push the conversion of a reaction that is limited by thermodynamic equilibrium while separating the reaction products. This is generally the case for condensation reactions that liberate water, such as esterification reactions. From the Le Chatelier principle, the use of a water permselective membrane in the reactor can shift the equilibrium, provided that the permeance of the membrane is larger than the reaction rate (Figure 6.19). Hence we can anticipate that hydrophilic MOF membranes could become good candidates [69].

6.5
Limitations

Regarding the formulation of MOFs, one of the main issues is their stability in water or in humid conditions. Robustness of the MOF membranes is essential for any industrial application. Willis and co-workers confirmed with a steam stability map for MOFs that the most resistant materials belong to the families of ZIFs and some carboxylate-based MOFs [70]. Clearly, the next membrane developments should be focused on the most stable phases while the stability of the MOF structures has to be more clearly understood.

Owing to the flexibility of the structures, MOFs can usually adsorb larger molecules than the pore size or cavity windows as they are defined by XRD. As an example, ZIF-8 can adsorb large amounts of i-C_4H_{10}, which has a kinetic diameter of 5.0 Å, much larger than the reported sodalite window of 3.0 Å (Figure 6.20). This "elasticity" will affect molecular sieving properties, lowering the expected selectivity of the MOF membrane.

With regard to the manufacture of large membrane modules, the secondary growth approach is difficult to control and scale up. On the other hand, we and others have demonstrated that direct nucleation growth on a porous support can be achieved for ZIF materials [40, 43, 44, 71]. Good reproducibility has been obtained at the laboratory scale. Nevertheless, procedures have to be further optimized to prepare very thin membranes less than 1 µm thick. Further, nucleation–growth mechanisms

Table 6.2 Summary of gas separation performances of MOF membranes.

MOF	Seeding technique	Synthesis procedure	Type of support	Single permeation (mol m^{-2} s^{-1} Pa^{-1}) H$_2$	CO$_2$	N$_2$	CH$_4$	Separation factor (50:50, 25 °C, 1 bar) H$_2$/CH$_4$	CO$_2$/CH$_4$	CO$_2$/N$_2$	Ref.
ZIF-8		Microwave	TiO$_2$ disc	6.04×10^{-8}	1.33×10^{-8}	5.20×10^{-9}	4.80×10^{-9}	11.2			[39]
ZIF-7	ZIF-7 in PEI dip-coating	Microwave	Al$_2$O$_3$ disc	7.40×10^{-8}	1.10×10^{-8}	1.10×10^{-8}	1.08×10^{-8}	5.9			[40]
HKUST-1		Solvothermal	Oxidized Cu net	1.25×10^{-6}	2.77×10^{-7}	2.72×10^{-7}	1.61×10^{-7}	5.9			[45]
MMOF	Rubbing	Solvothermal	PEI-modified Al$_2$O$_3$ disc	1.20×10^{-8}	5.00×10^{-9}	5.00×10^{-9}					[47]
MOF-5	In situ microwave	Solvothermal	Graphite-coated Al$_2$O$_3$ disc	8.00×10^{-7}	2.20×10^{-7}	2.40×10^{-7}	3.80×10^{-7}				[46]
MOF-5		Solvothermal	Al$_2$O$_3$ disc	1.20×10^{-6}	3.00×10^{-7}	4.00×10^{-7}	5.00×10^{-7}				[67]
ZIF-7	ZIF-7 in PEI dip-coating	Microwave	Al$_2$O$_3$ disc	4.55×10^{-8}	3.50×10^{-9}	2.20×10^{-9}	3.10×10^{-9}	14.0			[43]
ZIF-69		Solvothermal	Al$_2$O$_3$ disc	6.50×10^{-8}	2.50×10^{-8}		1.50×10^{-8}				[42]
ZIF-8	ZIF-8 rubbing	Solvothermal	Al$_2$O$_3$ tube		1.69×10^{-6}		2.42×10^{-6}		7.0		[59]
SIM-1		Solvothermal	Al$_2$O$_3$ tube							4.5	[44]

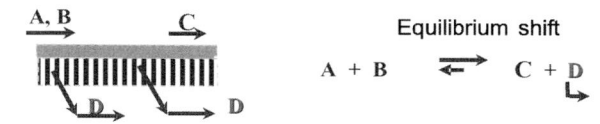

Figure 6.19 Equilibrium shift principle.

to inorganic interfaces must be better understood to allow the generalization of the synthesis to other MOFs.

Zeolites are stable up to 700 °C for high-Al zeolites, but the temperature can be raised up to 1300 °C for all silica zeolites. MOF materials cannot withstand this temperature range. We could expect around 500 °C for the most resistant, such as MIL-53 [72], TUDMOF-2 [73], UiO-66 [74], and ZIF-11 [41]. This is not a limitation for most separation processes, which operate under conditions close to ambient temperature. Nevertheless, attention has to be paid to thermal treatments. For supported membranes, there is always a mismatch, more or less pronounced, between the expansion coefficient of the support and the film. The changes in the thermal expansion coefficients of zeolites as a function of temperature can cause stress problems for the attachment of the zeolite layer to the support and also for the connection of the individual microcrystals within the zeolite layer. Unlike metal or ceramic supports, which expand with increase in temperature (expansion coefficients of $2-7 \times 10^{-6}$, 11×10^{-6}, and $19 \times 10^{-6}\ °C^{-1}$ for Al_2O_3, SiO_2, and stainless steel, respectively), the volume of the unit cell of most zeolites passes through a maximum when zeolites become heated. The transition from the low-temperature phase (near room temperature) with positive thermal expansion ($\sim 10 \times 10^{-6}\ °C^{-1}$) to the high-temperature phase with negative thermal expansion ($-10 \times 10^{-6}\ °C^{-1}$) occurs for MFI structures between 75 and 120 °C [75, 76]. This means that the

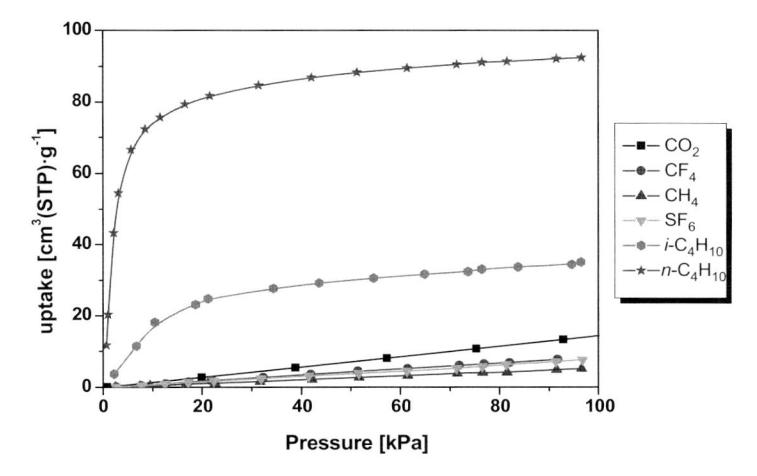

Figure 6.20 Adsorption isotherms of different gases at 303 K with ZIF-8.

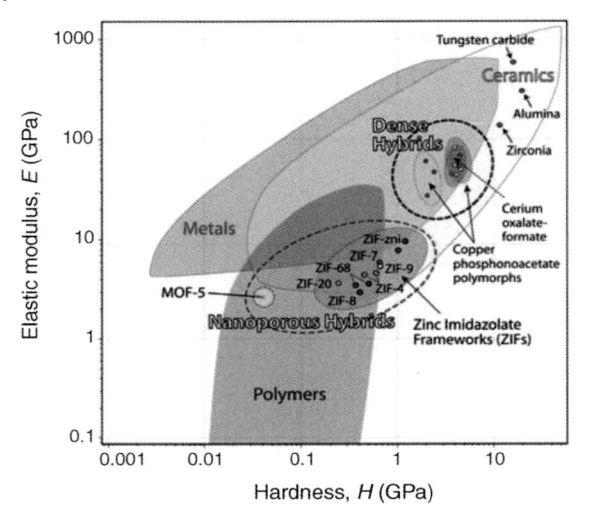

Figure 6.21 Elastic modulus versus hardness materials property map for dense and nanoporous hybrid framework materials shown alongside purely organic and inorganic materials [78].

support expands with increase in temperature, whereas the zeolite layer shrinks. In the case of MOF materials, the first studies indicate expansion coefficients that are neutral or slightly negative due to the framework flexibility [77]. Therefore, heat treatments of MOF membranes have to be considered with care –milder washing or vacuum treatments should be applied to liberate the solvent from the porous network.

On the other hand, MOF materials show mechanical properties that are intermediate between those of ceramics and plastics (Figure 6.21). This plastic feature could be an asset to maintain stability upon thermal cycles [78].

The cost of raw materials is usually an argument against MOF materials (see Chapter 14). However, the amount of MOF material needed to prepare membranes is very small. In addition, no templates are used, in contrast to zeolites, the cost of which is usually associated with the cost of the organic template.

6.6
Conclusions and Outlook

The first cases of separations using MOF membranes have been published. However, there is little insight into the mass transport and separation mechanism. This calls for systematic investigations of gas/vapor permeation under various conditions and *ab initio* modeling to understand better and then simulate permeation profiles, especially by taking into account the softness of the MOF and effect of framework flexibility

The reported membranes confirm the high permeability with respect to their large porosity. For gas separations, this might be at the expense of a lower selectivity, however. Hence the design of novel MOF membranes with higher adsorbate affinity shall be targeted. Fortunately, we can take advantage of the ability to functionalize MOF porous networks by post-modification (Chapter 2). Hence the engineering of the framework would enable one to tune adsorption properties, such as hydrophilic–hydrophobic, without significantly affecting the porous surface. In the most advanced designs, we could anticipate grafting chiral functions, which would allow enantioselective separations.

References

1 Bruschke, H. (1995) *Pure Appl. Chem.*, **65**, 993.
2 Scott, K. (1995) *Handbook of Indutrial Membranes*, Elsevier, Amsterdam.
3 INOPOR GmbH www.inopor.com August 2010
4 de Vos, R.M. and Verweij, H. (1998) *Science*, **279**, 1710.
5 Van der Bruggen, B., Mänttäri, M., and Nyström, M. (2008) *Sep. Purif. Technol.*, **63**, 251.
6 Breck, D.W. (1974) *Zeolite Molecular Sieves*, John Wiley & Sons, Inc., New York.
7 Mintova, S., Hedlund, J., Valtchev, V., Schoeman, B.J., and Sterte, J. (1998) *J. Mater. Chem.*, **8**, 2217.
8 van de Graaf, J.M., van der Bijl, E., Stol, A., Kapteijn, F., and Moulijn, J.A. (1998) *Ind. Eng. Chem. Res.*, **37**, 4071.
9 Miano, F. (1996) *Colloids Surf.*, **110**, 95.
10 Vavlitis, A.P., Ruthven, D.M., and Loughlin, K.F. (1981) *J. Colloid Interface Sci.*, **84**, 526.
11 Ciavarella, P., Moueddeb, H., Miachon, S., Fiaty, K., and Dalmon, J.-A. (2000) *Catal. Today*, **56**, 253.
12 Gump, C.J., Tuan, V.A., Noble, R.D., and Falconer, J.L. (2001) *Ind. Eng. Chem. Res.*, **40**, 565.
13 Lai, Z.P., Bonilla, G., Diaz, I., Nery, J.G., Sujaoti, K., Amat, M.A., Kokkoli, E., Terasaki, O., Thompson, R.W., Tsapatsis, M., and Vlachos, D.G. (2003) *Science*, **300**, 456.
14 Lai, Z.P. and Tsapatsis, M. (2004) *Ind. Eng. Chem. Res.*, **43**, 3000.
15 Xomeritakis, G., Lai, Z.P., and Tsapatsis, M. (2001) *Ind. Eng. Chem. Res.*, **40**, 544.
16 Xomeritakis, G. and Tsapatsis, M. (1999) *Chem. Mater.*, **11**, 875.
17 Noack, M. and Caro, J. (2002) In *Handbook of Porous Solids* (eds. F. Schüth, K. Sing, and J. Weitkamp), Wiley-VCH Verlag GmbH, Weinheim, p. 2433.
18 Tricoli, V., Sefcik, J., and McCormick, A.V. (1997) *Langmuir*, **13**, 4193.
19 Sano, T., Kiyozumi, Y., Kawamura, M., Mizukami, F., Takaya, H., Mouri, T., Inaoka, W., Toida, Y., Watanabe, M., and Toyoda, K. (1991) *Zeolites*, **11**, 842.
20 Li, Y., Chung, T.S., and Kulprathipanja, S. (2007) *AIChE J.*, **53**, 610.
21 Alfaro, S., Arruebo, M., Coronas, J., Menendez, M., and Santamaria, J. (2001) *Micropor. Mesopor. Mater.*, **50**, 195.
22 Kikuchi, E., Yamashita, K., Hiromoto, S., Ueyama, K., and Matsukata, M. (1997) *Micropor. Mater.*, **11**, 107.
23 Tavolaro, A. and Drioli, E. (1999) *Adv. Mater.*, **11**, 975.
24 Kapteijn, F., Bakker, W.J.W., Vandegraaf, J., Zheng, G., Poppe, J., and Moulijn, J.A. (1995) *Catal. Today*, **25**, 213.
25 Vroon, Z., Keizer, K., Gilde, M.J., Verweij, H., and Burggraaf, A.J. (1996) *J. Membr. Sci.*, **113**, 293.
26 Ramsay, J., Giroir-Fendler, A., Julbe, A., and Dalmon, J.A. (1994) French Patent 94-05562.
27 Lovallo, M.C., Gouzinis, A., and Tsapatsis, M. (1998) *AIChE J.*, **44**, 1903.
28 Kondo, M., Komori, M., Kita, H., and Okamoto, K. (1997) *J. Membr. Sci.*, **133**, 133.
29 Li, S.G., Tuan, V.A., Noble, R.D., and Falconer, J.L. (2001) *Ind. Eng. Chem. Res.*, **40**, 4577.

30 Balkus, K.J. and Scott, A.S. (1999) *Chem. Mater.*, **11**, 189.
31 Hermes, S., Schroder, F., Chelmowski, R., Woll, C., and Fischer, R.A. (2005) *J. Am. Chem. Soc.*, **127**, 13744.
32 Hermes, S., Zacher, D., Baunemann, A., Woll, C., and Fischer, R.A. (2007) *Chem. Mater.*, **19**, 2168.
33 Biemmi, E., Scherb, C., and Bein, T. (2007) *J. Am. Chem. Soc.*, **129**, 8054.
34 Scherb, C., Schödel, A., and Bein, T. (2008) *Angew. Chem. Int. Ed.*, **47**, 5777.
35 Arnold, M., Kortunov, P., Jones, D.J., Nedellec, Y., Karger, J., and Caro, J. (2007) *Eur. J. Inorg. Chem.*, 60.
36 Gascón, J., Aguado, S., and Kapteijn, F. (2008) *Microporous Mesoporous Mater.*, **113**, 132.
37 Yoo, Y. and Jeong, H.K. (2008) *Chem. Commun.*, 2441.
38 Cravillon, J., Munzer, S., Lohmeier, S.J., Feldhoff, A., Huber, K., and Wiebcke, M. (2009) *Chem. Mater.*, **21**, 1410.
39 Bux, H., Liang, F., Li, Y., Cravillon, J., Wiebcke, M., and Caro, J. (2009) *J. Am. Chem. Soc.*, **131**, 16000.
40 Li, Y., Liang, F., Bux, H., Feldhoff, A., Yang, W., and Caro, J. (2010) *Angew. Chem. Int. Ed.*, **49**, 548.
41 Park, K.S., Ni, Z., Cote, A.P., Choi, J.Y., Huang, R.D., Uribe-Romo, F.J., Chae, H.K., O'Keeffe, M., and Yaghi, O.M. (2006) *Proc. Natl. Acad. Sci. USA*, **103**, 10186.
42 Liu, Y.Y., Hu, E.P., Khan, E.A., and Lai, Z.P. (2010) *J. Membr. Sci.*, **353**, 36.
43 Li, Y.S., Liang, F.Y., Bux, H.G., Yang, W.S., and Caro, J. (2010) *J. Membr. Sci.*, **354**, 48.
44 Aguado, S., Nicolas, C.H., Moizan-Baslé, V., Nieto, C., Amrouche, H., Bats, N., Audebrand, N., and Farrusseng, D. (2011) *New J. Chem.*, **35**, 41. DOI: 10.1039/C0NJ00667J.
45 Guo, H., Zhu, G., Hewitt, I.J., and Qiu, S. (2009) *J. Am. Chem. Soc.*, **131**, 1646.
46 Yoo, Y., Lai, Z., and Jeong, H.K. (2009) *Micropor. Mesopor. Mater.*, **123**, 100.
47 Ranjan, R. and Tsapatsis, M. (2009) *Chem. Mater.*, **21**, 4920.
48 Takamizawa, S., Takasaki, Y., and Miyake, R. (2010) *J. Am. Chem. Soc.*, **132**, 2862.
49 Aguado, S., Bergeret, G., Pera-Titus, M., Moizan-Baslé, V., Nieto-Dragui, C., Bats, N., and Farrusseng, D. (2011) *New J. Chem.*, **35**, 546.
50 Yazaydin, A., Snurr, R.Q., Park, T.-H., Koh, K., Liu, J., LeVan, M.D., Benin, A.I., Jakubczak, P., Lanuza, M., Galloway, D.B., Low, J.J., and Willis, R.R. (2009) *J. Am. Chem. Soc.*, **131**, 18198.
51 Banerjee, R., Furukawa, H., Britt, D., Knobler, C., O'Keeffe, M., and Yaghi, O.M. (2009) *J. Am. Chem. Soc.*, **131**, 3875.
52 Choudhary, V.R. and Mayadevi, S. (1996) *Langmuir*, **12**, 980.
53 Zhang, J., Singh, R., and Webley, P.A. (2008) *Micropor. Mesopor. Mater.*, **111**, 478.
54 Barrer, R.M. and Goibbons, R.M. (1965) *Trans. Faraday Soc.*, **61**, 948.
55 Dunne, J.A., Rao, M., Sircar, S., Gorte, R.J., and Myers, A.L. (1996) *Langmuir*, **12**, 5896.
56 Dunne, J.A., Mariwala, R., Rao, M., Sircar, S., Gorte, R.J., and Myers, A.L. (1996) *Langmuir*, **12**, 5888.
57 Banerjee, R., Phan, A., Wang, B., Knobler, C., Furukawa, H., O'Keeffe, M., and Yaghi, O.M. (2008) *Science*, **319**, 939.
58 Kusakabe, K., Kuroda, T., Murata, A., and Morooka, S. (1997) *Ind. Eng. Chem. Res.*, **36**, 649.
59 Venna, S.R. and Carreon, M.A. (2010) *J. Am. Chem. Soc.*, **132**, 76.
60 Carreon, M.A., Li, S.G., Falconer, J.L., and Noble, R.D. (2008) *J. Am. Chem. Soc.*, **130**, 5412.
61 Bernal, M.P., Coronas, J., Menéndez, M., and Santamaría, J. (2004) *AIChE J.*, **50**, 127.
62 van den Bergh, J., Zhu, W., Gascon, J., Moulijn, J.A., and Kapteijn, F. (2008) *J. Membr. Sci.*, **316**, 35.
63 Li, S., Falconer, J.L., and Noble, R.D. (2004) *J. Membr. Sci.*, **241**, 121.
64 Li, S., Falconer, J.L., and Noble, R.D. (2006) *Adv. Mater.*, **18**, 2601.
65 Ramos-Fernandez, E.V., Garcia-Domingos, M., Juan-Alcañiz, J., Gascon, J., and Kapteijn, F. (2011) *Appl. Catal. A*, **391**, 261.
66 Aguado, S., Canivet, J., and Farrusseng, D. (2010) *Chem. Commun.*, 7999.

67 Liu, Y., Ng, Z., Khan, E.A., Jeong, H.-K., Ching, C.-B., and Lai, Z. (2009) *Micropor. Mesopor. Mater.*, **118**, 296.

68 Basu, S., Maes, M., Cano-Odena, A., Alaerts, L., De Vos, D.E., and Vankelecom, I.F.J. (2009) *J. Membr. Sci.*, **344**, 190.

69 Kusgens, P., Rose, M., Senkovska, I., Frode, H., Henschel, A., Siegle, S., and Kaskel, S. (2009) *Micropor. Mesopor. Mater.*, **120**, 325.

70 Low, J.J., Benin, A.I., Jakubczak, P., Abrahamian, J.F., Faheem, S.A., and Willis, R.R. (2009) *J. Am. Chem. Soc.*, **131**, 15834.

71 Farrusseng, D., Aguado, S., Nicolas, C.H., Siret, B., and Durecu, S. (2009) PCT09/04489.

72 Loiseau, T., Serre, C., Huguenard, C., Fink, G., Taulelle, F., Henry, M., Bataille, T., and Férey, G. (2004) *Chem. Eur. J.*, **10**, 1373.

73 Senkovska, I. and Kaskel, S. (2006) *Eur. J. Inorg. Chem.*, 4564.

74 Cavka, J., Jakobsen, S., Olsbye, U., Guillou, N., Lamberti, C., Bordiga, S., and Lillerud, K. (2008) *J. Am. Chem. Soc.*, **130**, 13850.

75 Caro, J., Noack, M., Kolsch, P., and Schafer, R. (2000) *Micropor. Mesopor. Mater.*, **38**, 3.

76 Miachon, S., Landrivon, E., Aouine, M., Sun, Y., Kumakiri, I., Li, Y., Prokopova, O.P., Guilhaurne, N., Giroir-Fendler, A., Mozzanega, H., and Dalmon, J.A. (2006) *J. Membr. Sci.*, **281**, 228.

77 Zhao, L. and Zhong, C.L. (2009) *J. Phys. Chem. C*, **113**, 16860.

78 Tan, J.C., Bennett, T.D., and Cheetham, A.K. (2010) *Proc. Natl. Acad. Sci. USA*, **107**, 9938.

7
Research Status of Metal–Organic Frameworks for On-Board Cryo-Adsorptive Hydrogen Storage Applications

Anne Dailly

7.1
Introduction – Research Problem and Significance

7.1.1
Challenges in Hydrogen Storage Technologies for Hydrogen Fuel Cell Vehicles

In the context of world population and vehicle growth, climate change, and dwindling oil reserves, the sustainability of personal mobility is a serious concern. Green sources of fuel and transportation are required to lower environmental and national security costs. Hydrogen fuel cell electric vehicles (FCEVs) are transformative relative to today's conventional internal combustion engine vehicles as they potentially offer the unique combination of zero petroleum consumption, zero greenhouse gas tailpipe emissions, comparable durability, range, and performance, and fast refueling. Hydrogen-powered FCEVs are a real technology. Compelling fuel cell vehicle performance with a conventional driving range (>300 miles) has been demonstrated by General Motors, Daimler, Toyota, and Honda. A critical challenge for transportation applications is to have a propulsion system based on hydrogen and fuel cells designed and validated that can go head-to-head with the internal combustion engine so that the new technology meets or exceeds customer expectations in terms of affordability, reliability, performance, and durability [1]. Realizing the widespread commercialization of hydrogen-fueled vehicles still requires overcoming technical barriers. One such barrier is the development of an on-board hydrogen storage system.

Hydrogen is an energy carrier with great potential to become a major fuel for both mobile and stationary applications. Hydrogen has the highest energy content per unit of weight of any known element. It is also the lightest element. As a result, it is characterized by low volume energy density. On a weight basis, hydrogen has nearly three times the energy content of gasoline (120 MJ kg^{-1} for hydrogen versus 44 MJ kg^{-1} for gasoline). However, on a volume basis, the situation is reversed (8 MJ L^{-1} for liquid hydrogen versus 32 MJ L^{-1} for gasoline). This presents significant challenges to storing the large quantities of hydrogen that are necessary in the hydrogen energy

Metal-Organic Frameworks: Applications from Catalysis to Gas Storage, First Edition. Edited by David Farrusseng.
© 2011 Wiley-VCH Verlag GmbH & Co. KGaA. Published 2011 by Wiley-VCH Verlag GmbH & Co. KGaA.

economy (5–13 kg of hydrogen are required to cover small to mid-sized cars). Today, hydrogen for transportation application includes 5000 psi (~35 MPa) and 10 000 psi (~70 MPa) compressed gas tanks that have been certified worldwide according to ISO 11 439 (Europe) and NGV-2 (United States) and approved by TUV (Germany) and The High Pressure Gas Safety Institute of Japan (KHK). Carbon fiber-reinforced high-pressure tanks are arguably the most mature current approach to automotive on-board storage and have been demonstrated in several prototype hydrogen-powered FCEVs. Hydrogen pressure vessels are becoming far more practical as the strength of carbon fiber composites makes increasingly high gas pressures conceivable. Nevertheless, driving ranges for compressed tanks remain inadequate and the energy consumed to compress hydrogen reduces the efficiency of this storage medium. The key issues facing containment in hydrogen pressure vessels revolve around public acceptance of their reactivity, the cost of manufacturing, and the time and complexity of the filling operation.

7.1.2
Current Status of Hydrogen Storage Options and R&D for the Future

Compressed gas is likely to be used as an interim solution and the research community is actively trying to develop advanced hydrogen solid-state storage alternatives for on-board applications, as presented in Figure 7.1, which are potentially robust and abuse tolerant and more energy efficient.

Each of these options has advantages and disadvantages compared with the 70 MPa high-pressure compressed gas hydrogen system which is the benchmark technology at present for on-board hydrogen storage. Vehicular hydrogen storage system must satisfy a number of requirements, as shown in Table 7.1, to come anywhere close to gasoline. The United States Department of Energy (DOE) has set criteria for the

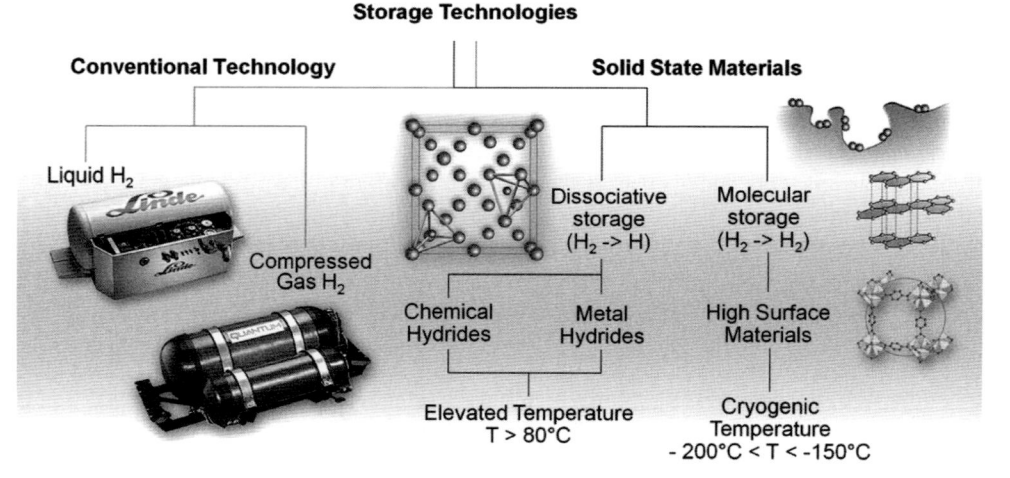

Figure 7.1 Hydrogen storage options for on-board applications.

Table 7.1 Hydrogen storage system requirements.

Parameter	Goal for 2015	Ultimate goal
System usable energy per unit mass	6.48 MJ kg^{-1} (1.8 kWh kg^{-1})	9 MJ kg^{-1} (2.5 kWh kg^{-1})
Net useful energy/maximum system mass	0.055 kg H_2/kg system	0.075 kg H_2/kg system
System usable energy per unit volume	4.68 MJ l^{-1} (1.3 kWh l^{-1})	8.28 MJ l^{-1} (2.3 kWh l^{-1})
Net useful energy/maximum system volume	0.040 kg H_2/l system	0.070 kg H_2/l system
H_2 tank capacity	4 kg (DOE)/~3–6 kg (General Motors)	
Operating temperature for H_2 tank	−40 to 60 °C	
H_2 delivery temperature	−40 to 85 °C (DOE)/~10 °C around stack coolant temperature (General Motors)	
Refueling time	<5 min	<5 min
Supply pressure to the fuel cell system	0.5 MPa	0.5 MPa

comparison of storage media by defining on-board hydrogen storage system performance targets [2]. The system includes the weight and volume of the hydrogen and its required fuel delivery support, such as the tank, pipes, pumps, and heat exchanger. These targets entail that the criteria for the storage materials be even higher.

The two most important criteria are the gravimetric energy density and the volumetric energy density. The other parameter targets are that the release temperature should be near the operating temperature of the fuel cell stack, the refueling time should be minimal, and the delivery pressure to the system fuel cell needs to be about 0.5 MPa. Solid-state hydrogen storage materials which are attracting attention because of their potential to provide compact and lightweight hydrogen storage systems, can be grouped roughly into two broad categories: physical storage, where molecular hydrogen is bonded to surfaces via weak van der Waals interactions (typically <10 kJ mol^{-1} H_2) in a process known as physisorption, and chemical storage, which usually involves the dissociation of H_2 into two hydrogen atoms that can be trapped in the materials bulk with strong chemical bonds (typically >10 kJ mol^{-1} H_2). Metal hydrides, complex hydrides, and chemical hydrides are an example. Chemical storage of such hydrides can be divided into two classes, either "reversible" or "nonreversible." which refers to a material's ability to rehydrogenate under reasonable temperature and pressure conditions. Nonreversible materials typically release or generate hydrogen on-board and leave dehydrogenated products that must be physically removed from the vehicle and recharged off-board. Reversible solid-state hydrogen storage on-board the vehicle can be considered the ultimate goal for automotive applications. Prior to the late 1990s, most of the focus on reversible storage was directed towards traditional metal hydrides such as $LaNi_5H_6$, $FeTiH_2$, and Mg_2NiH_4 that absorb atomic hydrogen into interstitial sites in the crystal structure. Significant interest was subsequently turned to complex hydrides such as alanate (AlH_4^-) materials that have the potential for higher gravimetric hydrogen capacities in the operational window than simple metal hydrides. Issues with complex metal hydrides include low hydrogen capacity and slow uptakes, release kinetics, and thermal management during refueling. In adsorptive hydrogen storage, that is, physisorption, an adsorbate is formed in which the density of the gas species is much higher than the gaseous state, provided that the interaction energy is sufficiently high. Molecular hydrogen storage typically shows fast kinetics and good reversibility, but since it relies on weak interactions one might expect significant hydrogen adsorption only at low temperatures such as liquid nitrogen temperature (77 K). Physisorptive processes typically require highly porous materials to maximize the surface area and the number of available binding sites for hydrogen.

Although no existing storage materials system has yet satisfied all the requirements given in Table 7.1, progress continues to be made thanks to worldwide efforts from both academia and industry. This chapter considers one specific aspect of that progress, namely the study of metal–organic frameworks (MOFs) as potential cryoadsorbents for hydrogen storage applications.

7.2
MOFs as Adsorptive Hydrogen Storage Options

Various adsorbents have been explored as potential hydrogen storage materials, among them zeolites and different types of carbon, including activated carbons and nanotubes. The intrinsic properties of MOFs, that is, their extraordinarily high surface areas, their high porosities, and their low densities, make them among the most promising adsorbent materials for hydrogen storage. MOFs have specific surface areas and micropore volumes that can exceed those of traditional adsorbents such as zeolites and activated carbons. A few MOFs possess Brunauer–Emmett–Teller (BET) specific surface areas over $4000\,m^2\,g^{-1}$. For instance, MIL-101 [3], NOTT-116 [4], and UCM-2 [5] display BET surface areas of 4230, 4664 and $5200\,m^2\,g^{-1}$, respectively. MOFs are hybrid inorganic–organic frameworks that are assembled by the connection of secondary building blocks (SBUs), usually consisting of metal ions or clusters, through rigid organic ligands. The variety of cations and molecular bridges that can be combined in the framework yields an extended range of materials with diverse pore sizes and functionalities. One of the greatest advantages of MOFs over traditional adsorbents such as inorganic zeolites and porous activated carbons is that their pore sizes can be tuned from several ångstroms to a few nanometers just by controlling the length of the organic linkers. Further, because of their crystalline nature, MOFs offer a well-defined structural platform from which, in principle, their properties can be studied more easily. Carboxylate-based MOFs have received by far the most attention because most of these carboxylate-based ligands are either commercially available, such as 1,4-benzenedicarboxylate (bdc^{2-}) and 4,4′,4″-benzene-1,3,5-triyltribenzoate (btb^{3-}), or are easily synthesized. Carboxylates also have a high acidity ($pK_a \approx 4$), which allows for facile *in situ* deprotonation. The bridging bidentate coordination ability of carboxylate groups facilitates a high degree of framework connectivity and robust metal–ligand bonds, so that MOFs can preserve the integrity of their structure upon desolvation and activation of the porosity. A variety of cations can participate in the frameworks. Zn^{2+} and Cu^{2+} are the most common transition metals used in carboxylate-based frameworks. Light main group metal ions such as Be^{2+}, Mg^{2+}, and Al^{3+} have been explored with the purpose of lightening the weight of the framework and ultimately increasing the gravimetric hydrogen uptake, but few of these frameworks have actually been reported to date. A second generation of MOFs has been based on functional nitrogen-based bridging ligands structurally analogous to carboxylate ligands. The linking organic units are, for instance, bdp (1,4-benzenedipyrazolate) [6], btt (1,3,5 benzenetristetrazolate) [7], and tpt-3tz [2,4,6-tri-*p*-(tetrazol-5-yl)phenyl-*s*-triazine] [8]. More recently, two sub-families of MOFs have been introduced: the zeolitic imidazolate frameworks (ZIFs) and the covalent organic frameworks (COFs). ZIFs are a new class of nanoporous compounds consisting of tetrahedral clusters of MN_4 (M = Co, Cu, Zn, etc.) linked by simple imidazolate ligands [9]. COFs are formed solely from light elements (H, B, C, N, and O) and their organic building units are held together by strong covalent bonds (C–C, C–O, B–O, and Si–C) rather than metal ions to produce materials with high porosity [10].

All the MOFs that have been considered as potential hydrogen cryo-adsorbents share the same attributes. In addition to being crystalline and three-dimensional and having well-defined macromolecular structure, they exhibit high porosity and high specific surface area. Extensive sets of MOFs have been synthesized at the laboratory scale and evaluated as hydrogen cryo-adsorbents in order to select the most promising adsorbents for selected (pressure, temperature) operating conditions. The purpose of this screening is also to understand the mechanisms and the interplay between the structure, the material stability, and the storage densities (per unit volume and per unit weight) for specific operating conditions. Figure 7.2 gives an overview of the organic linkers that have been combined with different SBUs in the studied MOFs.

A substantial number of MOFs have been based on aromatic carboxylate ligands such as bdc (1,4-benzenedicarboxylate) [11–14], btb (4,4′,4″-benzene-1,3,5-triyltribenzoate) [15, 16], ttdc (thieno[3,2-*b*]thiophene-2,5-dicarboxylate) [17], tcn (3,5,3′,5′-tetracarboxylatenaphthalene) [18], tcp (3,5,3′,5′ tetracarboxylatephenanthrene) [16], tptc (terphenyl 3,5,3′,5′ tetracarboxylate) [19], qptc (quaterphenyl 3,3‴,5,5‴ tetracarboxylate) [17], btc (1,3,5-benzenetricarboxylate) [20, 21], dobdc (2,5-dihydroxy-1,4-benzenedicarboxylate) [22], adip [5,5′-(9,10-anthracenediyl)diisophthalate] [23], betei [5,5′,5″-benzene-1,3,5-triyltris(1-ethynyl-2-isophthalate] [24], ntei {5,5′,5″-[4,4′,4″-nitrolotris(benzene-4,1-diyl)tris(ethyne-2,1-diyl)]triisophthalate} [22], and L^2 {5,5′,5″-[benzene-1,3,5-triyltris(ethyne-2,1-diyl)]triisophthalate} [2]. Nitrogen-based bridging

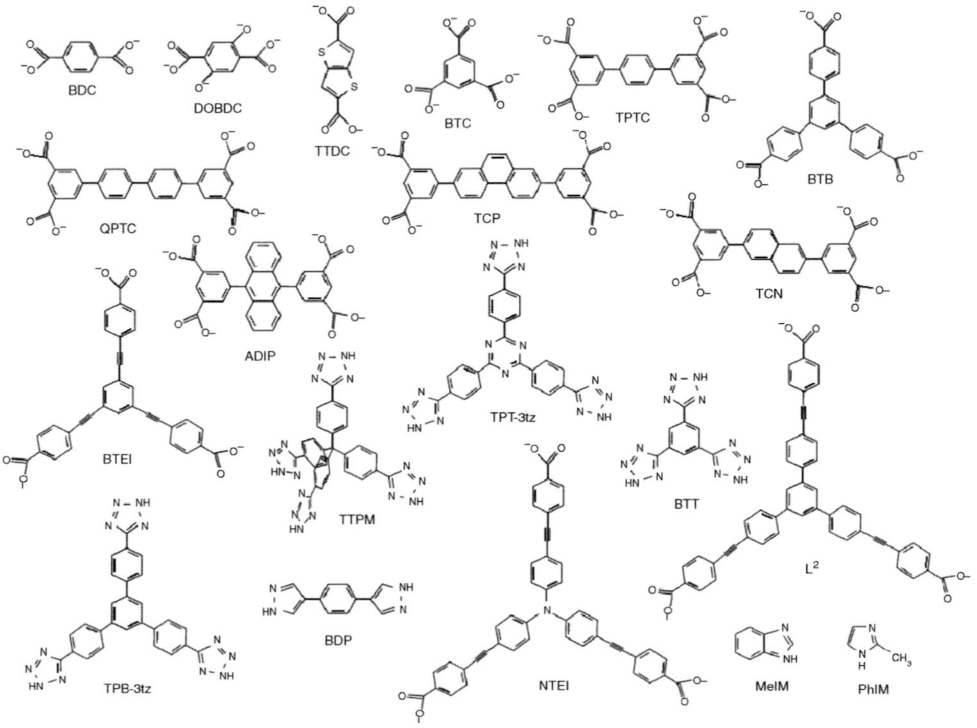

Figure 7.2 Examples of the organic linkers used in the coordination frameworks.

ligands were also examined, such as bdp, btt [5], tpb-3tz [1,3,5-tri-*p*-(tetrazol-5-yl) phenylbenzene] [6], tpt-3tz [6], and ttpm [tetrakis(4-tetrazolylphenyl) methane] [25]. MOFs based on metal imidazolates, PhIm (benzimidazolate) and MeIm (2-methylimidazolate), were also explored [26]. The versatility of organic ligands has provided many possibilities for the construction of porous MOFs with various topologies, which allows us to establish the key physical properties (chemistry, structure, surface area, etc.) that influence the hydrogen storage behavior of various new and promising microporous MOF materials for hydrogen storage.

7.3
Experimental Techniques and Methods for Performance and Thermodynamics Assessment of Porous MOFs for Hydrogen Storage

Accurate evaluations of the hydrogen storage capacity of MOF materials require reliable adsorption data over a wide range of temperature and pressure. Most measurements have focused on relatively low pressures (1 bar), well below full coverage (i.e., below saturation) of the surface area by hydrogen gas molecules. Low-pressure measurements offer a clear view of the low coverage interaction energy between hydrogen and the surface (the adsorption enthalpy), but may not accurately predict the maximum hydrogen storage capacity. Most of the reported hydrogen adsorption studies on MOFs were also performed at liquid nitrogen temperature and room temperature, as these are the most directly achievable conditions with common sorption instruments, but isotherm measurements at different temperatures can provide important information such as thermodynamics and mechanisms of adsorption. For the characterization of MOFs as potential hydrogen physisorbents, the hydrogen storage capacity can be determined on a gravimetric basis as the amount of hydrogen stored by unit weight (gravimetric capacity) or on a volumetric basis as the amount of hydrogen stored by unit volume (volumetric capacity). The gravimetric capacity can be measured experimentally using either gravimetric or volumetric methods. The gravimetric method determines the amount of hydrogen uptake by measuring the mass change of the material following exposure to a hydrogen atmosphere. The measured weight of the sample at each equilibrium point is then corrected, as appropriate, for buoyancy effects. A volumetric measurement, also known as Sieverts' method, determines the amount of hydrogen uptake by measuring changes in pressure within a system of fixed volume at a known temperature, following the real gas law, $PV = nZRT$, where Z is the hydrogen compressibility at a given temperature and pressure. Three definitions of adsorptions (excess, absolute, and total) are shown schematically in Figure 7.3. The performances of cryo-adsorption materials are evaluated and compared by assessing their excess

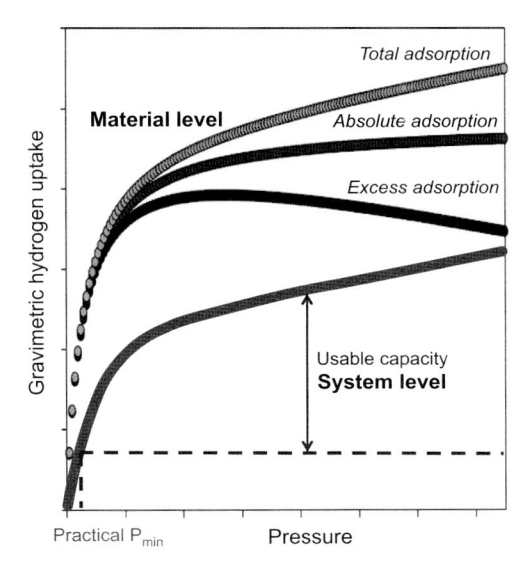

Figure 7.3 Schematic representations of the different gravimetric hydrogen adsorptions: excess isotherm (lower of the three top lines; e.g., most measurements), absolute isotherm (middle line), and total isotherm (upper line) at both material and system levels. The usable capacity is represented by the arrow to the bottom line.

hydrogen adsorption uptakes. The excess adsorption amount is the only quantity associated with the adsorbed phase that is readily accessible to measurements. This quantity is defined as the difference, under given temperature and pressure (P, T), between the amount of adsorbate stored in the porous volume of the adsorbent and the amount that would be present in an identical volume in the absence of solid–gas interactions. The presence in the isotherm of a maximum is a typical feature of excess adsorption isotherms. At higher pressures, the gas density in the material's pores starts to saturate whereas the bulk gas density continues to increase. The maximum and following rollover in the isotherm occur when the density of the coexisting bulk phase increases faster than the adsorbed phase. The absolute adsorption amount, which refers to the amount of hydrogen actually stored in the adsorbed phase, is relevant for the study and modeling of adsorption thermodynamics. Nevertheless, in practice, one must take into account the contribution from the bulk phase hydrogen stored in the adsorbent material's interparticular void volume and defects, which will have a detrimental effect on the volumetric storage capacity. The total hydrogen storage capacity of the material contrasts significantly with the excess adsorption and will naturally depend, in particular, on the amount of macropores and interparticular voids present in the materials. In this latter volume, the hydrogen does not interact with the adsorbent and the hydrogen density is a function of pressure and temperature only. Unlike amorphous materials such as activated carbons, the well-defined crystal structure of MOFs allows the determination of the total adsorption. An oversimplified estimate of a material's total adsorption capacity can be made from its excess gravimetric hydrogen uptake and its crystallographic density, which excludes all voids that are not a fundamental part of the molecular arrangement. The bulk density should be used instead, since it includes the contribution of the particle volume, the interparticulate void volume, and the internal pore volume. In contrast to the crystallographic density, the bulk density is not an intrinsic property of the material and it can change depending on how the material is handled. Hence it is essential in reporting the bulk density to specify how the determination was made.

It is important to consider the total amount of gas stored in the material in both the adsorbed and gas phases in order to compare the performance of an adsorbent with other hydrogen storage technologies (e.g., the various hydride classes). Total capacities can be reported at the materials level and on a system level, which takes into account the weight and volume of the auxiliary fuel system components such as the vessel for material containment, the thermal management equipment, the hydrogen fuel, pressure valves, pipes, and so on. The usable storage capacity of the adsorption based system is determined by subtracting the hydrogen that would remain in the vessel due to the delivery pressure, 0.5 MPa, which is imposed by the fuel cell. Very little work has been done on collecting engineering properties such as densification, packing density, thermal conductivity, and material compatibility on MOFs. For this reason, the usable hydrogen capacities for the system are rarely available and most studies focus on gravimetric storage uptake on a materials basis.

7.4
Material Research Results

7.4.1
Structure–Hydrogen Storage Properties Correlations

Since the first investigation of hydrogen adsorption at liquid nitrogen temperature at pressures up to 1 bar on the porous cubic carboxylate-based framework $Zn_4O(bdc)_3$ (named MOF-5 and then IRMOF-1) was reported [27], many more porous MOFs have been screened and evaluated as cryo-adsorbents. The maximum excess gravimetric uptake on Zn(bdc) was independently determined in different laboratories to reach 4.5–5 wt% [28–30]. However, it was also shown that the gas storage properties were strongly dependent on the methods used for the preparation and activation of the samples. Zn(bdc) represents one of the best examples of the sensitivity of several MOFs against moisture which leads to decomposition of the framework. When Zn(bdc) is protected from exposure to air and water, the maximum excess gravimetric uptake reaches 7.1 wt% at 77 K [31], with complete activation of the porosity. This final step is also decisive for the gas sorption performance of the sample. A high degree of connectivity and strong metal–ligand bonds appear to be necessary for maintaining the framework architecture while evacuating the solvent from the pores. The incomplete activation or partial collapse of the framework structure during the outgassing procedure can contribute to a loss of porosity and ultimately to a decline in the gas sorption properties of the material.

A complete set of data including excess isotherms of hydrogen adsorption and desorption, maximum hydrogen uptake, BET specific surface area (N_2), pore size distribution, and total pore volume (Ar porosimetry) have been measured on an extended range of MOF materials. As mentioned previously, the hydrogen uptake at low pressures correlates with the adsorption enthalpy [32]. This quantity is a measure of the solid–gas interaction and is determined by the nature of the chemical constituents and the geometric properties of the adsorbent. At higher pressures, that is, when the excess maximum is reached, hydrogen storage capacities can be related to the specific surface area and the pore volume. The BET specific surface areas provide a good indication of the extent of porosity and surface adsorption sites.

The maximum excess hydrogen uptakes correlate well with the BET specific surface areas, and a fairly linear trend between maximum hydrogen uptake and BET specific surface area is evidenced on Figure 7.4. The curve has a slope of $m_0 = 0.02 \pm 0.002 \, \text{mg m}^{-2}$ or $1 \pm 0.1 \, \text{wt\%}$ per $500 \, \text{m}^2 \, \text{g}^{-1}$. This trend is in reasonable agreement with Chahine's rule established for activated carbons and zeolites [33]. The correlation between BET specific surface area and maximum excess hydrogen uptake seems fairly independent of the chemical composition and structure of the MOFs. Functionality and chemistry have little impact on maximum hydrogen uptake. The present data were obtained on various porous frameworks built from the linking of different building blocks and organic units with different functionalities. The 10% error, estimated from the standard deviation, is in agreement with that inherent in the BET surface area calculations for microporous materials [34]. It can be associated

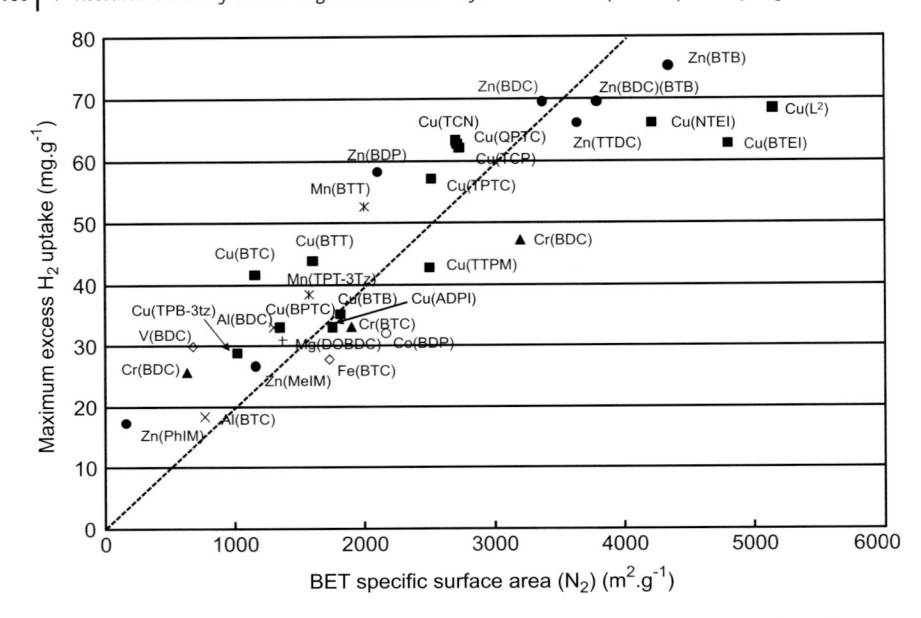

Figure 7.4 Maximum excess hydrogen storage capacity at 77 K versus BET specific surface area (N_2) measurements for different types of MOF (metal from SBU with the organic linker in parentheses).

in part with the presence of a relatively large number of very small pores. Because of their larger molecular size, the nitrogen molecules used for the BET measurement might not access the same surface sites that are available to hydrogen [35, 36]. Moreover, some variability in the trend may also occur because of micropore filling and differences in adsorbate packing densities. A generally linear relationship between BET specific surface area and maximum excess hydrogen uptake is evident. As a strategy to achieve high levels of hydrogen storage in MOFs, increasing the surface area of materials has been strongly pursued. Further, methods serving as useful tools for predicting the theoretically achievable surface areas in microporous coordination polymers have been developed. A geometric accessible surface area method in which values coincide well with grand canonical Monte Carlo (GCMC) theoretical values and with experimental BET determination was presented by Düren et al. [37] Geometric accessible surface area calculations were also used to build a predictive model that calculates the upper limit of surface area potential based on the molecular weight of building block components for a perfect crystal. The accessible surface areas of MOFs with the two most commonly employed metal clusters, Zn_4O $(CO_2R)_6$ and $Cu(CO_2R)_4$, revealed that the upper limit of conceptually organic-dominated frameworks is around $8000 \, m^2 \, g^{-1}$ [38]. Experimental surface areas can be affected by common experimental barriers such as interpenetration, incomplete guest removal during activation, framework collapse, and phase purity. With this in mind, and assuming a surface area of $\sim$1 wt% per $500 \, m^2 \, g^{-1}$, one can estimate an ideal upper limit for gravimetric hydrogen storage in MOFs of $\sim$16 wt% at 77 K.

Although they definitely provide exciting opportunities for the materials research community, the tailoring, synthesis, and discovery of MOFs for hydrogen storage applications present an exceedingly complex challenge. A great deal of progress has already been achieved over the last decade. As Férey established in a review [39]: in a few years hybrid solids passed from curiosity to strategic materials. Nevertheless, the maximum excess gravimetric hydrogen uptake appears to have plateaued at ~7 wt% even for materials showing BET specific surface areas of ~5000 $m^2\,g^{-1}$. Besides, increasing the gravimetric capacity should not be the only concern: for on-board applications, a suitable volumetric capacity must also be reached. This latter is typically calculated from the product of the material-based gravimetric capacity and the density of the adsorbent. However, high surface area materials tend to exhibit inherently low crystal density. In most cases, the low bulk density leads to a decrease in the volumetric hydrogen uptake of the MOF despite its high gravimetric uptake. Finding the best adsorbent for hydrogen storage implies optimizing both the specific surface area and the bulk density of the material; essentially, the surface area available per unit volume of the adsorbent should be maximized.

Similar correlations, although with slightly larger scatter, can be found if the maximum excess hydrogen uptakes are plotted against the total pore volumes. Figure 7.5 shows the linearity between the maximum adsorbed amounts and the measured total pore volumes obtained using argon porosimetry. A deviation from linearity is observed when the MOF materials contain a non-negligible proportion of mesopores, as is the case for Zn(bdc)(btb) with a pore size distribution around 1.4 and 3.1 nm and for Cr(bdc) with most of its pores around 2.7 and 3.2 nm. The adsorption

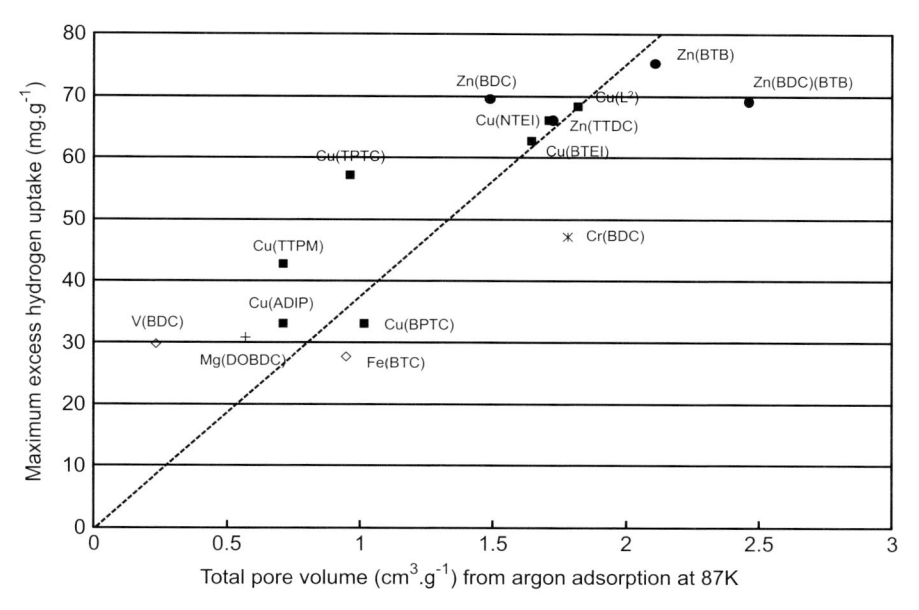

Figure 7.5 Maximum excess hydrogen storage capacity at 77 K versus total pore volume (Ar) measurements for different types of MOF (metal from SBU with the organic linker in parentheses).

properties of a porous solid depend primarily on the size of its pores, which are classified as micropores (<2 nm), mesopores (2–50 nm) and macropores (>50 nm). The MOFs which are considered for hydrogen storage applications are microporous solids. The adsorption properties of MOFs are strongly dependent on the size of the pores, as illustrated in a study by Lin *et al.*, who investigated hydrogen adsorption on isostructural Cu-based MOFs that differ in the length of their tetracarboxylate ligands [40]. By changing the length of the polyaromatic backbone, the influence of pore size could be assessed. Clearly, both the extent of the surface area and the accessible pore volume appear as determinant factors for the hydrogen storage capacity of a MOF.

7.4.2
Nature of the Adsorbed Hydrogen Phase

An important step in tuning the adsorbents for a specific application is to understand the relationship between the key properties of the material and the adsorbate storage mechanism. In order to gain further insights into the nature and behavior of the adsorbed phase and its influence on the correlation with the surface area, the structure and chemical functionality, the excess adsorption isotherms of some specific microporous materials were analyzed at specific (P, T) operating conditions. The interpretation of an adsorption process and the extraction of thermodynamic quantities in terms of a specific adsorption model require the determination of the dependence of the amount of hydrogen adsorbed in a MOF on the pressure and temperature under supercritical conditions. Until couple years ago, little was known in that regard. Part of the reason for this incomplete current understanding is the difficulty in performing adsorption measurements over a wide range of temperature, especially on small sample masses. Moreover, conducting experiments at cryogenic temperatures requires high temperature stability as the temperature error will cause uncertainties on measured quantities. A volumetric adsorption measurement method was developed at General Motors to conduct hydrogen storage experiments over 0–60 bar and 50–300 K. The system involves a continuous-flow cryostat operated under liquid helium that was adapted in-house to provide stable temperature control of the sample cell to within 0.01 K. The resulting excess hydrogen isotherms were fitted using the Dubinin–Astakhov (DA) micropore filling model to extract thermodynamic parameters such as the adsorbed phase density and the adsorption enthalpy as a function of coverage and temperature and the sorption mechanisms. When expressed in terms of the DA model, the excess hydrogen uptake takes the particular form

$$n_{\text{ex}} = \left(\varrho_a - \varrho_g\right) W_0 \exp\left(-\left\{\frac{RT}{E}\ln\left[\frac{(T/T_c)^\gamma P_c}{P}\right]\right\}^m\right) \quad (7.1)$$

where ϱ_a is the adsorbed phase density at saturation, W_0 is the adsorbed phase volume, E is a characteristic energy, R is the universal gas constant, m is an heterogeneity parameter, and ϱ_g is the hydrogen gas phase density determined from the NIST12 software package [41]. The pseudo-saturation pressure is defined using approaches

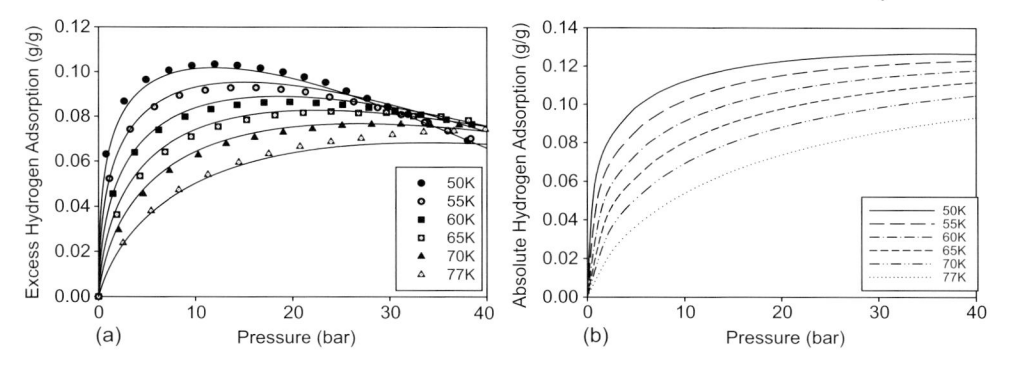

Figure 7.6 (a) Excess and (b) absolute hydrogen adsorption isotherms on Zn(btb) (MOF-177). The data points correspond to experimental measurements; the dashed lines were calculated using the DA model.

such as that proposed by Amankwah and Schwarz [42] and is related to the temperature by a parameter γ specific to the adsorbent–adsorbate pair, where P_c and T_c are the critical temperature and pressure of the adsorbate:

$$P_s = (T/T_c)^\gamma P_c \tag{7.2}$$

The DA model in the form described previously has been successfully applied to hydrogen adsorption isotherms on MOFs. It provides a formal basis to establish the dependence of the adsorbed amount on the (P, T) operating conditions. The parameters, ϱ_a, W_0, E, γ, and m can be determined by a simultaneous regression fit to all the experimental adsorption isotherms on the adsorbent over the studied (P, T) range. As a typical example, measured excess hydrogen adsorption isotherms on Zn(btb) (MOF-177) in the range 50–77 K are presented in Figure 7.6. The isotherms are strongly temperature dependent. The observed increasingly linear-like behavior of the isotherms as the temperature increases is common for physisorption.

The decreasing excess amount adsorbed at high pressure results from the conjunction of an adsorption process approaching saturation and a constantly increasing bulk gas density. This implies that the advantage of having an adsorbent is reduced when high densities are already achieved in the gaseous phase. The entire set of measured isotherms was analyzed using the DA model. The calculated isotherms (dashed lines) plotted against the measured isotherms show very good agreement. The corresponding absolute isotherms can then be calculated using the determined parameters.

The isotherms exhibit plateaus in the high filling regions, as expected from an absolute model since the contribution of the "compressed" gas in the adsorption volume is considered. The isotherms are increased relative to the measured excess isotherms, gaining up to 10–30% at the pressures corresponding to the excess maxima. The variation of the maximum uptake with temperature means that less hydrogen is required to fill the pores as the temperatures increases. The adsorbed phase densities and the adsorption enthalpies, key thermodynamic quantities

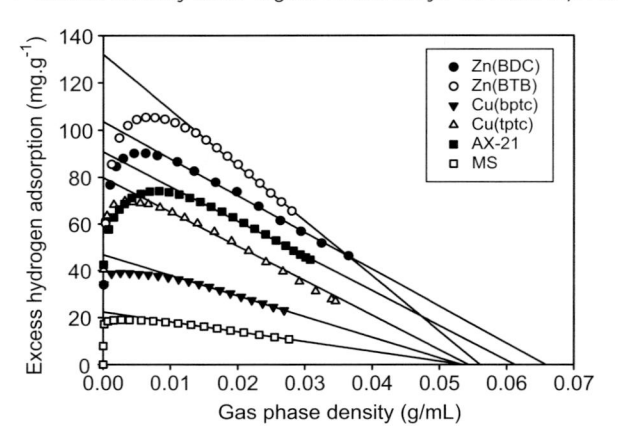

Figure 7.7 Plot of the measured hydrogen adsorption data at 50 K as a function of the gas-phase density.

characterizing the nature of the adsorption mechanism, are also extracted by modeling using the classical micropore filling DA model. In Figure 7.7, the measured hydrogen adsorption data at 50 K are plotted as a function of the gas-phase density for four MOFs, AX-21, a well known coconut shell KOH-activated carbon, and a Type Y molecular sieve (MS) for comparison. The saturation regime past the excess maxima was successfully fitted and extrapolated with linear plots.

This linearity suggests that the adsorbed phase behaves, past the characteristic excess maximum, like an incompressible fluid of density comparable to that of bulk liquid hydrogen (50–70 g l^{-1}) on all the tested samples. This observation demonstrates that the nature of the adsorbed phase, near saturation, may not be influenced by the geometric and chemical environment. The incompressibility associated with the liquid state near saturation could represent a physical limit to hydrogen storage in such materials.

As shown in Figure 7.8, the adsorption enthalpies, calculated from model parameters on both copper- and zinc-based MOFs, range from 3 to 9 kJ mol^{-1}, as expected for physisorption. The material with the smaller pore size, Cu(bptc), shows the highest enthalpy, in accordance with the steep increase in its isotherms as a function of pressure. Figure 7.8 also indicates that the higher enthalpies of Cu-based MOFs lead to a more stable adsorbed phase than Zn-based MOFs. The MOF material with the smallest pore size and with the highest adsorption enthalpy [Cu(bptc)] exhibits steadier maximum excess amounts as a function of temperature.

Reducing the pore size may not enhance the maximum adsorbed phase density but it favors the retention of hydrogen as the temperature increases as a result of stronger solid–gas interaction. The higher hydrogen affinity on the materials with smaller pores leads to displacement of the excess maximum to lower pressures and to a steadier adsorption capacity as a function of temperature. However, improvement in the hydrogen affinity does not necessarily lead to higher adsorbed phase density, as is sometimes supposed. The (P, T) operating adsorption conditions can be improved on

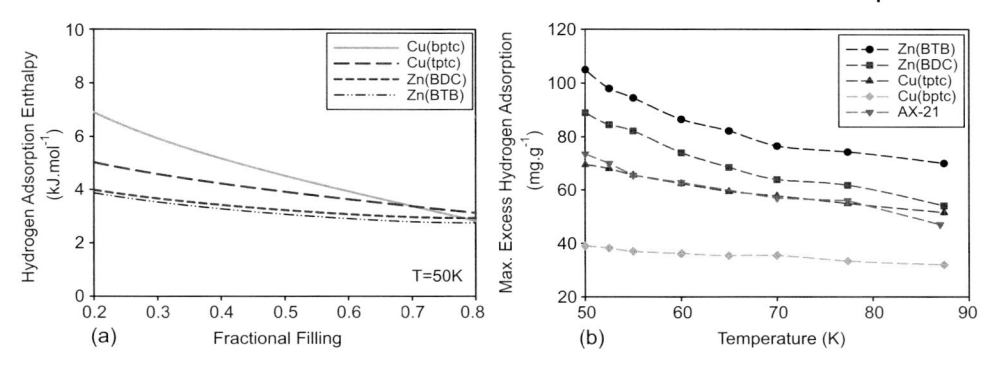

Figure 7.8 (a) Comparison of hydrogen adsorption enthalpy as a function of filling on Zn- and Cu-based MOFs and (b) profile of the excess adsorption maxima of the same MOFs as a function of temperature. The activated carbon AX-21 is also shown for comparison.

materials with small pores but at the cost of reductions in pores volumes and storage capacities.

Interesting paths to tailor adsorbents for specific (P, T) operating conditions might be achieved through the unusual behaviors from the flexible MOFs. For instance, the adsorption–desorption curve of Al(bdc), also known as MIL-53 Al, shows hysteresis that was recently explained by a structural transition between an open-pore to a closed-pore structure as a function of temperature [43]. Co(bdp) also shows a significant hydrogen uptake with wide hysteresis consistent with an accordion-type flexible behavior of the channel pores [44]. This type of breathing effect or gate mechanism could provide a kinetic trapping mechanism for storing hydrogen at modest pressures.

7.5
From Laboratory-Scale Materials to Engineering

To date, most of the studies and measurements on potential MOFs for hydrogen storage have been conducted on laboratory-scale samples. Conclusions regarding the adsorption mechanism, the relationships between key properties and the adsorbate, insights into possible limitations, and potential optimizations at the material level have been drawn from these results. However, if MOFs are to be used in a cryoadsorption storage vessel, kilogram-scale materials will be required. These materials are made mostly from hydrocarbons and relatively inexpensive metals and have the potential for low-cost mass production. However, almost all currently available MOFs are synthesized in batch reactors using the solvothermal approach, which is ineffective and expensive. Consider, for example, Zn(btb), known as MOF-177, which is one of the most promising adsorbents. 4,4′,4″-benzene-1,3,5-triyltribenzoate is the aromatic carboxylate ligand for MOF-177 and recently became commercially available for US$179 per gram. The yield of MOF-177 synthesis reaction based on H_3btb was ~32% [45]. This represents a minimum price of ~US$560 per gram of Zn(btb) before taking into account solvent costs. With such costs, the use of MOFs as hydrogen sorbents for vehicular applications seems highly compromised. Although many researchers in both academia and industry around the world are conducting MOFs research, to our knowledge no work is under way on cost and scalability. This issue will have to be addressed in order to develop and identify low-cost MOFs for hydrogen storage and ultimately a cost-effective solid-state hydrogen storage option for on-board applications.

Another challenge for the practical application of MOFs is to deliver them in a suitable shape. Usually, as a common industrial practice, the adsorbent powders are compacted under external pressure into pellets or monoliths in order to increase their packing density. The shape of the densified adsorbents or the monoliths can be easily adapted to any tank or filter geometry or other shapes and allow safe handling since dust formation is prevented. The first objective of the compaction is to ensure that the adsorbent occupies the container volume more efficiently, but also, in the specific case of hydrogen storage for on-board applications, to minimize the specific interparticle void volume without destroying the pore structure and with a minor reduction in the gravimetric storage capacity. As stated previously, large pores and voids are inefficient in adsorbing significant amounts of gas. The important aspect of the densification process and its impact on the properties and performances of the materials has hardly been considered in the literature and almost all of the published studies relating to the synthesis of adsorbents overlook this point.

Zacharia *et al.* [46] recently reported the volumetric hydrogen sorption capacity of monoliths of Zn(btb) prepared by mechanical densification, without the addition of any organic solvent or water-based binders. Excess, total gravimetric, and volumetric hydrogen storage capacities were measured for monoliths with bulk densities ranging from 0.39 to 1.40 g cm^{-3} and compared with the bulk powdered sample of Zn(btb). The monoliths showed diminishing excess gravimetric capacity with increasing density, which can be explained by their decreasing micropore volume. It was demonstrated that the latter is due to the progressive collapse of Zn(btb) crystals to an amorphous phase upon densification. However, a compression-induced enhancement of the excess and total volumetric hydrogen storage capacity was observed. It was found that by simply compressing Zn(btb) its excess volumetric hydrogen storage capacity at 77 K could be increased by 78% compared with the bulk powdered sample.

The hydrogen storage performance of industrial pilot-scale densified Zn(btb) obtained from BASF was also evaluated internally at General Motors on both a gravimetric and a volumetric basis. The first observation was that the compaction of the powder does not alter the nature of the solid–gas interaction in the material. The adsorbed phase volume and the limiting excess adsorption were found to be lower for the pellet than the corresponding powder because of the decrease in the pore volume upon densification. However, the same conclusion could be drawn regarding the liquid-like adsorbed phase. We also observed that the total amount of hydrogen stored on a volumetric basis could reach, depending on the packing density of the pellets, about 40 g l^{-1}, which is the 2015 DOE system target at specific (*P*, *T*) operating conditions [2]. Collecting total, both gravimetric and volumetric, capacities on structured materials is crucial to determine the performance of the system and to be able to make comparisons with the other solid-state hydrogen storage options and, most importantly, with the 70 MPa high-pressure compressed gas system that is the current benchmark technology.

7.6 Conclusion

Environmentally friendly hydrogen-powered fuel cell vehicles (FCEVs) will possibly move personal transportation away from internal combustion engines and fossil fuels in the near future. The problems related to hydrogen storage are of major concern to further development. High-pressure compressed hydrogen gas is likely to be used as an interim solution but the research community is actively trying to develop advanced solid-state hydrogen storage alternatives. Among a number of promising hydrogen storage materials that have been explored for on-board applications, MOFs have attracted wide interest. As a relatively new emerging class of materials, MOFs exhibit great potential because of their unique ability, compared with classical sorbents, to tune their pore size and pore wall functionality.

Although the DOE targets for hydrogen storage are set at the conditions of near ambient temperature and high pressures, the maximum excess hydrogen uptakes on MOFs at 77 K and pressures up to 60 bar have been widely investigated. The results are crucial in the early stage of exploration of MOFs since they provide insights into the key property relationships between the adsorbent and the adsorbate for specific (P, T) operating conditions. Our investigations have shown that the high-pressure adsorbed density of hydrogen on MOFs generally correlates linearly with the pore volume and the specific surface area of the adsorbent, whereas chemistry and functionality have little effect on the maximum hydrogen uptake but play an important role in the enthalpy of adsorption and ultimately in the specific (P, T) operating conditions. It was found that the hydrogen adsorbed phase, past the excess maximum, is limited to bulk liquid hydrogen densities. This behavior was also observed for classical microporous adsorbents such as activated carbons and zeolites, which suggests that the nature of the adsorbed phase is independent of the structure and chemical composition of the adsorbents. Furthermore, the incompressibility associated with the liquid state could represent a physical limit to the amount of hydrogen that can be stored in such materials.

Although the intense focus placed on laboratory-scale MOF materials has been appropriate, there is also a need to collect and assess data on materials engineering properties. Indeed, these drive the storage tank to be more or less efficient and the data are crucial to making comparisons with the other storage options.

A great deal of progress has been achieved over the last decade in the area of MOFs for hydrogen storage applications in terms of enhanced performances and better understanding of the adsorption mechanisms. Still further significant advances will be required for a viable MOF-based storage system that can meet or exceed the 70 MPa high-pressure compressed gas system that is the current benchmark technology. The tailoring, synthesis, and discovery of MOFs for specific applications such as hydrogen storage present a complex challenge that certainly requires the cooperation and synergy of different scientific communities.

References

1 Burns, L.D. (2005) On the future of cars at TED.
2 US Department of Energy. Targets for on-board hydrogen storage, https://www1.eere.energy.gov (accessed on April 7).
3 Férey, G., Mellot-Draznieks, C., Serre, C., Millange, F., Dutour, J., Surblé, S., and Margiolaki, I. (2005) *Science*, **309**, 2040–2042.
4 Yan, Y., Telepeni, I., Yang, S., Lin, X., Kockelmann, W., Dailly, A., Blake, A.J., Lewis, W., Walker, G.S., Allan, D.R., Barnett, S.A., Champness, N., and Schroder, M. (2010) *J. Am. Chem. Soc.*, **132**, 4092–4094.
5 Koh, K., Wong-Foy, A.G., and Matzger, A.J. (2009) *J. Am. Chem. Soc.*, **131**, 4184–4185.
6 Choi, H.J., Dinca, M., Dailly, A., and Long, J.R. (2010) *Energy Environ. Sci.*, **3**, 117–123.
7 Dinca, M., Dailly, A., Liu, Y., Brown, C.M., Neumann, D.A., and Long, J.R. (2006) *J. Am. Chem. Soc.*, **128**, 16876–16883.
8 Dinca, M., Dailly, A., Tsay, C., and Long, J.R. (2008) *Inorg. Chem.*, **47** (1), 11–13.
9 Banerjee, R., Phan, A., Wang, B., Knobler, C., Furukawa, H., O'Keeffffe, M., and Yaghi, O.M. (2008) *Science*, **319**, 939–943.
10 El-Kaderi, H.M., Hunt, J.R., Mendoza-Cortes, J.L., Cote, A.P., Taylor, R.E., O'Keeffe, M., and Yaghi, O.M. (2007) *Science*, **316**, 268–272.
11 Li, H., Eddaoudi, M., O'Keeffe, M., and Yaghi, O.M. (2008) *Nature*, **402**, 276–279.
12 Barthelet, K., Marrot, J., Riou, D., and Férey, G. (2002) *Angew. Chem. Int. Ed.*, **41** (2), 281–284.
13 Ferey, G., Mellot-Draznieks, C., Serre, C., Millange, F., Dutour, J., Surblé, S., and Margiolaki, I. (2005) *Science*, **309**, 2040–2042.
14 Loiseau, T., Serre, C., Huguenard, C., Fink, G., Taulelle, F., Henry, M., Bataille, T., and Férey, G. (2004) *Chem. Eur. J.*, **10**, 1373–1382.
15 Furukawa, H., Miller, M.A., and Yaghi, O.M. (2007) *J. Mater. Chem.*, **17** (30), 3197–3204.
16 Chen, B.L., Eddaoudi, M., Hyde, S.T., O'Keeffe, M., and Yaghi, O.M. (2001) *Science*, **291** (5506), 1021–1023.
17 Rowsell, J.L.C. and Yaghi, O.M. (2006) *J. Am. Chem. Soc.*, **128**, 1304–1315.
18 Yang, S.H., Lin, X., Dailly, A., Blake, A.J., Hubberstey, P., Champness, N.R., and Schroder, M. (2009) *Chem. Eur. J.*, **15** (19), 4829–4835.
19 Lin, X., Telepeni, I., Blake, A.J., Dailly, A., Brown, C.M., Simmons, J.M., Zoppi, M., Walker, G.S., Thomas, K.M., Mays, T.J., Hubberstey, P., Champness, N.R., and Schroder, M. (2009) *J. Am. Chem. Soc.*, **131** (6), 2159–2171.
20 Mueller, U., Schubert, M., Teich, F., Puetter, H., Schierle-Arndt, K., and Pastre, J. (2006) *J. Mater. Chem.*, **16**, 626–636.
21 Loiseau, T., Lecroq, L., Volkringer, C., Marrot, J., Férey, G., Haouas, M., Taulelle, F., Bourrelly, S., Llewellyn, P.L., and Latroche, M. (2006) *J. Am. Chem. Soc.*, **128** (31), 10223–10230.
22 Caskey, S.R., Wong-Foy, A.G., and Matzger, A.J. (2008) *J. Am. Chem. Soc.*, **190**, 10670–10671.
23 Ma, S., Sun, D., Simmons, J.M., Collier, C.D., Yuan, S., and Zhou, H.C. (2008) *J. Am. Chem. Soc.*, **130**, 1012–1016.
24 Zhao, D., Yuan, D., Sun, D., and Zhou, H.C. (2009) *J. Am. Chem. Soc.*, **131**, 9186–9188.
25 Dinca, M., Dailly, A., and Long, J.R. (2008) *Chem. Eur. J.*, **14** (33), 10280–10285.

26. Park, K.S., Ni, Z., Cote, A.P., Choi, J.Y., Huang, R., Uribe-Romo, F.J., Chae, H.K., O'Keeffe, M., and Yaghi, O.M. (2006) *Proc. Natl. Acad. Sci. USA*, **103** (27), 10186–10191.
27. Rosi, N.L., Eckert, J., Eddaoudi, M., Vodak, D.T., Kim, J., O'Keeffe, M., and Yaghi, O.M. (2003) *Science*, **300**, 1127–1129.
28. Dailly, A., Vajo, J.J., and Ahn, C.C. (2006) *J. Phys. Chem. B*, **100** (3), 1099–1101.
29. Panella, B., Hirscher, M., Putter, H., and Muller, U. (2006) *Adv. Funct. Mater.*, **16** (4), 520–524.
30. Wong-Foy, A.G., Matzger, A.J., and Yaghi, O.M. (2006) *J. Am. Chem. Soc.*, **128** (11), 3494–3495.
31. Kaye, S.S., Dailly, A., Yaghi, O.M., and Long, J.R. (2007) *J. Am. Chem. Soc.*, **129**, 14176–14177.
32. Frost, H., Düren, T., and Snurr, R.S. (2006) *J. Phys. Chem. B*, **110**, 9565–9570.
33. Chahine, R. and Bose, T.K. (1996) In *Hydrogen Energy Progress XI, Proceedings of the 11th World Hydrogen Energy Conference, Stuttgart, 23–28 June 1996* (eds T.N. Veziroğlu, C.J. Winter, J.P. Baselt and G. Kreysa), International Association for Hydrogen Energy, Coral Gables, FL, p. 1259.
34. Gregg, S.J. and Sing, K.S.W. (1982) *Adsorption, Surface Area and Porosity*, 2nd edn, Academic Press, New York.
35. Poirier, E., Chahine, R., and Bose, T.K. (2001) *Int. J. Hydrogen Energy*, **26**, 831–835.
36. Ahn, C.C., Ye, Y., Ratnakumar, B.V., Witham, C., Bowman, R.B., and Fultz, B. (1998) *Appl. Phys. Lett.*, **73** (23), 3378–3380.
37. Düren, T., Millange, F., Férey, G., Walton, K.S., and Snurr, R.Q. (2007) *J. Phys. Chem. C*, **111**, 15350–15356.
38. Schnobrich, J.K., Koh, K., Sura, K.N., and Matzger, A.J. (2010) *Langmuir*, **26** (8), 5808–5814.
39. Férey, G. (2008) *Chem. Soc. Rev.*, **37**, 191–214.
40. Lin, X., Jia, J., Zhao, X., Thomas, K.M., Blake, A.J., Walker, G.S., Champness, N.R., Hubberstey, P., and Schröder, M. (2006) *Angew. Chem. Int. Ed.*, **45**, 7358–7364.
41. Lemmon, E.W., Peskin, A.P., McLinden, M.O., and Friend, D.G. (2007) *NIST12 Thermodynamic and Transport Properties of Pure Fluids – NIST Standard Reference Database 23, Version 8.0*, US Department of Commerce, Washington, DC.
42. Amankwah, K.A.G. and Schwarz, J.A. (1995) *Carbon*, **33** (9), 1313–1319.
43. Liu, Y., Her, J.H., Dailly, A., Ramirez-Cuesta, A.J., Neumann, D., and Brown, C.M. (2008) *J. Am. Chem. Soc.*, **130** (35), 11813–11818.
44. Choi, H.J., Dinca, M., and Long, J.R. (2008) *J. Am. Chem. Soc.*, **130**, 7848–7850.
45. Chae, H.K., Siberio-Pérez, D.Y., Kim, J., Go, Y., Eddaoudi, M., Matzger, A.J., O'Keeffe, M., and Yaghi, O.M. (2004) *Nature*, **427**, 523–527.
46. Zacharia, R., Cossement, D., Lafi, L., and Chahine, R. (2010) *J. Mater. Chem.*, **20**, 2145–2151.

Part Three
Bulk Chemistry Applications

8
Separation of Xylene Isomers
Joeri F.M. Denayer, Dirk De Vos, and Philibert Leflaive

8.1
Xylene Separation: Industrial Processes, Adsorbents, and Separation Principles

Xylenes (dimethylbenzenes) are aromatic compounds of general formula C_8H_{10} (Figure 8.1). The industrial separation of xylene isomers takes place within the aromatic complex and aims to produce pure isomers from their mixture, called "mixed xylenes," which contains the three xylene isomers [*ortho* (oX), *meta* (mX), and *para* (pX) isomers] and ethylbenzene (EB) [1].

The aromatic complex consists of a combination of process units that are used to convert reformates and pyrolysis gasoline into the basic petrochemical intermediates, benzene, toluene, and xylenes (BTX). It can be divided in three parts. In a first step, catalytic reforming is used to convert petroleum refinery naphthas, typically having low octane ratings, into products called reformates, containing a large amount of aromatics which are components of high-octane gasoline. These reformates are then fed to the second step. Alternatively, pyrolysis gasoline, which is a by-product of the steam cracking of hydrocarbon feeds in ethylene crackers, can also be used. The main role of the second part of the aromatic complex is to produce three different streams: pure benzene, pure toluene, and a "mixed xylenes" stream. Benzene is a building block for over 250 products, including EB (for styrene production), cumene (for phenol production), and cyclohexane. Toluene is mainly used for the production of extra xylenes through disproportionation, a catalytic process which converts toluene into a benzene–xylenes mixture, and transalkylation with C_9-aromatics. The contributions of these four processes to the "mixed xylenes" stream are 83% for reforming, 4% for pyrolysis gasoline, and 13% for toluene disproportionation and transalkylation combined. The majority of "mixed xylenes" are processed further within the last step of the complex to produce one or more of the individual isomers. A typical "mixed xylenes" stream contains 20% oX, 42% mX, 18% pX, and 20% EB. As this repartitioning does not fit the market demand, this stream is processed into the so-called "C_8-aromatics loop," which consists first in the separation of the desired isomers (essentially pX and to a lesser extent oX), then the processing of the other compounds in the isomerization unit, which brings the isomers almost back to their

Metal-Organic Frameworks: Applications from Catalysis to Gas Storage, First Edition. Edited by David Farrusseng.
© 2011 Wiley-VCH Verlag GmbH & Co. KGaA. Published 2011 by Wiley-VCH Verlag GmbH & Co. KGaA.

8 Separation of Xylene Isomers

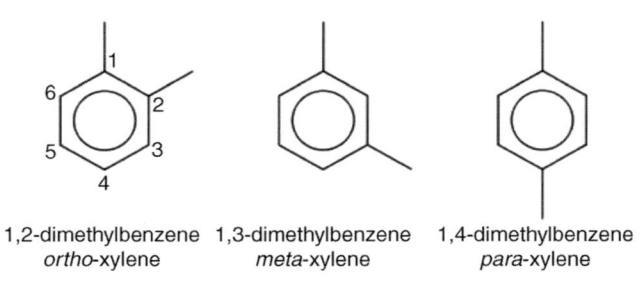

1,2-dimethylbenzene 1,3-dimethylbenzene 1,4-dimethylbenzene
   ortho-xylene   meta-xylene   para-xylene

Figure 8.1 Molecular structures of o-, m- and p-xylene.

thermodynamic equilibrium, and finally the recycling of the isomerate to the separation units.

The xylene isomer with the broadest commercial use is pX: more than 80% of the world production of xylenes is used for the production of pX. The pX production is almost all consumed by the polyester chain via one of the intermediates purified terephthalic acid (PTA) or dimethyl terephthalate (DMT). The main end uses are polyester fibers for clothing and poly(ethylene terephthalate) (PET) resins for beverage bottles. oX is oxidized to phthalic anhydride before being converted into plasticizers. mX is oxidized to isophthalic acid (IPA), which is used as a co-monomer with terephthalic acid in the manufacture of PET resins, where it lowers the melting point of the resin and decreases the rate of crystallization. The total world "mixed xylenes" production in 2008 was about 39.2 Mt, of which 33.0 Mt was used as pure pX, 3.6 Mt as oX, and 0.4 Mt as mX, the rest being the direct use of "mixed xylenes" mainly as solvents.

It is almost impossible to obtain pX product that meets commercial specifications (pX purity >99.7 wt%) by distillation from a C_8-aromatics mixture, as the relative volatility of pX and mX is only 1.02. Industrially, pX is recovered using crystallization or adsorption. The first commercial process for producing high-purity pX was crystallization, in the 1950s. These units were typically operated at temperatures of −65 to −60 °C and provided only 60–70% recovery of pX due to the presence of a eutectic point. In the 1970s, adsorption separation processes were preferred because of their significantly higher recovery (>96%) when feeds are at equilibrium (∼20–23% pX), which is usually the case. From the 1990s, crystallization was only considered for pX-rich feed (80%) from toluene disproportionation (a few units) [2]. Currently, about 25% of the pX is produced worldwide by crystallization, the remaining 75% being separated by adsorption. These two technologies can also be combined [3].

At present, all commercial adsorption-based methods for pX separation are carried out in the liquid phase using simulated moving bed (SMB) technology [4, 5]. SMB is a continuous chromatographic countercurrent process which consists of a set of adsorbent beds connected in series; the countercurrent flow of the solid phase is simulated by the periodic shifting of inlets (feed and desorbent) and outlets (extract and raffinate) in the direction of the fluid flow. Separation is accomplished by exploiting the differences in affinity of the adsorbent for pX, relative to the other C_8 isomers. The adsorbed pX is then removed from the adsorbent by displacement by a desorbent. Typical pX purity and recovery are 99.7–99.9% and 97–99%, respectively. The adsorbents used are alkali or alkaline earth metal-exchanged faujasite zeolites (generally BaX or KBaX zeolite) with a controlled water content (4–8 wt%) [6, 7]. The operating conditions are 140–180 °C and moderate pressures around 10 bar. The desorbent is generally *p*-diethylbenzene, or toluene in a few cases.

Concerning the other isomers, oX is generally separated from the "mixed xylenes" by fractional distillation, its boiling point being 5 °C higher than those of the other isomers. The first commercial process for mX production was based on the formation of a complex between mX (the most basic C_8-aromatic) and $HF–BF_3$. At present, adsorption-based processes using SMB technology are preferred. The adsorbents used are NaY and NaLiY zeolites with a controlled water content (0–8 wt%) depending on operating temperature [8, 9]. The operating conditions are 100–180 °C and 10 bar. The desorbent is generally toluene.

Regarding the mechanism of the separation xylene isomers on faujasites, it was shown that selectivity is determined by (i) the Si:Al ratio of the zeolite, (ii) the nature of the exchangeable cations, (iii) the composition of the xylene mixture, (iv) the degree of pore filling, and (v) the presence of small polar molecules such as water. The selectivity originates from the thermodynamics of adsorption and not from diffusion as in MFI-based membranes. This latter selectivity is currently not used industrially to produce high-purity pX from a "mixed xylene" stream due to the presence of EB. which has a critical diameter very close to that of pX.

According to Iwayama and Suzuki [10], the selective adsorption of xylene isomers on faujasite zeolites is affected by the ionic potential of the exchanged cation. The isomer with the highest base strength, mX, is selectively adsorbed on the more "acid" alkaline zeolite (NaY and NaLiY), whereas "basic" zeolites such as CsX and RbX selectively adsorb EB [11]. The pX isomer is only selectively adsorbed on "intermediate" zeolites, KY, BaY, BaX, and KBaY, and the selectivity then depends on both pore filling and the presence of water molecules. For the adsorption of pX and mX on BaY and NaY, Cottier *et al.* [12] observed two selective adsorption processes according to the degree of pore filling. This is in agreement with the powder neutron diffraction data of Mellot *et al.*, who observed a significant molecular rearrangement of adsorbed mX molecules when the coverage was increased above two molecules per supercage in BaX zeolite [13]. This molecular rearrangement is due to intermolecular interactions and has the goal of minimizing the intermolecular repulsion. It is also connected to the size of the cations and the presence of water molecules that create steric hindrance inside the zeolite cages. The presence of aromatic molecules adsorbed on non-cationic sites at high pore filling was also confirmed by Lachet *et al.* on KY using molecular simulation [14] and by Pichon *et al.* in BaX using neutron diffraction [15]. The origin of the selectivity for pX within alkali and alkaline earth metal-exchnged zeolites then results from a very complex energetically heterogeneous adsorption behavior due to confinement and the associated cation and molecular rearrangements.

Considering the use of metal–organic frameworks (MOFs) versus zeolites in xylene separation, the major limitations to be expected are probably the following:

- As stated above, pX selectivity seems mainly to rely on confinement and entropic effects. Selectivity for other isomers is then much more encountered within zeolites than pX selectivity, which will probably also be the case for MOFs.
- pX separation is by far the most important application of SMB technology. The quantity of adsorbent required for commercial operation is generally very high and is typically several hundred tons (up to 2000 t) for new units. The use of MOFs for this application would then require an extensive scale-up of MOF manufacture at reasonable cost.
- The use of very stable zeolites (both thermally and chemically) allows very long times on-stream (typically several years) before replacement is necessary. This should also be the same for MOFs if commercial application is targeted.

8.2
Properties of MOFs Versus Zeolites in Xylene Separations

As adsorbents, MOFs show many similarities to the traditional zeolite adsorbents used in the industrial separation of xylene isomers, but also differ in many aspects. Zeolites are inorganic solids, their rigid crystalline framework being essentially built up from SiO_4 and AlO_4 tetrahedra, while charge compensating extra-framework cations are present as mobile species in the zeolite's nanopores [16]. Although several views exist on the exact nature and role of the interactions between the extra-framework cations and the adsorbing xylene isomers, it is obvious that they play a dominant role in the separation mechanism. Zeolites are microporous, with pore sizes typically ranging between 3 and 13 Å. As a result, only a few aromatic molecules can be packed in the zeolitic cages or the pores. As a consequence, the size and shape of the adsorbing molecules are a governing factor in the adsorption equilibrium; molecules that fit better inside the micropores are adsorbed selectively; this property is commonly referred to as shape selectivity. At high pressure or in liquid-phase conditions, the pores of the zeolite are saturated with adsorbate molecules. The ease or efficiency with which molecules can be packed in the constrained environment of the zeolite pores affects the adsorption selectivity significantly [17]; this entropic effect often leads to large changes in selectivity with degree of pore filling. Many different pore topologies and architectures are known; zeolites with one-dimensional (1D) tubular pores, intersecting pores, or cages interconnected via windows have found their way into industrial applications [18]. In addition to affecting shape selectivity, pore topology also influences the diffusion rate of molecules inside the crystals. For example, in narrow 1D pores, molecules are transported via a single-file diffusion mechanism, which is much slower than three-dimensional (3D) intracrystalline transport in structures with open cages, connected to each other via several windows. With respect to industrial applicability, perhaps one of the most important properties of zeolites is their very high thermal and mechanical stability. Many zeolites can be heated to temperatures exceeding 500 °C, allowing their successive thermal regeneration in industrial catalytic reactors or adsorptive separation units.

Currently, faujasite-type zeolites are used in the separation of xylene isomers. These materials combine high selectivity due to the presence of cations, a high adsorption capacity due to the open-pore structure consisting of large cages, relatively fast intracrystalline transport due the presence of large windows between the zeolite cages, and high thermal and mechanical stability. When searching for alternatives for the successful replacement of faujasites by MOFs, all of these aspects have to be balanced against each other.

MOFs can be described as hybrid porous solids formed by a network of metal atoms, held together by bridging multidentate organic ligands. In this way a crystalline, 3D open framework can be formed. The metal ions (di-, tri-, or tetravalent) or ion clusters form the corners where the organic ligands come together. MOFs can accept almost all di-, tri-, and tetravalent cations, including rare earth elements. A large variety of different topologies have been observed, with pore structures ranging from inaccessible cavities to 3D intersecting channels. In general, very high specific surface areas and pore volumes are found for MOFs, exceeding those of zeolites. This is the result of the dimensions of the pore walls. In zeolites, the pore walls are constructed with a thickness of at least several Si, O, and Al atoms, creating dead space and an inaccessible surface. In contrast, almost all of the framework atoms of MOFs can be used as a surface since the pore walls are often only one carbon atom thick [19]. A specific surface area of $4500\,m^2\,g^{-1}$ and a pore volume of $1.59\,ml\,g^{-1}$ have been reported for MOF-177 [20]. This specific surface area is significantly higher than that for the material with the highest specific surface area in the group of disordered structures ($2030\,m^2\,g^{-1}$) [21]. In the group of ordered structures, this place is occupied by zeolite Y with a specific surface area of about $1000\,m^2\,g^{-1}$ and a pore volume of $0.3\,ml\,g^{-1}$ [22]. With respect to the tailor-made design of adsorbents for a given separation, probably the most interesting property of the MOFs is their exceptional tunability. Since they are built from metal ions, interconnected through organic ligands, changing these ligands allows the pore dimensions to be altered and also extra functionalities (chiral center, active site, change in electronic properties, etc.) to be introduced fairly easily without affecting the underlying topology. The completely different chemical nature of MOFs compared with zeolites (organic versus inorganic) is expected to result in a different adsorption behavior. The absence of extra-framework cations will lead to less strong interactions with, for example, aromatic molecules. On the other hand, organic linkers containing aromatic groups might also result in favorable interactions with adsorbing aromatics. A weaker but still selective interaction could even be beneficial for the design of an adsorptive separation process, since desorption would require less energy.

Whereas zeolites have a rigid structure, many MOFs have a flexible framework [23–27]. The structural changes found in MOFs arise from stretching motions around the connectors and/or linkers resulting from bond formation/cleavage, rotation around a single bond, or breathing (contraction/expansion) of the framework [28–32]. Adsorption or desorption of guest molecules sometimes introduces a structural change without loss of crystallinity. When during the transformation the topology of the framework is maintained, the structural change is referred to as breathing, which will always result in a change in the cell volume and is guest dependent [33]. When during the adsorption or exchange of guest molecules the topology of the framework is changed, the structural change is referred to as a deformation. Possible advantages of framework flexibility for separation processes are difficult to assess. It has been demonstrated that flexible MOFs allow "zero-order kinetics" in the release of bioactive compounds [34]. In certain cases, a phenomenon called "gate opening" causes a sharp step in adsorption and desorption isotherms [35]. Only a "gate opening" pressure do the pores of the MOF open, and adsorption occurs. Such an effect could be beneficial for the kinetics and energy consumption in pressure swing adsorption processes, as only a small decrease in pressure is needed to achieve a large change in amount adsorbed. An issue that needs further research is the possible change in crystal size during adsorption and desorption of molecules on flexible MOF structures. This potentially could lead to mechanical instability of adsorbent particles consisting of crystal agglomerates.

Thermal stability is one property for which MOFs cannot compete with zeolites. In MOFs, the building blocks are held together by coordination bonds, H-bonds, $\pi-\pi$ stacking, and van der Waals interactions, whereas the building units of zeolites are connected through covalent bonds, which are substantially stronger. The thermal stability of MOFs is often limited to 300–400 °C.

Overall, MOFs possess several interesting features for molecular separations, and xylene isomer separation in particular. The very large adsorption capacity and the high chemical tunability offer perspectives for the development of better adsorbents [36, 37].

8.3
Separation of Xylenes Using MIL-47 and MIL-53

MIL-47 and MIL-53 are two isostructural MOFs with interesting properties for the separation of aromatic compounds. MIL-53 is built from $CrO_4(OH)_2$, $AlO_4(OH)_2$, or $FeO_4(OH)_2$ octahedra, held together by the dicarboxylate groups of the terephthalate linkers [38]. In this way, a 3D microporous framework with 1D diamond shaped channels with a free internal diameter of about 0.85 nm is formed (Figure 8.2). MIL-47 differs from MIL-53 in that during calcination its V^{3+} atoms are oxidized to V^{4+}. As a result, MIL-47 has free-standing oxo groups instead of the free-standing hydroxyl groups found in the MIL-53 framework. The presence of the corner-sharing hydroxyl groups in the MIL-53 framework is responsible for the high flexibility of the structure upon hydration/dehydration. [39] The presence of aromatic rings in the framework of both materials might induce selectivity in the adsorption of mixtures of aromatic molecules. The constrained pore size could also enhance molecular separation, similarly to zeolites.

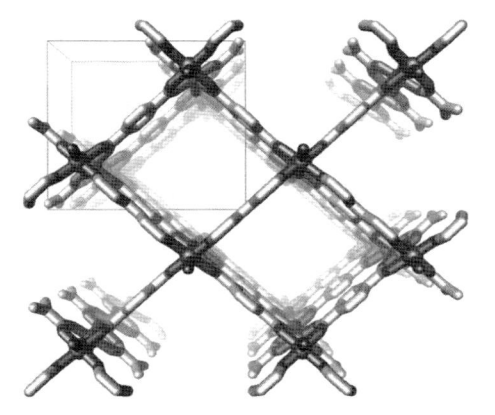

Figure 8.2 Structure of MIL-53. The box is the representation of one unit cell.

8.3.1
Low-Coverage Gas-Phase Adsorption Properties

At very low hydrocarbon partial pressures, MIL-47 and MIL-53 are not selective in the adsorption of C_8-alkylaromatics. This is shown in Figure 8.3, where chromatograms obtained after injection of the individual C_8-alkylaromatic isomers on a MIL-47 column at 270 °C are presented [40]. Only small differences were observed with retention times differing by a maximum of 27%. The order of elution of the C_8-alkylaromatic components on the MIL-47 column does not follow the order of boiling points (EB 136 < pX 138 < mX 139 < oX 144 °C) as would be the case on a typical aspecific analytical chromatographic column.

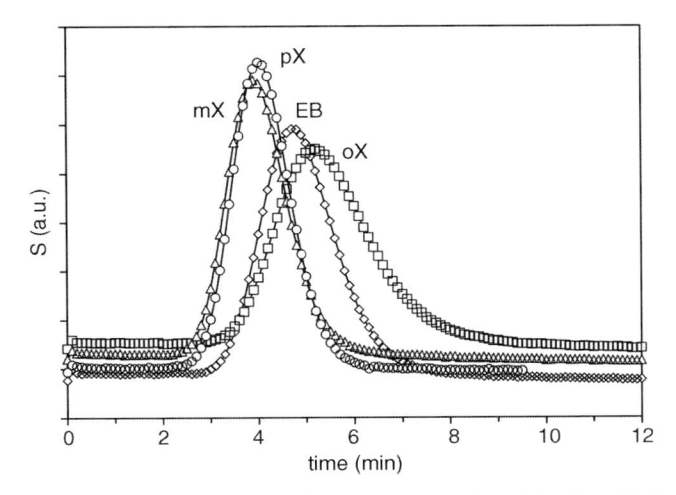

Figure 8.3 Chromatograms of EB, oX, mX, and pX on MIL-47 at 270 °C.

Table 8.1 Henry constants (at 270 °C) and adsorption enthalpies on MIL-47.

Compound	K'_{270} (mol kg^{-1} Pa^{-1})	$-\Delta H_0$ (kJ mol^{-1})
Ethylbenzene	$2.5 \times 10^{-4} \pm 9.6 \times 10^{-6}$	59.7 ± 1.8
o-Xylene	$2.9 \times 10^{-4} \pm 1.1 \times 10^{-5}$	59.6 ± 1.8
m-Xylene	$2.1 \times 10^{-4} \pm 7.9 \times 10^{-6}$	59.7 ± 1.8
p-Xylene	$2.2 \times 10^{-4} \pm 8.5 \times 10^{-6}$	61.2 ± 1.8

This nonselective behavior at low pressure also follows from the Henry constants and adsorption enthalpies, which are very similar for these four components (Table 8.1). oX, mX, pX, and EB interact with about the same strength with the MIL-47 framework. The adsorption enthalpy of the C$_8$-alkylaromatic components (about 60 kJ mol^{-1}) is significantly lower than that of n-octane (65.7 kJ mol^{-1}). This is in clear contrast to what is commonly observed, that is, a stronger interaction of the aromatic component compared with the saturated alkane. It therefore appears that the π-clouds of the aromatics are not dominating the interaction with the MIL-47 framework. Similar observations are made with MIL-53 [41].

8.3.2
Molecular Packing

Adsorption isotherms of the C$_8$-alkylaromatics and n-octane at 70 C on MIL-47 are shown in Figure 8.4. High adsorption capacities, up to almost 40 wt% at 70 °C, were measured, which significantly exceed those of zeolitic adsorbents. The largest adsorption capacity is measured for pX, followed in order by oX, mX, and EB. pX reaches its plateau at a partial pressure of 0.005 bar, but for the other isomers the amount adsorbed continues to increase over the whole experimental pressure range.

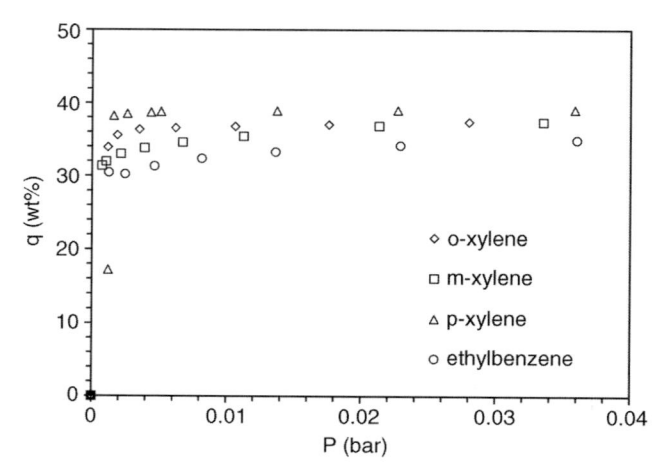

Figure 8.4 Adsorption isotherms of EB, pX, oX, and mX on MIL-47 at 70 °C.

The observed tendencies can be related to the modes of packing of the C_8-alkylaromatic molecules in the MIL-47 pore system, as elucidated in earlier work by Rietveld refinement of the X-ray powder diffraction patterns of MIL-47 samples saturated with each of the isomers at room temperature and in the liquid phase [42]. It was found that all isomers adsorb in pairs along the length of the pores, with two pairs in each unit cell. This is in fair agreement with the maximum adsorption capacity of 3.4 molecules per unit cell as determined by gravimetry at 70 °C. The major difference between the molecular arrangements of the three xylene isomers is the relative alignment of the aromatic planes of the molecules within a pair. For both mX and oX, a higher pressure is needed to achieve the same packing efficiency than for pX. Steric constraints in the pores caused by the ethyl group of EB prevent planar alignment of the aromatic rings, such that π–π interactions between EB molecules or EB molecules and the framework become insignificant. This implies an energy cost, which explains the significantly lower uptake of EB.

Adsorption isotherms on the flexible material MIL-53(Al) look different (Figure 8.5). oX, mX, and pX show a manifest two-step adsorption isotherm with hysteresis [43]. For EB, the appearance of an intermediate plateau is less obvious, but a discontinuity can still be appreciated at a pressure of about 0.003 bar.

For all alkylaromatics, the amount adsorbed, q, increases steeply to reach a first plateau at vapor pressures between 0.001 and 0.003 bar. At this first plateau, all components have rather similar capacities (see Figure 8.5); EB and oX show an adsorption capacity of 19 wt%. This first plateau in the isotherm is followed by a second increase in q. The onset of the second adsorption branch and the position of the second plateau are component dependent. oX shows a large and steep jump in capacity from 20 to 35 wt% at a partial pressure of 0.003 bar. After this jump, q gradually increases with increase in pressure to reach a final adsorption capacity of

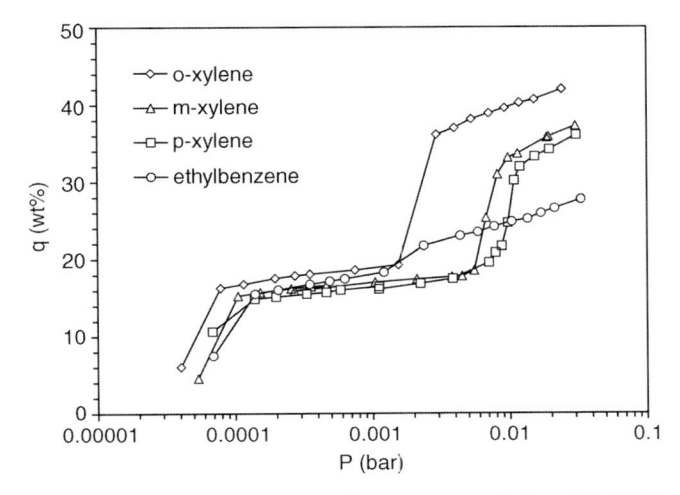

Figure 8.5 Adsorption isotherms of EB, oX, mX, and pX on MIL-53(Al) at 110 °C.

42 wt% at a pressure of 0.025 bar, corresponding to 3.3 molecules per unit cell. With respect to adsorption capacity, oX is followed by mX and pX with adsorption capacities of 37 and 36 wt%, respectively, at a pressure of 0.03 bar, corresponding to about 2.9 molecules per unit cell. The second step in the isotherm starts at 0.007 bar for pX, 0.005 bar for mX, and 0.003 bar for oX. Instead of a jump or step, rather a kink in the isotherm occurs with EB at a pressure of about 0.003 bar. An adsorption capacity of only 28 wt% is reached at a pressure of 0.033 bar, corresponding to 2.2 molecules per unit cell.

Similar two-step isotherm profiles have been reported for several microporous coordination polymers. The occurrence of steps in the adsorption isotherm is accepted to be caused by the flexibility of the framework (see above) [44]. It has been shown that the sudden increase in the amount adsorbed at higher pressure is due to structural changes induced in the crystal framework leading to the opening of the pores [45, 46]. A similar mechanism is expected to prevail in the adsorption of the C_8-alkylaromatic components in the MIL-53 structure. Adsorption of alkylaromatic compounds in the MIL-53(Al) MOF in liquid-phase conditions induces a significant change in the framework lattice parameters [47]. More specifically, interactions between the methyl groups on the aromatic ring of the adsorbed molecules and the framework carboxyl groups appear to be responsible for framework deformation. Rietveld refinement of X-ray powder diffraction patterns of MIL-53(Al) samples with different oX loadings demonstrated a structural transition of the framework. At low partial pressure, a structure with strongly reduced pore diameter is observed. This MIL-53iX(Al) structure has unit cell dimensions of $a = 18.51$, $b = 6.64$, and $c = 9.60$ Å. At high partial pressure and thus degree of pore filling, the pores are reopened and the unit cell becomes larger: $a = 6.63$, $b = 15.90$, and $c = 14.02$ Å (Figure 8.6). At intermediate pressures, mixtures of the two phases are observed.

Together with the structural transition from the MIL-53iX(Al) to the MIL-53ht(Al) form, the molecular siting mechanism changes. The comparable adsorption capacity of all C_8-alkylaromatic components at the first plateau in the isotherm is explained by the molecular packing mechanism in the "closed" pores of the MIL-53iX(Al) framework. Experimentally, a capacity of 1.4–1.5 molecules per unit cell is found for the first plateau. This corresponds to a single file of molecules, adsorbed along the length of the pores (see Figure 8.6). Given the small pore diameter (6.6 × 9.6 Å) in the MIL-53iX(Al) form, only one molecule can be adsorbed in the cross section of the pore, regardless of the position of the methyl side chains.

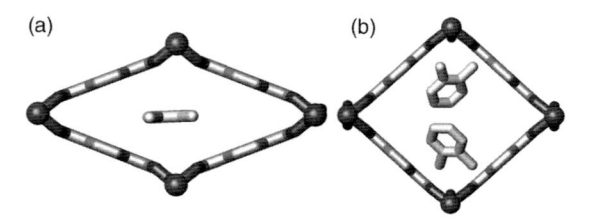

Figure 8.6 Pore structure of MIL-53(Al) at (a) low and (b) high xylene pressure.

8.3 Separation of Xylenes Using MIL-47 and MIL-53

In order to achieve a higher molecular density in the pores, a large energy input is required, given the large increase in pressure needed to induce additional uptake of molecules (Figure 8.5). As soon as the transition pressure is reached, the amount adsorbed doubles for all alkylaromatic components apart from EB, in a small pressure interval. These observations point to a complete molecular reorganization of the adsorbed phase.

The pressure or energy required to trigger the rearrangement from a single file of adsorbed molecules to a double file, in which two molecules are present per channel segment, is related to the stability of the adsorbed phase at high degrees of pore filling. Steric restrictions imposed on the adsorption of a pair of alkylaromatic molecules in one pore segment, together with the energetic interactions between the adsorbed molecules and the host, govern the adsorption equilibrium. As evidenced by Rietveld refinement, oX has the strongest interaction with the framework at full pore occupancy. Therefore, this component shows a transition from a single file of adsorbed molecules to a double file at the lowest pressure. mX, having a lower energetic interaction, needs a higher pressure to rearrange the adsorbed phase. pX, having the lowest interaction among the xylene isomers at full pore occupancy, has the largest transition pressure. EB is expected to interact via only one methyl group with a carboxylate group in the framework, similarly to pX. Moreover, as shown before for the adsorption of EB in MIL-47, the ethyl group causes steric hindrance and prevents optimal pairwise organization in the pores. This translates into a much lower uptake of EB, even at high pressures.

Another remarkable feature of the isotherms on MIL-53(Al) is the occurrence of a hysteresis loop between adsorption and desorption (Figure 8.7). This hysteresis can also be explained by the transitions between the different MIL-53(Al) structures. The occurrence of a first plateau followed by a second increase in the adsorption isotherm can be seen as a "delay" in the adsorption caused by the transition from both

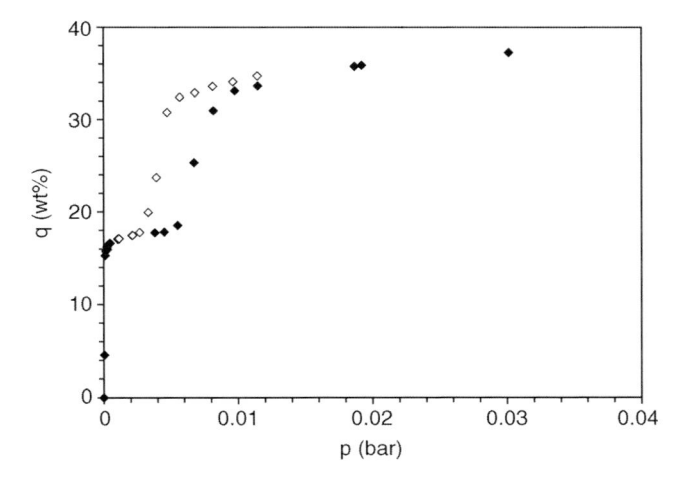

Figure 8.7 Adsorption (solid symbols) and desorption (open symbols) isotherms of mX on MIL-53 (Al) at 110 °C.

the small-pore MIL-53(Al) form to the open MIL-53(Al) form and from a single-file adsorption to a double-file adsorption. Thus, during adsorption, an extra energy input is required to reach full saturation. During desorption, at hydrocarbon pressures between the first and the second plateau, the material is already in the MIL-53*ht*(Al) form. As the MIL-53*ht*(Al) form is stabilized by the adsorbate–adsorbent interactions, the transition from the MIL-53*ht*(Al) back to the MIL-53*iX*(Al) form and from double-file to single-file adsorption during desorption will take place at a pressure lower than the onset of the second increase in the adsorption isotherm. At present, it is not fully understood how such a hysteresis loop will affect the performance of a cyclic adsorption–desorption process. It could be argued that more energy will be needed for the desorption step, since the pressure has to be lowered significantly before desorption occurs, but experimental and theoretical work are needed to obtain a deeper insight into this effect.

8.3.3
Separation of Xylene-Mixtures

Pulse chromatographic experiments show that neither MIL-47 nor MIL-53 is suitable for separating xylene isomers at very low pressure. Pure component isotherms, on the other hand, show differences between the components at higher pressure. Separation is therefore expected to depend on pressure. Figure 8.8 shows breakthrough profiles obtained by flowing an equimolar vapor mixture of EB and oX through a column packed with MIL-53(Al) adsorbent, at different hydrocarbon partial pressures. A clear evolution of the shape of the breakthrough profiles with increasing partial hydrocarbon pressure is observed. At a pressure of 0.0009 bar, which is well below the pressure at which the transition from contracted to open pores occurs (Figure 8.5), both components elute simultaneously. No separation of the components could therefore be achieved under this condition. This is in conformity with the measured adsorption isotherms, where EB and oX are adsorbed to the same extent at low pressure.

At the highest partial hydrocarbon pressure, a classical breakthrough profile is observed with a significant difference in breakthrough time between EB and oX. Pure EB elutes from the column. Already from the start of the experiment, the loading of the pores corresponds to the second plateau on the isotherm. As mentioned before, oX is selectively adsorbed in this regime because of its more efficient packing and higher energetic interaction compared with EB. An average separation factor of 6.4 is obtained.

At an intermediate pressure of 0.005 bar, which is slightly above the pressure at which transition from the narrow-pore to the open-pore form occurs, three different phases can be observed after the initial phase in which both components are fully adsorbed. The occurrence of this unconventional breakthrough profile and the appearance of the distinct phases can be rationalized by a transition from aselective adsorption in the single-file adsorption mode in the closed form of the pores to selective adsorption in the double-file adsorption mode in the open form of the pores.

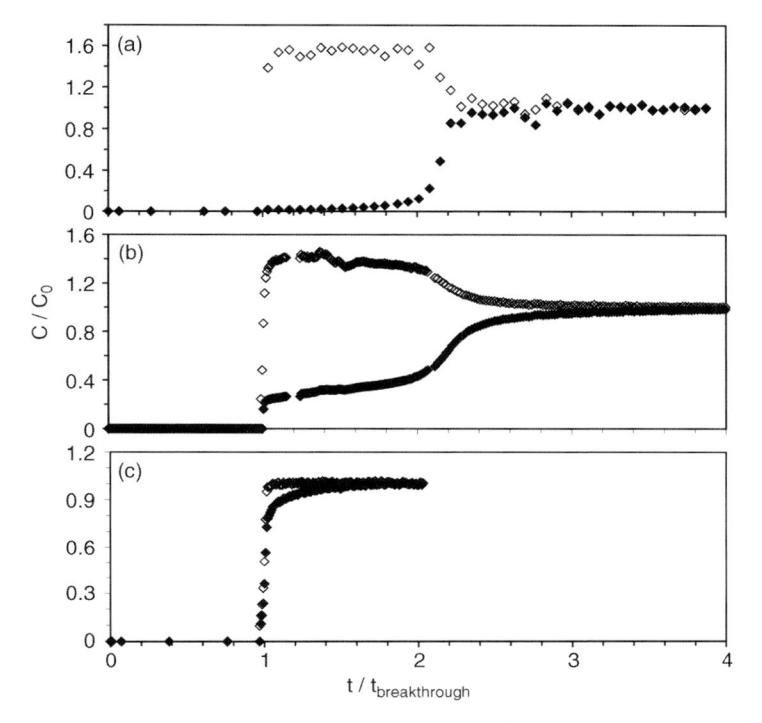

Figure 8.8 Breakthrough curves for the separation of EB (open symbols)–oX (solid symbols) mixtures on MIL-53(Al) at different total partial hydrocarbon vapor pressures and 110 °C. (a) 0.056; (b) 0.005; (c) 0.0009 bar.

These experiments clearly indicate that the separation mechanism of MIL-47 and MIL-53 relies on molecular packing in the cross-section of the 1D pores of these MOFs. The greater the degree of pore filling, the higher is the selectivity. This is confirmed in liquid-phase separation experiments, where the concentration is significantly higher than in vapor-phase experiments. Under these conditions, a clear separation between, for example, EB and oX is obtained (Figure 8.9).

Figure 8.10 shows the vapor-phase separation of an equimolar mixture of pX, mX, oX, and EB on MIL-47 at 110 °C. These four C_8-alkylaromatics are separated from each other, which demonstrates the potential of MOFs for molecular separation processes with aromatic mixtures.

8.4
Conclusions

The MIL-47 and MIL-53 MOFs allow the rather difficult separation of xylene isomers to be achieved. The advantages of these materials over zeolites are their high adsorption capacity and lower adsorption enthalpy, which is advantageous with

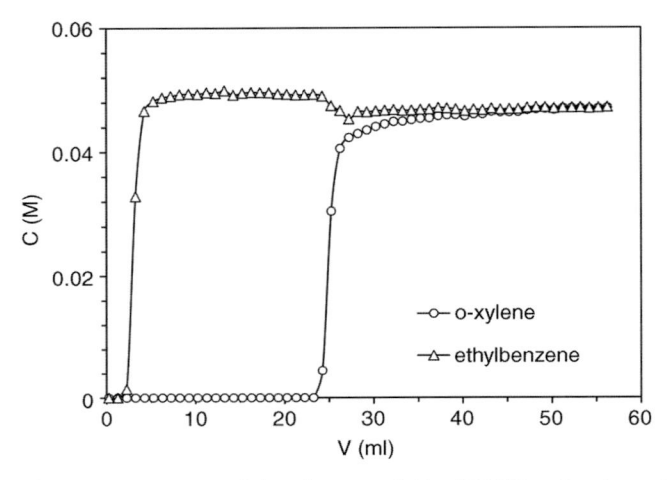

Figure 8.9 Separation of a liquid mixture of oX and EB (diluted in *n*-hexane) on a column filled with MIL-53 crystallites at 298 K.

respect to regeneration of the adsorbent. Disadvantages are the lower thermal and mechanical stability. Nevertheless, the much larger tunability of MOFs compared with zeolites will undoubtedly lead to new applications in the field of molecular separation. Figure 8.11 shows a flow scheme for the separation of C_8- and C_9-alkylaromatics in individual components using different MOFs in adsorptive separation units. This demonstrates that the subtle differences in selectivity of the different materials can be exploited to achieve complex molecular separations.

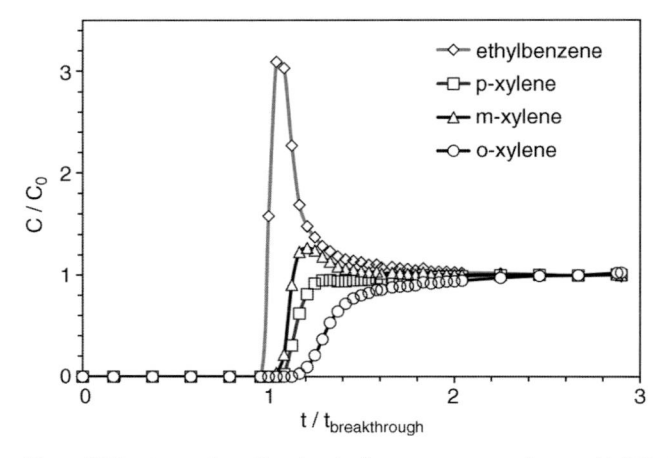

Figure 8.10 Separation of equimolar four-component mixture with MIL-47 at 110 °C and a total hydrocarbon pressure of 0.05 bar.

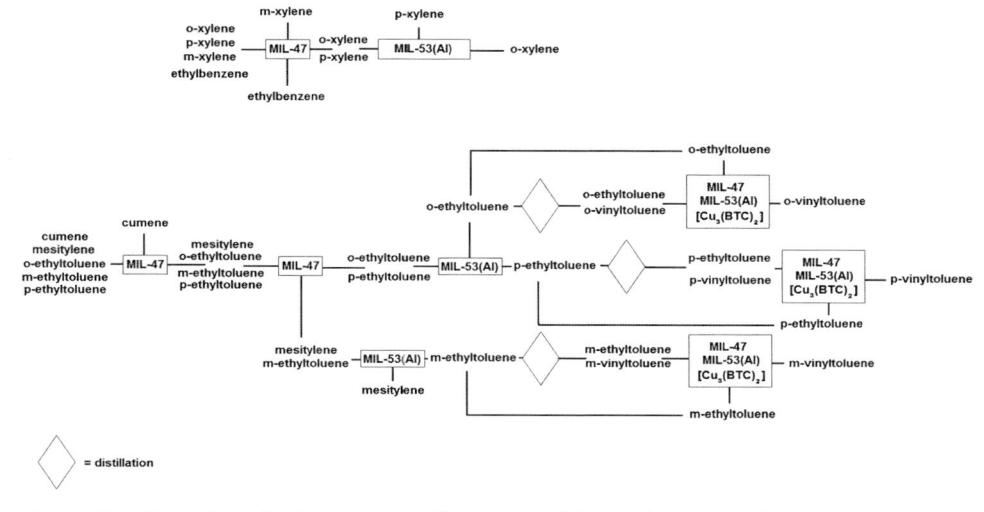

Figure 8.11 Flow scheme for the separation of C_8- and C_9-alkylaromatics using MOFs.

Acknowledgments

This work was performed in the frame of the IAP Functional Supramolecular Systems of the Belgian Federal Government. The authors thank FWO Vlaanderen and IAP for funding this research.

References

1 Canella, W.J. (2000) Xylenes and ethylbenzene. In (ed *Kirk–Othmer*) *Encyclopedia of Chemical Engineering*, John Wiley & Sons, Inc., New York, 101–172.
2 Starkey, D.R., Andrews, J.L., Luo, S.L., and Knob, K.J. (2009) PX max with crystallization. An integrated process for high purity *para*-xylene production. Presented at the Asian Refining Technology Conference Annual Meeting, Kuala Lumpur, 10–12 March 2009.
3 Hotier, G. and Methivier, A. (2002) Debottlenecking of existing aromatic production loop: the hybrid eluxyl process offers an attractive alternative solution. Presented at the AIChE Spring Meeting, New York, 10–14 March 2002.
4 Guiochon, G., Shirazi, D.G., Felinger, A., and Katti, A.M. (2006) Chapter 17, *Fundamentals of Preparative and Nonlinear Chromatography*, Academic Press, New York.
5 Wauquier, J.P., Trambouze, P., and Favennec, J.P. (1995) *Petroleum Refining: Separation Processes*, vol. 2, Editions Technip, Paris.
6 Plee, D. and Methivier, A. (2008) Agglomerated zeolitic adsorbents, their process of preparation and their uses. US Patent 7 452 840.
7 Cheng, L.S. and Johnson, J.A. (2009) Adsorbent with improved mass transfer properties and their use in the adsorptive separation of *para*-xylene. US Patent Application 2009/0326311.
8 Kulprathipanja, S. (1995) Process for the adsorptive separation of *meta*-xylene from aromatic hydrocarbons. US Patent 5 382 747.

9 Leflaive, P. and Barthelet, K. (2008) Process for separating *meta*-xylene from a feed of aromatic hydrocarbons by liquid phase adsorption. US Patent 7,468,468.

10 Iwayama, K. and Suzuki, M. (1994) Adsorption of C8 aromatic isomers on faujasite zeolite. *Stud. Surf. Sci. Catal.*, **83**, 243–250.

11 Barthomeuf, D. (1986) US Patents 4 593 149 and 4 613 725.

12 Cottier, V., Bellat, J.P., Simonot-Grange, M.H., and Methivier, A. (1997) Adsorption of *p*-xylene/*m*-xylene gas mixtures on BaY and NaY zeolites. Coadsorption equilibria and selectivities. *J. Phys. Chem. B*, **101**, 4798–4802.

13 Mellot, C., Simonot-Grange, M.H., Pilverdier, E., Bellat, J.P., and Espinat, D. (1995) Adsorption of gaseous *p*- or *m*-xylene in BaX zeolite: correlation between thermodynamic and crystallographic studies. *Langmuir*, **11**, 1726–1730.

14 Lachet, V., Boutin, A., Tavitian, B., and Fuchs, AH. (1999) Molecular simulation of *p*-xylene and *m*-xylene adsorption in Y zeolites. Single components and binary mixtures study. *Langmuir*, **15**, 8678–8685.

15 Pichon, C., Methivier, A., Simonot-Grange, M.H., and Baerlocher, C. (1999) Location of water and xylene molecules adsorbed on prehydrated zeolite BaX. A low-temperature neutron powder diffraction study. *J. Phys. Chem. B*, **103**, 10197–10203.

16 Frising, T. and Leflaive, P. (2008) Extraframework cation distributions in X and Y faujasite zeolites: a review. *Micropor. Mesopor. Mater.*, **114** (1–3), 27–63.

17 Schenk, M., Calero, S., Maesen, T.L.M., van Benthem, L.L., Verbeek, M.G., and Smit, B. (2002) Understanding zeolite catalysis: inverse shape selectivity revised. *Angew. Chem. Int. Ed.*, **114**, 2610–2612.

18 Kulprathipanja, S. (2010) *Zeolites in Industrial Separation and Catalysis*, Wiley-VCH Verlag GmbH, Weinheim.

19 Kitagawa, S., Kitaura, R., and Noro, S. (2004) Functional porous coordination polymers. *Angew. Chem. Int. Ed.*, **43**, 2334–2375.

20 Chae, H.K., Siberio, P., Kim, J., Go, Y., Eddaoudi, M., Matzger, A.J., O'Keeffe, M., and Yaghi, O.M. (2004) A route to high surface area, porosity and inclusion of large molecules in crystals. *Nature*, **427**, 523–527.

21 Nijkamp, M.G., Raaymakers, J.E., van Dillen, A.J., and de Jong, K.P. (2001) Hydrogen storage using physisorption – materials demands. *Appl. Phys. A: Mater.*, **72**, 619–623.

22 Chester, A.W., Clement, P., and Han, S. (2000) C US Patent 6 136 291, assigned to Mobile Oil Corporation.

23 Ghosh, S.K., Zhang, J.P., and Kitagawa, S. (2007) Reversible topochemical transformation of a soft crystal of a coordination polymer. *Angew. Chem. Int. Ed.*, **46**, 7965–7968.

24 Kubota, Y., Takata, M., Matsuda, R., Kitaura, R., Kitagawa, S., and Kobayashi, T.C. (2006) Metastable sorption state of a metal–organic porous material determined by *in situ* synchrotron powder diffraction. *Angew. Chem. Int. Ed.*, **45**, 4932–4936.

25 Lee, E.Y., Jang, S.Y., and Suh, M.P. (2005) Multifunctionality and crystal dynamics of a highly stable, porous metal-organic framework [$Zn_4O(NTB)_2$]. *J. Am. Chem. Soc.*, **127**, 6374–6381.

26 Mellot-Draznieks, C., Serre, C., Surblé, S., Audebrand, N., and Férey, G. (2005) Very large swelling in hybrid frameworks, a combined computational and powder diffraction study. *J. Am. Chem. Soc.*, **127**, 16273–16278.

27 Zhang, J.P., Lin, Y.Y., Zhang, W.X., and Chen, X.M. (2005) Temperature- or guest-induced drastic single-crystal-to-single-crystal transformations of a nanoporous coordination polymer. *J. Am. Chem. Soc.*, **127**, 14162–14163.

28 Fletcher, A.J., Thomas, K.M., and Rosseinsky, M.J. (2005) Flexibility in metal–organic framework materials, impact on sorption properties. *J. Solid State Chem.*, **178**, 2491–2510.

29 Horike, S., Matsuda, R., Tanaka, D., Matsubara, S., Mizuno, M., Endo, K., and Kitagawa, S. (2006) Dynamic motion of building blocks in porous coordination polymers. *Angew. Chem. Int. Ed.*, **45**, 7226–7230.

30 Serre, C., Mellot-Draznieks, C., Surblé, S., Audebrand, N., Filinchuk, Y., and Férey, G. (2007) Role of solvent–host interactions that lead to very large swelling of hybrid frameworks. *Science*, **315**, 1828–1831.

31 Kitagawa, S., Kitaura, R., and Noro, S. (2004) Functional porous coordination polymers. *Angew. Chem. Int. Ed.*, **43**, 2334–2375.

32 Uemura, K., Matsuda, R., and Kitagawa, S. (2005) Flexible microporous coordination polymers. *J. Solid State Chem.*, **178**, 2420–2429.

33 Mellot-Draznieks, C., Serre, C., Surblé, S., Audebrand, N., and Férey, G. (2005) Very large swelling in hybrid frameworks, a combined computational and powder diffraction study. *J. Am. Chem. Soc.*, **127**, 16273–16278.

34 Horcajada, P., Serre, C., Maurin, G., Ramsahye, NA., Balas, F., Vallet-Regi, M., Sebban, M., Taulelle, F., and Férey, G. (2008) Flexible porous metal–organic frameworks for a controlled drug delivery. *J. Am. Chem. Soc.*, **130** (21), 6774–6780.

35 Chun, H. and Seo, J. (2009) Discrimination of small gas molecules through adsorption: reverse selectivity for hydrogen in a flexible metal–organic framework. *Inorg. Chem.*, **48** (21), 9980–9982.

36 Gu, Z.Y., Jiang, D.Q., Wang, H.F., Cui, X.Y., and Yan, X.P. (2009) Adsorption and separation of xylene isomers and ethylbenzene on two Zn-terephthalate metal–organic frameworks. *J. Phys. Chem. C*, **114**, 331–336.

37 Nicolau, M.P.M., Barcia, P.S., Gallegos, J.M., Silva, J.A.C., Rodrigues, A.E., and Chen, B. (2009) Single- and multicomponent vapor-phase adsorption of xylene isomers and ethylbenzene in a microporous metal–organic framework. *J. Phys. Chem. C*, **113**, 13173–13179.

38 Loiseau, T., Serre, C., Huguenard, C., Fink, G., Taulelle, F., Henry, M., Bataille, T., and Férey, G. (2004) A rationale for the large breathing of the porous aluminum terephthalate (MIL-53) upon hydration. *Chem. Eur. J.*, **10**, 1373–1382.

39 Serre, C., Millange, F., Thouvenot, C., Nogues, M., Marsolier, G., Louer, D., and Férey, G. (2002) Very large breathing effect in the first nanoporous chromium(III)-based solids, MIL-53 or $Cr^{III}(OH)\{O_2C–C_6H_4–CO_2\}\{HO_2C–C_6H_4–CO_2H\}_x H_2O_y$. *J. Am. Chem. Soc*, **124**, 13519–13526.

40 Finsy, V., Verelst, H., Alaerts, L., De Vos, D., Jacobs, P.A., Baron, G.V., and Denayer, J.F.M. (2008) Pore-filling-dependent selectivity effects in the vapor-phase separation of xylene isomers on the metal–organic framework MIL-47. *J. Am. Chem. Soc.*, **130** (22), 7110–7118.

41 Finsy, V., Calero, S., Garcia-Perez, E., Merkling, P.J., Vedts, G., De Vos, D.E., Baron, G.V., and Denayer, J.F.M. (2009) Low-coverage adsorption properties of the metal–organic framework MIL-47 studied by pulse chromatography and Monte Carlo simulations. *Phys. Chem. Chem. Phys.*, **11** (18), 3515–3521.

42 Alaerts, L., Kirschhock, C.E.A., Maes, M., van der Veen, M.A., Finsy, V., Depla, A., Martens, J.A., Baron, G.V., Jacobs, P.A., Denayer, J.F.M., and De Vos, D.E. (2007) Selective adsorption and separation of xylene isomers and ethylbenzene with the microporous vanadium(IV) terephthalate MIL-47. *Angew. Chem. Int. Ed.*, **46**, 4293–4297.

43 Finsy, V., Kirschhock, C.E.A., Vedts, G., Maes, M., Alaerts, L., De Vos, D.E., Baron, G.V., and Denayer, J.F.M. (2009) Framework breathing in the vapour-phase adsorption and separation of xylene isomers with the metal–organic framework MIL-53. *Chem. Eur. J.*, **15** (31), 7724–7731.

44 Coudert, FX., Mellot-Draznieks, C., Fuchs, AH., and Boutin, A. (2009) Double structural transition in hybrid material MIL-53 upon hydrocarbon adsorption: the thermodynamics behind the scenes. *J. Am. Chem. Soc.*, **131**, 3442–3444.

45 Zhao, X., Xiao, B., Fletcher, A.J., Thomas, K.M., Bradshaw, D., and Rosseinsky, M.J. (2004) Hysteretic adsorption and desorption of hydrogen by nanoporous

metal–organic frameworks. *Science*, **306**, 1012–1015.

46 Tanaka, D., Nakagawa, K., Higuchi, M., Horike, S., Kubota, Y., Kobayashi, T.C., Takata, M., and Kitagawa, S. (2008) Kinetic gate-opening process in a flexible porous coordination polymer. *Angew. Chem. Int. Ed.*, **47**, 3914–3918.

47 Alaerts, L., Maes, M., Giebeler, L., Jacobs, P.A., Martens, J.A., Denayer, J.F.M., Kirschhock, C.E.A., and De Vos, D.E. (2008) Selective adsorption and separation of *ortho*-substituted alkylaromatics with the microporous aluminum terephthalate MIL-53. *J. Am. Chem. Soc.*, **130**, 14170–14178.

9
Metal–Organic Frameworks as Catalysts for Organic Reactions

Lik Hong Wee, Luc Alaerts, Johan A. Martens, and Dirk De Vos

9.1
Introduction

The design of metal–organic frameworks (MOFs) for catalytic applications has emerged as an important research field. In this chapter, the state of the art of MOF materials engineering specifically in relation to catalysis is presented. In addition to the structural design of MOF catalysts, critical issues with respect to catalysis in the liquid phase are the solubility and stability of MOF materials under reaction conditions, and their operation as truly heterogeneous catalysts rather than as reservoirs providing dissolved active species that are leached from the MOF. The chapter is divided into five parts according to the concept through which the MOF is functionalized catalytically: (1) MOFs with active metal nodes; (2) MOFs with catalytic functions introduced via grafting on linkers; (3) homochiral MOFs; (4) MOF-encapsulated catalytically active guest molecules or clusters; and (5) mesoporous MOFs. Advantages and limitations of the different types of MOF catalysts and current challenges are presented.

9.2
MOFs with Catalytically Active Metal Nodes in the Framework

Many metal ions found in MOFs are well known for their activity in homogeneous Lewis acid–base and oxidation catalysis. Their incorporation in a solid MOF lattice can be seen as a strategy to generate self-supporting catalysts. Metal sites as integral parts of the MOF structure present a high concentration of active sites provided that they can be prevented from leaching. Their incorporation into solid particles allows easy catalyst separation from the reaction products. Although the high selectivity obtained with a number of MOF catalysts is attractive and invites their further exploration, the activity of the open metal sites in the current generation of MOFs in many cases is low compared with the activity of ion-exchanged transition metal cations in zeolites [1, 2]. In MOFs, the metal ions play an important structural role as node points of the framework, which partly or in most cases completely impedes their availability for catalysis. This may also suggest an inverse relationship between the activity of the metal ions and the stability of the MOF material.

By accepting electron pairs from reactant molecules, open metal sites function as Lewis acids. Although MOFs with accessible metal ions are continuously being developed, for instance with lanthanide metal centers that easily enlarge their coordination sphere, there is clearly a parallel need for strategies to increase the degree of exposure of the metal sites, for instance by the deliberate introduction of structural defects. Strategies to tune the Lewis acidity of the metal centers could also be further developed, for instance by the use of functionalized linkers or even mixed linkers. The development of MOFs with mixed-metal nodes would also be interesting. A trend towards MOFs based on elements with low toxicity has already been discerned from the increasing number of MOF materials containing Al, Mg, Fe, Ti, or Zr.

9.2.1
Transition Metal Nodes

A typical example of a familiar metal–organic compound to be incorporated in MOFs is the Cu paddle-wheel cluster. The best known MOF constructed from this SBU is [$Cu_3(btc)_2$], or HKUST-1. It has an open structure with easy access to the open metal Cu sites inside the large pores [3]. The Lewis acid properties of these Cu sites were first exemplified using cyanosilylation reactions (Figure 9.1) [4] and were thoroughly investigated with organic isomerization reactions of α-pinene oxide, (+)-citronellal, and the ethylene ketal of α-bromopropiophenone. In these reactions, the selectivities to the desired products obtained with [$Cu_3(btc)_2$] as catalyst, 84, 66, and 75%,

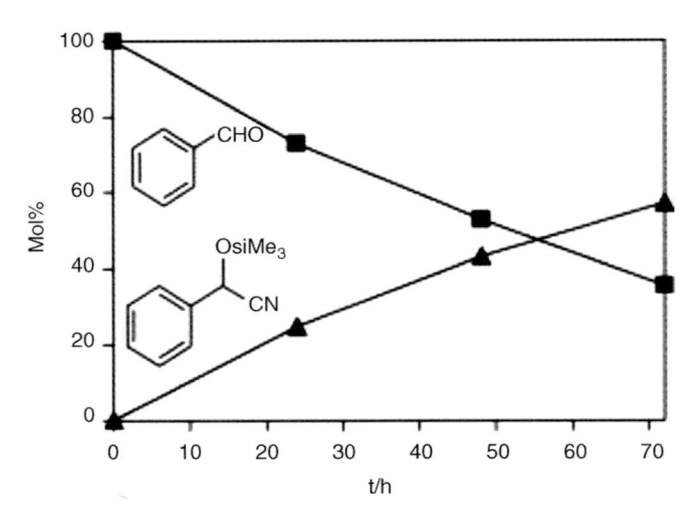

Figure 9.1 Educt and product concentration of the $Cu_3(btc)_2$-catalyzed cyanosilylation of benzaldehyde (solvent, pentane; temperature, 40 °C) [4].

respectively, were equal to or higher than those obtained with a series of Cu- and Zn-based reference catalysts, and this allows the identification of the Cu sites as fairly hard Lewis acids. These findings were confirmed by IR spectroscopy on adsorbed carbon monoxide [5]. The activity of [$Cu_3(btc)_2$] in these reactions is fairly low as a consequence of a strong shielding effect by the four oxygen atoms surrounding the active sites [6]. [$Cu_3(btc)_2$] was resistant to leaching in these reactions. Catalyst deactivation was induced by progressive formation of deposits inside the pores. As MOFs cannot withstand a calcination treatment, an efficient alternative washing protocol was developed to restore the activity of the Cu catalyst to close to its initial value [5].

The very first demonstration of the catalytic potential of MOFs was made in 1994 for a two-dimensional Cd bipyridine lattice that catalyzes the cyanosilylation of aldehydes [7]. A continuation of this work in 2004 for imines showed that the hydrophobic surroundings of the framework enhance the reaction in comparison with homogeneous Cd(pyridine) complexes [8]. The activity of MIL-101(Cr) is much higher than that of the Cd lattices, but in consecutive reaction runs the activity decreases. In contrast with [$Cu_3(btc)_2$], the catalytic sites are not prone to reduction by benzaldehyde [9]. The MOF $Mn_3[(Mn_4Cl)_3(btt)_8(CH_3OH)_{10}]_2$, with two different types of open Mn^{II} sites, catalyzes the cyanosilylation of aromatic aldehydes and ketones with a remarkable reactant shape selectivity. This MOF also catalyzes the more demanding Mukaiyama aldol reaction [10].

Acetalization of benzaldehyde with trimethyl orthoformate can be carried out with a series of MOFs constructed from In and bdc or btc ligands with open In sites. The catalysts are even stable in aqueous medium and can be reused without loss of activity. The reaction takes place only at the outer surface of the crystals [11]. In another MOF constructed from In and 4,4'-(hexafluoroisopropylidene)bis(benzoic acid), the same reaction takes place inside the pores [12].

When open metal sites in MOFs are used for oxidation catalysis, not only the coordination number but also the oxidation number of the metal may change. However, in some MOFs constructed from Cu paddle-wheel SBUs, instead of the framework metal atoms, some of the ligated species change their oxidation state. Upon oxidation of a flexible Cu *trans*-1,4-cyclohexanedicarboxylate MOF with H_2O_2, the Cu paddle-wheels are disconnected, rotated, and again linked to each other by peroxo bridges. This material is a heterogeneous and reusable oxidation catalyst for various aliphatic and aromatic alcohols with H_2O_2 with selectivities above 99% [13, 14]. A similar type of peroxo species might be formed on the Cu paddle-wheel sites of [$Cu_3(btc)_2$] in the oxidation of benzylic compounds such as xanthene with TBHP [15]. The intrinsic potential of Cu paddle-wheel clusters to catalyze oxidation reactions is also found in a MOF constructed from Cu^{II} and 5-methylisophthalate ligands used for CO oxidation. The activities are similar to or higher than those of CuO and CuO/Al_2O_3 reference catalysts, with full conversion of CO at 200 °C. The porous lattice has a stable activity and remains intact after catalytic reaction; no CuO is formed [16].

Other types of MOF Cu sites have also been explored for oxidation catalysis. [Cu(H_2btec)(bpy)] was used for the epoxidation of cyclohexene and styrene with TBHP with yields of 65 and 24%, respectively, after 24 h of reaction at 75 °C. TOFs as high as 79 h^{-1} for cyclohexene illustrate the efficiency of the catalyst. The mechanism begins with an expansion of the coordination number of the Cu^{II} center by coordination of TBHP. After a nucleophilic attack of the alkene substrate on this species, a concerted oxygen transfer takes place, leading to the departure of *tert*-butanol. Finally, the rupture of the Cu–O bond produces the epoxide and the regenerated catalyst [17]. [Cu(bpy)(H$_2$O)$_2$(BF$_4$)$_2$(bpy)] was used for the allylic oxidation of cyclohexene with molecular oxygen. Under close to ambient conditions and without solvent, the selectivity to cyclohexene-3-hydroperoxide reached 90%. The active Cu^{II} sites are located at the surface of the framework and both bpy and water are involved in the active complex. Removal of the structural water at 100 °C in vacuum opens the pores of this catalyst but eliminates the oxidation activity due to the direct coordination of all bpy ligands to Cu^{2+} ions [18].

In the oxidation of sulfides with H$_2$O$_2$ with a series of MOFs containing Sc, Y, or La, it is clear that the better ability of Y and La to accommodate an extra oxygen atom of the proposed peroxo derivative in the metal coordination environment is essential for their good performance. All catalysts are fully heterogeneous and can be reused without loss of selectivity. All catalysts are as selective as their homogeneous counterparts. However, almost stoichiometric amounts of the catalysts were needed [19–21]. Better results were obtained with a series of Sc and Y naphthalenedisulfonates [22].

In a MOF constructed from Pd and 2-hydroxypyrimidinolate, the framework might be able to accommodate the change in the coordinative demand of Pd from tetra- to dicoordination, for instance in the Suzuki–Miyaura coupling between phenylboronic acid and 4-bromoanisole. This MOF can also catalyze the aerobic oxidation of cinnamyl alcohol to cinnamylaldehyde and the hydrogenation of alkenes. Reactant shape and size selectivity are observed in the reaction of cyclododecene compared with the reaction of 1-octene [23]. This shows that the catalytic reactions are heterogeneous and are occurring within the MOF. At least for hydrogenation, catalysis could alternatively also take place at the surface of palladium nanoparticles inside the MOF or at defect sites [24].

9.2.2
Coordinatively Unsaturated Metal Nodes

The role of structural defects at surfaces or inside the pores of MOFs has only recently been elucidated for catalysis. For instance, although the Zn atoms in intact MOF-5 are inaccessible for ligation, catalytic activity has been reported for this material, for instance for esterification and transesterification reactions or for *para*-alkylation of large polyaromatic compounds [25–28]. In this last example [28], the preference for *para*-alkylation products over *ortho*-products suggests an intraporous reaction. Most probably Zn–OH defects are created inside the pores as a consequence of adsorption of moisture [29]. In many MOF structures obviously lacking open metal sites, significant catalytic activity was observed. For instance, a series of Zn-MOFs having different supramolecular network structures (diamondoid, lonsdaleite, and PtS) was

Table 9.1 Alkylation of toluene and biphenyl catalyzed by Zn-MOF with different supramolecular network structures after 2 h at 170 °C.

Catalyst	Toluene				Biphenyl			
	p-	o-	Di-	Conversion (%)	p-	o-	Di-	Conversion (%)
H-BEA	72	28	0	60	55	22	23	60
AlCl$_3$	46	54	0	60	51	38	11	60
Diamondoid	99.3	0.7	0	99.6	100	0	0	66.1
Lonsdaleite	99.4	0.6	0	88	100	0	0	30
PtS	100	0	0	18.1	100	0	0	5.3
None	0.2	0.2		0.5				0.2

Data taken from [30].

tested for alkylation and benzylation reactions [30]. The Zn-MOF with a hexagonal diamondoid structure showed high selectivity (99.3%) towards the *para*-oriented product at 99.6% toluene conversion (Table 9.1). Other examples are the unexplained activity of commercially available [Al$_2$(bdc)$_3$] for the reduction of carbon–carbon multiple bonds with hydrazine [31] and the *N*-methylation of aromatic amines [32].

The deliberate introduction of such defects is a challenging task, owing to the low chemical stability of MOF lattices and, even more, the difficult characterization of the defects, not only concerning their exact chemical nature but also their distribution inside the porous matrix. Pioneering work in this field has been carried out on MOF-5 for the monoalkylation of biphenyl with *tert*-BuCl. Two approaches were followed: a fast precipitation and the partial substitution of bdc linkers by monofunctionalized carboxylic acid linkers such as 2-methyltoluic acid. Unfortunately, the complexity of the catalysts impeded an assessment of the exact contribution of these introduced modifications to the catalytic activity [33].

9.3
Catalytic Functionalization of Organic Framework Linkers

9.3.1
Porphyrin Functional Groups

Attempts have been made to circumvent the limited catalytic activity of structural metal ions by functionalizing the organic ligands constituting the lattice. While many possibilities could be envisaged using various organic reactions, this section is devoted to the use of metal-bearing porphyrin-based ligands, well known as homogeneous catalysts and as prosthetic group of metalloenzymes. Regarding their incorporation in MOFs, some structural issues need to be considered, for instance, the susceptibility to framework collapse or catenation as very large cages are inherently formed with these ligands. Another issue is a possible interference of the porphyrin complexation site with the MOF formation, via competitive coordination of ligands to the metals already present there, but also by taking away metal ions that were intended for the MOF lattice if the porphyrins are added in their non-metalated form [24]. In practice, the majority of the reported materials are synthesized with the same metal at the porphyrinic sites and at the structural nodes. The main advantage of heterogeneous porphyrinic catalysts is the prevention of oxidative self-degradation of these structures by their physical separation.

In the MOF PIZA-3, Mn^{III} is found both in the porphyrin ligands and as a structural metal node. The framework is structurally stable and is used for the oxidation of cyclic alkanes and alkenes with iodosylbenzene or peracetic acid as the oxidant. Reaction takes place at the outer surface, for which the authors blamed the unfavorable hydrophilic pore interior. Yields were similar to those obtained with Mn^{III} porphyrins in homogeneous systems or immobilized inside inorganic supports as heterogeneous catalysts. Less than 0.1 µM of metalloporphyrin or degradation products were observed in the reaction mixtures. With peracetic acid, no loss of oxidation activity was observed in a second run [34].

The Zn-loaded porphyrin ligands in ZnPO-MOF are connected to a Zn paddle-wheel lattice with H_4tcpb as a second ligand providing structural rigidity. The dense network of the paddle-wheel–tcpb clusters, moreover, prevents catenation. ZnPO-MOF was used for acyl transfer reactions, for instance, between N-acetylimidazole and 3-pyridylcarbinol. The catalytic results revealed a 2400-fold rate enhancement compared with the control experiment. This was interpreted by proposing that the Lewis acid action of the porphyrin Zn sites is strongly enhanced by reagent preconcentration inside the MOF cages [35].

A MOF with Rh^{II} paddle-wheel clusters linked to porphyrinic ligands loaded with Cu^{II}, Ni^{II}, or Pd^{II} was synthesized by ligand exchange of the preloaded porphyrin ligands with $Rh_2(CH_3COO)_4$ complexes. Here an interesting synergetic behavior was observed in the hydrogenation of alkenes: the hydride species at the Rh center is transferred to the alkene coordinated to a metal ion in the center of the porphyrin ring to form an alkyl species, and next this alkyl species reacts with a hydride species activated at the Rh center to form the alkane [36].

9.3.2
Amine and Amide Functions Incorporated via Grafting

Grafting of amide functional groups to the unsaturated metal sites of MOFs has attracted considerable attention because amides can act either as hydrogen bond donors or acceptors. This also implies that bonds among the amides themselves could inhibit the interaction with guest molecules. Kitagawa and co-workers [37] successfully demonstrated the use of a tridentate ligand (4-btapa) for the construction of a MOF, $\{[Cd(4\text{-btapa})_2(NO_3)_2]\cdot 6H_2O\cdot 2dmf\}_n$, possessing free amide groups as guest-accessible functional organic sites. The catalytic activity and size selectivity of the material were evaluated in the Knoevenagel condensation of benzaldehyde in benzene with three different reactants (malononitrile, ethyl cyanoacetate, and tert-butyl cyanoacetate) having different molecular sizes. When malononitrile was used as a substrate, a high conversion of 98% was obtained after 12 h, whereas almost no activity was observed in the conversion of ethyl and *tert*-butyl cyanoacetate, suggesting that the catalyst was size selective. The observation of reactant shape selectivity revealed that the catalytic turnovers take place in the MOF channels rather than at the outer surface of the MOF particles. Upon desolvation, for example, by removal of the H_2O or dmf molecules from the structure, the crystalline material loses its porosity; it then shows low activity in the Knoevenagel condensation (<20% conversion), mainly due to the absence of accessible channels. The catalyst was truly heterogeneous and well recyclable. In solventless conditions, 100% conversion of malononitrile was achieved after a 2.5 h reaction time. Interestingly, this MOF catalyst changes to an amorphous state upon heating under reduced pressure but regains its crystallinity when immersed in methanol or exposed to methanol vapor.

Functionalization of dehydrated MIL-101 through grafting of amine molecules (ed, deta or aps) was reported by Hwang *et al.* [38]. The catalytic performance of the catalyst was evaluated using the Knoevenagel condensation of benzaldehyde and ethyl cyanoacetate in cyclohexane solvent at 80 °C. A high conversion was obtained with ed-, deta- and aps-MIL-101 with high selectivity for *trans*-ethyl cyanocinnamate (99.3%) (Table 9.2). The unmodified hydrated MIL-101 exhibited much lower catalytic activity.

Gascon *et al.* [39] reported that IRMOF-3 and amino-functionalized MIL-53 were effective as base catalysts in the Knoevenagel condensation of benzaldehyde and ethyl cyanoacetate in DMF at 60 °C. The IRMOF-3 catalyst outperformed the aniline reference homogeneous catalyst, with a conversion of 90% compared with 55%. Only a minor activity loss of IRMOF-3, ascribed to incomplete catalyst recovery, was observed in a second catalytic test. After removal of the IRMOF-3 catalyst from the liquid reaction product mixture, no further conversion was observed, demonstrating the stability of the heterogeneous catalyst. IRMOF-3 was synthesized from two different solvents (DEF and DMF) and denoted IRMOF-3$_{DEF}$ and IRMOF-3$_{DMF}$, respectively, with the former being somewhat more active. The observed difference in catalytic activity was ascribed to a larger specific surface area for IRMOF-3$_{DEF}$ (3683 $m^2 g^{-1}$) compared with IRMOF-3$_{DMF}$ (3130 $m^2 g^{-1}$), resulting in a better accessibility of the amino groups. Amino-MIL-53 catalyst was found to be less active in this reaction, probably because of too strong adsorption and slow diffusion of the

Table 9.2 Catalytic properties of amine-grafted MIL-101 and of the amine-grafted ordered mesoporous silica SBA-15 in the Knoevenagel condensation of benzaldehyde and ethyl cyanoacetate.

Catalyst	Conversion (%)	Selectivity (%)	TOF (h^{-1})
MIL-101	31.5	91.4	—
ED-MIL-101	97.7	99.1	328
DETA-MIL-101	97.7	99.3	190
APS-MIL-101	96.3	99.3	160
SBA-15	2.6	93.0	—
APS-SBA-15	74.8	93.5	32

Data taken from [38].

reaction products inside the pores of this particular MOF. Farrusseng and co-workers [40] reported the preparation of catalytically active IRMOF-3 and ZnF (Am$_2$taz) via a two-step post-modification method. First, the MOF materials were functionalized with pyridine groups to enhance the hydrophobicity, followed by grafting of nicotinoyl group. Both functionalized IRMOF-3 and ZnF(Am$_2$taz) outperformed MCM-41-NH$_2$ catalyst in solvent-free aza-Michael condensation and transesterification reactions.

9.4
Homochiral MOFs

The wealth of molecular tools for designing catalytically active MOFs has inspired researchers to attempt the development of chiral catalysts. Asymmetric catalysis necessitates chirality in the immediate environment of the active sites. Generally, MOFs with active metal sites connected via chiral linkers show modest enantiomeric excess (*ee*) values because of the too remote position of the chiral organic groups with respect to the active site. The alternative approach, involving incorporation of chiral metalloligands known from homogeneous catalysis in a MOF, is more effective. In such catalysts, the catalytically active metal centers have no structural role in the MOF lattice but are coordinated in a chiral reaction environment inside the pore. The high cost of such sophisticated heterogeneous chiral catalysts renders a large-scale application unlikely. Critical issues discussed in Section 9.2 are also relevant for chiral MOF design, such as the compromise between flexibility of the lattice and stability. A mixed chiral and nonchiral ligand approach may be such compromise.

9.4.1
MOFs with Intrinsic Chirality

This section focuses on catalytic reactions taking place at metal centers inside chiral MOF cages. In a MOF constructed from Zn, L-lactic acid, and terephthalic acid, sulfoxides are adsorbed with modest *ee* values between 20 and 27%, but no asymmetric induction was found for the oxidation of sulfides to sulfoxides [41]. In a series of homochiral lanthanide MOFs tested for cyanosilylation of aldehydes and ring opening of *meso*-carboxylic anhydrides, negligible *ee* values were observed, again illustrating the challenge of designing highly enantioselective MOF catalysts via remote influence by chiral environments of the open channels [42].

In a MOF consisting of 5,5'-dicarboxy-substituted binol ligands connected via Cu paddle-wheels, modest *ee* values (maximum 50%) are obtained for the asymmetric, Cu-catalyzed ring opening of epoxides with amines. This weak chiral induction could again be explained by the distance between the catalytic site and the asymmetric moiety of the lattice. Free (*S*)-binol ligands were not active at all for this reaction. The catalyst could be recovered by filtration and recycled without affecting reactivity and enantioselectivity. However, the three-dimensional structure is reversibly destroyed upon desolvation of the MOF [43].

9.4.2
Chiral Organic Catalytic Functions

In the MOFs presented in this section, catalytic activity is induced via pendant organic groups inside the cages. In the cages of the MOF D-POST-1, based on Zn and the ligand shown in Figure 9.2a, catalytically active pyridyl groups are present together with chiral tartrate-derived ligands. Half of the pyridine groups are not involved in the framework building, and are freely exposed in the channels. The reaction takes place inside the pores, as revealed by the occurrence of reactant shape selectivity, but the relatively large distance between the two moieties results in a low enantioselective induction (ee 8%) in the transesterification between 2,4-dinitrophenyl acetate and 1-phenyl-2-propanol [44]. POST-1 with N-methylated pyridine rings showed excellent catalytic properties in the transesterification reaction between 2,4-dinitrophenyl acetate and ethanol (Figure 9.2b). Similar observations were made with [Cu(L-asp)$_2$(bipy)]. In this MOF, Brønsted acid sites are first introduced via treatment with HCl, resulting in the protonation of the carboxylate groups of the linker. In the methanolysis of cis-2,3-epoxybutane, moderate yields (32–65%) but low ee values (up to 17%) were obtained. Again, a reactant shape selectivity study showed that the reaction does take place inside the pores [45].

An interesting approach was followed for the post-synthetic modification of MIL-101(Cr). By grafting the open metal sites of this highly porous MOF with an L-proline-bearing ligand, asymmetric aldol condensations can be carried out with yields

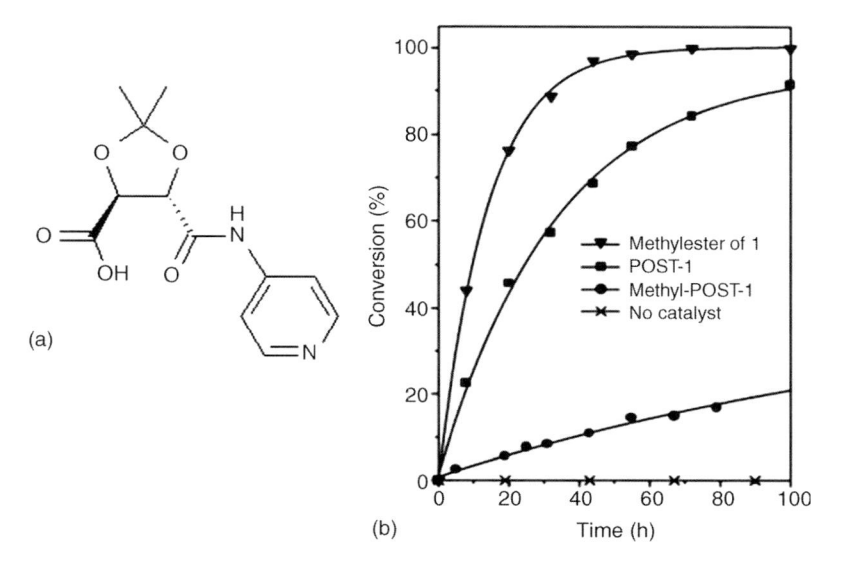

Figure 9.2 (a) Ligand employed in POST-1 synthesis and (b) transesterification of 2,4-dinitrophenyl acetate with ethanol in carbon tetrachloride at 27 °C in the presence of POST-1. The reaction was carried out in the absence of catalyst, and in the presence of 0.1 equiv. (based on the number of pyridyl units exposed to channels) of POST-1, methylated POST-1, and the methyl ester of the ligand in (a). Data taken from [44].

between 60 and 90% and *ee* values between 55 and 80% for the *R*-isomers. The relatively high chiral induction could be due to the restricted molecular movement inside the pores, combined with the very large number of chiral groups present as many open Cr sites are available in the large cages for docking of the proline ligands. The ee values obtained are even higher than those of the corresponding homogeneous catalysts. Size selectivity studies with bulky aldehydes prove that the reactions take place inside the pores [46].

9.4.3
Metalloligands

The incorporation of chiral metalloligands in MOFs is a way to combine an active site and a chiral center. The chiral functionalities can be introduced into the ligands before or after the formation of the framework. When introduced before, these functionalities should of course be compatible with the other framework constituents. To achieve an effective loading of the chiral complexing sites into a MOF, a low affinity for the structural metal ions is a strict requirement. As metalloligands are large and flexible molecules, extra precautions need to be taken to ensure structural stability. In a Zn-based MOF with Mn(salen) ligands, the metalloligands are anchored to sheets composed of Zn paddle-wheels connected by 4,4′-biphenyldicarboxylate ligands. These sheets have a structural role only and are not involved in catalysis. This MOF was applied in the asymmetric epoxidation of 2,2-dimethyl-2*H*-chromene. The MOF was less active than the homogeneous salen complex, but more stable over time, as the immobilization of the Mn(salen) complexes precludes oxidation of one metal center by another. The immobilization decreases the flexibility of the ligands, explaining the slightly lower enantioselectivity compared with that of the free complex (*ee* values of 82 and 88%, respectively). A small loss (4–7%) of catalytically inactive Mn was observed after each reaction run, whereas the enantioselectivity remained constant. Upon prolonged use of this catalyst, the outer layers of the MOF crystals were gradually damaged by oxidation. This degradation gave rise to an enhanced substrate size selectivity as the catalytic reaction proceeded [47].

A similar system was encountered in a MOF constructed from Cd^{II} and binol-type ligands. In this system, one-third of the dihydroxybinaphthyl groups are available for post-synthetic coordination. After reaction of $Ti(O^iPr)_4$ with the OH groups of the ligands, a heterogeneous Ti–binol catalyst is obtained that catalyzes the addition of diethylzinc to aromatic aldehydes to yield chiral secondary alcohols with *ee* values exceeding 80% for most substrates. Bulky aldehyde molecules were less reactive on this catalyst, suggesting the occurrence of reagent shape selectivity and the presence of catalytic activity mainly inside the pores of the MOF. The catalytic inactivity of a mixture of $Cd(pyridine)_2(H_2O)_2$ and $Ti(O^iPr)_4$ in solution illustrated the heterogeneous nature of the MOF catalyst [48, 49]. Recently, three-dimensional chiral MOFs based on tetracarboxylate bridging ligands and dizinc secondary building units were synthesized by the same group [50]. Post-treatment of these materials with $Ti(O^iPr)_4$ resulted in the formation of intermolecular [Ti(binolate)$_2$] complexes. Interestingly, chiral MOFs constructed from the ligand (*R*)-2,2′-dihydroxy-1,1′-binaphthyl-4,4′,6,6′-

tetrabenzoate possess inter single-crystal cross-linking. However, low *ee* values of the aromatic aldehyde product (29.8%) were obtained due to steric influence of the naphthyl rings at the Ti center.

9.5
MOF-Encapsulated Catalytically Active Guests

9.5.1
Polyoxometalates (POMs)

POMs with strong Brønsted acidity are proven homogeneous catalysts for many acid-catalyzed organic reactions, such as alkene hydration, esterification, and alkylation. Encapsulation of POMs into MOFs as a host lattice has been attempted as a means of heterogenization of these homogeneous catalysts. Incorporation of POMs in the cavities of $Cu_3(btc)_2$ [51–53] and MIL-101 [54] has been reported. $Cu_3(btc)_2$ encapsulating $[H_3PW_{12}O_{40}]^{5-}$ polyanions [HPW/$Cu_3(btc)_2$] showed high catalytic activity in the hydrolysis of esters in aqueous media depending on the size of the ester substrate molecule with regard to the window diameter of the cages of the MOF material [55]. The conversion of methyl acetate with dimensions 4.87×3.08 Å was about 64% after 5 h at 80 °C and reached a maximum of >95% after 7 h. In contrast, for a large ester such as 4-methylphenyl propionate with dimensions 10.61×4.04 Å, almost no conversion was observed under similar reaction conditions. The authors claimed that small ester substrates with high hydrophilicity can diffuse swiftly through the pores and reach active sites, whereas the diffusion of large esters with hydrophobic groups is impeded. The POMs did not leach out of the MOF. In consecutive runs with a recycled catalyst, the catalytic activity was maintained. Recently, catalytically active nanocrystals of HPW/$Cu_3(btc)_2$ were prepared by Martens and co-workers [56] via rapid room temperature synthesis in combination with a freeze-drying approach. The as-synthesized nanocatalysts outperformed ultrastable Y zeolite and micron-sized HPW/$Cu_3(btc)_2$ in esterification of acetic acid with 1-propanol (60 °C, 7 h).

MIL-101 encapsulating $[H_3PW_{12}O_{40}]^{5-}$ was evaluated for Knoevenagel condensation of benzaldehyde with ethyl cyanoacetate, esterification of acetic acid with *n*-butanol, and dehydration of methanol to dimethyl ether [57]. MIL-101 with 10 and 20 wt% impregnated $[H_3PW_{12}O_{40}]^{5-}$ showed >99% conversion with 100% selectivity for Knoevenagel condensation of benzaldehyde with ethyl cyanoacetate after 2 h at 40 °C in DMF or toluene. For the esterification of acetic acid with *n*-butanol, the highest conversion (70%) was achieved for MIL-101 with 20 wt% $[H_3PW_{12}O_{40}]^{5-}$ after 5 h at 110 °C. No acid leaching or deactivation of the MOF catalysts was observed for consecutive runs.

Encapsulation of the Ti- and Co-monosubstituted Keggin heteropolyanions $[PW_{11}CoO_{39}]^{5-}$ and $[PW_{11}TiO_{40}]^{5-}$ in the matrix of MIL-101 was reported by Maksimchuk *et al.* [58]. The as-synthesized catalysts were evaluated in (1) α-pinene oxidation with molecular oxygen at 50 °C or with H_2O_2 at 30 °C, (2) caryophyllene

Table 9.3 Results of alkene oxidation with H_2O_2 and O_2 catalyzed by MIL-101-encapsulated heteropolyacids.

Catalyst	Oxidant	Conversion (%)	Selectivity (%)	
		(α-pinene)	(verbenol)	(verbenone)
None	O_2	15	33	27
MIL-101		14	14	7
$(PW_{11}CoO_{39})^{5-}$		45	36	24
MIL-101/$(PW_{11}CoO_{39})^{5-}$		45	29	27
None	H_2O_2	16	7	7
MIL-101		18	8	7
$(PW_{11}TiO_{40})^{5-}$		42	25	18
MIL-101/$(PW_{11}TiO_{40})^{5-}$		40	32	30
		(caryophyllene)	(caryophyllene oxide)	
None	H_2O_2	13	13	
MIL-101		40	41	
$(PW_{11}TiO_{40})^{5-}$		60	52	
MIL-101/$(PW_{11}TiO_{40})^{5-}$		71	80	
		(cyclohexene)	(cyclohexenol + cyclohexenone)	(cyclohexanediol)
MIL-101	H_2O_2	8	60	10
$(PW_{11}TiO_{40})^{5-}$		7	7	32
MIL-101/$(PW_{11}TiO_{40})^{5-}$		39	32	46

Data taken from [58].

oxidation with H_2O_2 at 50 °C, and (3) cyclohexene oxidation with H_2O_2 at 70 °C. Both Ti- and Co-POM/MIL-101 catalysts showed fairly good activity and selectivity towards α-pinene oxidation with 40 and 45% conversion, respectively, with an average selectivity of 30% for both verbenol and verbenone as compared with unmodified MIL-101 (14–18% conversion) (Table 9.3). Higher conversion (88%) with 100% selectivity was achieved for caryophyllene epoxidation catalyzed by Ti-POM/MIL-101. In α-pinene oxidation with O_2, the Co-POM/MIL-101 catalyst maintained its activity and selectivity in up to five catalyst reuse cycles [58]. No structural changes were observed in the XRD pattern. However, a 20% decrease in the specific surface area and pore volume were noted.

Recently, the immobilization of polyoxotungstates $[PW_4O_{24}]^{3-}$ (PW_4) and $[PW_{12}O_{40}]^{3-}$ (PW_{12}) in the nanocages of MIL-101 was reported by the same group [59]. The catalytic activity of the resulting functionalized MOF materials was evaluated in cyclohexene epoxidation. MIL-101 with a 5 wt% $[PW_{12}O_{40}]^{3-}$ loading gave a maximum cyclohexene conversion of 84% with 80% selectivity to the epoxide. The TOF value of the heterogeneous PW_{12} catalyst approached the value for the homogeneous PW_{12} catalyst (0.40 min^{-1}). The MOF catalyst was stable in up to five reuse cycles and retained its activity and selectivity.

A POM-containing MOF having lanthanide-based nanocages as an SBU was synthesized with the general formula $\{[Ho_4(dpdo)_8(H_2O)_{16}BW_{12}O_{40}](H_2O)_2\}^{7+}$. This material showed catalytic activity for phosphodiester bond cleavage in bis(4-nitrophenyl)phosphate at 50 °C. The authors claimed that the hydrolytic cleavage of the phosphodiester was catalyzed by the lanthanide MOF rather than by the POM based on the observation that the free $[BW_{12}O_{40}]^{5-}$ anions did not exhibit catalytic activity [60].

9.5.2
Metalloporphyrins

The encapsulation of metalloporphyrins in the cavities of an indium imidazoledicarboxylate-based *rho*-zeolite-like metal–organic framework (*rho*-ZMOF) has been demonstrated [61]. The catalytic activity of this material was assessed by cyclohexane oxidation with TBHP as the oxidant. The cyclohexane conversion reached 91.5% after 24 h at 65 °C (Figure 9.3). Cyclohexanol and cyclohexanone were the only observed products, suggesting that the investigated oxidation reaction is selective towards the desired products. Furthermore, upon reuse of the catalyst no loss of crystallinity, reactivity, and selectivity in up to 11 cycles was observed. There was no leaching of the

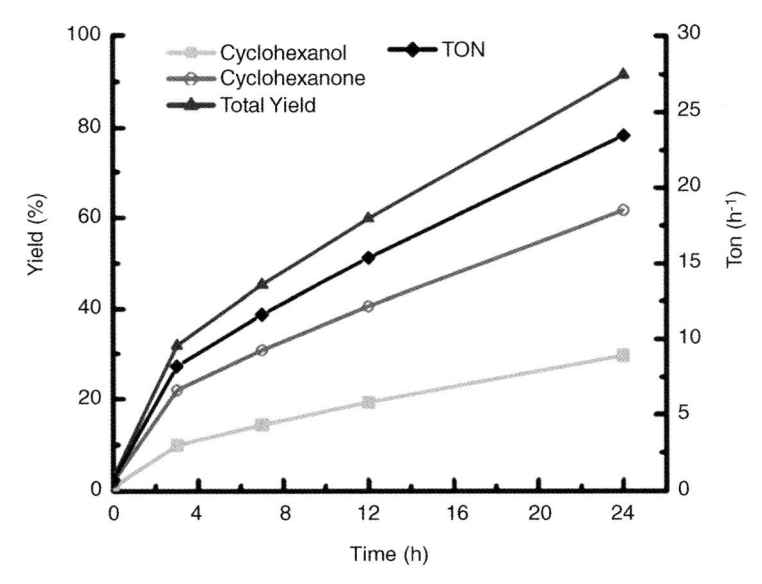

Figure 9.3 Cyclohexane catalytic oxidation catalyzed by Mn-metalated porphyrin encapsulated in *rho*-ZMOF [61].

encapsulated metalloporphyrin into the product solution, as evidenced by the UV–Vis spectra.

9.5.3
Metal Nanoparticles

Nanoparticle deposition into MOFs has also emerged as an interesting but challenging approach to heterogeneous catalysis. For example, Kaskel and co-workers [62] described incipient wetness infiltration of Pd metal precursors into MOF-5 followed by reduction to produce occluded nanoparticles. The resulting Pd/MOF-5 material was tested for hydrogenation of styrene, 1-octene, and *cis*-cyclooctene for 24 h at 35 °C. After 12 h, >99.7% of ethylbenzene was obtained from styrene hydrogenation (Figure 9.4), whereas for the hydrogenation of *cis*-cyclooctene, <13 wt% of cyclooctane was obtained after 12 h. The authors proposed that the low conversion of *cis*-cyclooctene catalyzed by Pd/MOF-5 could probably be due to pore accessibility or different activation energies for the hydrogenation of styrene as compared with that of *cis*-cyclooctene, or to the larger kinetic diameter of *cis*-cyclooctene as compared with that of styrene, which may lead to diffusion limitations. In addition, deactivation of the Pd/MOF-5 catalyst was observed for styrene hydrogenation in two consecutive runs. It should be noted that the Pd/MOF-5 catalyst is not stable upon contact with water [62]. Later, MIL-101 was used as a support for Pd; advantages of MIL-101 are a very high specific pore volume and long-term stability in an air atmosphere [9]. In contrast to the instability of the Pd/MOF-5, Pd/MIL-101 catalyst is stable in an air atmosphere, as proven by repeated nitrogen physisorption measurements on Pd/MIL-101 stored in air for 2 months.

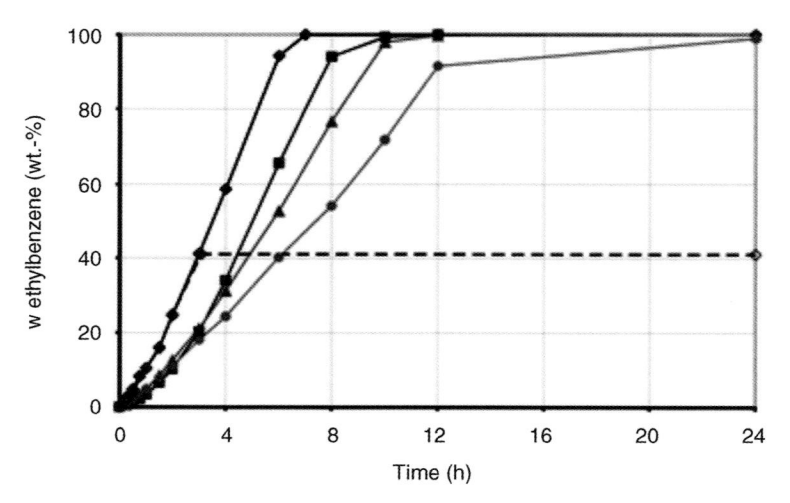

Figure 9.4 Hydrogenation of styrene with different supported palladium catalysts (1 wt%): ◆, Pd/MIL-101; ■, Pd/MOF-5; ▲, Pd/Norit A; ●, Pd/C; the broken line represents the filtration test [9].

In addition, the stability of Pd/MIL-101 was tested further under more severe reaction conditions. The catalyst was used in the gas-phase hydrogenation of an acetylene–ethylene mixture in a continuous fixed-bed reactor operated in repeated temperature cycles between 40 and 100 °C. The results obtained showed superior activity: 80–100% of acetylene was converted to ethane along with smaller amounts of ethylene as compared with Pd/ZnO or Pd/Al$_2$O$_3$ having the same total amount of Pd (0.05 wt%). A very slow deactivation was observed after a prolonged 145 h time-on-stream, indicating the stability of the catalyst [9].

IRMOF-3 [Zn$_4$O(sita-AuCl$_2$)$_x$(ata)$_{3-x}$] encapsulating Au nanoparticles was synthesized by Corma and co-workers [63]. The catalyst was tested on domino coupling and cyclization of N-tosyl-protected ethynylaniline, piperidine, and paraformaldehyde at 40 °C for 14 h with 1,4-dioxane as solvent. The results showed that 93% ethynylaniline conversion was achieved for Au/IRMOF-3 as compared with AuCl$_3$, Au/ZrO$_2$, and Au^{III} Schiff base complex, which gave conversions of 60, 36, and 38%, respectively. The catalytic activity of Au/IRMOF-3 was also studied for hydrogenation of 1,3-butadiene either in H$_2$ at 250 °C. A higher 1,3-butadiene conversion (about 96%) was obtained with Au/MOF-3 than with the reference catalyst (Au/TiO$_2$), which gave a low conversion of about 9% (Figure 9.5). The TOF (540 h^{-1}) for Au/IRMOF-3 calculated on the basis of total weight is one order of magnitude higher than that of the Au/TiO$_2$ catalyst (50.4 h^{-1}). Also, Au/IRMOF-5 shows very high selectivity (97%) for hydrogenation of 1,3-butadiene, giving 1-butene and (E)-2-butene as the main products, and only 3% of butane.

MOF-177 [Zn$_4$O(1,3,5-benzenetribenzoate)$_2$] with embedded Pt nanoparticles in a size regime of 2–5 nm in the host lattice was prepared via hydrogen reduction

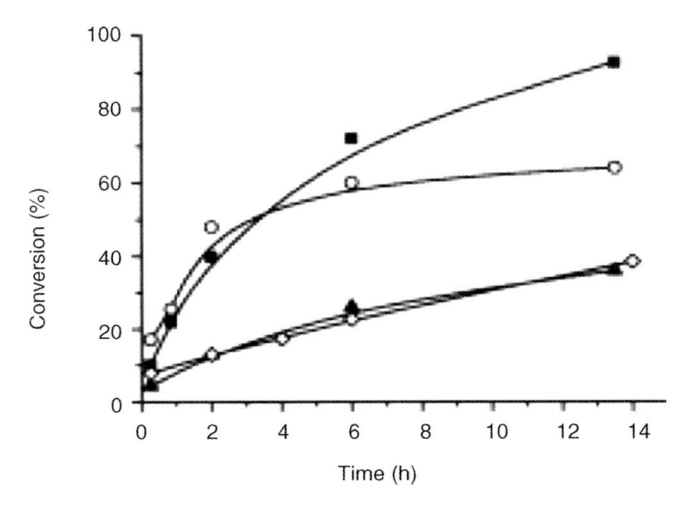

Figure 9.5 Comparison of catalytic hydrogenation over IRMOF-3-SI-Au (■, gold: 0.0008 mmol), Au/ZrO$_2$ (▲, gold: 0.0014 mmol), homogeneous Au^{III} Schiff base complex (◇, gold: 0.0008 mmol), and AuCl$_3$ (○, gold: 0.025 mmol) for domino coupling and cyclization of ethynylaniline (0.10 mmol), piperidine (0.12 mmol), and paraformaldehyde (0.20 mmol) with 1,4-dioxane (1.0 ml) as solvent at 40 °C [63].

(100 bar, 100 °C) for 24 h using [Me$_3$PtCp] as a platinum precursor by taking into account that MOF-177 has the highest hydrogen adsorption capacity [64]. The as-synthesized Pt@MOF-177 was tested for room temperature oxidation of allylic and aliphatic alcohols. The results obtained showed that allylic alcohols were readily oxidized (>99% conversion) with high selectivity; however, the host lattice of MOF-177 collapsed after the first recycling, as evidenced by XRD characterization. The reason is that water was produced as a by-product in the oxidation reaction.

MOF-5 samples loaded with Cu and ZnO nanoparticles were prepared by Muhler et al. [65]. Cu/MOF-5 (13.8 wt% Cu loading) was tested for methanol synthesis. A stable and reproducible rate of 70 µmol$_{MeOH}$ g$_{cat}^{-1}$ h^{-1} was obtained at 1 atm pressure for freshly prepared Cu/MOF-5. The catalytic properties of Cu/ZnO-MOF-5 (1.4 wt% of Cu and 9.9 wt% of ZnO) were also evaluated for the same reaction under similar conditions. An extremely high activity of 212 µmol$_{MeOH}$ g$_{cat}^{-1}$ h^{-1} was measured after 1 h on-stream, but this gradually decreased to 12 µmol$_{MeOH}$ g$_{cat}^{-1}$ h^{-1} after 12 h. XRD characterization of the tested catalyst showed sintering of the ZnO phase and also collapse of the MOF-5 structure. According to the TEM images, the initial particle size of the Cu, with a size distribution between 1 and 3 nm increased to 15–20 nm after catalysis.

Ru nanoparticles embedded inside the pore structure of MOF-5 (Ru/MOF-5) were prepared via gas-phase loading with the volatile [Ru(cod)(cot)], followed by subsequent hydrogenolysis [66]. Ru/MOF-5 was tested for oxidation of benzyl alcohol to benzaldehyde at 80 °C. Prior to catalysis, the Ru/MOF-5 catalyst was converted to RuO$_x$/MOF-5 via oxidation with diluted O$_2$ gas. However, relatively low conversion (25%) was obtained after 48 h. XRD characterization revealed collapse of the MOF-5 structure, which is as expected since the oxidation reaction produces water, and MOF-5 is a highly water-sensitive material.

MIL-101 encapsulating nanoparticles (Pd, Pt, and Au) in its mesoporous cages was prepared by Férey and co-workers [38, 67] via a series of steps (functionalization, neutralization, and reduction). The catalyst (1 wt% Pd) was then tested on the coupling reaction of acrylic acid with iodobenzene at 120 °C with added N,N-dimethylacetamide as a solvent. Superior activity with 100% conversion was achieved in less than 1 h, which is comparable to results obtained with the commercially available Pd/C catalyst (1.09 wt% Pd) (Figure 9.6) [38].

9.6
Mesoporous MOFs

The design of MOFs with mesoporous cavities has lately attracted a lot of attention for overcoming the potential diffusion and mass transfer limitations in heterogeneous catalysis. The microporous MOF materials MIL-100(Fe) [68] and MIL-101(Cr) [55] with mesoporous cages were first synthesized by Férey's group. The catalytic activity of MIL-100(Fe) was evaluated for Friedel–Crafts alkylation between benzene and benzyl chloride to form diphenylmethane at 70 °C [68]. The results obtained showed

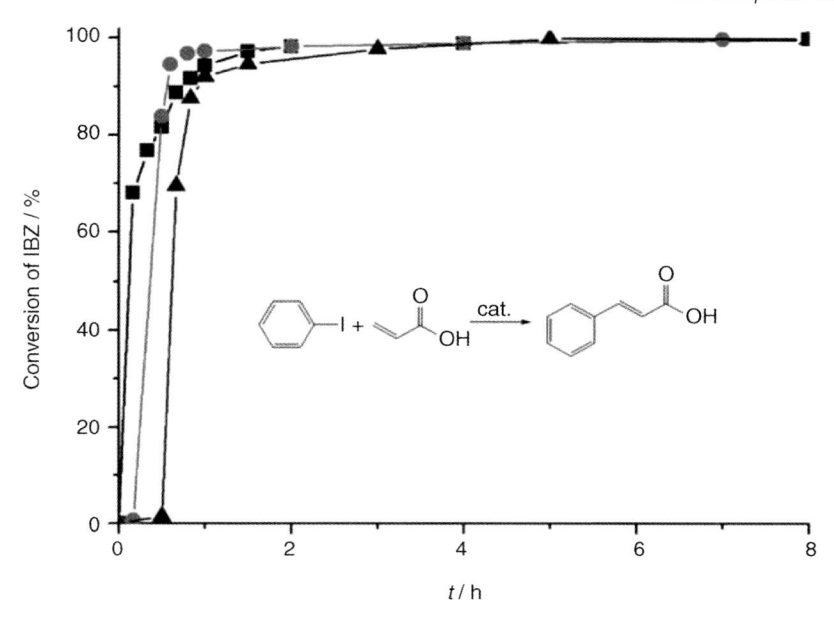

Figure 9.6 Catalytic activities of various palladium-loaded catalysts as a function of reaction time in the Heck coupling reaction. ■, Pd/C (1.09 wt% Pd); ●, Pd/APS-MIL-101 (0.93 wt% Pd); ▲, Pd/EDMIL-101 (0.95 wt% Pd). Reaction was carried out with 1.0 mmol of iodobenzene, 1.5 mmol of acrylic acid, 1.5 mmol of triethylamine, and 50 mg of catalyst in 25 ml of N,N-dimethylacetamide as solvent at 120 °C [38].

100% benzyl chloride conversion with ~100% diphenylmethane selectivity after 5 min. In comparison, MIL-100(Cr), HBEA, and HY catalysts were less reactive under the same reaction conditions, giving 42, 43 (after 5 h), and 54% (after 30 h) conversion, respectively. The results clearly demonstrated that iron is more suitable than chromium as a catalytically active site for this typical reaction, which could be due to the redox properties of trivalent iron species ($Fe^{3+} + e^- \rightleftharpoons Fe^{2+}$). The short induction period (5 min) observed in the benzylation is ascribed to the inhibition effect by the water present in the catalyst or the diffusion limitation of the molecules to and from the active site in the pores. However, it been has been reported recently that for MIL-53(Ga), the isolated Lewis acid sites and the appropriate large micropores allowing fast diffusion of the reactants are the key factors explaining its higher activity in comparison with microporous zeolites (H-BEA and H-MOR) for akylation of biphenyl. Although a mild acidity (Ga-OH) was probed by *in situ* FTIR studies based on CO adsorption, the tilted OH evident from DFT calculations induces much stronger stabilization of the positively charged intermediates by π–π stacking interactions between the aromatic intermediates and benzene ring of the linker [69].

The catalytic activity of MIL-101(Cr) was assessed in the cyanosilylation of benzaldehyde with the addition of trimethylsilylcyanide at 40 °C [9]. A 98.5% conversion was achieved after 3 h, showing no structural deformation of the catalyst

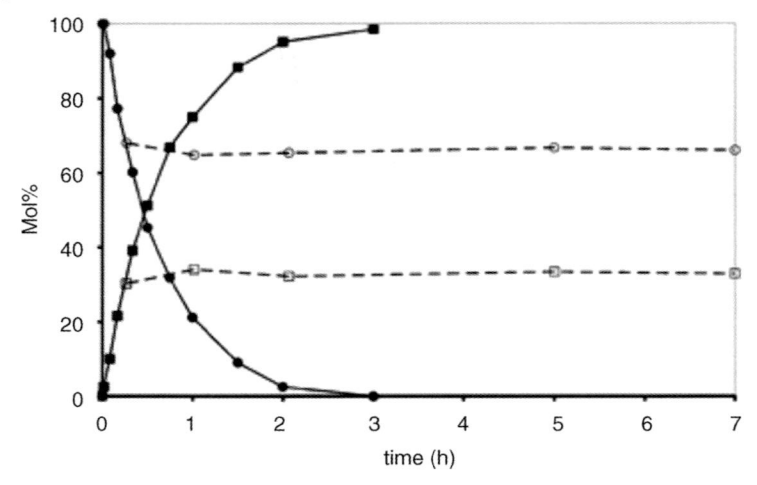

Figure 9.7 Reactant and product concentration during the MIL-101 = catalyzed cyanosilylation of benzaldehyde, Solid line: ●, concentration of benzaldehyde; ■, concentration of product. Broken line: filtration test; ○, concentration of benzaldehyde; □, concentration of product [9].

after use as evidenced by XRD. A slight decrease in catalytic activity was observed after three consecutive cycles. MIL-101 was also tested on sulfoxidation of aryl sulfides with H_2O_2 at room temperature with acetonitrile added as a solvent. A maximum conversion (99%) with 100% selectivity was achieved for the conversion of diphenyl sulfide after 15 h (Figure 9.7) [70]. Relative reaction rates of various substrates were determined in competitive experiments. Electron-releasing or -withdrawing groups in the oxygenation of p-X-phenyl methyl sulfides (X–Ph–S–CH$_3$, X = CH$_3$, H, NO$_2$, CN) were found to influence the oxidation reactivity strongly, as proven by a large negative ϱ value (ϱ = − 1.8) obtained from a linear free-energy relationship. It is important to mention that no catalyst deactivation or leaching of chromium species from the host lattice was observed after up to five cycles.

A crystalline mesoporous MOF (MesMOF-1) with an MTN-type framework and two meso-cages of diameter 3.9 and 4.7 nm (S and L cages, respectively) was prepared by Park et al. [71]. The material was functionalized via subsequent gas-phase loading with nickelocene and reduction under an N_2 flow, resulting in Ni nanoparticles (1.4–1.9 nm) embedded in the lattice matrix of MesMOF-1, as evidenced by TEM. [72]. The catalytic activity of the MesMOF-1-immobilized Ni nanoparticles was examined in the reduction of styrene to ethylbenzene (H_2 in methanol) and of nitrobenzene to aniline (NaBH$_4$ in methanol) at room temperature. Over 99% conversion was achieved for styrene (after 4 h) and nitrobenzene (after 15 min), whereas the non-loaded MesMOF-1 material showed no reaction. Removal of the catalyst showed no further conversion after 12 h, as evidenced by ^{1}H NMR spectroscopy, confirming no leaching of the nanoparticles into the reaction mixture. No deactivation of the catalyst was observed after three cycles of the reaction.

9.7 Conclusions

Although catalytic studies with MOFs are still in their infancy, MOFs have already shown catalytic activity and selectivity for a wide variety of reactions ranging from acid–base to redox catalysis. In some cases, recycling of the MOF catalysts is possible without significant catalyst deactivation or structural deterioration. In MOFs, not only the nature of the active site can be tailored, but also the size, dimensionality, and even chirality of the pores. A rich chemistry is also opened by encapsulating guest species in MOF hosts. This opens up new opportunities for many important applications, especially in the field of base chemicals and fine chemicals production. To achieve this goal, the development of efficient protocols for the synthesis of chemically and thermally stable MOFs is highly desirable.

List of Abbreviations

Am	amino
aps	3-aminopropyltrialkoxysilane
ata	2-aminoterephthalate
bdc	1,4-benzenedicarboxylate
4-btapa	1,3,5-benzenetricarboxytris[*N*-(4-pyridyl)amide]
bpy	4,4′-bipyridine
btc	1,3,5-benzenetricarboxylate
btec	1,2,4,5-benzenetetracarboxylate
btt	1,3,5-benzenetristetrazol-5-yl
cod	1,5-cyclooctadiene
cot	1,3,5-cyclooctatriene
Cp	cyclopentadienyl
DEF	*N*,*N*-diethylformamide
deta	diethylenetriamine
DFT	density functional theory
dmf	dimethylformamide (ligand)
DMF	*N*,*N*-dimethylformamide
dpdo	4,4′-bipyridine-*N*,*N*′-dioxide
ed	ethylenediamine
ee	enantiomeric excess
FTIR	Fourier transform infrared
MOF	metal–organic framework
POM	polyoxometalate
SBU	secondary building unit
sita	2-salicylidenimine terephthalate
taz	triazolate
TBHP	*tert*-butyl hydroperoxide
tcpb	1,2,4,5-tetrakis(4-carboxyphenyl)benzene

TEM   transmission electron microscopy
TOF   turnover frequency
XRD   X-ray diffraction

Acknowledgments

The authors are grateful to the Flemish Government for long-term structural support in the Casas Methusalem project and to the EU for supporting the FP-7 projects Macademia and Nanomof. D.D.V. is grateful to FWO (Belgium) for project support.

References

1 Ogura, M., Nakata, S., Kikuchi, E., and Matsukata, M. (2001) *J. Catal.*, **199**, 41.
2 Concepción-Heydorn, P., Jia, C., Herein, D., Pfänder, N., Karge, H.G., and Jentoft, F.C. (2000) *J. Mol. Catal. A: Chem.*, **162**, 227.
3 Chui, S., Lo, S., Charmant, J., Orpen, A., and Williams, I. (1999) *Science*, **283**, 1148.
4 Schlichte, K., Kratzke, T., and Kaskel, S. (2004) *Micropor. Mesopor. Mater.*, **73**, 81.
5 Alaerts, L., Seguin, E., Poelman, H., Thibault-Starzyk, F., Jacobs, P., and De Vos, D. (2006) *Chem. Eur. J.*, **12**, 7353.
6 Prestipino, C., Regli, L., Vitillo, J., Bonino, F., Damin, A., Lamberti, C., Zecchina, A., Solari, P., Kongshaug, K., and Bordiga, S. (2006) *Chem. Mater.*, **18**, 1337.
7 Fujita, M., Kwon, Y., Washizu, S., and Ogura, K. (1994) *J. Am. Chem. Soc.*, **116**, 1151.
8 Ohmori, O. and Fujita, M. (2004) *Chem. Commun.*, 1586.
9 Henschel, A., Gedrich, K., Kraehnert, R., and Kaskel, S. (2008) *Chem. Commun.*, 4192.
10 Horike, S., Dinca, M., Tamaki, K., and Long, J. (2008) *J. Am. Chem. Soc.*, **130**, 5854.
11 Gomez-Lor, B., Gutierrez-Puebla, E., Iglesias, M., Monge, M., Ruiz-Valero, C., and Snejko, N. (2005) *Chem. Mater.*, **17**, 2568.
12 Gandara, F., Gomez-Lor, B., Gutierrez-Puebla, E., Iglesias, M., Monge, M., Proserpio, D., and Snejko, N. (2008) *Chem. Mater.*, **20**, 72.
13 Kato, C., Hasegawa, M., Sato, T., Yoshizawa, A., Inoue, T., and Mori, W. (2005) *J. Catal.*, **230**, 226.
14 Kato, C. and Mori, W. (2007) *C. R. Chim.*, **10**, 284.
15 Dhakshinamoorthy, A., Alvaro, M., and Garcia, H. (2009) *J. Catal.*, **267**, 1.
16 Zou, R., Sakurai, H., Han, S., Zhong, R., and Xu, Q. (2007) *J. Am. Chem. Soc.*, **129**, 8402.
17 Brown, K., Zolezzi, S., Aguirre, P., Venegas-Yazigi, D., Paredes-Garcia, V., Baggio, R., Novak, M., and Spodine, E. (2009) *Dalton Trans.*, 1422.
18 Jiang, D., Mallat, T., Meier, D.M., Urakawa, A., and Baiker, A. (2010) *J. Catal.*, **270**, 26.
19 Miller, S., Wright, P., Serre, C., Loiseau, T., Marrot, J., and Férey, G. (2005) *Chem. Commun.*, 3850.
20 Perles, J., Iglesias, M., Ruiz-Valero, C., and Snejko, N. (2004) *J. Mater. Chem.*, **14**, 2683.
21 Perles, J., Iglesias, M., Martin-Luengo, M., Monge, M., Ruiz-Valero, C., and Snejko, N. (2005) *Chem. Mater.*, **17**, 5837.
22 Perles, J., Snejko, N., Iglesias, M., and Monge, M. (2009) *J. Mater. Chem.*, **19**, 6504.
23 Llabres, F., Xamena, I., Abad, A., Corma, A., and Garcia, H. (2007) *J. Catal.*, **250**, 294.

24 Lee, J., Farha, O., Roberts, J., Scheidt, K., Nguyen, S., and Hupp, J. (2009) *Chem. Soc. Rev.*, **38**, 1450.

25 Müller, U., Schubert, M., Teich, F., Pütter, H., Schierle-Arndt, K., and Pastré, J. (2006) *J. Mater. Chem.*, **16**, 626.

26 Zhou, Y., Song, J., Liang, S., Hu, S., Liu, H., Jiang, T., and Han, B. (2009) *J. Mol. Catal. A: Chem.*, **308**, 68.

27 Kleist, W., Jutz, F., Maciejewski, M., and Baiker, A. (2009) *Eur. J. Inorg. Chem.*, 3552.

28 Ravon, U., Domine, M., Gaudillere, C., Desmartin-Chomel, A., and Farrusseng, D. (2008) *New J. Chem.*, **32**, 937.

29 Huang, L., Wang, H., Chen, J., Wang, Z., Sun, J., Zhao, D., and Yan, Y. (2003) *Micropor. Mesopor. Mater.*, **58**, 105.

30 Thallapally, P., Fernandez, C., Motkuri, R., Nune, S., Liu, J., and Peden, C. (2010) *Dalton Trans.*, **39**, 1692.

31 Dhakshinamoorthy, A., Alvaro, M., and Garcia, H. (2009) *Adv. Synth. Catal.*, **351**, 2271.

32 Dhakshinamoorthy, A., Alvaro, M., and Garcia, H. (2010) *Appl. Catal. A: Gen.*, **378**, 19.

33 Ravon, U., Savonnet, M., Aguado, S., Domine, M., Janneau, E., and Farrusseng, D. (2010) *Micropor. Mesopor. Mater.*, **129**, 319.

34 Suslick, K., Bhyrappa, P., Chou, J., Kosal, M., Nakagaki, S., Smithenry, D., and Wilson, S. (2005) *Acc. Chem. Res.*, **38**, 283.

35 Shultz, A., Farha, O., Hupp, J., and Nguyen, S. (2009) *J. Am. Chem. Soc.*, **131**, 4204.

36 Sato, T., Mori, W., Kato, C., Yanaoka, E., Kuribayashi, T., Ohtera, R., and Shiraishi, Y. (2005) *J. Catal.*, **232**, 186.

37 Hasegawa, S., Horike, S., Matsuda, R., Furukawa, S., Mochizuki, K., Kinoshita, Y., and Kitagawa, S. (2007) *J. Am. Chem. Soc.*, **129**, 2607.

38 Hwang, Y.K., Hong, D.-Y., Chang, J.-S., Jhung, S.H., Seo, Y.-K., Kim, J., Vimont, A., Daturi, M., Serre, C., and Férey, G. (2008) *Angew. Chem. Int. Ed.*, **47**, 4144.

39 Gascon, J., Aktay, U., Hernandez-Alonso, M.D., van Klink, G.P.M., and Kapteijn, F. (2009) *J. Catal.*, **261**, 75.

40 Sovonnet, M., Aguado, S., Ravon, U., Bazer-Bachi, D., Lecocq, V., Bats, N., Pinel, C., and Farrusseng, D. (2009) *Green Chem.*, **11**, 1729.

41 Dybtsev, D., Nuzhdin, A., Chun, H., Bryliakov, K., Talsi, E., Fedin, V., and Kim, K. (2006) *Angew. Chem.*, **118**, 930; *Angew. Chem. Int. Ed.*, **45**, 916.

42 Evans, O., Ngo, H., and Lin, W. (2001) *J. Am. Chem. Soc.*, **123**, 10395.

43 Tanaka, K., Oda, S., and Shiro, M. (2008) *Chem. Commun.*, 820.

44 Seo, J.S., Whang, D., Lee, H., Jun, S.I., Oh, J., Jeon, Y.J., and Kim, K. (2000) *Nature*, **404**, 982.

45 Ingleson, M., Barrio, J., Bacsa, J., Dickinson, C., Park, H., and Rosseinsky, M. (2008) *Chem. Commun.*, 1287.

46 Banerjee, M., Das, S., Yoon, M., Choi, H., Hyun, M., Park, S., Seo, G., and Kim, K. (2009) *J. Am. Chem. Soc.*, **131**, 7524.

47 Cho, S., Ma, B., Nguyen, S., Hupp, J., and Albrecht-Schmitt, T. (2006) *Chem. Commun.*, 2563.

48 Wu, C., Hu, A., Zhang, L., and Lin, W. (2005) *J. Am. Chem. Soc.*, **127**, 8940.

49 Wu, C. and Lin, W. (2007) *Angew. Chem.*, **119**, 1093; *Angew. Chem. Int. Ed.*, **46**, 1075.

50 Ma, L., Wu, C.-D., Wanderley, M.M., and Lin, W. (2010) *Angew. Chem. Int. Ed.*, **49**, 8244.

51 Hundal, G., Hwang, Y.K., and Chang, J.-S. (2009) *Polyhedron*, **28**, 2450.

52 Yang, L., Naruke, H., and Yamase, T. (2003) *Inorg. Chem. Commun.*, **6**, 1020.

53 Bajpe, S.R., Kirschhock, C.E.A., Aerts, A., Breynaert, E., Absillis, G., Parac-Vogt, T.N., Giebeler, L., and Martens, J.A. (2010) *Chem. Eur. J.*, **16**, 3926.

54 Férey, G., Mellot-Draznieks, C., Serre, C., Millange, F., Dutour, J., Surblé, S., and Margiolaki, I. (2005) *Science*, **309**, 2040.

55 Sun, C.-Y., Liu, S.-X., Liang, D.-D., Shao, K.-Z., Ren, Y.-H., and Su, Z.-M. (2009) *J. Am. Chem. Soc.*, **131**, 1883.

56 Wee, L.H., Bajpe, S.R., Janssens, N., Hermans, I., Houthoofd, K., Kirschhock, C.E.A., and Martens, J.A. (2010) *Chem. Commun.*, **46**, 8186.

57 Juan-Alcañiz, J., Ramos-Fernandez, E.V., Lafont, U., Gascon, J., and Kapteijn, F. (2010) *J. Catal.*, **269**, 229.

58 Maksimchuk, N.V., Timofeeva, M.N., Melgunov, A.N., Chesalov, Y.A., Dybtsev, D.N., Fedin, V.P., and Kholdeeva, O.A. (2008) *J. Catal.*, **257**, 315.

59 Maksimchuk, N.V., Kovalenko, K.A., Arzumanov, S.S., Chesalov, Y.A., Melgunov, M.S., Stepanov, A.G., Fedin, V.P., and Kholdeeva, O.A. (2010) *Inorg. Chem.*, **49**, 2920.

60 Dang, D., Bai, Y., He, C., Wang, J., Duan, C., and Niu, J. (2010) *Inorg. Chem.*, **49**, 1280.

61 Alkordi, M.H., Liu, Y., Larsen, R.W., Jarrod, J.F., and Eddaoudi, M. (2008) *J. Am. Chem. Soc.*, **130**, 12639.

62 Sabo, M., Henschel, A., Fröde, H., Klemm, E., and Kaskel, S. (2007) *J. Mater. Chem.*, **17**, 3827.

63 Zhang, X., Llabrés i Xamena, F.X., and Corma, A. (2009) *J. Catal.*, **265**, 155.

64 Proch, S., Herrmannsdörfer, J., Kempe, R., Kern, C., Jess, A., Seyfarth, L., and Senker, J. (2008) *Chem. Eur. J.*, **14**, 8204.

65 Müller, M., Hermes, S., Kähler, K., van den Berg, M.W.E., Muhler, M., and Fischer, R.A. (2008) *Chem. Mater.*, **20**, 4576.

66 Schröder, F., Esken, D., Cokoja, M., van den Berg, M.W.E., Lebedev, O.I., Tendeloo, G.V., Walaszek, B., Buntkowsky, G., Limbach, H.-H., Chaudret, B., and Fischer, R.A. (2008) *J. Am. Chem. Soc.*, **130**, 6119.

67 Hong, D.-Y., Hwang, Y.K., Serre, C., Férey, G., and Chang, J.-S. (2009) *Adv. Funct. Mater.*, **19**, 1537.

68 Horcajada, P., Surblé, S., Serre, C., Hong, D.-Y., Seo, Y.-K., Chang, J.-S., Grenèche, J.-M., Margiolaki, I., and Férey, G. (2007) *Chem. Commun.*, 2820.

69 Ravon, U., Chaplais, G., Chizallet, C., Seyyedi, B., Bonino, F., Bordigo, S., Bats, N., and Farrusseng, D. (2010) *ChemCatChem*, **2**, 1235.

70 Hwang, Y.K., Hong, D.-Y., Chang, J.-S., Seo, H., Yoon, M., Kim, J., Jhung, S.H., Serre, C., and Férey, G. (2009) *Appl. Catal. A: Gen.*, **358**, 249.

71 Park, Y.K., Choi, S.B., Kim, H., Kim, K., Won, B.-H., Choi, K., Choi, J.-S., Ahn, W.-S., Won, N., Kim, S., Jung, D.H., Choi, S.-H., Kim, G.-H., Cha, S.-S., Jhon, Y.H., Yang, J.K., and Kim, J. (2001) *Angew. Chem. Int. Ed.*, **46**, 8230.

72 Park, Y.K., Choi, S.B., Nam, H.J., Jung, D.-Y., Ahn, H.C., Choi, K., Furukawa, H., and Kim, J. (2010) *Chem. Commun.*, 3086.

Part Four
Medical Applications

10
Biomedical Applications of Metal–Organic Frameworks
Patricia Horcajada, Christian Serre, Alistair C. McKinlay, and Russell E. Morris

10.1
Introduction

The large range of possible compositions (metal, linker) and the structural diversity (pore size, structure, etc.) of nontoxic porous crystalline hybrid inorganic–organic solids or metal–organic frameworks (MOFs) make them attractive for use in biomedicine. Most MOFs are biodegradable, as confirmed by *in vitro* experiments, and their stability ranges from less than a few hours up to weeks as a function of their composition and/or structure, which reduces their accumulation in the body. Iron oxo-centered polycarboxylate-based MOFs are nontoxic after intravenous (i.v.) administration of high doses in rats.

The dual hydrophilic/hydrophobic character of MOFs and the presence of accessible Lewis acidic metal sites make these materials suitable candidates to host active therapeutic molecules. A few MOFs possess very large loading capacities of therapeutic drug molecules in comparison with other carriers (polymers, liposomes, zeolites, etc.), and can release them in a controlled manner under physiological conditions. MOFs can also exhibit specific therapeutic activity by incorporating directly an active molecule within their framework: either (1) as the linker, with a release of the bioactive molecule when the solid degrades, or (2) as an active metal (Gd, Mn, Fe, etc.), with interesting imaging properties for theranostics or as contrast agents.

Finally, MOFs can be prepared as (1) pellets or tablets, (2) thin films for patches, (3) composite materials based on polymers or inorganic matrices (silica-covered MOFs, patches, etc.), or (4) stable solutions of nanoparticles with modified surfaces. Stealth, addressing, or imaging properties of MOF nanoparticles are introduced through surface modifications with organic biomolecules [poly(ethylene glycol) (PEG), chitosan, etc.], at the different stages of the synthesis. A short discussion of the still remaining challenges in this new domain of research is also proposed at the end of this chapter.

Metal-Organic Frameworks: Applications from Catalysis to Gas Storage, First Edition. Edited by David Farrusseng.
© 2011 Wiley-VCH Verlag GmbH & Co. KGaA. Published 2011 by Wiley-VCH Verlag GmbH & Co. KGaA.

10.2
MOFs for Bioapplications

10.2.1
Choosing the Right Composition

The science of porous solids has so far dealt mostly with crystalline silica-based zeolites, metal phosphates, and activated carbons. The latest class of porous solids concerns coordination polymers or MOFs [1–4]. The advantage of MOFs over other types of porous solids concerns the possibility of easily changing the metal and/or the organic linker. Indeed, the list of available linkers, including aliphatic or aromatic organic spacers with carboxylates, phosphonates, sulfonates, imidazolates, amines, pyridyl complexing functions, and so on is almost endless. As a result, hundreds of porous MOFs with various pore sizes, shapes (tunnels, cages, etc.) and organic functionalities (polar, apolar) have been reported. To use such solids for a given biomedical application, it is nevertheless important to focus on toxicologically acceptable MOFs. Apart from the first results dealing with the *in vivo* toxicity study at the preclinical level of several porous iron carboxylate-based MOFs, there has been no report on the toxicity of MOFs. Therefore, one has to rely on toxicity data already reported for the metals and linkers themselves. Each metal possesses its own degree of toxicity, ranging from a few $\mu g\,kg^{-1}$ up to more than $1\,g\,kg^{-1}$ (calcium). Hence, in principle, all metals and linkers could be used for biomedical applications but at doses below their degree of toxicity. One also has to consider the daily requirements of the human body to select the right metal due to the potential repetitive administration of the MOFs. For instance, calcium, magnesium, zinc, and iron, whose oxides possess $LD_{50}s$ over a few $g\,kg^{-1}$ (see Table 10.1) are of particular interest and, except for calcium where no porous MOFs have been reported so far, several porous coordination polymers have been described for each of these metals (Figure 10.1). For titanium and zirconium, the situation is different with very low daily doses (0.8 mg kg^{-1} per day), but as these metals do not accumulate in the body, they are not considered as being toxic ($LD_{50} > 3\,g\,kg^{-1}$).

Taking all these considerations together, examples of MOFs of interest for biomedical applications are zinc MOFs such as ZIF-8 (ZIF = zinc imidazolate framework) [5], Mg coordination polymers such as CPO-27(Mg) (CPO = coordination polymer from Oslo) [6], iron carboxylates such as MIL-100(Fe) [7], tetravalent

Table 10.1 LD_{50} and daily requirements of selected metals. LD_{50} for chromium(III) chloride [9], titanium dioxide [10], copper(II) sulfate [11], manganese(II) chloride [12], iron(II) chloride [13], zinc chloride [14], magnesium chloride [15], and calcium chloride [16].

Parameter	Cr	Ti	Cu	Mn	Fe	Zn	Mg	Ca
LD_{50} (mg kg^{-1})	1800 (Cr^{III})	>25 000	25	1500	450	350	8100	1000
Daily dose (mg)	0.12	0.8	2	5	15	15	350	1000

| ZIF-8 | CPO-27 | MIL-100 | MIL-125 | UiO-66 |
| (Zn) | (Mg) | (Fe) | (Ti) | (Zr) |

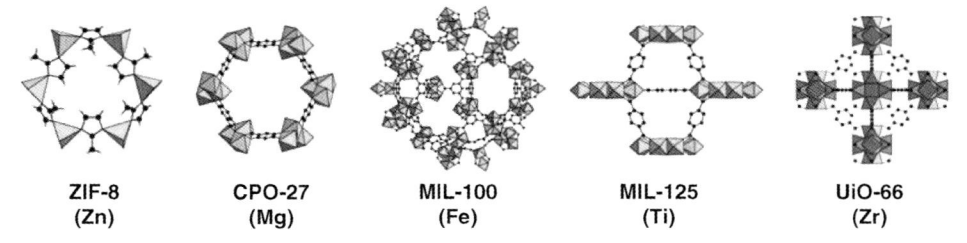

Figure 10.1 View of the structures of a few topical non toxic MOFs, with the corresponding labeling and metals part of the framework below. Metal polyhedra and carbon atoms are in gray and black, respectively.

Table 10.2 Toxicity data for polycarboxylic linkers.

Linker	Terephthalic acid [17]	Trimesic acid [18]	2,6-Naphthalenedicarboxylic acid [19]	1-Methylimidazole [20]
LD_{50} (g kg^{-1})	5	8.4	5	1.13

dicarboxylate solids such as MIL-125(Ti) (MIL = material from Institut Lavoisier), and UiO-66(Zr) (UiO = University of Oslo) [8]. These solids exhibit large pore sizes (4–29 Å) and BET surface areas between 1200 and 2200 m^2 g^{-1}, respectively.

Once the metal has been selected, one has to consider the choice of the linker. Most ligands used so far for the construction of porous MOFs are aromatic or aliphatic polycarboxylates. Two types of linkers have to be considered, however. First are exogenous linkers, which are not naturally part of any body cycles. Once introduced into the body, the molecule might be either metabolized or eliminated through urine or feces. Here again, some toxicity data are available (LD_{50}) (Table 10.2) that show, for instance, that typical polycarboxylic linkers are not very toxic.

However, such studies are often based on oral (rat) administration and no in-depth toxicity study of all the possible exogenous linkers of interest is available. Therefore, the best strategy might be the use of endogenous linkers to build up new porous coordination polymers.

So far, although there are numerous MOFs with endogenous linkers, only a few of them are really porous, that is, capable of encapsulating biological molecules. Most of them are built up from simple dicarboxylic or amino acid linkers (Figure 10.2).

10.2.2
The Role of Flexibility

Another property of a few selected MOFs is the possibility of keeping the same structure type while increasing the size of the organic spacer, thus enlarging the pore size, which is denoted isoreticularity [3, 21, 22] For a given structure, these pores can

| Fe fumarate | Mg formate | Bi citrate | Cu aspartate |

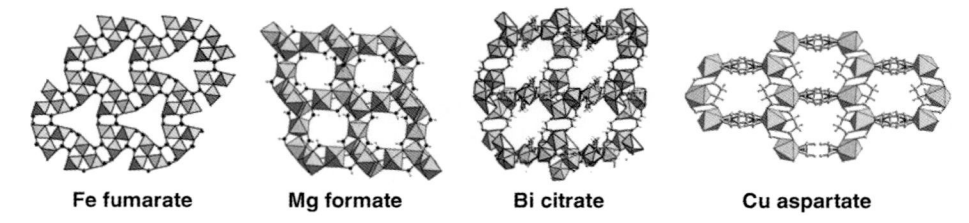

Figure 10.2 View of the structures of a few MOFs based on endogenous linkers, with names of the corresponding metals and linkers below. Metal polyhedra and carbon atoms are in gray and black, respectively.

also be functionalized through grafting of a polar or apolar functional group by substituting a proton from the linker or through post-synthetic modification in order to tune its properties [22].

Most MOFs possess a rigid framework, but a few of them are flexible and breathe (or swell) in the presence of a change in pressure or temperature or during the adsorption of a fluid (breathing = flexible or dynamic porosity, able to be adapted to the guest adsorbed molecule) [4d, 23]. Typical examples are the metal terephthalates denoted MIL-53(Al, Cr, Fe, etc.) (Figure 10.3) [24] or MIL-88(Fe, Cr) [25]. They exhibit a reversible and selective pore opening, which leads to drastic changes in the unit cell volumes of the solvated and the dried forms, from 40 to 230%.

Such large and selective variations might lead in the near future to new applications in terms of adsorption, storage, and facile delivery of gases or biological molecules [26, 27].

Finally, the question of the stability of MOFs under physiological conditions is raised. It has been shown previously that the hydrothermal stability of MOFs depends strongly on the metal, linker (complexing function and spacer), the inorganic secondary building unit and the structure, leading from highly unstable MOFs in the presence of air moisture to hydrothermally stable solids [28]. Such strong discrepancies between the various candidates for bioapplications are nevertheless advantageous, since one could choose an adequate solid with a stability suitable for a given application and its corresponding administration conditions (cutaneous, *in vivo*, etc.).

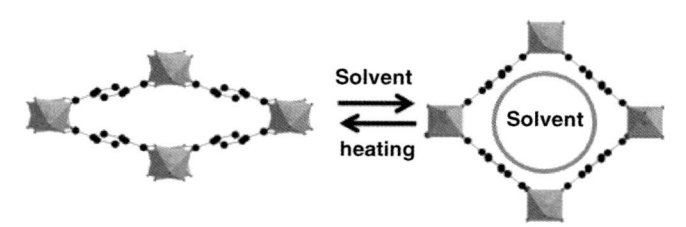

Figure 10.3 The breathing behavior of the iron terephthalate MIL-53(Fe). Iron(III) octahedra and carbon atoms are in white and gray, respectively.

10.2.3
The Role of Functionalization

As mentioned above, the choice of organic linkers to build up porous MOFs is almost infinite. This represents a major advantage of MOFs over their purely organic (carbons) or inorganic (zeolites, metal phosphates) counterparts. Thus, for biomedical applications, there is a real possibility of tuning the host–guest interactions between the active molecule and the pores of the hybrid solid, in order to control the delivery of the drug molecule. For instance, several porous metal terephthalates based on iron or zinc have been reported with a series of modified terephthalate linkers that bear polar or apolar functional groups such as amino, nitro, chloro, bromo, carboxylate, methyl, or perfluoro (Figure 10.4) [22c]. In addition, introducing functional groups in flexible MOFs not only modifies their polarity but also drastically affects their flexibility, which might be of the utmost importance for the loading of a bioactive guest molecule [27].

10.2.4
Biodegradability and Toxicity of MOFs

For many potential industrial applications related to MOFs, stability of the porous solid is a pertinent issue. This concern has been widely raised due to the low water stability of the well-known microporous zinc terephthalate MOF-5 [29]. A tentative rationalization of the relative hydrothermal stability of MOFs was reported by Low *et al.* (Figure 10.5) [28]. Thus, the charge and coordination number of the metal, chemical functionality of the organic linker, framework dimensionality, and interpenetration are the main parameters controlling the degree of stability of a MOF. For bioapplications, MOFs have to be stable enough to realize their function and then to be removed/recycled once in the body, preventing endogenous accumulation and its

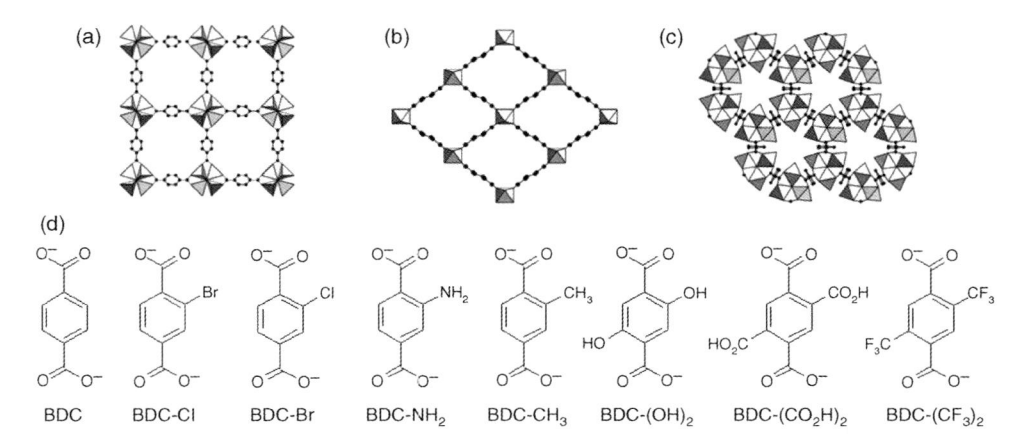

Figure 10.4 Structures of (a) MOF-5, (b) MIL-53, and (c) MIL-88B. (d) Selected modified terephthalic acids. BDC = 1,4-benzenedicarboxylic acid.

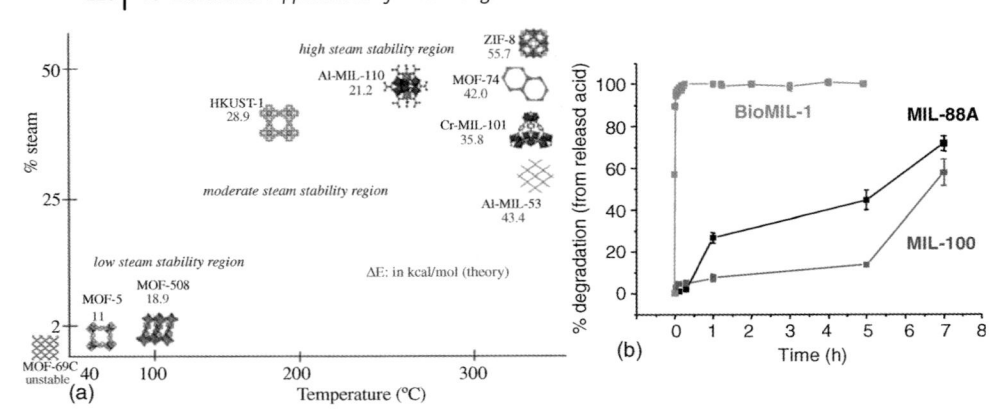

Figure 10.5 (a) Hydrothermal stability of different MOFs. From Low et al., [28]. (b) In vitro degradation tests on the BioMIL-1 and the nanoparticles of MIL-100 and MIL-88A under simulated physiological conditions (PBS, pH = 7.4, 37 °C). Modified from Horcajada et al. [33].

potential toxic effects. The wide range of possible compositions, functionalization of the linker [22c], and structures offer theoretically quasi-infinite possibilities to tune the kinetics of biodegradation, according to the desired application.

Interestingly, as mentioned below, the first studies have confirmed that MOFs possess different stabilities in body fluids according to their composition and structure.

The aminoterephthalate MIL-101(Fe) [30] and the iron nicotinate Bio-MIL-1 [31] degrade rapidly within a few hours in simulated body fluid (PBS). Thus, to prolong the half-live of such MOFs, an alternative consists in protecting the nanoparticles by a silica- cover. In contrast, Morris and co-workers showed that the metal(II) hydroxoterephthalates M-CPO-27 (Ni, Co, etc.) are stable under physiological simulated conditions [phosphate-buffered saline (PBS) and bovine serum, at 37 °C] for days, showing only low degradation after 1 week of assay (around 15 wt% of metal was released) [32]. Similarly, in tablet form, the flexible microporous iron terephthalate MIL-53(Fe) is stable under a simulated body fluid (SBF; same inorganic composition as human plasma, 37 °C) for more than 3 weeks [27].

Finally, despite a higher surface-to-volume ratio compared with micrometric particles, nanoparticles of some porous iron carboxylate MOFs have also been reported to exhibit acceptable stability, with partial degradation over a few days in PBS (Figure 10.5) [33].

As mentioned above, the choice of the MOFs was based first on the toxicity data for their separate components (linker and metal). However, the first in vitro and in vivo toxicities of a few selected porous iron polycarboxylates have been determined by Horcajada et al. [33].

The cytotoxicity of two iron carboxylates as nanoparticles was first studied through MTT assay. The IC_{50} (half maximum inhibitory concentration) of the iron fumarate MIL-88A nanoparticles on mouse macrophages was comparable to those of currently available nanocarrier systems [34], thus confirming their lack of cytotoxicity. The same method has been used for the iron trimesate MIL-100 nanoparticles on human peripheral blood mononuclear cells (PBMCs) with no sign of cytotoxic effects even at high doses.

In a second step, the first acute and subacute *in vivo* toxicity assays have been carried out [33]. Three iron carboxylates (nanoparticles) were selected: (1) an microporous iron fumarate (MIL-88A) based on the hydrophilic endogenous linker (Krebs cycle), fumaric acid, (2) the mesoporous iron trimesate (MIL-100) based on an exogenous hydrophilic aromatic linker, and (3) the microporous iron tetramethylterephthalate (MIL-88B-4CH$_3$) based on a hydrophobic exogenous aromatic linker. Important doses of these iron carboxylate nanoparticles (up to 220 mg kg^{-1}) were administered i.v. to rats groups, evaluating different parameters after 1, 3, 7, 14, 30, 60, and 90 days of administration (weight, serum parameters, inflammatory and immune reactions, organ histology, hepatic enzymatic activity, etc.). Remarkably, no sign of severe toxicity was evidenced, even after the administration of very high doses, which were considerably higher than those of usual carriers. Note that some oral iron supplements, currently approved, have the same chemical composition as the iron fumarate MIL-88A solid (e.g., Ferrets®).

10.3
Therapeutics

10.3.1
Drug Delivery

Controlled drug release technology has expanded rapidly since the 1970s. These systems offer many advantages over conventional drug administration, such as low toxicity issues and higher efficiency values due to targeting. Additionally, many interesting therapeutic molecules suffer from drawbacks, such as instability in biological media, low solubility, and/or poor ability to cross natural barriers, which could be solved by using drug carriers [35]. The most commonly used drug delivery devices are micelles, liposomes, and polymeric systems [36–39]. Although these technologies are now on the market, they are not able to satisfy fully all the requirements for the administration of many therapeutic molecules. Therefore, there is still a need to develop new delivery carriers. Among drug delivery systems, ordered porous materials have emerged as a new alternative. Great attention has been focused on the development of stable uniform porous structures, with narrow pore size distributions, high surface areas and pore volumes, and tunable structures and compositions. Inorganic porous materials such as zeolites [40–43] or ordered mesoporous silica [44–47] fulfill most of these requisites. However, zeolites have shown lower drug capacities and toxicity restrictions [48]. Ordered mesoporous silica materials have been widely studied for drug delivery, with interesting results [45, 49–51]. Nevertheless, mesoporous silicas are frequently organically surface modified for better control of drug release [52], but sometimes have controversial toxicity [53, 54]. Therefore, an alternative might be the use of porous

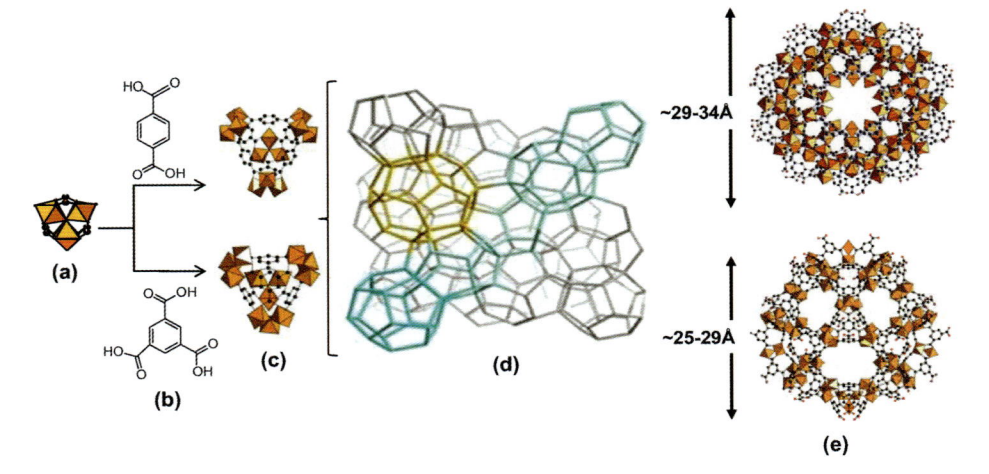

Figure 10.6 Schematic view of the MIL-100 and MIL-101 structures. (a) Metal trimer; (b) terephthalic acid (top) and trimesic acid (bottom); (c) hybrid supertetrahedra built from metal trimers and carboxylate linker; (d) schematic view of the cubic unit cell, representing the larger cage in yellow and the smaller cage in blue; (e) Larger (top) and smaller cages (bottom) of MIL-100.

MOFs, which combine a highly regular porosity with an easily tunable, crystalline hybrid inorganic–organic framework [55].

The first evaluation of the use of porous MOFs in controlled delivery of drugs was tested first [55], using the anti-inflammatory and analgesic drug ibuprofen {2-[4-(2-methylpropyl)phenyl]propanoic acid} as a model drug, and both the model mesoporous rigid MOFs MIL-100(Cr) and MIL-101(Cr) [55] and the porous flexible MOFs MIL-53(Cr) and MIL-53(Fe) [27]. The toxic mesoporous chromium solids MIL-100(Cr) [56] and MIL-101(Cr) [57] were chosen as model solids due to their large pores (diameter 25–34 Å), large BET surface areas (2000–4400 $m^2 g^{-1}$) and high aqueous stability (Figure 10.6). In addition, they exhibit accessible Lewis acid sites, which are good candidates to interact with active molecules.

The second type of hybrid materials concerns the microporous metal carboxylate MIL-53(Cr, Fe) solids [24], which exhibit a highly flexible framework able to modulate the pore size reversibly upon the adsorption of guest molecules, inducing a variation in cell volume of up to 40%. Ibuprofen was loaded into the dehydrated hybrid solids by soaking in hexane or ethanol drug-containing solutions. Table 10.3 shows that exceptionally high drug loadings were achieved, up to 1.4 g of ibuprofen per gram of MIL-101(Cr), corresponding to about 56 ibuprofen molecules in each "small" cage and about 92 molecules in the "large" cage, in agreement with Monte Carlo predictions [58]. This is roughly four times larger than the capacity obtained using the usual mesoporous silica materials [59].

MIL-100(Cr) exhibits a much lower capacity than MIL-101(Cr) (0.35 $g g^{-1}$). probably due to the smaller pentagonal windows (5 Å), which rule out the loading of ibuprofen within the small cages.

Table 10.3 Pore size and pore volume of different porous solids, and encapsulation/release results for ibuprofen (the release was achieved in SBF at 37 °C).

Porous solid	Pore diameter (Å)	Pore volume (cm^3 g^{-1})	g ibuprofen g^{-1} dehydrated solid	Time release (days)
MIL-101(Cr)	29–34[a]	1.9–2.3	1.38	6
MIL-100(Cr)	25–29[b]	1.2	0.35	3
MIL-100(Fe)	25–29[b]	1.1	0.35	3
MIL-53(Cr)	8	0.5	0.22	21
MIL-53(Fe)[c]	8	0.5	0.21	21
Faujasite	11	0.3	0.16	7
MCM-41	36	1.0	0.34	2
MCM-41-NH$_2$	28	0.4	0.22	5

a) Accessible by microporous pentagonal (∼12 Å) and hexagonal (∼16 Å) windows. Smaller cages only accessible by pentagonal windows.
b) Accessible by microporous pentagonal (∼5 Å) and hexagonal (∼8.5 Å) windows. Smaller cages only accessible by pentagonal windows.
c) MIL-53(Fe), although nonporous in its dehydrated form, possesses in its open form, that is, in the presence of gases, vapors, or liquids, an equivalent pore volume of about 0.5 g cm^{-3}.

The use of the flexible MOF led to a lower drug loading (∼20 wt%) than with the mesoporous MIL-101/100(Cr), in agreement with its lower pore volume (Table 10.3). Interestingly, as observed previously with small gases [60, 61], liquids [62], or vapors [63], MIL-53 solids also breath upon ibuprofen adsorption [27]. An intermediate pore opening, that is, with sizes between those of the narrow and large pores forms, is present (Figure 10.7). Furthermore, using experimental cell parameters, the conformation of the drug inside the pores could be calculated from computer modeling.

Drug delivery from ibuprofen-containing MOFs pellets was performed using SBF under continuous stirring at 37 °C (Figure 10.8). Interestingly, all solids completely release the ibuprofen to the media by an exchange process. However, the mesoporous and rigid MIL-100(Cr) and MIL-101(Cr) solids release their cargo after 3 or 6 days, respectively, with three different stages: (1) first delivery of the ibuprofen located not in direct contact with the pore walls (∼5 h) and a slower release of the ibuprofen

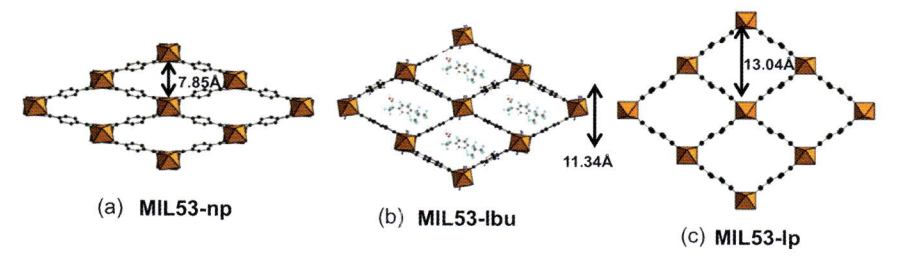

Figure 10.7 Pore openings of the MIL-53 solid: water (a), ibuprofen (b), and open form (c). Metal octahedra, oxygen and carbon atoms are in orange, red, and black, respectively. From Horcajada et al. [27].

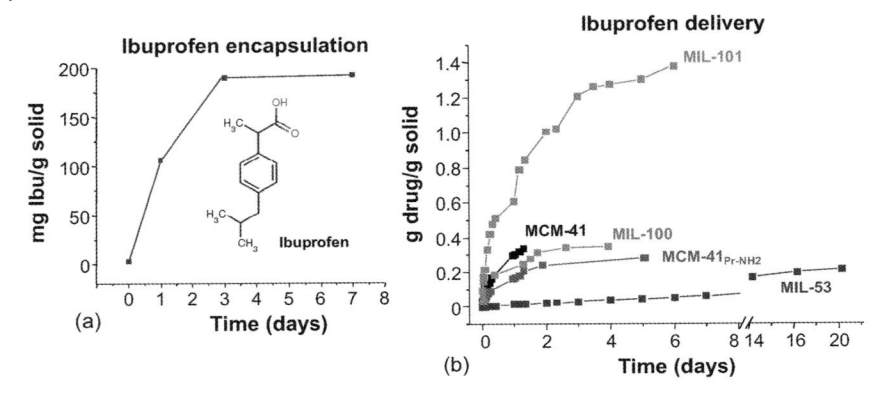

Figure 10.8 (a) Encapsulation kinetics of ibuprofen in MIL-53 solids. (b) Delivery kinetics of ibuprofen from several porous carriers (SBF, 37 °C).

fraction located close to the pore wall into the (2) larger and (3) smaller cages. Considering the stability of these solids under the release conditions, which was confirmed by high-performance liquid chromatography (HPLC), porosimetry, and X-ray powder diffraction (XRPD), drug release here is governed only by a diffusion process for the first step and by a combination between diffusion and drug–matrix interactions for the second and third steps [55].

^{1}H solid-state NMR spectroscopy indicates that ibuprofen is deprotonated, suggesting that a large amount of drug molecules are coordinated through its carboxylate group to the Lewis acid metal sites, with a DFT-estimated heat of adsorption of -73.17 kJ mol^{-1}, while the remaining active molecules are in interaction with the aromatic linkers. Finally, the slower release in MIL-101 compared with MIL-100 might arise from a higher proportion of aromatic rings [Ar/Cr = 1.0 (MIL-101), 0.66 (MIL-100)].

The flexible MIL-53(Fe, Cr) solids slowly release the ibuprofen (up to 3 weeks) with exceptional zero-order kinetics, which makes the process predictable and independent of the drug concentration in the solid [27]. Solid state NMR showed that (1) ibuprofen confinement is stronger in MIL-53 due to its smaller pore size and (2) the carboxylic group from ibuprofen is here protonated, in agreement with hydrogen bonds between this carboxylic group and the hydroxyl group from MIL-53. DFT calculation gives an overall heat of adsorption of about -57 kJ mol^{-1} Additionally, the zero-order kinetics could be due to the maintenance of a local environment throughout the whole delivery process, as confirmed by XRPD. Hence flexible MOFs are excellent candidates for controlling drug release by the optimization of both diffusion process and drug–matrix interactions through the adaptation of the pore dimensions to the API size.

Noteworthily, similar encapsulation and delivery tests were performed using iron analogs of MIL-53 and MIL-100 structures.

Finally, MOFs are on the whole better, in terms of loading capacity and kinetics of release, than microporous zeolites (faujasite, pore diameter $\sim$7.4 Å) [64] and ordered mesoporous silica MCM-41 [65]. Grafting aminopropyl groups on the surface of the mesoporous silica MCM-41 leads to an increase in the delivery time but with an unfortunate decrease of around 1.5-fold in the drug capacity [66]. Thus MOFs, by combining a regular porosity and the presence of organic groups within the framework, achieve both high loadings and longer releases.

More recently, encapsulation and delivery tests on drugs with greater therapeutic interest have been carried out. An et al. prepared a 3D anionic zinc adeninate framework with a one-dimensional (1D) channel system, Zn_8(adeninate)$_4$(biphenyldicarboxylate)$_6O_2Me_2NH_2 \cdot 8dmf \cdot 11H_2O$ (BioMOF-1), able to absorb the cationic antiarrythmic drug procainamide [67]. After 15 days of exchange in a procainamide solution, this solid showed important drug loadings (up to 22 wt%). This means 3.5 procainamide-H^+ molecules per formula unit, estimating that 2.5 molecules are located inside the pores and one molecule on the external surface. Procainamide release depends strongly on the medium, since the completed exchange is achieved after 3 days in PBS (Figure 10.9) compared with 20% (probably out-surface) in nanopure water.

Highly challenging drugs related to economically and societally important fields such as cancer or AIDS have also recently been encapsulated in different porous MOFs. For instance, Taylor-Pashow et al. [68] reported the successful encapsulation of a cisplatin prodrug (ethoxysuccinatocisplatin or ESCP) with loadings up to 12.8 wt% in the mesoporous iron terephthalate MIL-101 (Table 10.2). This solid was previously modified with 17 mol% of the aminoterephthalate linker in order to increase interactions with the drug and allow the grafting of a contrast imaging agent

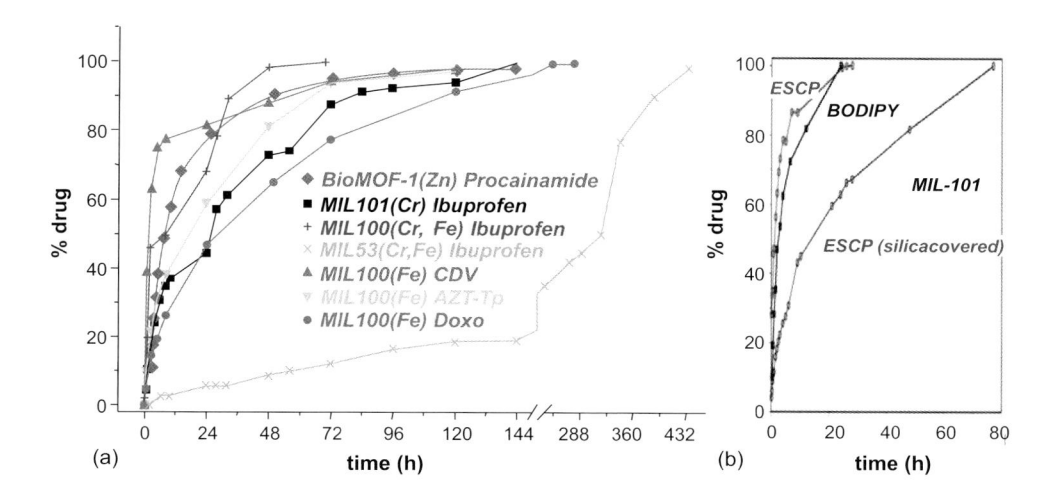

Figure 10.9 (a) Drug delivery kinetics of (i) ibuprofen from pellets of MIL-100(Cr or Fe), MIL-101(Cr), and MIL-53(Cr or Fe); (ii) cidofovir, azidothymidine triphosphate, and doxorubicin from nanoparticles of MIL-100 (Fe); and (iii) procainamide from particles of BioMOF-1(Zn). (b) Delivery of BODIPY, ESCP from MIL-101, and silica-covered MIL-101. From Taylor-Pashow et al. [68].

[1,3,5,7-tetramethyl-4,4-difluoro-8-bromomethyl-4-bora-3a,4a-diaza-s-indacene (Br-BODIPY)]. However, owing to the high instability of this compound in biological media, nanoparticles were coated with a thin silica layer, resulting in a total release of the prodrug under PBS physiological conditions after 14 h in the original nanoMIL-101 solid and after 3 days from the silica-coated nanoparticles (Figure 10.4).

The antitumoral molecules busulfan (Bu) and doxorubicin (Doxo) and the antiviral azidothymidine triphosphate (AZT-Tp) and cidofovir (CDV) were also encapsulated from nanoparticles of porous iron carboxylates [55]. These drugs suffer from important drawbacks such as poor solubility and/or stability in aqueous biological media [69–71], often resulting in short half-lives and low bioavailabilities [72, 73], and/or limited bypass of physiological barriers [74, 75]. The available nanocarriers have shown very poor loadings (never exceeding 1–5 wt%) and also fast releases ("burst effect") [76–79]. or require the use of toxic solvents (Bu) [80]. MOFs could load much higher amounts of busulfan, from 8.0 to 25.5 wt% (Table 10.4), compared with those of existing nanocarriers (0.4–5 wt%) [81].

Similar very high loadings of AZT-Tp, CDV, and Doxo were achieved using nanoparticles of the mesoporous iron trimesate (nanoMIL-100) [55], up to 21, 16, and 9 wt%, respectively, and an unprecedented capacity of 42 wt% was reached for AZT-Tp and CDV using MIL-101-NH_2 nanoparticles (Table 10.4). This is far better than the less than 1 wt% loading reported previously using other carriers and these drugs [82]. Furthermore, AZT-Tp, CVD, and Doxo were progressively released from MIL-100 nanoparticles under simulated physiological conditions after 3–5 days for AZT-Tp and CDV, and up to 2 weeks for the hydrophobic Doxo, showing no "burst effect" (Figure 10.9). It is assumed here that the controlled release is mainly governed by a combination of host–guest interactions and a diffusion process. It is likely also

Table 10.4 Drug capacities of different drugs on several iron carboxylate nanoparticles.

	MIL-89	MIL-88A	MIL-100	MIL-101-NH_2	MIL-53
Flexible solid	Yes	Yes	No	No	Yes
Pore size (Å)	11	6	25 (5.6)	29 (12)	8.6
			29 (8.6)	34 (16)	
Particle size (nm)	50–100	150*	200	120	350*
Ethoxysuccinatocisplatin	—	—	—	12.8	—
Busulfan	9.8	8.0	25.5	—	14.3
Azidothymidine triphosphate	—	0.60	21.2	42.0	0.24
Cidofovir	14	2.6	16.1	41.9	—
Doxorubicin	—	—	9.1	—	—
Ibuprofen	—	—	33	—	22
Caffeine	—	—	24.2	—	23.1
Urea	—	—	69.2	—	63.5
Benzophenone 4	—	—	15.2	—	5
Benzophenone 3	—	—	1.5	—	—

a) Modified from Horcajada et al. [55].
* Bimodal distribution of sizes, with presence of larger particles.

that degradation of the MOF does not significantly affect the release of the cargo since only 10% of MIL-100(Fe) is degraded under similar conditions.

Finally, other cosmetic or active molecules with different chemical structures and properties have been successfully encapsulated into MOFs, supporting the "universality" of these drug nanocarriers [55]. In conclusion, the exceptional loadings of drugs in MOFs whatever the polarity and structure of the active molecule originates from their adapted internal microenvironment (amphiphilic: polar metal, oxygen atoms from the linker, and nonpolar organic spacer) suitable for a large number of host molecules [83].

10.3.2
BioMOFs: the Use of Active Linkers

MOFs are built up from the connections of an organic linker and a metallic subunit (polyhedral, clusters, chains, etc.). In addition, the most obvious potential use of porous MOFs in biomedicine would be to use their porosity for the encapsulation of biomolecules. However, once administered and the cargo delivered, it is very likely that the MOFs will be degraded within the body, leading to a release of the metal and the linker issued from the MOF itself, introducing corresponding additional toxicity issues. Hence the use of endogenous linkers is an acceptable solution and indeed several amino acids and endogenous Zn-based MOFs that exhibit a therapeutic activity have been reported recently [84]. A further step could be the preparation of biologically active MOFs. So far, drug delivery using MOFs has involved typical encapsulation/impregnation/release methods. The loading capacity and how fast the release of their cargo is depend both on the porosity of the MOFs and host–guest interactions. Hence this requires a porous MOF with a release of the biomolecule related to the host–guest interactions and also the stability of the MOF. An alternative way would be to use directly the biomolecule as the linker to react with a nontoxic metal to make the MOF (Figure 10.10). There will be no need to make a porous framework any longer. The release of the drug will be achieved through the degradation of the MOF, with no toxicity concerns issuing from the use of exogenous

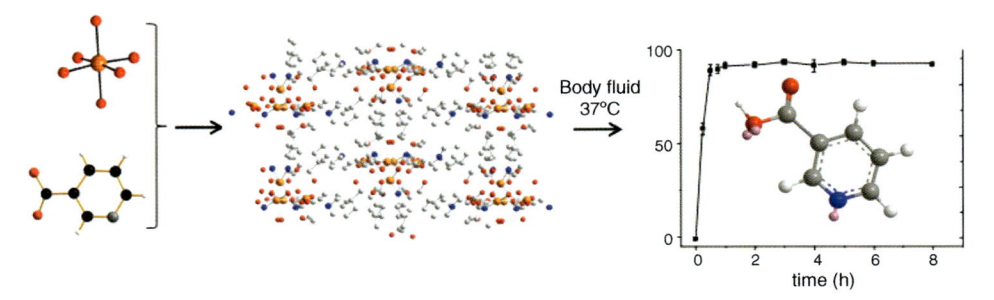

Figure 10.10 Schematic view of the formation of a BioMOF (Bio-MIL-1) built up from bioactive linker and its delivery. Here the bioactive linker is nicotinic acid. Iron, oxygen, nitrogen, and carbon atoms are in orange, red, gray, and black, respectively.

linkers. Other MOFs built up from bioactive linkers have been reported, such as peptide- and adenine-based MOFs, but so far none of them has been used for controlled released of the active molecule [85]. Yu *et al.* nevertheless made a bioactive layered MOF based on the drug exonacin combined with manganese, but did not report anything related to its use for controlled release of the drug through the degradation of the coordination polymer [86].

The proof of principle for the use of bioactive MOFs to deliver drugs has been reported through the synthesis of a three-dimensional (3D) iron(II/III) nicotinate. Nicotinic acid (pyridine-3-carboxylic acid or niacin) corresponds to vitamin B_3 [31], which is an endogenous acid with pellagra-curative, vasodilating, and antilipemic properties. The MOF denoted Bio-MIL-1 or $Fe^{III}_2Fe^{III}_{1-x}Fe^{II}_xO_{1-y}(OH)_y[O_2C-C_5H_4N]_5[O_2CCH_3]$ ($x \approx 0.15$) (Bio-MIL for Bioactive Material from Institut Lavoisier) is a rare example of bioactive MOFs.

The release of nicotinic acid, using SBF at 37 °C, is very rapid, within 1 h [87]. The bioactive MOF is degraded, as indicated by XRPD and HPLC (release of linker). Such a rapid degradation under physiological conditions has been observed previously with the iron aminoterephthalate MIL-101(Fe) [88]. Other porous iron carboxylates based on the same inorganic building units, such as the iron fumarate MIL-88A and the iron trimesate MIL-100 solids, however, exhibit much longer degradation times, over 1 week. Hence the stability of a MOF depends not only on its building unit but also on its structure, composition (metal, spacer, complexing group, etc.). Depending on the nature of the bioactive linker, one would expect future bioactive MOFs to present different stabilities. For a given application, this means that the choice of the metal and structure will be crucial in order to find the best conditions for a suitable controlled release of the bioactive molecule through the degradation of the bioactive MOFs. Finally, as most drugs possess one or several polar groups (e.g., carboxylates, phosphates, phosphonates, amines, heterocycles) that are able to form strong ionocovalent bonds with nontoxic metals, this approach could be considered as a complementary method for the controlled released of biomolecules using biomaterials with known toxicity and activity.

10.3.3
Release of Nitric Oxide

In 1998, three American scientists (F. Murad, R. Furchgott and L. Ignarro) won the Nobel Prize in Physiology for discovering the uses of NO in the human body and in particular its action as a muscle relaxant [89, 90]. A family of enzymes are now known to produce NO in the body, collectively called nitric oxide synthase (NOS). As NO is a very small, highly reactive species, it diffuses very easily in the body, and its impact is therefore dependent upon the concentration of NO produced and where its production occurs. The NO requirement is a delicate balance as too little or too much can lead to life-threatening diseases and can even be fatal, as illustrated in Figure 10.11.

Since its discovery as an important muscle relaxant in the body, interest in this small radical has increased exponentially. A wide variety of roles have now been shown to be controlled by NO in the body and indeed improvements in certain

NO Concentration	
Too Low	**Too High**
Hypertension	Septic Shock
Atherosclerosis	Hypotension
Parkinsons disease	Excessive bleeding
Fibrosis	Meningitis
Sexual Impotence	Rheumatoid Arthritis
	Ischemia
	Cancer
	Diabetes
	Asthma

Figure 10.11 Diseases related to insufficient or excess NO in the body.

functions of the body are now linked to an increase in NO, whether it be administered as pure NO or generated by taking supplements that encourage its generation (L-arginine and L-citrulline) [91–93]. As a result, NO-related therapies are attracting considerable interest, utilizing a wide variety of delivery methods depending on the intended application. One of the most heavily researched areas of NO delivery concerns its wound-healing abilities [94, 95]. It has been shown that increasing the concentration of NO at the site of a wound greatly decreases the wound healing time in diabetic patients [96]. Therefore, numerous novel NO delivery methods are now being investigated. One of the most promising methods of delivering NO is to use porous materials, specifically MOFs, to store and release NO controllably when required.

Certain porous solids, upon complete removal of occluded solvent molecules, contain coordinatively unsaturated metal sites (CUSs) [97–99]. which are the key in the storage and delivery of NO from such materials. Morris's group of St. Andrews began storing and releasing NO in zeolites [100–102] and then tried storing NO in copper trimesate (HKUST-1) [103]. Owing to the larger surface area and high number of coordinatively unsaturated metal sites, HKUST-1 displays a higher uptake of NO than the zeolites previously investigated. However, the release of NO from HKUST-1 is two orders of magnitude less than the amount actually adsorbed. Despite this, NO-loaded HKUST-1 completely inhibits platelet aggregation, showing that the small amount of NO released is still of biological significance.

The M-CPO-27 Ni and Co solids, which are 1D metal(II) dihydroxyterephthalate microporous compounds, were synthesized and activated to make free their large number of free Lewis acid sites (>6 mmol g^{-1}). The activated MOF sample was then exposed to NO. The results were remarkable. Whereas HKUST-1, which possesses copper CUS sites upon activation, had an NO uptake of 3 mmol g^{-1}, both Ni and Co

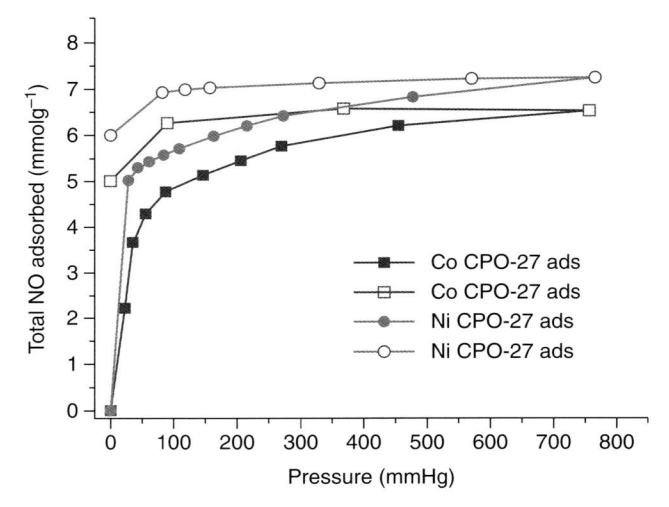

Figure 10.12 NO adsorption/desorption isotherm for Ni and Co CPO-27 at 298 K showing the exceptional NO uptake of these MOFs.

CPO-27 showed NO capacities of between 6 and 7 mmol g^{-1} activated MOF (Figure 10.12) [104]. This is twice as much as the previous MOF studied and about five times greater than that for the best zeolite. As can also be seen from the isotherm, the desorption arm shows a very distinct hysteresis, which is indicative of the NO binding strongly to the CUSs. The amount of NO adsorbed is about one NO molecule per available CUS. Therefore, the maximum adsorption capacity should be 6.4 mmol g^{-1} of NO; however, there are extra NO molecules weakly adsorbed (physisorbed) on the pore walls of the MOF. The actual amount of NO chemisorbed is 4.99 mmol g^{-1} for Co CPO-27 and 5.99 mmol g^{-1} for Ni CPO-27, which is very close to the theoretical maximum of 6.4 mmol g^{-1}.

This large capacity of NO storage is extremely significant. However, it is only useful for potential applications if the gas can be recovered. In this case, a very simple trigger is used to start the release of NO from the samples. Water or moisture has a greater affinity for the metal in the MOF and therefore, when triggered, the NO gas should be released from the metal and replaced with water molecules. Investigations into the release ability of NO from preloaded Ni and Co CPO-27 were performed. The activated materials show a distinct color change when exposed to NO. The Co CPO-27 (already a much darker color from the activation step) turns almost black and Ni CPO-27 turns from yellow to very dark green. These color changes can be seen in Figure 10.13. This color change is indicative of the NO interacting with the available CUSs.

Release profiles for both Ni and Co CPO-27 are shown Figure 10.14. All the NO adsorbed was released, up to ∼7 mmol g^{-1}, in ∼20 h, as evidenced by the reversible change in color (red for Co CPO-27 and yellow for Ni CPO-27). This release behavior is in complete contrast to that which occurs in HKUST-1, where only 2 μmol g^{-1} is released [105].

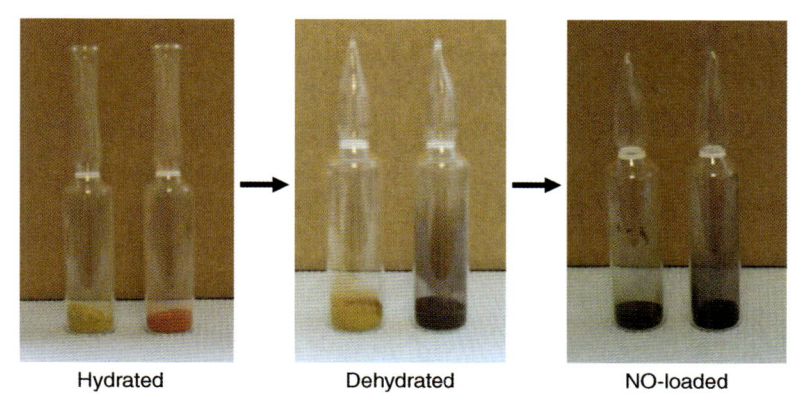

Figure 10.13 Photographs of Ni CPO-27 (always on the left) and Co CPO-27 (always on the right) showing the color changes when samples are dehydrated and NO loaded.

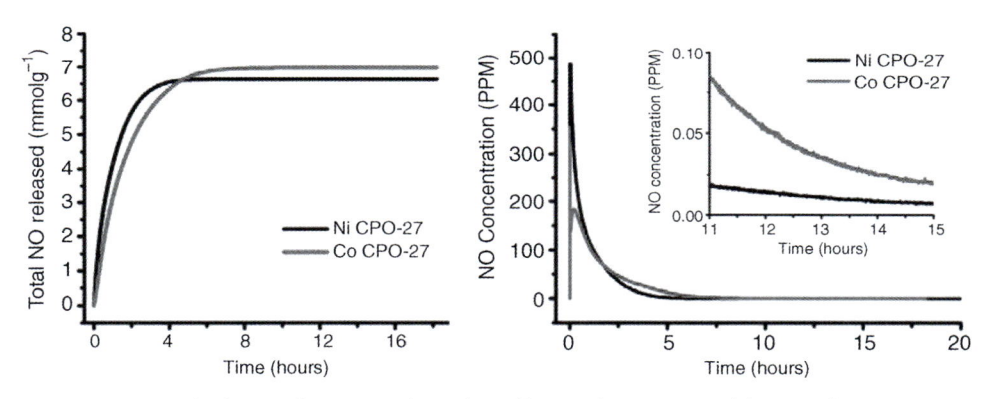

Figure 10.14 Graphs showing the amount of NO released by Ni and Co CPO-27 and the time taken.

This complete cycle of activation, storage, and delivery (see Figure 10.15) is essential for a successful gas storage material. This is extremely significant and represents a milestone for the use of MOFs as possible delivery vehicles for biologically active gases for medicinal applications.

10.3.4
Activity Tests

10.3.4.1 Activity of Drug-Containing MOFs
Some carriers that showed good drug delivery performances failed, however, in *in vitro* and *in vivo* conditions, since they were are not able to release the active cargo, bypass the organic barriers, or remain stable enough. For a better understanding, it is therefore necessary to study the *in vitro* and *in vivo* activity of the drug-loaded carrier. Using MOFs or other carriers, *in vitro* assays using physiological simulated

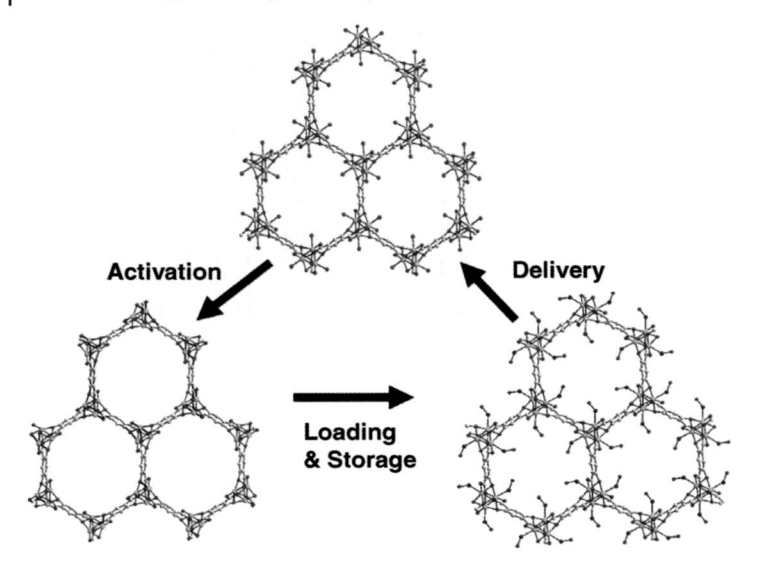

Figure 10.15 The required cycle of activation, loading, storage, and delivery for a prospective candidate material suitable for gas storage. The activation stage is extremely important as great care must be taken to ensure all guest solvent molecules are removed to allow all CUSs available for NO binding. Color key: cyan, Ni/Co; red, oxygen; gray, carbon; pink, oxygen of coordinated water molecules. The hydrogen atoms and uncoordinated water molecules have been omitted for clarity.

conditions are required. The *in vitro* cancer cell cytotoxicity activity of silica-covered nanoparticles of terbium-based MOF (mouse DL_{50} oral administration (v.o.) 3.6 g kg^{-1} [106]; intraperitoneal (i.p.) 0.3 g kg^{-1} [107]) and the antitumoral prodrug NCP-1′ [$Tb_2(dscp)_3(H_2O)_{12}$, where dscp = disuccinatocisplatin] were evaluated using an angiogenic human colon carcinoma cell line (HT-29) (Figure 10.16) [108]. The prodrug was released through the degradation of the solid, with a release of 1, 5.5,

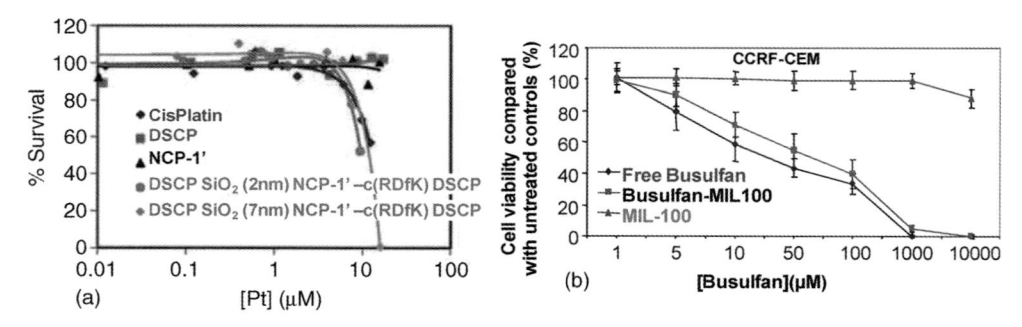

Figure 10.16 *In vitro* cytotoxicity of cisplatin prodrug-loaded NCP-1′ nanoparticles (a) and of Busulfan-charged MIL-100 nanoparticles (b). From (a) Rieter *et al.* [108] and (b) Horcajada *et al.* [55].

and 9 h for coverage with a 0, 2 and 7 nm silica layer, respectively. These silica-coated nanoparticles showed no antitumoral activity after 72 h of incubation, presumably because DSCP was not able to penetrate the cell. Thus, silica-covered nanoparticles were superficially grafted with a c(RGDfK) peptide to target the $\alpha_v\beta_3$ integrins overexpressed on the membrane of many angiogenic cancer cells, and showed an IC_{50} (50% inhibitory concentration) of 10, which is slightly smaller than that of free cisplatin itself (13 μM). This suggests that the coated nanoparticles are probably internalized into the cells and then release the prodrug within the cell, where it could be further reduced by endogenous biomolecules from Pt^{IV} to the active Pt^{II} species. In contrast, *in vitro* cytotoxicity assays performed on another carcinoma cell line (human breast carcinoma cell MCF-7) showed similar activities to free cisplatin, confirming the activity of the formulation.

Lin and co-workers also evaluated the *in vitro* anticancer efficacy of silica-covered MIL-101 nanoparticles loaded with the prodrug ethoxisuccinate cisplatin on the human colon carcinoma HT-29 line [68]. The *in vitro* cytotoxicity was slightly lower than that of free cisplatin (IC_{50} = 29 and 20 μM for drug loaded silica-covered MIL-101 and free cisplatin, respectively). The authors have observed cytotoxicity values similar to those for free cisplatin through increasing the cell uptake by functionalizing the surface with the c(RGDfK) peptide. This means that cisplatin prodrug-MOF is released and reduced to its active form with no loss of cytotoxicity activity.

Finally, promising results were obtained with the antiretroviral AZT-Tp and antitumoral Bu using the mesoporous iron trimesate MIL-100. *In vitro* activity tests on both drug-containing nanoparticles were performed [55]. The cytotoxicity of Bu-loaded iron trimesate MIL-100 nanoparticles was evaluated on human leukemia (CCRF-CEM) and human multiple myeloma (RPMI-8226) cell lines, resulting in a similar activity to that of the free drug, which indicates that Bu is released in its active form without any degradation (Figure 10.16). The anti-HIV activity of the AZT-Tp-containing MIL-100 nanoparticles was also investigated using HIV-1 LAI-infected peripheral blood mononuclear cells (PBMCs). A significant anti-HIV activity was observed only for the drug-loaded nanoparticles (90% inhibition of HIV replication for 200 nM of AZT-Tp in MIL-100). Considering that the highly hydrophilic AZT-Tp is not able to penetrate the cell, this not only proves that drug cargo is delivered in its active form, but also that a cellular uptake of the nanoparticles occurs, allowing the release of the active drug into the HIV-infected cells.

10.3.4.2 Activity of NO-Loaded Samples

As shown above, MOFs can release their store of NO under PBS conditions which mimic the body's own physiological environment. As has been shown already, NO-loaded MOFs completely inhibit the formation of platelet aggregation in blood [105]. To see the effect of an NO-loaded MOF on tissue, an initial biological experiment was carried out using a pressed pellet of Ni CPO-27 (5 mg) on pre-contracted pig coronary arteries *in vitro* (Figure 10.17) [104]. Placement of pellets at a distance of 2 mm from the vessel in a 10 ml organ bath resulted in rapid 100% relaxation of the vessel. The pellet could be seen to generate bubbles of gas for ∼10 min of submersion, although the relaxation remained maximal for longer than 1 h. In some experiments

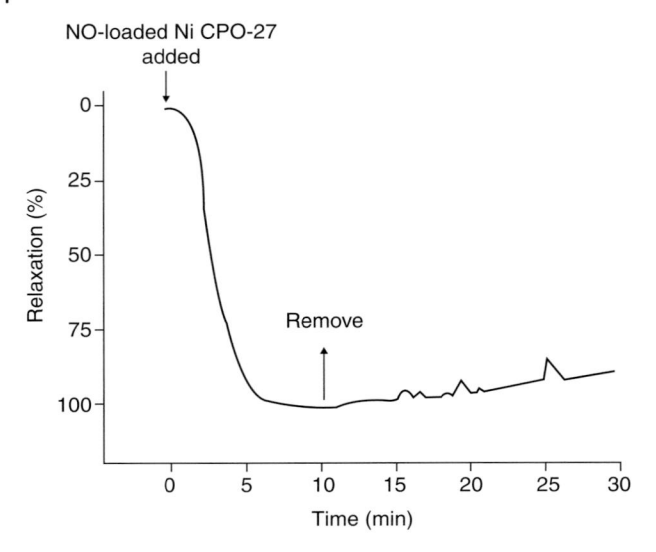

Figure 10.17 Representative trace showing the vasodilatory effect of NO-loaded Ni CPO-27 in a pre-contracted porcine coronary artery. A pellet (~5 mg) of NO-loaded Ni CPO-27 was placed in the organ bath (10 ml) and removed as indicated. Within 4 min of the addition of the pellet, 100% relaxation was achieved, and the relaxation slowly recovered upon pellet removal. In experiments in which the pellet was left in the bath, recovery was not seen for up to 1 h after pellet addition.

(Figure 10.17), the pellet was removed from the bath after 10 min and the relaxation was seen gradually to recover. Parallel control experiments with NO-free Ni CPO-27 MOFs failed to cause relaxation and did not generate bubbles on submersion, indicating that NO release from the loaded MOF is responsible for the relaxing effect.

10.3.4.3 Activity of Silver Coordination Polymers

Silver is an increasingly important antibacterial agent, which shows excellent effectiveness against many different organisms. Although no porous MOFs have been studied for their antibacterial properties, nonporous silver coordination polymers with various structures and linkers have been intensively studied during the last two decades [109]. It has been shown that these solids exhibit antimicrobial properties in addition to their capacity to form silver nanoparticles [110]. For instance, the reaction between silver nitrate and an isonicotinic derivative of bis(ethylene glycol) results in a layered coordination polymer comprised of 1D silver–ligand chains bridged by nitrates. Dental restorative materials such as a gold alloy was then coated with the silver coordination polymer. Using a flow chamber mimicking the oral cavity, it was shown that bacteria still adhere to the coated surface but were killed upon contact with the substrate (Figure 10.18) [111]. In the present case, activity is due to the release of silver from the coating.

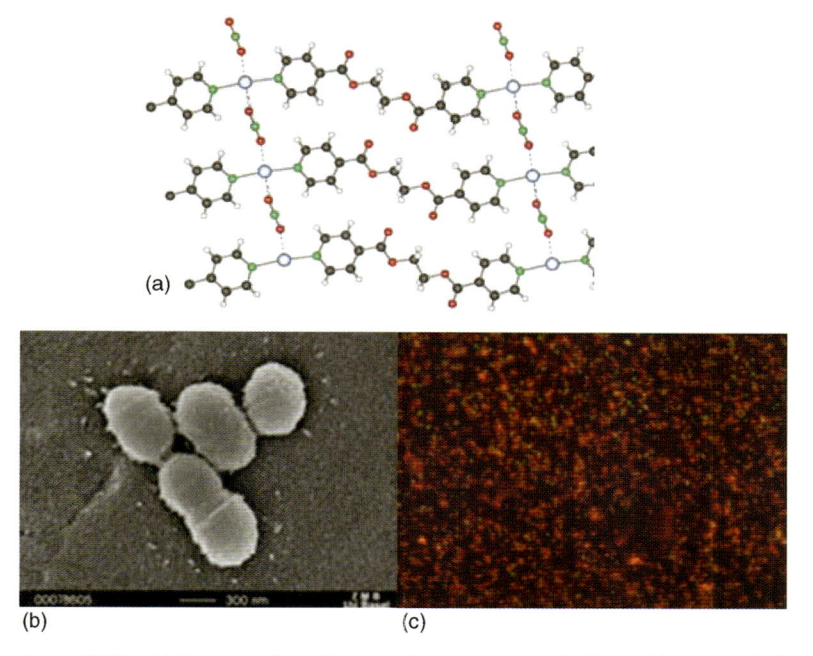

Figure 10.18 (a) Structure of the silver coordination polymer. (b) View of the bacterial adhesion to gold alloy. (c) Optical micrograph of dead bacteria (with fluorescent marker) on a gold alloy coated with silver coordination polymer. According to ref. [34].

10.4 Diagnostics

10.4.1 Magnetic Resonance Imaging

Another interesting application is to use MOFs based on paramagnetic metals as contrast agents for magnetic resonance imaging (MRI), which was reported in 2006 by Lin and co-workers through the use of nanoscale Gd^{3+}-based MOFs (Figure 10.19) [116]. These gadolinium terephthalate or trimellitate materials exhibited extraordinarily large longitudinal r_1 and transversal r_2 relaxivities due to the presence of a very high content of paramagnetic Gd^{3+} centers within each particle. Other nanoMOFs (nanoparticles of MOFs) as contrast agents were later reported also based on Gd and benzenehexacarboxylate [112]. However, the poor chemical stability of these Gd-MOF nanoparticles rules out their practical use due to the high toxicity of gadolinium.

This approach was later extended to less toxic Mn^{2+} carboxylate-coated MOFs with very high *in vivo* r_1 relaxivities by binding to intracellular proteins [113]. However, silica was used to coat the nanoparticles to prevent rapid degradation, which led to modest r_1 values, not enough to be efficient T_1-weighted contrast agents.

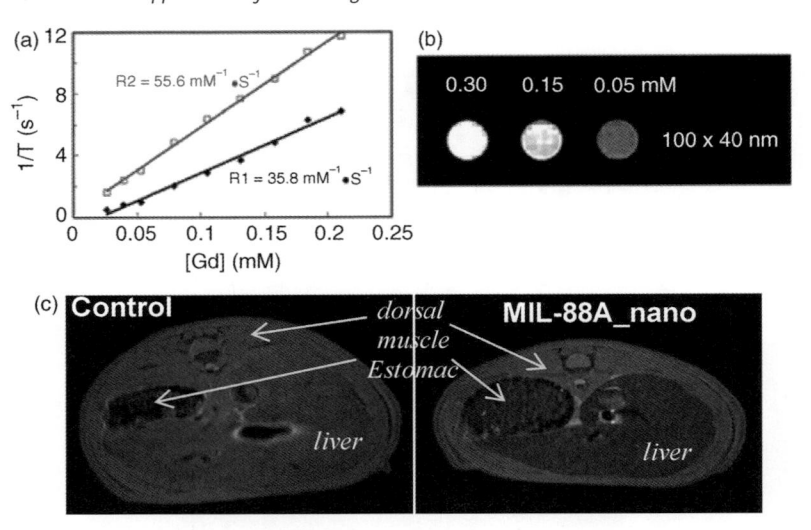

Figure 10.19 (a) r_1 and r_2 relaxivity curves of the Gd carboxylate nanoparticles of dimensions 100 × 40 nm and (b) T_1-weighted MRI images of suspensions of Gd^{3+} terephthalate MOF in water containing 0.1% xanthan gum. From Rieter et al. [108]. (c) MRI images of spin echo sequence of the control and the MIL-88A-nano-treated rats 30 min after injection. From Horcajada et al. [55].

If one considers typical "superparamagnetic" agents, such as nanoparticles of magnetite (Fe_3O_4) [114], their relaxivity values r_1 and r_2 are much higher than those of Gd- or Mn-based MOFs [108]. An alternative is to use iron(III)-based MOFs as contrast agents. Recently, nanoparticles of several nontoxic porous iron(III) carboxylates have shown significant relaxivity times, r_2 from 56 to 73 s^{-1} mM^{-1} for MIL-88A and MIL-100 nanoparticles, respectively, considered to be sufficient for *in vivo* use [55]. After intravenous administration to rats, the spleen and liver were clearly darker than with the control, in agreement with the accumulation of the nanoparticles within these target organs (Figure 10.19). Imaging properties are ascribed to both the paramagnetic oxo-centered trimers of iron(III) octahedra and an interconnected 3D porous network with free and coordinated water molecules, which can easily exchange their protons. Hence these nanoparticles may act as efficient drug nanocarriers, associating therapeutics and diagnostics suitable for theranostics.

10.4.2
Optical Imaging

In vitro fluorescence has also been studied using different functionalized MOFs. For instance, the amino groups present at the surface of nanoparticles of the iron aminoterephthalate MIL-101 were used for the covalent attachment of the biologically relevant fluorophore BODIPY (1,3,5,7-tetramethyl-4,4-difluoro-8-bromomethyl-4-bora-3a,4a-diaza-s-indacene) [68]. An anticancer drug (loading of 12.8 wt%)

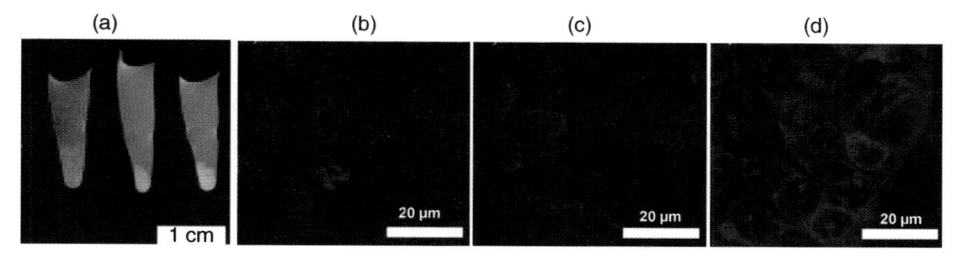

Figure 10.20 (a) *In vitro* MRI images of HT-29 cells incubated without Mn trimesate nanoparticles (left), Mn(II) trimesate nanoparticles (middle), and c(RGDfK)–Mn(II) trimesate nanoparticles (right). (b–d) Confocal images of HT-29 cells incubated without Mn trimesate nanoparticles (b), with Mn trimesate nanoparticles (c), and with c(RGDfK)–Mn trimesate nanoparticles (d). The blue color is from DRAQ5 used to stain the cell nuclei and the green color is from rhodamine. From Taylor and coworkers [113]).

was then introduced to associate *in vitro* optical imaging (fluorophore) and anticancer activity (drug). The *in vitro* efficacy of silica-coated BODIPY-grafted MIL-101 nanoparticles as optical contrast agents was evaluated on HT-29 human colon adenocarcinoma cells. Despite the presence of fluorescence from the BODIPY particles after incubation, it was difficult to assess the effective internalization of the nanoparticles within the cells since nanoparticles are nonemissive due to the d–d luminescence quenching from the Fe^{III} centers. This would signify that fluorescence is due to the released BODIPY moieties, occurring either (1) after the cell uptake of the nanoparticles, confirming the penetration of the nanoparticles, or (2) before, so that BODIPY detection does not give any information about the particle location. Similar results were obtained using Mn^{II} trimesate nanoparticles surface functionalized with rhodamine B (fluorescent) and c(RGDfK) (targeting agent) [113]. These nanoparticles combine the imaging properties of the delivered Mn^{2+} with the optical imaging of the rhodamine B. As shown below, c(RGDfK) surface-modified nanoparticles are taken up well by human colon cancer HT-29 cells (Figure 10.20). Finally, nanoparticles of several lanthanide terephthalate MOFs were grafted with a Tb complex and dipicolinic acid (fluorophore) for imaging purposes [115].

10.5
From Synthesis of Nanoparticles to Surface Modification and Shaping

10.5.1
Synthesis of Nanoparticles

The synthesis of nanoparticles is required for the administration of the drug carrier system by intravenous (i.v.), intraperitoneal (i.p.), subcutaneous (s.c.), intranasal (i.n.), intraocular (i.o.), and so on routes, in order to avoid embolism. For i.v. administration of MOFs, homogeneous nanoparticles smaller than 200 nm are necessary to prevent aggregation, reactivity, or precipitation. The synthesis of

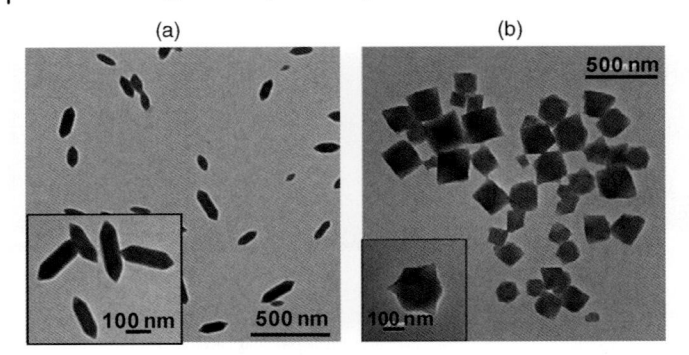

Figure 10.21 Scanning and transmission electron microscopy images of nanoparticles of MIL-88B-NO$_2$ (a) and MIL-101-NH$_2$ (b) made from microwave synthesis.

MOFs is usually performed under hydro/solvothermal conditions by tuning the temperature, pressure, the nature of the solvent, pH, stoichiometry, and so on. Recently, synthetic methods such as (1) reverse emulsions [112, 116] and (2) microwave-assisted hydro/solvothermal routes [55, 117, 118] have been proposed for the synthesis of narrow distributed nanoparticles of MOFs (Figure 10.21).

Reversed-phase emulsions consist in producing a microemulsion from the metal source and linker in the presence of a cationic surfactant [cetyltrimethylammonium bromide (CTAB)]–isooctane–1-hexanol–water system. This simple and easily generalized method has been applied to different Ln-based nonporous MOFs with imaging properties [108, 112] and to nanorods of manganese polycarboxylates [113]. The main drawbacks of such a method are its low yield and the very important volumes of toxic components required, which would rule out their practical *in vivo* use {CTAB, LD$_{50}$ (i.v. rats) = 44 mg kg^{-1} [119, 120]; isooctane, LD$_{50}$ (oral rats) = 3310 mg kg^{-1} [121]; 1-hexanol, LD$_{50}$ (oral rats) = 720 mg kg^{-1}) [122]}.

Microwave-assisted hydro/solvothermal synthesis is a convenient method to obtain high yields of nanoMOFs with good control of the particle size and a narrow size distribution. Several examples have been reported recently, such as the mesoporous chromium terephthalate MIL-101 (22 nm) [117], the microporous zinc imidazolate ZIF-8 (40 nm) [123], the flexible microporous iron terephthalate MIL-53 (350 × 1000 nm), the mesoporous iron trimesate MIL-100 (200 nm) [55], and the iron aminoterephthalate MIL-101-NH$_2$ (120 nm) [68].

Additionally, nanoparticles can also be obtained through traditional hydro/solvothermal routes by tuning the different synthetic parameters such as the synthesis time, temperature, concentration, pH, or the use if inhibitors (acetic acid, hydroxybenzoic acid, etc.). This led to nanoMOFs of the porous zinc terephthalate MOF-5 [124], the flexible porous iron muconate MIL-89 (30 nm) [125], and the iron fumarate MIL-88A (150 nm) [55]. However, lower yields and/or a higher degree of polydispersity of the particle size are usually obtained compared with the microwave hydro/solvothermal method.

10.5.2
Surface Engineering

For stealth, addressing, or bioadhesion purposes [126–128], it is necessary to graft various functional groups at the surface of MOFs. In general, MOF particles are covered with various reactive groups (metals, counteranions, functional groups from the linker, etc.), which can interact weakly or strongly with grafting groups.

Surface engineering might also be helpful in preventing aggregation in a formulation of within the biological medium. Electrostatic or by hindrance repulsion is a suitable method to prevent aggregation. Additionally, this can help to stabilize the nanoparticles further in the biological media by slowing their degradation. Lin and co-workers [68, 108, 113] applied the method developed by van Blaaderen and co-workers [129] to several MOF nanoparticles. As indicated earlier, covering nano-MOFs, previously modified with polyvinylpyrrolidone (PVP), with silica shells of variable thickness using a sol–gel method, avoided the fast leaching of toxic metals. The same method was applied to an Mn trimesate nanoMOF, which increased its half-live in water from 3.5 to 7.5 h and in PBS from 18 min to 1.44 h (both at 37 °C) [113]. This also allowed a slower release of drug, as shown also for the cisplatin silica-covered iron terephthalate particles [68]. However, the *in vivo* toxicity of silica nanoparticles is still an issue.

The external surface of nanoMOFs can also be modified, either through post-synthetic functionalization or directly during the synthesis of MOFs, using organic moieties well known to provide stealth or addressing properties, such as described above through different silica coatings of nanoMOFs using a targeting peptide overexpressed in angiogenic cancers [$\alpha_v\beta_3$ integrin c (RGDfK)] and/or imaging agents (BODIPY [68], rhodamine [113], dipicolinic acid [108]).

Finally, several porous iron(III) polycarboxylate nanoparticles were superficially modified, by post-synthetic or direct methods, using PEG, to avoid rapid sequestration of the nanoMOFs by the reticuloendothelial system (RES) [130–132]. Chitosan and dextran–fluorescein–biotin polymers were also grafted on the surface of these particles to obtain bioadhesive or addressing properties [55].

In conclusion, surface engineering in MOFs if still in its infancy, and can improve their stability, controlled release, or imaging properties, or provide them with addressing or stealth properties.

10.5.3
Shaping

Achieving real applications requires both suitable synthesis scale-up and shaping. Although most MOFs are synthesized at the milligram laboratory scale through conventional solvothermal synthesis, there are now several examples of MOFs produced at the industrial scale and that are commercially available {[e.g., copper trimesate (HKUST-1 or Basolite™ C300), aluminum terephthalate [MIL-53(Al) or Basolite™ A100], magnesium formate (Basosiv™ M050), zinc 2-methylimidazole (Basolite™ Z1200), and iron trimesate (Basolite™ F300)} [133].

Further, using MOFs for biomedical application requires the preparation of specific formulations with optimal performances (tablets, pellets, extrudates, packed columns, etc.). Here, the choice of the formulation depends mainly on the administration route (topical, enteral, or parenteral). Whatever the route, it is vital to control the particle size and distribution, which affect the process conditions and the stability of formulations. Therefore, if one can succeed in good control of the key parameters, such as rheology, packing, and reactivity, then reproducible and stable formulations might be obtained.

Homogeneous powders, pellets, or tablets could be also used for enteral administration (Figure 10.21). In this case, one the following parameters should be studied: (1) stability of the MOFs in stomachal (acidic) or intestinal (basic) conditions; (2) surface charge of particles, to avoid agglomeration through electrostatic repulsion; and (3) mechanical stability under pressure. For instance, tablets of ibuprofen containing iron carboxylates (MIL-53, MIL-100) were obtained using low pressures without any additives [27]. However, addition of a binder might improve the shaping or the stability of pellets or tablets of MOFs. Noteworthily, despite the lack of experimental shaping data for MOFs, several commercialized MOFs are now produced as pellets or extrudates using binders [134]. For bioapplications, many biologically relevant excipients are available and could help in the near future to achieve optimal shaping.

Making thin films of MOFs could also represent an alternative to produce patches or composites. So far, homogeneous thin films of MOFs have been prepared from stable colloidal solutions by a simple dip or spin coating method (Figure 10.22) [135–137]. This allows control of the film thickness by multilayer deposition or by controlling the particle size and/or deposition conditions. This could be extended in the near future to the preparation of cream/ointment formulations of MOFs. This concept has already been applied to zeolites [138].

Finally, composite materials based on MOFs and organic polymers and/or silica have been prepared. Three different approaches have been considered so far:

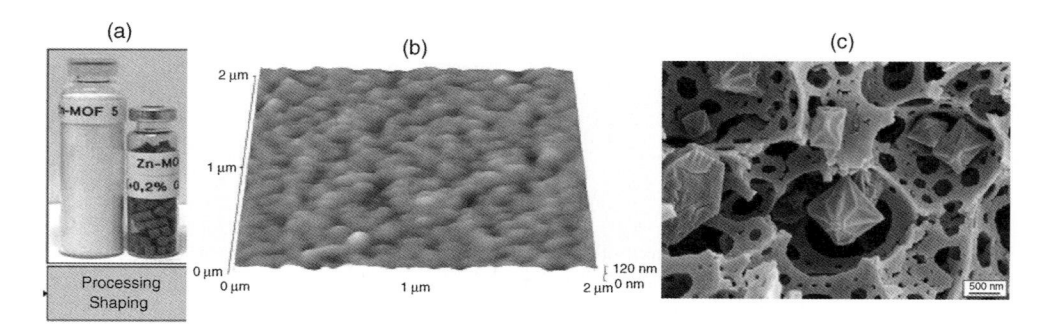

Figure 10.22 (a) Photograph of powder and pellets of MOFs. (b) Atomic force microscopy image of the surface of an MIL-101 thin film. (c) Scanning electron microscopy image of HKUST-1 particles entrapped into a macroporous polyHIPE. (a, c) From Mueller et al. [133] and (b) from Demessence et al. [136] (center).

(1) coating (core–shell) with, for instance, a PVP and silica cover shell to control the degradation and delivery; (2) composite materials, by mixing the MOF with a nonmiscible material, and (3) integrating the MOF directly into a polymer matrix [e.g., the copper trimesate HKUST-1 within a monolithic macroporous hydrophilic polymer (polyHIPE)] (Figure 10.22) [139].

The delivery of NO to the skin, for wound-healing purposes, cannot be performed directly using powders, so composites based on MOF particles and hydrocolloids have been prepared. Such patches are required particularly for patients with diabetic wounds (wounds that have very long repair times and are extremely susceptible to infection). In addition, there is need to find a suitable method for storing powders for a practical formulation that can be sold commercially in a patient-friendly manner. Hydrocolloids are widely used dressings in medicine [140]. Here, combinations of a polymer and NO-loaded MOF together with a binder such as cellulose have been combined as a wound-healing and antibacterial dressing. The polymer, cellulose, and NO-loaded MOF were mixed under an inert atmosphere at elevated temperature, resulting in the production of a hydrocolloid with the ability to release NO over several days. For this study, Ni CPO-27 was chosen as the MOF. Figure 10.23 shows the hydrocolloid prepared in the laboratory.

Chemiluminescence measurements of NO release were made on the hydrocolloid and the resulting curves are illustrated in Figure 10.24. Although releasing only ~2.5 mmol g^{-1} (about half of the amount released by the powder), it released this store over almost 10 days, which is extremely significant. Further to this, the amount of NO released is much higher than that released by powdered zeolites.

Apart from the exciting release properties, however, it is important to demonstrate that no leaching of toxic nickel arises from the degradation of the MOF structure. Previous studies carried out using bovine serum have shown that the amount of nickel that leaches out from the framework is minimal [141]. The prepared hydrocolloid–MOF composite and a pellet of Ni CPO-27 were placed in PBS at 37 °C for several days. The hydrocolloid retains the Ni from the MOF significantly better than

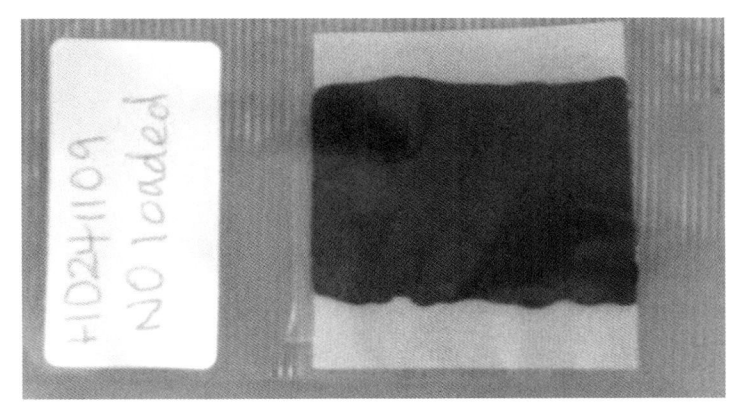

Figure 10.23 Photograph of an NO-loaded hydrocolloid in a sealed vacuum pack.

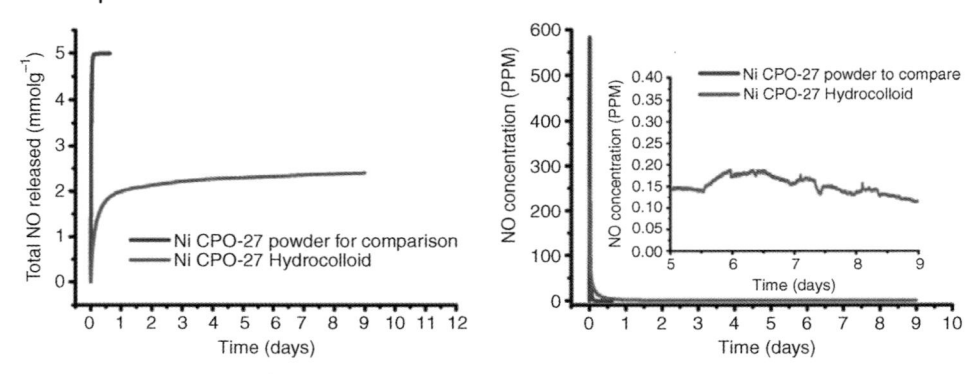

Figure 10.24 NO release curves for the hydrocolloid shown compared with a powder sample.

the pellet sample of Ni CPO-27. This confirms that when mixed with polyisobutylene (PIB), the MOF degrades more slowly and almost no nickel is leached out (Figure 10.25).

10.6
Discussion and Conclusion

MOFs are a fascinating class of crystalline porous hybrid solids. Their composition, structure, pore size, and volume are easy tunable through the choice of the metal (transition metal, lanthanide, calcium/magnesium, etc.) and the ligand (carboxylates, phosphonates, imidazolates, etc.) or through organic functionalization. This represents a major advantage in order to optimize their adsorption properties, compared with zeolites or activated carbons. The toxicity of a few porous iron

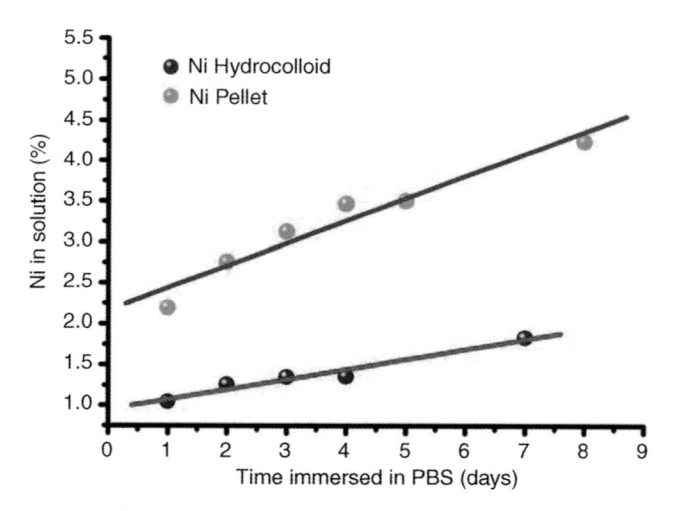

Figure 10.25 Atomic adsorption measurements carried out on a prepared hydrocolloid and a Ni CPO-27 pellet for comparison.

carboxylate MOFs has also been assessed at the preclinical level. Doses as high as 220 mg kg^{-1} have been injected i.v. into rats through a single dose with no sign of toxicity. In addition, iron polycarboxylate MOFs are biodegradable solids with differences in terms of degradation kinetics as a function of their composition and structure, from a few hours to weeks. Another advantage of MOFs over other carriers is their higher loading capacity for drug molecules or biological gases, up to 1.4 g of drug per gram of solid or 7 mmol g^{-1} of NO, together with controlled release in most cases. Hence this paves the way for the encapsulation of highly active drugs with huge loading capacities, followed in some cases by their controlled release into the environment. The bioactivity in MOFs can also be introduced directly by choosing a bioactive molecule (e.g., drug) as the linker and/or an active metal (Gd, Fe, etc.) as the inorganic counterpart with, for instances (Gd, Mn, Fe), imaging properties useful for theranostics or as contrast agents. Finally, the synthesis of MOF nanoparticles can be achieved through various techniques such as reverse emulsion techniques, microwave-assisted hydro- or solvothermal conditions, and sonothermal methods, while surface modification or shaping can be tuned directly through the synthesis or post-synthesis treatments.

However, despite these useful features compared with well-established carriers (liposomes, polymers, etc.), the use of MOFs for practical biomedical applications is still in its infancy and requires several critical issues to be resolved. First, even if very long releases are achieved in some cases, there are cases where burst release is (partially) observed. This is related to the absence of real control of the kinetics of delivery for a given drug molecule and MOF structure. Changing the structure, using organic functionalization or coating of particles, is a suitable approach used so far to tune the delivery rate of a biomolecule. However, there is still no rational approach for the control of the kinetics of release of drugs in MOFs and there is clearly a need for additional experimental data in order to predict the best MOF for a given active molecule and its administration route. One way would be to undertake systematic encapsulation/release studies of model drugs including adequate characterizations and modeling techniques. Second, unlike their inorganic or organic counterparts, it is difficult to obtain stable monodisperse MOF nanoparticles while avoiding their rapid degradation or particle aggregation. Part of the problem can be circumvented through surface modification, but strong efforts are still needed to resolve these issues. Stealth, addressing, or imaging properties of MOF nanoparticles can, in principle, be ensured through surface modifications with various organic biomolecules (PEG, chitosan, etc.), during or after the synthesis. However, here again, very little work has been carried out in this domain so far, and the corresponding surface modifications have been tested only for a few selected MOFs. There is a need to extend these methods further to other MOFs of interest in the near future. Similar issues are important in the design of MOFs for the controlled release of medical gases such as NO. In fact, as NO is such a small, rapidly diffusing gas molecule, these challenges can be all the more difficult. It is here that control over the formulation and shaping of the material becomes paramount in designing the properties of the therapy with sufficient control to obtain optimum delivery. Early studies in this area are very promising, and it is important to continue to develop new methods of combining the obvious advantages of MOFs with practical methods of delivering a therapy.

Another important issue concerns the very large differences in biodegradability from one MOF to another. This is critical if one wants to "choose" or to "design" a specific MOF with a suitable stability for a given bioapplication. Very little information is available so far (under physiological conditions) and, again, this is related to a lack of experimental data. The same situation arises in the production of films or composites, where the presence of organic and inorganic moieties in MOFs might be helpful for mixing MOFs and organic polymers for their potential use as patches, creams, and related applications. However, very little has been done in this area so far and many experimental studies will be required in order to establish the most appropriate processing conditions.

Absence of any toxicity at the preclinical level has been demonstrated for a few selected iron carboxylates. Additional toxicity data are required, however, either to assess the toxicity of other compositions or structures or to deepen the understanding of the toxicity of these iron-based MOFs, through i.v administration or other routes.

Acknowledgments

P.H and C.S. thank the EU for funding via the ERC-2007-209241-BioMOFs, Prof. G. Férey, Prof. P. Couvreur, Dr. R. Gref, Dr. J.S. Chang, co-workers and students, related to bioapplications of MOFs, from the Institut Lavoisier, Université Paris Sud, and Krict (Daejon, Korea). R.E.M.and A.C.McK. acknowledge the EPSRC for funding. R.E.M. is a Royal Society Wolfson Merit Award Holder.

References

1 (a) Clearfield, A. (1998) *Prog. Inorg. Chem.*, **47**, 371–510; (b) Férey, G. (2008) *Chem. Soc. Rev.*, **37**, 191.
2 See special issues: *Acc. Chem. Res.*, 2005, **38**, 215–378; *J. Solid State Chem.*, 2005, **178**, 2409–2574; *Chem. Soc. Rev.*, 2009, **38**, 1201–1508.
3 Eddaoudi, M., Moler, D.B., Li, H., Chen, B., Reineke, T.M., O'Keeffe, M., and Yaghi, O.M. (2001) *Acc. Chem. Res.*, **34**, 319–330.
4 (a) Some major references in the field: Yaghi, O.M., O'Keeffe, M., Ockwig, N.W., Chae, H.K., Eddaouddi, M., and Kim, J. (2003) *Nature*, **423**, 705–714; (b) Rowsell, J.L.C. and Yaghi, O.M. (2005) *Angew. Chem. Int. Ed.*, **44**, 4670–4679; (c) Rao, C.N.R., Natarajan, S., and Vaidhyanathan, R. (2004) *Angew. Chem. Int. Ed.*, **43**, 1466–1496; (d) Kitagawa, S., Kitaura, R., and Noro, S.-H. (2004) *Angew. Chem. Int. Ed.*, **43**, 2334–1375; (e) James, S.L. (2003) *Chem. Soc. Rev.*, **32**, 276–288; (f) Maspoch, D., Ruiz-Molina, D., and Veciana, J. (2007) *Chem. Soc. Rev.*, **36**, 770–818; (g) Cheetham, A.K. and Rao, C.N.R. (2006) *Chem. Commun.*, 4780–4795; (h) Kepert, C.J. (2006) *Chem. Commun.*, 695–700.
5 Park, K.S., Ni, Z., Côté, A.P., Choi, J.Y., Huang, R., Uribe-Romo, F.J., Chae, H.K., O'Keeffe, M., and Yaghi, O.M. (2006) *Proc. Natl. Acad. Sci. USA*, **103**, 10186–10191.
6 Dietzel, P.D.C., Blom, R., and Fjellvag, H. (2008) *Eur. J. Inorg. Chem.*, **23**, 3624–3632.

7 Horcajada, P., Surblé, S., Serre, C., Hong, D.-Y., Seo, Y.-K., Chang, J.-S., Grenèche, J.-M., Margiolaki, I., and Férey, G. (2007) *Chem. Commun.*, 2820.

8 (a) Hafizovic Cavka, J., Jakobsen, S., Olsbye, U., Guillou, N., Lamberti, C., Bordiga, S., and Lillerud, K.P. (2008) *J. Am. Chem. Soc.*, **130**, 13850; (b) Dan-Hardi, M., Serre, C., Frot, T., Rozes, L., Maurin, G., Sanchez, C., and Férey, G. (2009) *J. Am. Chem. Soc.*, **131**, 10857–10859.

9 Safety data for chromium chloride, 6-hydrate; Safety Officer in Physical Chemistry at Oxford University; updated on February 21, 2005, http://msds.chem.ox.ac.uk/CH/chromium_III_chloride_hexahydrate.html.

10 Material Safety data Sheet 7616; Imperial Supplies LLC; updated in 2010, http://www.imperialinc.com/msds0076160.shtml.

11 TOXNET (1975–1986) National Library of Medicine's Toxicology Data Network, Hazardous Substances Data Bank (HSDB). Public Health Service. National Institute of Health, US Department of Health and Human Services. National Library of Medicine, Bethesda, MD.

12 Safety data for manganese (II) chloride tetrahydrate; Safety Officer in Physical Chemistry at Oxford University; updated on October 20, 2003, http://msds.chem.ox.ac.uk/MA/manganese_II_chloride_tetrahydrate.html.

13 Material Safety data Sheet F1678 for Ferrous Chloride, 4-Hydrate; Mallinckrodt Baker Inc.; updated on August 20, 2008, http://www.jtbaker.com/msds/englishhtml/F1678.htm.

14 Material Safety data Sheet for Zinc Chloride; Safety Officer in Physical Chemistry at Oxford University; updated on April 20, 2009, http://msds.chem.ox.ac.uk/ZI/zinc_chloride.html.

15 Material Safety data Sheet MRD-108 for Magnesium Chloride Hexahydrate 64%; Mineral Research and Development; updated on March 31, 2008, http://www.mrdc.com/html/pdf/msds/Magnesium%20Chloride%2064%25%20Soln.%20MSDS.pdf.

16 Material Safety data Sheet for Calcium Chloride; Safety Officer in Physical Chemistry at Oxford University; updated on February 9, 2006, http://msds.chem.ox.ac.uk/CA/calcium_chloride_anhydrous.html.

17 Sids Initial Assessment Profile, UNEP Publications, Paris June 2001, http://www.chem.unep.ch/irptc/sids/OECDSIDS/100-21-0.pdf (2006)

18 Chemicalland 21, 2006, http://www.chemicalland21.com/specialtychem/perchem/TRIMESIC%20ACID.htm.

19 Johnson, W. (1991) Acute Oral Toxicity Study of 2,6-Naphthalene Dicarboxylic Acid (2,6-NDA) in Rats, Project No. 1659. Amoco, Chicago.

20 Safety data sheet for 1-methylimidazole, updated on October 17, Safety Officer in Physical Chemistry at Oxford University, 2006, http://msds.chem.ox.ac.uk/ME/1-methylimidazole.html.

21 (a) Serre, C., Millange, F., Surblé, S., and Férey, G. (2004) *Angew. Chem. Int. Ed.*, **43**, 6285–6289; (b) Surblé, S., Serre, C., Mellot-Draznieks, C., Millange, F., and Férey, G. (2006) *Chem. Commun.*, 284–286.

22 (a) Hwang, Y.K., Hong, D.-Y., Chang, J.S., Jhung, S.H., Seo, Y.-K., Kim, J., Vimont, A., Daturi, M., Serre, C., and Férey, G. (2008) *Angew. Chem. Int. Ed.*, **47**, 4144; (b) Wang, Z. and Cohen, S.M. (2008) *Angew. Chem.*, **120** (25), 4777–4780; Wang, Z. and Cohen, S.M. (2008) *Angew. Chem. Int. Ed.*, **47** (25), 4699–4702; (c) Devic, T., Horcajada, P., Serre, C., Salles, F., Maurin, G., Moulin, B., Leclerc, H., Heurtaux, D., Vimont, A., Clet, G., Daturi, M., Grenèche, J.M., le Ouay, B., Moreau, F., Magnier, E., Filinchuk, Y., Marrot, J., and Férey, G. (2010) *J. Am. Chem. Soc.*, **132**, 1127–1136.

23 (a) Kitagawa, S. and Uemura, K. (2005) *Chem. Soc. Rev.*, **34**, 109–119; (b) Uemura, K., Matsuda, R., and Kitagawa, S. (2005) *J. Solid State Chem.*, **178**, 2420–2429; (c) Férey, G. and Serre, C. (2009) *Chem. Soc. Rev.*, **38** (5), 1380–1399.

24 (a) Serre, C., Millange, F., Thouvenot, C., Noguès, M., Marsolier, G., Louër, D., and Férey, G. (2002) *J. Am. Chem. Soc.*, **124**, 13519–13526; (b) Loiseau, T., Serre, C.,

Huguenard, C., Fink, G., Taulelle, F., Henry, M., Bataille, T., and Férey, G. (2004) *Chem. Eur. J.*, **10**, 1373–1382; (c) Millange, F., Guillou, N., Walton, R.I., Grenèche, J.-M., Margiolaki, I., and Férey, G. (2008) *Chem. Commun.*, 4732–4734.

25. (a) Mellot-Draznieks, C., Serre, C., Surblé, S., Audebrand, N., and Férey, G. (2005) *J. Am. Chem. Soc.*, **127**, 16273–16378; (b) Serre, C., Mellot-Draznieks, C., Surblé, S., Audebrand, N., Fillinchuk, Y., and Férey, G. (2007) *Science*, **315**, 1828–1831.

26. (a) Fletcher, A.J., Thomas, K.M., and Rosseinsky, M.J. (2005) *J. Solid State Chem.*, **178**, 2491–2510; (b) Maji, T.K., Matsuda, R., and Kitagawa, S. (2007) *Nat. Mater.*, **6**, 142–148; (c) Bourrelly, S., Llewellyn, P.L., Serre, C., Millange, F., Loiseau, T., and Férey, G. (2005) *J. Am. Chem. Soc.*, **127**, 13519–13521; (d) Serre, C., Bourrelly, S., Vimont, A., Ramsahye, N., Maurin, G., Llewellyn, P.L., Daturi, M., Filinchuk, Y., Leynaud, O., Barnes, P., and Férey, G. (2007) *Adv. Mater.*, **19**, 2246–2251; (e) Millange, F., Serre, C., Guillou, N., Férey, G., and Walton, R.I. (2008) *Angew. Chem. Int. Ed.*, **47**, 4100–4105; (f) Trung, T.K., Trens, P., Tanchoux, N., Bourrelly, S., Llewellyn, P.L., Loera-Serna, S., Serre, C., Loiseau, T., Fajula, F., and Férey, G. (2008) *J. Am. Chem. Soc.*, **130** (50), 16926–16932; (g) Hamon, L., Llewellyn, P., Devic, T., Ghoufi, A., Clet, G., Guillerm, V., Pirngruber, G., Maurin, G., Serre, C., Driver, G., Van Beek, W., Jolimaître, E., Vimont, A., Daturi M., and Férey G. (2009) *J. Am. Chem. Soc.*, **131** (47), 17490–17499.

27. Horcajada, P., Serre, C., Maurin, G., Ramsahye, N.A., Vallet-Regí, M., Sebban, M., Taulelle, F., and Férey, G. (2008) *J. Am. Chem. Soc.*, **130**, 6774–6780.

28. Low, J.J., Benin, A.I., Jakubczak, P., Abrahamian, J.F., Faheem, S.A., and Willis, R.R. (2009) *J. Am. Chem. Soc.*, **131**, 15834–15842.

29. Li, H., Eddaoudi, M., O'Keeffe, M., and Yaghi, O.M. (1999) *Nature*, **402**, 276–279.

30. Taylor-Pashow, K.M.L., Rocca, J.D., Xie, Z., Tran, S., and Lin, W. (2009) *J. Am. Chem. Soc.*, **131** (40), 14261–14263.

31. Miller, S.R., Heurtaux, D., Baati, T., Horcajada, P., Grenèche, J.-M., and Serre, C. (2010) *Chem. Commun*, **46**, 4526–4528.

32. Hinks, N.J., McKinlay, A.C., Xiao, B., Wheatley, P.S., and Morris, R.E. (2010) *Micropor. Mesopor. Mater.*, **129** (3), 330–334.

33. Horcajada, P., Chalati, T., Serre, C., Gillet, B., Sebrie, C., Baati, T., Eubank, J.F., Heurtaux, E., Clayette, P., Kreuz, C., Chang, J.-S., Hwang, Y.K., Marsaud, V., Bories, P.-N., Cynober, L., Gil, S., Férey, G., Couvreur, P., and Gref, R. (2010) *Nature*, **9**, 172–178.

34. Soma, C.E., Dubernet, C., Barratt, G., and Benita, S. (2000) *J. Control. Release*, **68**, 283–289.

35. Mainardes, R.M. and Silva, L.P. (2004) *Curr. Drug Targets*, **5**, 449–455.

36. Qiu, L.Y. and Bae, Y.H. (2006) *Pharm. Res.*, **23** (1), 1–30.

37. Samad, A., Sultana, Y., and Aqil, M. (2007) *Curr. Drug Deliv.*, **4** (4), 297–305.

38. See special issue and related references: *Adv. Drug Deliv. Rev.*, 2008, **60**, 1203–1504.

39. Peer, D. et al. (2007) *Nat. Nanotechnol.*, **2**, 751–760.

40. Dyer, A., Morgan, S., Wells, P., and Williams, C. (2000) *J. Helminthol.*, **74** (2), 137–141.

41. Horcajada, P., Márquez-Alvarez, C., Rámila, A., Pérez-Pariente, J., and Vallet-Regi, M. (2006) *Solid State Sci.*, **8** (12), 1459–1465.

42. Arruebo, M., Fernández-Pacheco, R., Irusta, S., Arbiol, J., Ibarra, M.R., and Santamaría, J. (2006) *Nanotechnology*, **17**, 4057–4064.

43. (a) Uglea, C.V., Albu, I., Vatajanu, A., Croitoru, M., Antoniu, S., Panaitescu, L., and Ottenbrite, R.M. (1994) *J. Biomater. Sci. Polym. Ed.*, **12**, 633; (b) Pavelic, K., Hadzija, M., Bedrica, L., Pavelic, J., Dikic, I., Katic, M., Kralj, M., Bosnar, M.H., Kapitanovic, S., Poljak-Blazi, M., Krizanac, S., Stojkovic, R., Jurin, M., Subotic, B., and Colic, M. (2001) *J. Mol. Med.*, **78**, 708–720.

44. Slowing, I.I., Trewyn, B.G., Giri, S., and Lin, V.S.-Y. (2007) *Adv. Funct. Mater.*, **17**, 1225–1236.

45 Manzano, M., Colilla, M., and Vallet-Regí, M. (2009) *Expert Opin. Drug Deliv.*, **6** (12), 1383–1400.

46 Vallet-Regi, M., Balas, F., and Arcos, D. (2007) *Angew. Chem. Int. Ed.*, **46**, 7548–7558.

47 Giri, S., Trewyn, B.G., and Lin, V.S.Y. (2007) *Nanomedicine*, **2** (1), 99–111.

48 Levy, M.H. and Wheelock, E.F. (1975) *J. Immunol.*, **115**, 41–48.

49 Vallet-Regi, M., Ramila, A., del Real, R.P., and Perez-Pariente, J. (2001) *Chem. Mater.*, **13**, 308–311.

50 Huo, Q., Liu, J., Wang, L.Q., Jiang, Y., Lambert, T.N., and Fang, E. (2006) *J. Am. Chem. Soc.*, **128**, 6447–6453.

51 Slowing, I.I., Trewyn, B.G., Giri, S., and Lin, V.S.-Y. (2007) *Adv. Funct. Mater.*, **17**, 1225–1236.

52 Horcajada, P., Rámila, A., Férey, G., and Vallet-Regí, M. (2006) *Solid State Sci.*, **8**, 1243–1249.

53 Xie, G., Sun, J., Zhong, G., Shi, L., and Zhang, D. (2010) *Arch. Toxicol.*, **84** (3), 183–190.

54 Di Pasqua, A.J., Sharma, K.K., Shi, Y.-L., Toms, B.B., Ouellette, W., Dabrowiak, J.C., and Asefa, T. (2008) *J. Inorg. Biochem.*, **102**, 1416–1423.

55 Horcajada, P., Serre, C., Vallet-Regi, M., Sebban, M., Taulelle, F., and Férey, G. (2006) *Angew. Chem. Int. Ed.*, **45**, 5974–5978.

56 Férey, G., Serre, C., Mellot-Draznieks, C., Millange, F., and Surblé, S. (2004) *Angew. Chem. Int. Ed.*, **116**, 6456–6461.

57 Férey, G., Mellot-Draznieks, C., Serre, C., Millange, F., Dutour, J., Surblé, S., and Margiolaki, I. (2005) *Science*, **309**, 2040–2042.

58 Babarao, R. and Jiang, J. (2009) *J. Phys. Chem. C*, **113** (42), 18287–18291.

59 Vallet-Regi, M., Ramila, A., del Real, R.P., and Perez-Pariente, J. (2001) *Chem. Mater.*, **13**, 308–311.

60 Bourrelly, S., Llewellyn, P.L., Serre, C., Millange, F., Loiseau, T., and Férey, G. (2005) *J. Am. Chem. Soc.*, **127**, 13519–13521.

61 Llewellyn, P.L., Horcajada, P., Maurin, G., Devic, T., Rosenbach, N., Bourrelly, S., Serre, C., Vincent, D., Loera-Serna, S., Filinchuk, Y., and Férey, G. (2009) *J. Am. Chem. Soc.*, **131**, 13002–13008.

62 Millange, F., Serre, C., Guillou, N., Férey, G., and Walton, R.I. (2008) *Angew. Chem. Int. Ed.*, **47**, 4100–4105.

63 Trung, T.K., Trens, P., Tanchoux, N., Bourrelly, S., Llewellyn, P.L., Loera-Serna, S., Serre, C., Loiseau, T., Fajula, F., and Férey, G. (2008) *J. Am. Chem. Soc.*, **130** (50), 16926–16932.

64 Horcajada, P., Marquez-Alvarez, C., Ramila, A., Perez-Pariente, J., and Vallet-Regi, M. (2006) *Solid State Sci.*, **8** (12), 1459–1465.

65 Vallet-Regi, M., Rámila, A., del Real, R.P., and Pérez-Pariente, J. (2001) *Chem. Mater.*, **13**, 308–311.

66 Munoz, B., Ramila, A., Perez-Pariente, J., Diaz, I., and Vallet-Regi, M. (2003) *Chem. Mater.*, **15** (2), 500–503.

67 An, J., Geib, S.J., and Rosi, N.L. (2009) *J. Am. Chem. Soc.*, **131** (24), 8376–8377.

68 Taylor-Pashow, K.M.L., Rocca, J.D., Xie, Z., Tran, S., and Lin, W. (2009) *J. Am. Chem. Soc.*, **131** (40), 14261–14263.

69 Vassal, G., Gouyette, A., Hartmann, O., Pico, J.L., and Lemerle, J. (1989) *Cancer Chemother. Pharmacol.*, **24**, 386–390.

70 Vassal, G., Deroussent, A., Challine, D., Hartmann, O., Koscielny, S., Valteau-Couanet, D., Lemerle, J., and Gouyette, A. (1992) *Blood*, **79**, 2475–2479.

71 Slattery, J.T., Sanders, J.E., Buckner, C.D., Schaffer, R.L., Lambert, K.W., Langer, F.P., Anasetti, C., Bensinger, W.I., Fisher, L.D., and Appelbaum, F.R. (1995) *Bone Marrow Transplant.*, **16**, 31–42.

72 Xiaoling, L. and Chan, W.K. (1999) *Adv. Drug Deliv. Rev.*, **39**, 81–103.

73 Hillaireau, H., Le Doan, T., Besnard, M., Chacun, H., Janin, J., and Couvreur, P. (2006) *Int. J. Pharm.*, **324** (1), 37–42.

74 Loke, S.L. et al. (1989) *Proc. Natl. Acad. Sci. USA*, **86**, 3474–3478.

75 Kukhanova, M., Krayevsky, A., Prusoff, W., and Cheng, Y.C. (2000) *Curr. Pharm. Des.*, **6**, 585–598.

76 Layre, A., Couvreur, P., Chacun, H., Richard, J., Passirani, C., Requier, D., Benoit, J.P., and Gref, R. (2006) *J. Control. Release*, **111**, 271–280.

77 Layre, A., Gref, R., Richard, J., Requier, D., and Couvreur, P. (2004) Nanoparticules polimériques composites, French Patent Application 04 07569.

78 Hasan, A.S., Socha, M., Lamprecht, A., El Ghazouan, F., Sapin, A., Hoffman, M., Maincent, P., and Ubrich., N. (2007) *Int. J. Pharm.*, **344**, 53–61.

79 Hassan, Z., Nilsson, C., and Hassan, M. (1998) *Bone Marrow Transplant.*, **22**, 913–918.

80 Madden, T., de Lima, M., Thapar, N., Nguyen, J., Roberson, S., Couriel, D., Pierre, B., Shpall, E.J., Jones, R.B., Champlin, R.E., and Andersson, B.S. (2007) *Biol. Blood Marrow Transplant.*, **13** (1), 56–64.

81 (a) Hasan, A.S., Socha, M., Lamprecht, A., El Ghazouan, F., Sapin, A., Hoffman, M., Maincent, P., and Ubrich., N. (2007) *Int. J. Pharm.*, **344**, 53–61; (b) Vassal, G., Gouyette, A., Hartmann, O., Pico, J.L., and Lemerle, J. (1989) *Cancer Chemother. Pharmacol.*, **24**, 386–390; (c) Hassan, Z., Nilsson, C., and Hassan, M. (1998) *Bone Marrow Transplant.*, **22**, 913–918.

82 (a) Xiaoling, L. and Chan, W.K. (1999) *Adv. Drug Deliv. Rev.*, **39**, 81–103; (b) Loke S.L., Stein, C.A., Zhang, X.H., Mori, K., Nakanishi, M., Subasinghe, C., Cohen, J.S., and Neckers, L.M. (1989) *Proc. Natl. Acad. Sci. USA*, **86**, 3474–3478.

83 Alaerts, L., Kirschhock, C.E.A., Maes, M., van der Veen, M.A., Finsy, V., Depla, A., Martens, J.A., Baron, G.V., Jacobs, P.A., Denayer, J.F.M., and De Vos, D.E. (2007) *Angew. Chem. Int. Ed.*, **46**, 4293–4297.

84 Xie, Y., Yu, Z., Huang, X., Wang, Z., Niu, L., Teng, M., and Li, J. (2007) *Chem. Eur. J.*, **13**, 9399–9405.

85 (a) Srivatsana, S.G., Parvezb, M., and Vermaa, S. (2003) *J. Inorg. Biochem.*, **97**, 340–344; (b) Mantion, A., Massüger, L., Rabu, P., Palivan, C., McCusker, L.B., and Taubert, A. (2008) *J. Am. Chem. Soc.*, **130**, 2517–2526.

86 Yua, L.-C., Chen, Z.-F., Zhoua, C.-S., Lianga, H., and Li, Y. (2005) *J. Coord. Chem.*, **58**, 1681–1687.

87 Miller, S.R., Heurtaux, D., Baati, T., Horcajada, P., Grenèche, J.-M., and Serre, C. (2010) *Chem. Commun.*, **46**, 4526–4258.

88 Xie, Y., Yu, Z., Huang, X., Wang, Z., Niu, L., Teng, M., and Li, J. (2007) *Chem. Eur. J.*, **13**, 9399–9405.

89 Palmer, R.M.J., Ferrige, A.G., and Moncada, S. (1987) *Nature*, **327**, 524.

90 Moncada, S., Palmer, R., and Higgs, E. (1991) *Pharmacol. Rev.*, **43**, 109.

91 Shekhter, A.B., Serezhenkov, V.A., Rudenko, T.G., Pekshev, A.V., and Vanin, A.F. (2005) *Nitric Oxide*, **12**, 210.

92 Barbul, A., Lazarou, S.A., Efron, D.T., Wasserkrug, H.L., and Efron, G. (1990) *Surgery*, **108**, 331.

93 Roberts, J.D., Polaner, D.M., Lang, P., and Zapol, W.M. (1992) *Lancet*, **340**, 818.

94 Maria, B.W. and Adrian, B. (2002) *Am. J. Surg.*, **183**, 406.

95 Zhu, H., Wei, X., Bian, K., and Murad, F. (2008) *J. Burn Care Res.*, **29**, 804.

96 Miller, C.C., Miller, M.K., Ghaffari, A., and Kunimoto, B. (2004) *J. Cutan. Med. Surg.*, **8**, 233.

97 Dietzel, P.D.C., Morita, Y., Blom, R., and Fjellvag, H. (2005) *Angew. Chem. Int. Ed.*, **44**, 6354.

98 Dietzel, P.D.C., Panella, B., Hirscher, M., Blom, R., and Fjellvag, H. (2006) *Chem. Commun.*, 959.

99 Chui, S.S.-Y., Lo, S.M.-F., Charmant, J.P.H., Orpen, A.G., and Williams, I.D. (1999) *Science*, **283**, 1148.

100 Wheatley, P.S., Butler, A.R., Crane, M.S., Rossi, A.G., Megson, I.L., and Morris, R.E. (2005) In *Molecular Sieves: from Basic Research to Industrial Applications. Studies in Surface Science and Catalysis*, vol. 158A and B (eds. J. Čejka, N. Žilková, and P. Nachtigall), Elsevier, Amsterdam, pp. 2033–2040.

101 Wheatley, P.S., Butler, A.R., Crane, M.S., Fox, S., Xiao, B., Rossi, A.G., Megson, I.L., and Morris, R.E. (2006) *J. Am. Chem. Soc.*, **128**, 502.

102 Xiao, B., Wheatley, P.S., and Morris, R.E. (2008) In *Zeolites and Related Materials: Trends Targets and Challenges. Studies in Surface Science and Catalysis*, Vol. 174, Part 1 (eds. A. Gedeon, P. Massiani, and F. Babonneau), Elsevier, Amsterdam, p. 902.

103 Xiao F B., Wheatley, P.S., Zhao, X.B., Fletcher, A.J., Fox, S., Rossi, A.G., Megson, I.L., Bordiga, S., Regli, L., Thomas, K.M., and Morris, R.E. (2007) *J. Am. Chem. Soc.*, **129**, 1203.

104 McKinlay, A.C., Xiao, B., Wragg, D.S., Wheatley, P.S., Megson, I.L., and Morris, R.E. (2008) *J. Am. Chem. Soc.*, **130**, 10440.

105 Xiao, B., Wheatley, P.S., Zhao, X.B., Fletcher, A.J., Fox, S., Rossi, A.G., Megson, I.L., Bordiga, S., Regli, L., Thomas, K.M., and Morris, R.E. (2007) *J. Am. Chem. Soc.*, **129**, 1203.

106 EQSSDX *Environmental Quality and Safety, Supplement.* (Stuttgart, Fed. Rep. Ger.) (1975), V.1–5, 1975–76.

107 Heldref Pub., AEHLAU *Archives of Environmental Health.* 4000 Albemarle St., NW, Washington, DC 20016, 1962, **5**, 437.

108 Rieter, W.J., Pott, K.M., Taylor, K.M.L., and Lin, W. (2008) *J. Am. Chem. Soc.*, **130** (35), 11584–11585.

109 Fromm, K.M. (2008) *Coord. Chem. Rev.*, **252**, 856–885.

110 (a) Belser, K., Vig Slenters, T., Pfumbidzai, C., Upert, G., Mirolo, L., Fromm, K.M., and Wennemers, H. (2009) *Angew. Chem. Int. Ed.*, **48**, 3661–3664; (b) Dorn, T., Fromm, K.M., and Janiak, C. (2006) *Aust. J. Chem.*, **59**, 22–25.

111 Slenters, T.V., Sagué, J.L., Brunetto, P.S., Zuber, S., Fleury, A., Mirolo, L., Robin, A.Y., Meuwly, M., Gordon, O., Landmann, R., Daniels, A.U., and Fromm, K.M. (2010) *Materials*, **3**, 3407–3429.

112 Taylor, K.M.L., Jin, A., and Lin, W. (2008) *Angew. Chem. Int. Ed.*, **47**, 7722–7725.

113 Taylor, K.M.L., Rieter, W.J., and Lin, W. (2008) *J. Am. Chem. Soc.*, **130**, 14358–14359.

114 Bonnemain, B. (1998) *J. Drug. Target.*, **6** (3), 167–174.

115 Rieter, W.J., Taylor, K.M.L., and Lin, W. (2007) *J. Am. Chem. Soc.*, **129**, 9852–9853.

116 Rieter, W.J., Taylor, K.M.L., An, H., and Lin, W. (2006) *J. Am. Chem. Soc.*, **128**, 9024–9025.

117 Jhung, S.H., Lee, J.-H., Yoon, J.W., Serre, C., Férey, G., and Chang, J.-S. (2006) *Adv. Mater.*, **19** (1), 121–124.

118 Demessence, A., Horcajada, P., Serre, C., Boissière, C., Grosso, D., Sanchez, C., and Férey, G. (2009) *Chem. Commun.*, 7149–7151.

119 Isomaa, B., Reuter, J., and Djupsund, B.M. (1976) *Arch. Toxicol.*, **35** (2), 91–96.

120 Safety data sheet for hexadecyltrimethylammonium bromide, Safety Officer in Physical Chemistry at Oxford University, updated on March 25, 2008, http://msds.chem.ox.ac.uk/HE/hexadecyltrimethyl-ammonium_ bromide.html (2010)

121 Safety data sheet IRMM-442 for isooctane, Institute for Reference Materials and Measurements, 2006, https://irmm.jrc.ec.europa.eu/html/reference_materials_catalogue/catalogue/attachements/IRMM-442_msds.pdf (2010)

122 Safety data sheet for 1-hexanol, updated on October 31, Safety Officer in Physical Chemistry at Oxford University, 2007, http://msds.chem.ox.ac.uk/HE/1-hexanol.html (2010)

123 Cravillon, J., Munzer, S., Lohmeier, S.-J., Feldhoff, A., Huber, K., and Wiebcke, M. (2009) *Chem. Mater.*, **21** (8), 1410–1412.

124 Hermes, S., Witte, T., Hikov, T., Zacher, D., Bahnmüller, S., Langstein, G., Huber, K., and Fischer, R.A. (2007) *J. Am. Chem. Soc.*, **129**, 5324.

125 Horcajada, P., Serre, C., Grosso, D., Boissière, C., Perruchas, S., Sanchez, C., and Férey, G. (2009) *Adv. Mater.*, **21**, 1931–1935.

126 Gupta, P.K. (1990) *J. Pharm. Sci.*, **79**, 11, 949–962.

127 Gref, R., Minamitake, Y., Peracchia, M.T., Trubetskoy, V., Torchilin, V., and Langer, R. (1994) *Science*, **263**, 1600–1603.

128 Gabizon, A. (2001) *Clin. Cancer Res.*, **7**, 223–225.

129 Graf, C., Vossen, D.L.J., Imhof, A., and van Blaaderen, A. (2003) *Langmuir*, **19**, 6693.

130 Gref, R., Domb, A., Quellec, P., Blunk, T., Müller, R.H., Verbavatz, J.M., and Langer, R. (1995) *Adv. Drug. Deliv. Rev.*, **16**, 215–233.

131 Lehr, C.M., Bouwstra, J.A., Schacht, E.H., and Junginger, H.E. (1992) *Int. J. Pharm.*, **78**, 43–48.

132 Chen, L., Schechter, B., Arnon, R., and Wilchek, M. (2000) *Drug Dev. Res.*, **50**, 3–4, 258–271.

133 Mueller, U., Schubert, M., Teich, F., Puetter, H., Schierle-Arndt, K., and Pastre, J. (2006) *J. Mater. Chem.*, **16**, 626–636.

134 Finsy, V., Ma, L., Alaerts, L., De Vos, D.E., Baron, G.V., and Denayer, J.F.M. (2009) *Micropor. Mesopor. Mater.*, **120**, 221–227.

135 Horcajada, P., Serre, C., Grosso, D., Boissière, C., Perruchas, S., Sanchez, C., and Férey, G. (2009) *Adv. Mater.*, **21**, 1931–1935.

136 Demessence, A., Horcajada, P., Serre, C., Boissière, C., Grosso, D., Sanchez, C., and Férey, G. (2009) *Chem. Commun.*, 7149–7151.

137 Demessence, A., Boissière, C., Grosso, D., Horcajada, P., Serre, C., Férey, G., Soler-Illia, G.J.A.A., and Sanchez, C. (2010) *J. Mater. Chem.*, **20**, 7676–7681.

138 Mowbray, M., Tan, X.J., Wheatley, P.S., Morris, R.E., and Weller, R.B. (2008) *J. Invest. Dermatol.*, **128**, 352–360.

139 Schwab, M.G., Senkovska, I., Rose, M., Koch, M., Pahnke, L., Jonschker, G., and Kaskel, S. (2008) *Adv. Eng. Mater.*, **10**, 1151–1155.

140 Cho, C.Y. and Lo, J.S. (1998) *Dermatol. Clin.*, **16**, 25.

141 Hinks, N.J., McKinlay, A.C., Xiao, B., Wheatley, P.S., and Morris, R.E. (2010) *Micropor. Mesopor. Mater.*, **129**, 330.

11
Metal–Organic Frameworks for Biomedical Imaging
Joseph Della Rocca and Wenbin Lin

11.1
Introduction

Despite tremendous advances in our understanding of the fundamental biology behind many diseases, we have yet to achieve comparable clinical benefits. A major reason for this is that many diseases are not diagnosed until advanced stages, beyond the point where therapeutic intervention is effective. Advances in biomedical imaging techniques are therefore essential for achieving early diagnosis, leading to a greater chance of a patient's survival and recovery after therapy. One of the major limitations of most medical imaging technologies is the low inherent contrast of many tissues, so exogenous contrast agents are frequently administered. Most of these contrast agents are small molecules, which suffer from nonspecific biodistribution and low contrast enhancement capability, and must be administered in high doses to achieve effective contrast. It is highly desirable to create agents that offer specific biodistribution and high image contrast. Nanomaterials offer one possible solution. Nanoparticle imaging agents offer several advantages, including tunable size, tailorable surface properties, high agent loading, biocompatibility, and improved pharmacokinetics [1]. Nanomaterials can be targeted to specific parenchymal sites by conjugation with targeting ligands to enhance accumulation. A large number of nanoparticle platforms for imaging and therapeutic applications have received clinical approval or are in clinical trials [2].

Metal–organic frameworks (MOFs), also known as coordination polymers, are hybrid materials constructed from metal connecting points and organic bridging ligands [3]. MOFs are typically synthesized by the spontaneous self-assembly of metal ions or clusters and bridging ligands under mild conditions. These materials can be constructed from an almost limitless combination of metals and ligands, allowing the resulting material to be engineered for many applications. MOFs are well studied materials for gas storage [4], nonlinear optics [5], sensing [6], catalysis [7], and magnetism [8]. An emerging area of MOF research is for biomedical applications, particularly in imaging and drug delivery [9]. MOFs possess many desirable characteristics for biomedical applications, including high agent loading through several

---

Metal-Organic Frameworks: Applications from Catalysis to Gas Storage, First Edition. Edited by David Farrusseng.
© 2011 Wiley-VCH Verlag GmbH & Co. KGaA. Published 2011 by Wiley-VCH Verlag GmbH & Co. KGaA.

methods (direct incorporation into the framework, covalent post-synthetic modification, noncovalent encapsulation), intrinsic biodegradability from labile metal–ligand bonds, high porosity for loading/release of encapsulated agents, and versatile modification methodologies. Recently, these materials have been scaled down to the nano regime to form nanoscale metal–organic frameworks (NMOFs), which may be more suitable for systemic administration [10]. This chapter discusses the development of MOFs as biomedical imaging contrast agents and their preliminary *in vitro* and *in vivo* applications.

MOFs have been developed as potential contrast agents for several different imaging modalities, which are briefly discussed below. Magnetic resonance imaging (MRI) is a noninvasive imaging technique based on the detection of nuclear spin reorientations in a magnetic field [11]. The longitudinal relaxation time (T_1) and transverse relaxation time (T_2) of the proton spins and their densities determine the signal intensity. MRI allows for high spatial resolution, high soft-tissue contrast, and large penetration depth, but has low intrinsic sensitivity. Contrast agents are administered with approximately 35% of all MRI scans to improve sensitivity [12]. Most of the currently used MRI contrast agents are small-molecule gadolinium chelates, which act as T_1-weighted contrast agents, providing increased signal intensity [13]. The efficacy of a contrast agent is expressed by the longitudinal (r_1) and transverse relaxivity (r_2) values, which are the slope of a plot of $1/T_1$ or $1/T_2$ against the agent concentration. Gd chelates only have modest r_1 values, requiring large doses to provide sufficient contrast between normal and diseased tissues. Additionally, free Gd^{3+} ions leached from clinically used Gd chelates can cause nephrogenic systemic fibrosis (NSF), which affects the skin and other organs [14]. Superparamagnetic iron oxides (SPIOs) are clinically used T_2-weighted contrast agents, leading to negative contrast enhancement. [15] Manganese ions are used preclinically as T_1-weighted contrast agents and are less toxic than free Gd^{3+} [13a, 16].

Optical imaging has been widely used in biomedical research. Visible light in the 400–600 nm range is used to excite fluorophores within tissues, which subsequently fluorescence or phosphoresce at longer wavelengths [11a, 17]. Optical imaging is ubiquitous for *in vitro* and *ex vivo* analysis, but has severe limitations for *in vivo* applications due to poor penetration. Tissue penetration can be improved by using photons in the near-infrared region of the spectrum, where absorption by water and endogenous biomolecules is minimized. Nanomaterials such as quantum dots have shown great promise as superior optical contrast agents [18].

Computed tomography (CT) is a powerful imaging technique, based on X-ray attenuation by a specimen. CT can provide three-dimensional images with excellent spatial resolution [19]. CT contrast agents contain elements with a high Z number, such as iodine, barium, and bismuth. Current clinically approved CT contrast agents are iodinated aromatic molecules and barium sulfate. In order to provide adequate contrast, extremely large doses (tens of grams) must be administered. The use of nanoparticle contrast agents can potentially alleviate this issue, since the particles can be delivered more specifically to the targeted tissue and thus satisfactory image contrast can be obtained with a smaller dose of contrast agent.

In addition to imaging of specific tissues, NMOFs may be used for biological sensing of specific analytes or pathogens. In an ideal scenario, the sensor is off until a specific analyte is introduced. Additionally, the material would need to provide rapid detection of the analyte over a large linear range with a low limit of detection and no signal from non-target species.

11.2
Gadolinium Carboxylate NMOFs

Lin and co-workers first showed the potential of NMOFs as MRI contrast agents with the development of Gd carboxylate materials [20]. Crystalline nanorods of Gd$(bdc)_{1.5}(H_2O)_2$ (**1**) (bdc = 1,4-benzenedicarboxylate) were prepared in a reverse microemulsion. The nanoparticle morphology was found to be dependent on the w value (water-to-surfactant molar ratio) of the microemulsion. Nanorods synthesized at $w = 10$ were 1–2 μm in length and 100 nm in diameter (Figure 11.1a), whereas at $w = 5$ the nanoparticles were 100–125 nm in length and 40 nm in diameter. X-ray powder diffraction (XRPD) measurements showed that **1** matched a known crystalline lanthanide–bdc phase. Nanoplates of $Gd(btc)(H_2O)_3$ (**2**) (btc = benzene-1,2,4-tricarboxylate) were synthesized at $w = 15$ using the same microemulsion system. Compound **2** was 100 nm in diameter and 35 nm thick (Figure 11.1b). As these nanomaterials contained large amounts of Gd^{3+} ions that are at or near the surfaces, **1** and **2** were evaluated as potential MRI contrast agents (Table 11.1). Relaxivity measurements showed very high r_1 values for **1** and **2** compared with OmniScan, a clinically used small-molecule chelate. The nanoparticles could also be

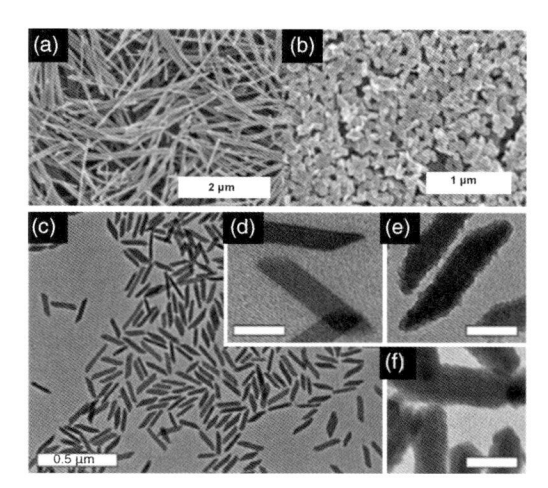

Figure 11.1 (a) SEM image of **1** synthesized at $w = 10$; (b) SEM image of **2** synthesized at $w = 15$; (c and e) TEM images of **1**·silica (2–3 nm coating); (d) TEM image of PVP-coated **1**; (f) TEM image of **1**·silica (8–9 nm coating). TEM scale bars for (d)–(f) are 50 nm. Reproduced with permission from [20, 21].

Table 11.1 Relaxivity values of Gd carboxylate NMOFs at 3 T on a per mM Gd basis [20].

NMOF (dimensions)	r_1 (mM^{-1} s^{-1})	r_2 (mM^{-1} s^{-1})
1 (100 × 40 nm)	35.8	55.6
1 (400 × 70 nm)	26.9	49.1
1 (1000 × 100 nm)	20.1	45.7
2 (100 × 35 nm)	13.0	29.4

used as T_2-weighted contrast agents when a different pulse sequence was employed. The r_1 values of **1** showed a dependence on particle size, with larger particles possessing lower relaxivity. This dependence was attributed to the reduced surface area of larger nanoparticles, which limited the accessibility of bulk water to the metal centers.

One of the drawbacks of using **1** or **2** in a clinical setting is that they readily break down under physiological conditions, releasing toxic Gd^{3+} ions. Therefore, these NMOFs must be stabilized to prevent premature Gd^{3+} release. Lin and co-workers were able to encapsulate **1** within a shell of amorphous silica [21]. Silica shells impart several advantages to **1**, including biocompatibility, increased water dispersibility, and covalent functionalization with silyl-derived molecules. The surface of **1** was first treated with polyvinylpyrrolidone (PVP) to form more dispersible PVP-coated **1** (Figure 11.1d) and then coated with a silica shell using tetraethyl orthosilicate (TEOS) under basic conditions (Figure 11.1c,e,f). The shell thickness could be controlled by the amount of TEOS added and the reaction time. For example, a reaction time of 3 h led to a 2–3 nm silica shell (Figure 11.1c,e), whereas extending the reaction time to 7 h led to an 8–9 nm thick coating (Figure 11.1f). Scanning electron microscopy (SEM) and XRPD studies showed that **1**@silica maintained its original morphology and crystallinity. NMOF dissolution studies demonstrated that the silica shell was able to retard the breakdown of **1**. The silica shell could then be further functionalized with a variety of silyl-derived molecules.

Lin and co-workers created luminescent analogues of **1** by adding a small percentage of Tb^{3+} or Eu^{3+} to the microemulsion, leading to NMOFs which were luminescent in different spectral regions [20]. Eu-doped **1** (**1**-Eu) was further developed as a sensor for dipicolinic acid (DPA) [21]. DPA is a major component of bacterial spores, constituting up to 15 wt% of the spore mass. **1**-Eu was coated with a silica shell (**1**-Eu·silica) and then functionalized with a silyl-derived Tb–EDTA monoamide (Figure 11.2a). After excitation of functionalized **1**-Eu·silica, only Eu luminescence was detected as the Tb–EDTA complex was nonemissive. However, upon DPA binding, the Tb luminescence was detected, the intensity of which increased linearly with DPA concentration until binding site saturation (Figure 11.2b). Eu^{3+} luminescence acts as a calibration standard as it does not interfere with Tb^{3+}-based luminescence. The limit of detection for this system was 48 nM and selectively detected DPA in the presence of amino acids.

NMOF **1** was coated with biocompatible polymers by Boyes and co-workers [22]. A copolymer of poly(N-isopropylacrylamide)-co-poly(N-acryloxysuccinimide)-co-

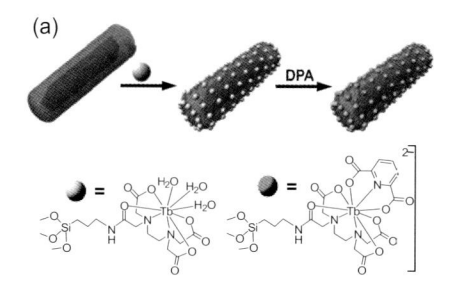

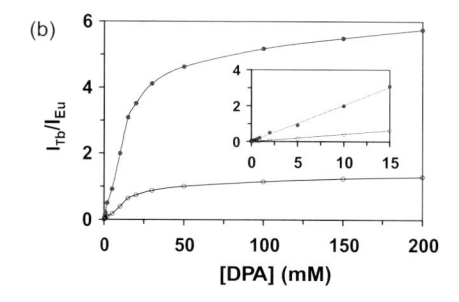

Figure 11.2 (a) Generalized scheme showing the functionalization of **1**-Eu·silica with silyl-derived Tb–EDTA monoamide and subsequent sensing of DPA. (b) Graph of the dependence of the ratio of Tb/Eu emission intensity as a function of DPA concentration (the top line = 544/592 nm and the bottom line = 544/615 nm). The inset shows the linear relationship at low [DPA]. Reproduced with permission from [21].

poly(N-acryloxysuccinimide)-co-poly(fluorescein O-methacrylate) (PNIPAM-co-PNAOS-co-PFMA) was synthesized by RAFT polymerization [22a]. The copolymer contains an optical imaging contrast agent (fluorescein) and the succinimide group allowing for the conjugation of a chemotherapeutic (methotrexate) and a targeting peptide (GRGDS). Thiol endgroups on the polymer were used to bind to free Gd ions on the surface of **1** (**1a**). An estimated 25 000 polymer chains were bound to the surface of the nanoparticle and dissolution studies showed that the polymer coating impeded Gd^{3+} release compared with the uncoated nanoparticles, but did not prevent it. Relaxivity measurements were taken at 1.5 T: the polymer modification tripled the r_1 relaxivity ($r_1 = 33.4 \, mM^{-1} s^{-1}$) compared with the unmodified nanoparticle ($r_1 = 9.86 \, mM^{-1} s^{-1}$). Polymer modification also increased the r_2 relaxivity ($r_2 = 47.2 \, mM^{-1} s^{-1}$) compared with the uncoated nanoparticle ($r_2 = 17.9 \, mM^{-1} s^{-1}$). Cytotoxicity assays with methotrexate-loaded **1a** showed cytotoxicity comparable to that of the free drug against canine FITZ-HSA cells. Targeted and untargeted **1a** were also evaluated as optical contrast agents against FITZ-HSA cells. Only the GRGDS-targeted nanoparticles showed fluorescence, as the untargeted nanoparticles were washed away during sample preparation.

One of the more interesting results from the work mentioned above is that the polymer modification could increase the relaxivity of **1**. Boyes and co-workers undertook a systematic study on how polymer modification could modify the relaxivity of **1** [22b]. Several different hydrophilic polymers {poly[N-(2-hydroxypropyl)methacrylamide]] (PHPMA), poly(N-isopropylacrylamide) (PNIPAM), poly[poly(ethylene glycol) methyl ether acrylate] (PPEGMEA), and poly(acrylic acid) (PAA)} and one hydrophobic polymer, polystyrene (PS), were synthesized at several molecular weights. The polymers were conjugated to **1** through thiol endgroups coordinating to surface metal centers and the relaxivity was measured at 1.5 T (Table 11.2). Several trends were observed: conjugation of hydrophilic polymers increased r_1 values and decreased the r_2 values. This observation was hypothesized as a result of increased water retention by the polymer, allowing for faster exchange with

Table 11.2 Relaxivity values of polymer-modified 1 taken at 1.5 T on a per mM Gd basis; NMOF 1 is 55 × 125 nm in size.

Polymer	MW	r_1 (mM^{-1}s^{-1})	r_2 (mM^{-1}s^{-1})
PHPMA	5327	17.8	25.8
	10281	32.9	44.8
	19370	105.4	129.6
PNIPAM	5690	20.27	29.73
	8606	46.99	64.10
	17846	62.51	79.9
PDMAEA	15120	37.2	54.2
PPEGMEA	19542	59.93	81.5
PAA	10888	21.3	31.82
PS	4802	1.17	141.16
	8972	1.2	25.75
	15245	3.91	123.4

the metal centers. Modification with hydrophobic PS led to very poor r_1 relaxivity but high r_2 relaxivity; this was due to decreased water retention by the polymer, with all of the contrast enhancement arising from water molecules within the framework structure.

The reverse microemulsion strategy was successful for synthesizing **1** and **2**; however, for many metal–ligand combinations this method leads to highly polydisperse nanoparticles with poorly defined size distributions. Lin and co-workers developed a surfactant-mediated hydrothermal method to synthesize Gd_2(bhc)$(H_2O)_6$ (**3**) NMOFs (bhc = benzenehexacarboxylate) [23]. NMOF **3** was formed by heating a microemulsion of $[NMeH_3]_6$(bhc) and $GdCl_3$ at 120 °C in a Parr reactor, and adopted a block-like morphology with dimensions of 25 × 50 × 100 nm by SEM and transmission electron microscopy (TEM) (Figure 11.3a). XRPD of **3** matched a known lanthanum bulk phase of $[La_2$(bhc)$(H_2O)_6]$. Luminescent analogues of **3** were created by doping with either Eu^{3+} or Tb^{3+}. The doped NMOFs were of the same size and morphology as **3**, but demonstrated intense luminescence under UV excitation. The crystalline phase and morphology of the Gd(bhc) NMOF could be tuned by changing the pH value of the microemulsion. Nanoparticles obtained by the reaction of $GdCl_3$ and benzenehexacarboxylic acid under microwave irradiation at 60 °C had a different crystalline structure to **3**. The nanoparticles obtained had the

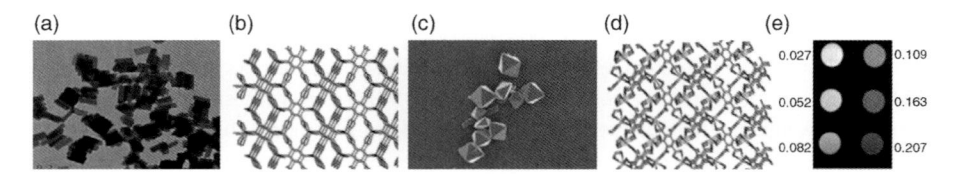

Figure 11.3 (a) TEM image of **3**. (b) Crystal structure of **3** showing packing of layers. (c) SEM image of **4a**. (d) Crystal structure of **4** showing packing of layers. (e) MRI phantoms of **3** taken at 9.4 T. Reproduced with permission from [23].

formula [Gd$_2$(bhc)(H$_2$O)$_8$](H$_2$O)$_2$] (**4**) and were rods 100–300 nm in diameter and several microns in length. Increasing the reaction temperature to 120 °C changed the morphology of **4** from nanorods to octahedra (**4a**), but the crystalline phase was maintained (Figure 11.3c). The difference between **3** and **4** was differing metal–ligand coordination modes. In **3**, the Gd centers were nine-coordinate, with each metal bound to two bridging and two chelating carboxylates from four bhc units (Figure 11.3b). Water molecules occupied the vacant coordination sites. In **4**, each metal was coordinated to two chelating carboxylates and one monodentate carboxylate from three different bhc units, with water occupying the remainder of the coordination sites (Figure 11.3d). Compound **3** was evaluated as a potential MRI contrast agent at 9.4 T (Figure 11.3e). It showed low r_1 relaxivity (1.5 mM^{-1} s^{-1}) but very high r_2 relaxivity (122.6 mM^{-1} s^{-1}), indicating that the nanoparticles may act as potential T_2-weighted contrast agents.

11.3
Manganese Carboxylate NMOFs

Lin and co-workers developed manganese carboxylate NMOFs as a potential multimodal contrast agent [24]. Nanorods of Mn(bdc)(H$_2$O)$_2$ (**5**) were synthesized in a reverse microemulsion, giving particles with diameters of 50–100 nm and lengths ranging from 750 nm to several microns. Microwave heating during the synthesis of **5** did not change the particle morphology. The relaxivity values of **5** were determined at 3 T (Table 11.3); only modest r_1 relaxivity but high r_2 relaxivity were displayed. NMOFs were also obtained by reacting MnCl$_2$ and trimesic acid [1,3,5-benzenetricarboxylic acid (tcb)]. In a reverse microemulsion, spiral rod NMOFs 50–100 nm in diameter and 1–2 μm in length were obtained and had the formula Mn$_3$(tcb)$_2$(H$_2$O)$_6$ (**6**). Microwave heating of MnCl$_2$ and tcb resulted in nanoparticles **6a**, which were of the same crystalline phase as **6**, but possessed a different morphology. Compound **6a** possesses a block-like morphology with lengths ranging from 50 to 300 nm in all directions. The relaxivity values of **6a** were determined at 3 and 9.4 T (Table 11.3). Compound **6a** also displayed low r_1 relaxivity and high r_2 relaxivity, suggesting that the intact NMOFs would not act as efficient T_1-weighted contrast agents. However, NMOFs are intrinsically biodegradable, so the nanoparticles would work by releasing Mn^{2+}, which would become effective T_1-weighted contrast agents after binding to endogenous biomolecules.

Table 11.3 Relaxivity values of Mn carboxylate NMOFs on a per mM Mn basis.

NMOF	Field strength (T)	r_1 (mM^{-1} s^{-1})	r_2 (mM^{-1} s^{-1})
5	3	5.5	80
6a	3	7.8	70.8
6a	9.4	4.6	141.2
6a·silica	9.4	4.0	112.8

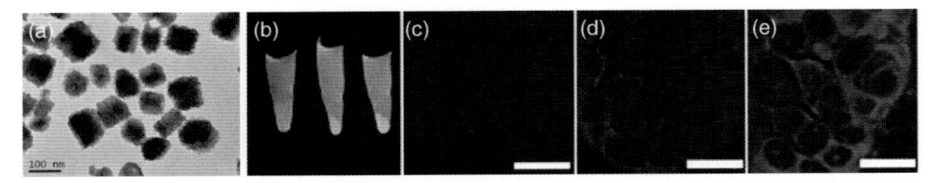

Figure 11.4 (a) TEM image of **6a**·silica. (b) MR phantoms of HT-29 cells taken with no particles (left), **6a**·silica (center), and cRGD-targeted **6a**·silica (right). Confocal microscopy images showing HT-29 cells incubated with no particles (c), **6a**·silica (d), and cRGD-targeted **6a**·silica (e). Blue fluorescence is from the DRAQ5 nuclear stain and green fluorescence is from rhodamine fluorophore used to functionalize **6a**·silica. The scale bars represent 20 μm for the confocal images. Reproduced with permission from [24].

NMOF **6a** was evaluated *in vitro* as a multimodal contrast agent for cancer cell imaging [24]. It was coated with a thin shell of amorphous silica (**6a**·silica) and functionalized with a silyl-derived rhodamine, and targeting peptide, cRGD (Figure 11.4a). The cRGD peptide binds tightly to integrins found on the surface of many angiogenic cancers. Dissolution studies of **6a**·silica showed that silica coating increased the half-life to 1.44 h, compared with 18 min for the uncoated nanoparticle, in PBS. Relaxivity measurements of **6a**·silica at 9.4 T showed a slight reduction in both r_1 and r_2 relaxivity compared with **6a** (Table 11.3). The **6a**·silica was evaluated as an optical and MR contrast agent *in vitro* against HT-29 human colon cancer cells. MR imaging showed increased uptake of targeted nanoparticles, which was confirmed by inductively coupled plasma mass spectrometry (ICP-MS) of the cell pellets (Figure 11.4b). Confocal microscopy images demonstrated that RGD targeting enhanced the uptake of the nanoparticles (Figure 11.4c–e). These results suggest that NMOFs **5** and **6** could act as potential multimodal contrast agents.

11.4
Iron Carboxylate NMOFs: the MIL Family

Among the tens of thousands of known MOFs, the MIL family, built from trivalent metal centers and carboxylate bridging ligands, has attracted a great deal of attention owing to their high stability, enormous porosity, and very large pores [3b]. For example, MIL-101(Cr) with the formula $Cr_3F(H_2O)_2O(bdc)_3 \cdot nH_2O$ (where n is ~25) has a Langmuir surface area of up to 5900 m^2 g^{-1} and contains two types of mesoporous cages with internal free diameters of ~29 and 34 Å [25]. The combination of these features makes the MIL family a unique candidate for storage and controlled release of biologically important molecules. Bulk-phase MIL materials were previously demonstrated as drug carriers [26]. These materials showed very high agent loadings by noncovalent incorporation and showed slow agent release.

Lin and co-workers developed an iron carboxylate MOF as a carrier for biomedically relevant agents [27]. Iron(III) NMOFs of the formula $Fe_3(\mu_3\text{-}O)Cl(H_2O)_2(bdc)_3$

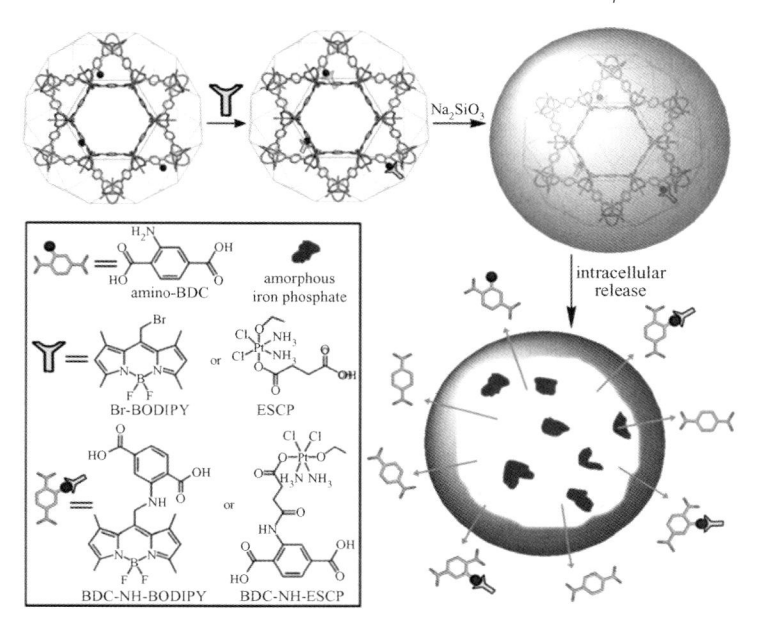

Figure 11.5 Generalized schematic showing the post-synthetic modification of **8** to create **8**-BODIPY or **8**-ESCP, subsequent silica coating, and intracellular release of the active agents.

(**7**) were synthesized by a microwave-assisted solvothermal method. XRPD studies showed that **7** matched the MIL-101 phase and were octahedral nanoparticles 200 nm in diameter. Compound **7** could be modified by adding 2-aminobenzenedicarboxylic acid (ABDC) to the precursor solution. It was found that the MIL-101 topology was maintained up to 17.4 mol% ABDC (**8**). At higher percentages of ABDC incorporation, the nonporous MIL-88B phase was obtained. With free amine groups located throughout the framework structure, NMOF **8** could be post-synthetically modified to contain a biomedically relevant cargo. As a proof of concept, an optical imaging agent, 3,3-difluoro-8-bromomethyl-4-bora-3a,4a-diaza-*s*-indacene (Br-BODIPY), and a chemotherapeutic, $cis,cis,trans$-Pt(NH$_3$)$_2$Cl$_2$(OEt)(O$_2$CCH$_2$CH$_2$CO$_2$H) (ESCP), were covalently loaded into the framework (Figure 11.5). The BODIPY-loaded **8** (**8**-BODIPY) contained 5.6–11.6 wt% fluorophore. **8**-BODIPY was not fluorescent due to iron quenching, but showed strong fluorescence upon framework decomposition and dye release. The nanoparticles were coated with a thin layer of amorphous silica and targeted using silyl-derived cRGD. Confocal microscopy assays demonstrated that **8**-BODIPY-silica was internalized by HT-29 colon cancer cells. Peptide targeting of **8**-BODIPY-silica increased the cellular localization and internalization. Control experiments with the free dye showed no cellular internalization, suggesting that the NMOF was internalized, then broke down intracellularly. The cisplatin prodrug ESCP was similarly loaded into **8** (**8**-ESCP) at loadings up to 12.9 wt%. Cytotoxicity assays using HT-29 cells revealed that RGD-targeted **8**-ESCP-silica showed comparable cytotoxicity to cisplatin.

Table 11.4 Relaxivity values for pegylated and nonpegylated iron carboxylate NMOFs at 9.4 T.

NMOF	r_2 (mM^{-1} s^{-1})
10	56
10-PEG	95
11	73
11-PEG	92

Horcajada and co-workers developed iron carboxylate NMOFs for T_2-weighted MRI imaging [28]. Highly porous iron(III) carboxylate NMOFs were synthesized by reacting iron salts with a variety of nontoxic carboxylates. The NMOF sizes ranged from 50 to 350 nm depending on the synthesis conditions. These nanoparticles corresponded to known bulk-phase materials MIL-89 (**9**), MIL-88A (**10**), MIL-100 (**11**), MIL-101-NH$_2$ (**12**), MIL-88Bt (**13**), and MIL-53 (**14**). The NMOFs were modified with biocompatible polymers, including PEG, chitosan, and dextran. The presence of large amounts of iron centers prompted the evaluation of the NMOFs as T_2-weighted MRI contrast agents. The r_2 relaxivity values of **10**, **10**-PEG, **11**, and **11**-PEG were reported at 9.4 T (Table 11.4). Pegylation of the NMOFs increased the relaxivity, which was attributed to either the pegylation increasing the size and lowering the tumbling rate of the NMOFs, or the pegylation decreasing the tendency of the particles to aggregate and settle out of aqueous solution. *In vivo* studies using **10** on Wistar rats showed T_2-weighted enhancement of the spleen and liver after injection (Figure 11.6). The contrast enhancement was lost after 3 months, indicating the accumulation and slow clearance of **10**.

11.5
Iodinated NMOFs: CT Contrast Agents

Lin and co-workers developed two iodinated NMOFs as potential CT contrast agents [29]. Five bulk MOFs were synthesized by combining Cu^{2+} or Zn^{2+} with 2,3,5,6-tetraiodo-1,4-benzenedicarboxylic acid. Single-crystal X-ray diffraction studies showed one-dimensional structures for all five MOFs. Two NMOFs containing Cu^{2+} (**14**) or Zn^{2+} (**15**) were synthesized. Compound **14** was synthesized within a reverse microemulsion and had a plate-like-morphology with a diameter of 300 nm and a thickness of 50 nm (Figure 11.7a). The morphology of **14** could be tuned by changing the microemulsion conditions to generate longer rod-like particles. Compound **15** was synthesized by a rapid precipitation technique to create crystalline rods 10–30 µm in length. Smaller particles **15a** were obtained by concentrating the precursor solutions. SEM of **15a** showed partially crystalline cubic NMOFs of 200–600 nm (Figure 11.7b). CT phantom studies were performed on dispersions of **14** and **15a** and compared with iodixanol, a clinically used CT contrast agent (Figure 11.7c–e). Both **14** and **15a** showed slightly higher X-ray attenuation than iodixanol, which was due to contributions from the metal centers.

11.6 Iodinated NMOFs: CT Contrast Agents | 261

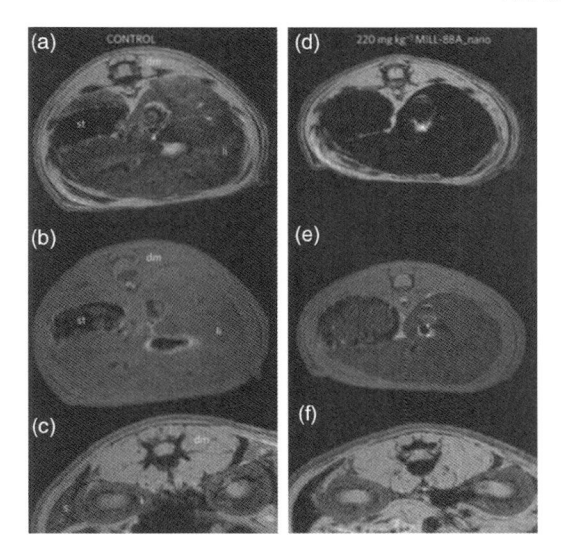

Figure 11.6 MR images of Wistar rats injected with no particles (a–c) or 220 mg kg^{-1} **10** (d–f). The images were acquired with either gradient-echo (a,c,d,f) or spin-echo (b,e) sequences. The images display either spleen (c,f) or liver (a,b,d,e) regions 30 min after injection; dm = dorsal muscle, k = kidney, li = liver, s = spleen, st = stomach. Reproduced with permission from [28].

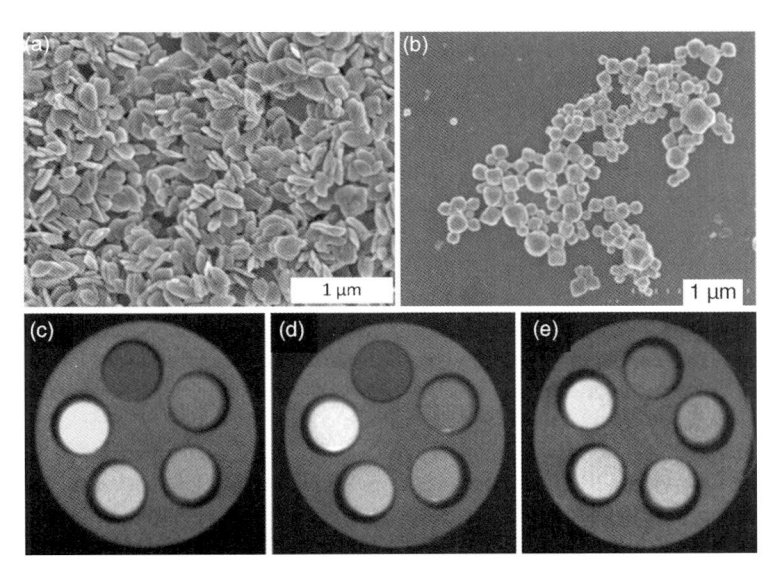

Figure 11.7 SEM images of (a) **14** and (b) **15a**. CT phantom images of (c) **14** and (d) **15a** dispersed in ethanol and (e) iodixanol in water. From the top, clockwise, the slots have [I] = 0, 0.075, 0.150, 0.225, and 0.300 M. Reproduced with permission from [29].

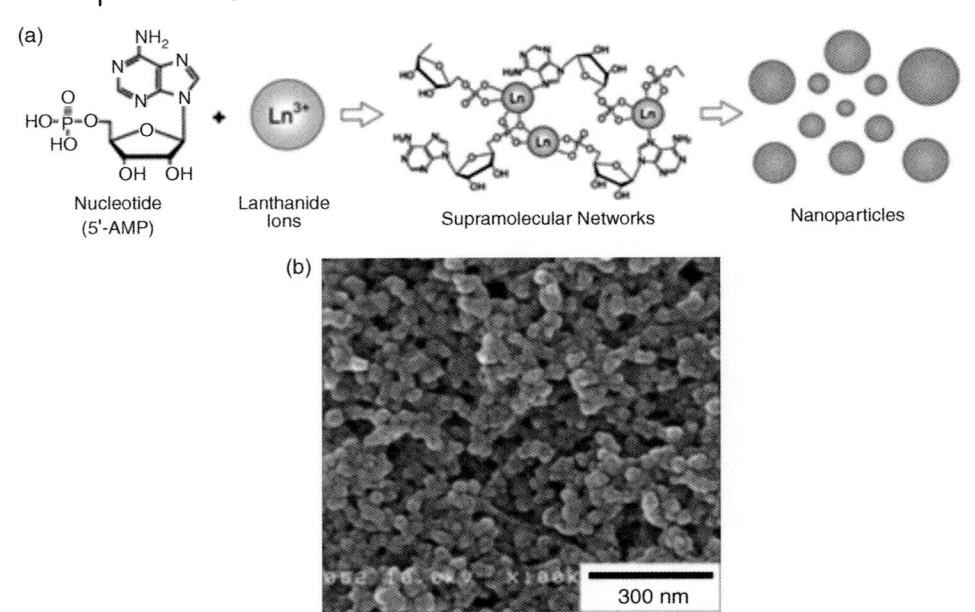

Figure 11.8 Formation of lanthanide–nucleotide NMOFs and SEM image of 17. Reproduced with permission from [30a].

11.6
Lanthanide Nucleotide NMOFs

Kimizuka and co-workers developed an NMOF system based on nucleotides and lanthanides, for a number of different biomedical applications [30]. Two MRI contrast agents were created by mixing equimolar amounts of $GdCl_3$ and a nucleotide in HEPES buffer [30a]. These amorphous MOFs were 30–180 nm in diameter depending on the nucleotide (Figure 11.8a). The Gd^{3+} ion was proposed to coordinate to both the nucleobase and the phosphate group present in the nucleotide. 5′GMP/Gd (**16**) and 5′AMP/Gd (**17**) NMOFs were evaluated as T_1-weighted MRI contrast agents at 0.3 T. Compounds **16** and **17** exhibited r_1 values of 13.4 and 12 $mM^{-1} s^{-1}$, respectively. Additionally, there was an inverse dependence between size and relaxivity, which was attributed to a greater proportion of Gd^{3+} ions at or near the NMOF surface.

Lanthanide–nucleotide NMOFs were shown to be effective materials for the encapsulation of biomedically relevant guest species [30a,b,d]. For example, water-soluble anionic dyes were encapsulated into lanthanide–nucleotide NMOFs by adding the dye during synthesis. The dye coordinated to the metal centers within the framework; cationic dyes could not be incorporated into the framework by this method. Dye encapsulation increased the fluorescence relative to the free dye. The anionic dye perylene-3,4,9,10-tetracarboxylic acid was doped within **17** (**17**-Perylene) to create fluorescent NMOFs. Confocal microscopy using HeLa cervical cancer cells showed uptake of **17**-Perylene into cellular lysosomes, while the anionic free dye remained outside the cells. The biodistribution of **17**-Perylene was studied *ex vivo* in a murine model by fluorescence and ICP-MS analysis. **17**-Perylene accumulated rapidly in the liver and remained there for up to 48 h after injection. No fluorescence was detected in the lungs, kidney, or spleen.

An NMOF created from 5′GMP/Tb^{3+} (**18**) inherently possessed intense green luminescence. This luminescence was attributed to strong coordination between the Tb^{3+} ions and the nucleobase, leading to energy transfer from the guanine base to coordinated metal. Analogous nanoparticles created from Tb^{3+} and other nucleotides did not show luminescence because of poor Tb–nucleobase interactions.

Kimizuka and co-workers took advantage of the non-luminescent properties of most nucleotide/Tb^{3+} NMOFs to create a switchable sensor [30c]. Spherical NMOFs of 5′AMP/Tb^{3+} (**19**) were not emissive because of quenching by coordinated water. If 2-hydroxypicolinic acid (2HPA) was added during the synthesis of **19**, a green luminescent NMOF was obtained (**19**-2HPA) as 2HPA strongly coordinates to Tb^{3+} to create a luminescent complex.

11.7
Guest Encapsulation within NMOFs

There have been two recent examples of the encapsulation of biomedically relevant agents within NMOF structures. Maspoch and co-workers created blue fluorescent nanospheres of Zn^{2+} and 1,4-bis(imidazol-1-ylmethyl)benzene (**20**) by a rapid precipitation technique [31]. SEM analysis showed that the size of **20** could be tailored from 100 to 1500 nm depending of the reaction conditions. Iron oxide nanoparticles (10 nm) were entrapped within **20** by adding iron oxide during the NMOF synthesis to create core–shell nanoparticles. Magnetization studies showed that the iron oxide maintained its magnetic properties after encapsulation and maintained the stability and fluorescent properties of unmodified **20**. This methodology was extended to entrap quantum dots, fluorescent dyes, and chemotherapeutics.

Li and co-workers developed peptide–polyoxometalate (POM) spheres for guest encapsulation. [32] POMs of phosphotungstic acid were treated with cationic dipeptides to form colloidal spheres **21** ∼150 nm in diameter. Analysis of **21** showed strong binding between the POMs and the peptides, with the peptides linking individual POMs through noncovalent interactions. Compound **21** was stable under acidic conditions, but rapidly disassembled under neutral or basic conditions. Cationic and anionic fluorescent dyes could be incorporated within **21** by adding the dye during the synthesis. The dye's charge made little difference to the encapsulation efficiency. Hydrophilic polymers, nanoparticles, and hydrophobic drugs could also be encapsulated within **21**.

11.8
Conclusion

NMOFs have received increasing attention as potential biomedical imaging agents. NMOF contrast agents generally have superior performance to the analogous small-molecule contrast agents and can be easily stabilized and functionalized to optimize their properties for biomedical applications. The *in vitro* and *in vivo* efficacy of these agents was demonstrated in multiple models. However, much more work needs to be performed before the future applications of NMOFs in the clinic can be realized.

References

1 (a) Ferrari, M. (2005) *Nat. Rev. Cancer*, **5**, 161–171; (b) Alexis, F., Pridgen, E., Molnar, L.K., and Farokhzad, O.C. (2008) *Mol. Pharm.*, **5**, 505–515; (c) Li, S.D. and Huang, L. (2008) *Mol. Pharm.*, **5**, 496–504.

2 (a) Cho, K., Wang, X., Nie, S., Chen, Z., and Shin, D.M. (2008) *Clin. Cancer Res.*, **14**, 1310–1316; (b) Peer, D., Karp, J.M., Hong, S., Farokhzad, O.C., Margalit, R., and Langer, R. (2007) *Nat. Nanotechnol.*, **2**, 751–760; (c) Torchilin, V.P. (2005) *Nat. Rev. Drug Discov.*, **4**, 145–160; (d) Pan, D., Lanza, G.M., Wickline, S.A., and Caruthers, S.D. (2009) *Eur. J. Radiol.*, **70**, 274–285; (e) Minchin, R.F. and Martin, D.J. (2010) *Endocrinology*, **151**, 474–481.

3 (a) Bradshaw, D., Warren, J.E., and Rosseinsky, M.J. (2007) *Science*, **315**, 977–980; (b) Férey, G., Mellot-Draznieks, C., Serre, C., and Millange, F. (2005) *Acc. Chem. Res.*, **38**, 217–225; (c) Hill, R.J., Long, D.L., Champness, N.R., Hubberstey, P., and Schroder, M. (2005) *Acc. Chem. Res.*, **38**, 335–348; (d) Yaghi, O.M., O'Keeffe, M., Ockwig, N.W., Chae, H.K., Eddaodi, M., and Kim, J. (2003) *Nature*, **423**, 705–714; (e) Kitagawa, S., Kitaura, R., and Noro, S.I. (2004) *Angew. Chem. Int. Ed.*, **43**, 2334–2375; (f) Moulton, B. and aworotko, M.J. (2001) *Chem. Rev.*, **101**, 1629–1658.

4 (a) Chen, B., Zhao, X., Putkham, A., Hong, K., Lobkovsky, E.B., Hurtado, E.J., Fletcher, A.J., and Thomas, K.M. (2008) *J. Am. Chem. Soc.*, **130**, 6411–6423; (b) Kesanli, B., Cui, Y., Smith, M.R., Bittner, E.W., Bockrath, B.C., and Lin, W. (2005) *Angew. Chem. Int. Ed.*, **44**, 72–75;

(c) Ma, L., Mihalcik, D.J., and Lin, W. (2009) *J. Am. Chem. Soc.*, **131**, 4610–4612; (d) Rosi, N.L., Eckert, J., Eddaoudi, M., Vodak, D.T., Kim, J., O'Keeffe, M., and Yaghi, O.M. (2003) *Science*, **300**, 1127–1129.

5 Evans, O.R. and Lin, W. (2002) *Acc. Chem. Res.*, **35**, 511–522.

6 (a) Chen, B.L., Wang, L.B., Zapata, F., Qian, G.D., and Lobkovsky, E.B. (2008) *J. Am. Chem. Soc.*, **130**, 6718–6719; (b) Xie, Z., Ma, L., Dekrafft, K.E., Jin, A., and Lin, W. (2010) *J. Am. Chem. Soc.*, **132**, 922–923.

7 (a) Dybtsev, D.N., Nuzhdin, A.L., Chun, H., Bryliakov, K.P., Talsi, E.P., Fedin, V.P., and Kim, K. (2006) *Angew. Chem. Int. Ed.*, **45**, 916–920; (b) Ma, L., Abney, C., and Lin, W. (2009) *Chem. Soc. Rev.*, **38**, 1248–1256; (c) Wu, C., Hu, A., Zhang, L., and Lin, W. (2005) *J. Am. Chem. Soc.*, **127**, 8940–8941; (d) Wu, C., and Lin, W. (2007) *Angew. Chem. Int. Ed.*, **46**, 1075–1078.

8 (a) Kuppler, R.J., Timmons, D.J., Fang, Q.R., Li, J.R., Makal, T.A., Young, M.D., Yuan, D.Q., Zhao, D., Zhuang, W.J., and Zhou, H.C. (2009) *Coord. Chem. Rev.*, **253**, 3042–3066; (b) Maspoch, D., Ruiz-Molina, D., and Veciana, J. (2004) *J. Mater. Chem.*, **14**, 2713–2723; (c) Maspoch, D., Ruiz-Molina, D., and Veciana, J. (2007) *Chem. Soc. Rev.*, **36**, 770–818.

9 (a) Della Rocca, J. and Lin, W. (2010) *Eur. J. Inorg. Chem.*, 2010, 3725–3724; (b) Huxford, R.C., Della Rocca, J., and Lin, W. (2010) *Curr. Opin. Chem. Biol.*, **14**, 262–268; (c) Taylor-Pashow, K.M.L., Della Rocca, J., Huxford, R.C., and Lin, W. (2010) *Chem. Commun.*, **46**, 5832–5849;

(d) McKinlay, A.C., Morris, R.E., Horcajada, P., Férey, G., Couvreur, P., and Serre, C. (2010) *Angew. Chem. Int. Ed.*, **49**, 6260–6266.

10 (a) Lin, W., Rieter, W.J., and Taylor, K.M.L. (2009) *Angew. Chem. Int. Ed.*, **48**, 650–658; (b) Catala, L., Volatron, F., Brinzei, D., and Mallah, T. (2009) *Inorg. Chem.*, **48**, 3360–3370; (c) Spokoyny, A.M., Kim, D., Sumrein, A., and Mirkin, C.A. (2009) *Chem. Soc. Rev.*, **38**, 1218–1227.

11 (a) Brindle, K. (2008) *Nat. Rev. Cancer* **8**, 94–107; (b) Hashemi, R.H. and Bradley, W.G. Jr. (1997) *MRI: the Basics*, Williams & Wilkins, Baltimore.

12 Aime, S., Crich, S.G., Gianolio, E., Giovenzana, G.B., Tei, L., and Terreno, E. (2006) *Coord. Chem. Rev.*, **250**, 1562–1579.

13 (a) Thunus, L. and Lejeune, R. (1999) *Coord. Chem. Rev.*, **184**, 125–155; (b) Caravan, P., Ellison, J.J., McMurry, T.J., and Lauffer, R.B. (1999) *Chem. Rev.*, **99**, 2293–2352.

14 (a) Ersoy, H. and Rybicki, F.J. (2007) *J. Magn. Reson. Imaging*, **26**, 1190–1197; (b) Lin, S.P. and Brown, J.J. (2007) *J. Magn. Reson. Imaging*, **25**, 884–899.

15 (a) Gao, J.H., Gu, H.W., and Xu, B. (2009) *Acc. Chem. Res.*, **42**, 1097–1107; (b) Lin, W.B., Hyeon, T., Lanza, G.M., Zhang, M.Q., and Meade, T.J. (2009) *MRS Bull.*, **34**, 441–448; (c) Laurent, S., Boutry, S., Mahieu, I., Vander Elst, L., and Muller, R.N. (2009) *Curr. Med. Chem.*, **16**, 4712–4727; (d) Laurent, S., Forge, D., Port, M., Roch, A., Vander Elst, L., and Nuller, R.N. (2008) *Chem. Rev.*, **108**, 2064–2110.

16 Silva, A.C., Lee, J.H., Aoki, L., and Koretsky, A.R. (2004) *NMR Biomed.*, **17**, 532–543.

17 (a) Kosaka, N., Ogawa, M., Choyke, P.L., and Kobayashi, H. (2009) *Fut. Oncol.*, **5**, 1501–1511; (b) Licha, K. (2002) *Top. Curr. Chem.*, **222**, 1–29.

18 (a) Smith, A.M., Duan, H., Mohs, A.M., and Nie, S. (2008) *Adv. Drug Deliv. Rev.*, **60**, 1226–1240; (b) Smith, A.M., Dave, S., Nie, S., True, L., and Gao, X. (2006) *Expert Rev. Mol. Diagn.*, **6**, 231–244.

19 (a) Kalender, W.A. (2006) *Phys. Med. Biol.*, **51**, R29; (b) Yu, S. and Watson, A.D. (1999) *Chem. Rev.*, **99**, 2353–2378.

20 Rieter, W.J., Taylor, K.M.L., An, H., Lin, W., and Lin, W. (2006) *J. Am. Chem. Soc.*, **128**, 9024–9025.

21 Rieter, W.J., Taylor, K.M.L., and Lin, W.B. (2007) *J. Am. Chem. Soc.*, **129**, 9852–9853.

22 (a) Rowe, M.D., Thamm, D.H., Kraft, S.L., and Boyes, S.G. (2009) *Biomacromolecules*, **10**, 983–993; (b) Rowe, M.D., Chang, C.C., Thamm, D.H., Kraft, S.L., Harmon, J.F., Vogt, A.P., Sumerlin, B.S., and Boyes, S.G. (2009) *Langmuir*, **25**, 9487–9499.

23 Taylor, K.M.L., Jin, A., and Lin, W. (2008) *Angew. Chem. Int. Ed.*, **47**, 7722–7725.

24 Taylor, K.M.L., Rieter, W.J., and Lin, W. (2008) *J. Am. Chem. Soc.*, **130**, 14358–14359.

25 Férey, G., Mellot-Draznieks, C., Serre, C., Millange, F., Dutour, J., Surblé, S., and Margiolaki, I. (2005) *Science*, **309**, 2040–2042.

26 (a) Horcajada, P., Serre, C., Maurin, G., Ramsahye, N.A., Balas, F., Vallet-Regi, M., Sebban, M., Taulelle, F., and Férey, G. (2008) *J. Am. Chem. Soc.*, **130**, 6774–6780; (b) Horcajada, P., Serre, C., Vallet-Regi, M., Sebban, M., Taulelle, F., and Férey, G. (2006) *Angew. Chem. Int. Ed.*, **45**, 5974–5978.

27 Taylor-Pashow, K.M.L., Della Rocca, J., Xie, Z., Tran, S., and Lin, W. (2009) *J. Am. Chem. Soc.*, **131**, 14261–14263.

28 Horcajada, P., Chalati, T., Serre, C., Gillet, B., Sebrie, C., Baati, T., Eubank, J.F., Heurtaux, D., Clayette, P., Kreuz, C., Chang, J.S., Hwang, Y.K., Marsaud, V., Bories, P.N., Cynober, L., Gil, S., Férey, G., Couvreur, P., and Gref, R. (2010) *Nat. Mater.*, **9**, 172–178.

29 Dekrafft, K.E., Xie, Z.G., Cao, G.H., Tran, S., Ma, L.Q., Zhou, O.Z., and Lin, W.B. (2009) *Angew. Chem. Int. Ed.*, **48**, 9901–9904.

30 (a) Nishiyabu, R., Hashimoto, N., Cho, T., Watanabe, K., Yasunaga, T., Endo, A., Kaneko, K., Niidome, T., Murata, M., Adachi, C., Katayama, Y., Hashizume, M., and Kimizuka, N. (2009) *J. Am. Chem. Soc.*, **131**, 2151–2158; (b) Nishiyabu, R., Aime, C., Gondo, R., Noguchi, T., and Kimizuka, N. (2009) *Angew. Chem. Int. Ed.*, **48**, 9465–9468; (c) Aime, C., Nishiyahu, R., Gondo, R., and

Kimizuka, N. (2010) *Chem. Eur. J.*, **16**, 3604–3607; (d) Nishiyabu, R., Aime, C., Gondo, R., Kaneko, K., and Kimizuka, N. (2010) *Chem. Commun.*, **46**, 4333–4335.

31 (a) Imaz, I., Hernando, J., Ruiz-Molina, D., and Maspoch, D. (2009) *Angew. Chem. Int. Ed.*, **48**, 2325–2329; (b) Imaz, I., Rubio-Martinez, M., Garcia-Fernandez, L., Garcia, F., Ruiz-Molina, D., Hernando, J., Puntes, V., and Maspoch, D. (2010) *Chem. Commun.*, **46**, 4737–4739.

32 Yan, X.H., Zhu, P.L., Fei, J.B., and Li, J.B. (2010) *Adv. Mater.*, **22**, 1283–1287.

Part Five
Physical Applications

12
Luminescent Metal–Organic Frameworks
John J. Perry IV, Christina A. Bauer, and Mark D. Allendorf

12.1
Introduction

Luminescence is a commonly reported property of metal–organic frameworks (MOFs), in large part because these structures typically contain aromatic or conjugated linker groups that readily fluoresce, or incorporate luminescent lanthanide ions. Because the moieties are largely immobilized by the crystal structure, the lifetimes and quantum efficiencies may be higher than those exhibited by the isolated luminophore in solution. MOFs are not completely rigid, however; their well-known structural flexibility can lead to dramatic changes in the local coordination environment and corresponding variations in emissive properties. The nanoporosity of these structures also creates new opportunities to impart light-emitting properties that are not available to traditional inorganic complexes. In particular, infiltrated guest molecules that are themselves emitters or sensitizers can be immobilized in close proximity to luminescent centers. This can influence the emission properties of the neat MOF in the form of wavelength shifts, intensity changes, or new emission resulting from excimer or exciplex formation. Since fluorescence spectrometers are readily available, many investigators report the excitation and emission spectra of newly synthesized MOF structures. As a result, more than 300 reports of luminescent MOFs are now available in the literature. Relatively few studies delve into the underlying spectroscopy or use luminescence as a tool for probing electronic structure and energy transfer. Nevertheless, it is apparent that luminescence holds considerable potential not only to serve as a diagnostic tool for probing fundamental aspects of MOF behavior and structure, but also to show potential for a variety of applications. As might be expected, chemical sensing is a logical possibility, but as will be seen later in this chapter, the potential extends well beyond this to include radiation detection, nonlinear optical effects, and solid-state lighting.

The purpose of this chapter is first to introduce the reader to the basic concepts of luminescence relevant to much of the current MOF literature. We then discuss luminescence associated with individual components of MOFs, each of which displays characteristic features. Examples from the recent literature are provided that illustrate the multidimensional character of MOF luminescence. However, a

Metal-Organic Frameworks: Applications from Catalysis to Gas Storage, First Edition. Edited by David Farrusseng.
© 2011 Wiley-VCH Verlag GmbH & Co. KGaA. Published 2011 by Wiley-VCH Verlag GmbH & Co. KGaA.

comprehensive review is not presented, as this was recently done [1]. As will be seen, a detailed understanding of luminescence phenomena in MOFs is beginning to emerge. Although still in its early stages, the results demonstrate that the unique characteristics of these supramolecular materials provide a platform for understanding light emission with both unparalleled synthetic flexibility and a highly ordered, inherently quantifiable, structure. From a practical point of view, it is also becoming clear that luminescence spectroscopy is a superb tool for assessing the structural integrity and purity of MOFs, with sensitivity far exceeding that of powder X-ray diffraction (PXRD). The chapter concludes with an overview of recent developments concerning applications of luminescent MOFs.

12.2
Luminescence Theory

In general, luminescence is the result of radiative decay (i.e., emission of light) by electronically excited molecules following the absorption of energy. The absorbed energy can originate from a variety of sources, which leads to different sub-classifications of luminescence. When the energy is absorbed from electromagnetic radiation in the form of photons (light), the subsequent emission of light is defined as *photoluminescence*. The nonthermal production of light by a chemical reaction is termed *chemiluminescence* [2–4], with the special case involving living organisms being referred to as *bioluminescence* [5–7]. In some cases, the presence of an electric field can cause electron–hole recombination in a material, producing *electroluminescence* [8, 9]. If the energy absorbed by the molecules is due to the material being bombarded with high-energy ionizing radiation (>10 eV), the emission of light is referred to as *radioluminescence*, while a beam of high-energy electrons imparted upon a material may lead to *cathodoluminescence*. Additionally, mechanical forces [10–12] acting on a material may also lead to the emission of light in the form of *triboluminescence* (mechanical stress, scratching, or fracture) or *piezoluminescence* (pressure). Although there appear to be several methods to excite a molecule or group of molecules in a MOF electronically to yield emission of light, to date only a few have been investigated, the most prevalent of which has been photoluminescence (PL).

12.2.1
Photoluminescence

Although many excellent texts are available concerning the theory of luminescence in general [13–15] and PL [16, 17] in particular, a general, albeit simplified, overview of the concepts is warranted. For the sake of simplicity in describing the photophysical processes in question, let us consider only the organic molecule, which acts as a linker or strut in the formation of the MOF. Before a molecule is excited by the absorption of light, the molecule resides in its lowest electronic energy level or ground state. Typically, an organic molecule has a singlet ground state (S_0), but in rare cases a triplet ground state (T_0) is possible. Higher energy electronic states exist (S_n or

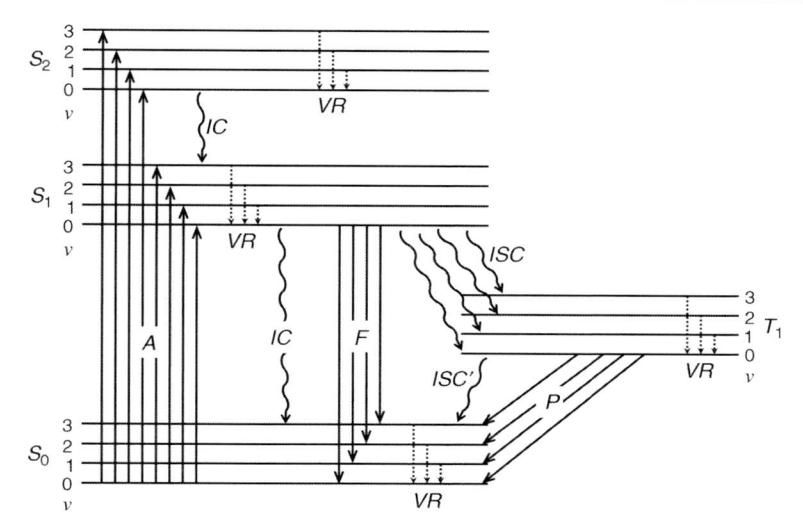

Figure 12.1 Jablonski diagram depicting the photophysical processes of a conjugated organic molecule. Absorption (A) of energy is followed by fast, non-radiative processes such as vibrational relaxation (VR) and internal conversion (IC) down to the lowest vibration level of the first excited state ($S_{1,0}$). From here there are three process for the deactivation of $S_{1,0}$; the non-radiative processes of internal conversion (IC) and singlet-triplet intersystem crossing (ISC), as well as the radiative process of fluorescence (F). If the deactivation process leads to singlet-triplet ISC, further relaxation from the lowest excited triplet state ($T_{1,0}$) can proceed through either radiative (Phosphorescence, P) or non-radiative (triplet-singlet intersystem crossing, ISC') deactivation to the ground state.

T_n; $n \geq 1$), and it is these elevated states to which an electron is promoted upon the absorption of energy. Additionally, each electronic state is associated with several vibrational energy levels and together the electronic and vibrational levels constitute a vibronic state ($S_{n,v}$: $v = 0, 1, 2, 3, \ldots$). A Jablonski diagram (Figure 12.1) is a simplified graphical representation of the vibronic states for a given molecule wherein higher energy levels are stacked vertically over lower ones. Jablonski diagrams are useful for depicting the various photophysical processes involved in the absorption and eventual loss of excess energy by the molecule, but do not account for geometric distortions. The vertical nature of the transitions is reflective of the fact that absorption occurs in a time frame that is too short for significant relocation of nuclei (Franck–Condon principle) and therefore rearrangement of electronic energy states is negligible.

In the ground state, nearly all of the molecules exist in the lowest totally symmetric (a_1) vibrational level ($S_{0,0}$) at room temperature, and so it is from this vibronic state that absorption of energy leads to excitation to elevated energy levels. In order to energize a molecule sufficiently and promote it to a higher electronic state, the use of ultraviolet (UV) or visible light is required. Light of lower energy (and longer wavelength) such as infrared (IR) is typically not energetic enough to cause promotion to higher electronic states, but can be useful for the excitation of higher vibrational levels of the ground electronic state. Upon excitation, the molecule begins rapidly to lose excess vibrational and electronic energy to the surrounding medium (when in a condensed phase) in a nonradiative thermal deactivation process. In the first phase, the molecule loses excess vibrational energy through vibrational relaxation (VR), in which the molecule undergoes a radiationless transition from a higher to a lower vibrational level within the same electronic state. This is followed by a radiationless transition between electronic states of the same multiplicity called internal conversion (IC) and possibly further VR within the new electronic state. Taken together, VR and IC constitute thermal deactivation of the molecule carrying it rapidly to the lowest vibrational level of the lowest excited state (10^{-13}–10^{-12} s). From the $S_{1,0}$ excited state, the molecule can undergo deactivation via one of three first-order processes. For many molecules it is possible to become deactivated by nonradiative internal conversion (rate constant k_{ic}). However, for some molecules containing relatively large gaps in energy between the lowest excited and ground states, this process is unfavorable, and the majority of these surrender their excess energy via the spin-allowed emission of light known as fluorescence (F, k_f). The third possibility is the formally spin-forbidden intersystem crossing (ISC, k_{isc}) to the lowest excited triplet state T_1. The excited triplet state must also undergo deactivation and can do so through either spin-forbidden radiative emission known as phosphorescence (P, k_p) or though nonradiative spin-forbidden ISC ($k_{isc'}$). As there are several competing pathways for the deactivation of the lowest excited singlet state, the number of molecules that fluoresce is only a fraction of the total population of excited molecules. This ratio between the number of molecules which fluoresce and the total excited population is the fluorescence efficiency or *quantum yield of fluorescence* (Φ_f). The quantum yield is a value between 0 (no fluorescence) and 1 (all excited molecules fluoresce) and is related to the first-order rate constants for the deactivation pathways (Equation 12.1). Similarly, the ratio of the number of molecules to undergo phosphorescence to the number of molecules to be deactivated in any of the other processes is given by the related Equation 12.2.

$$\Phi_f = \frac{k_f}{k_{ic} + k_f + k_{isc}} \tag{12.1}$$

$$\Phi_p = \frac{k_p k_{isc}}{(k_{isc'} + k_p)(k_{ic} + k_f + k_{isc})} \tag{12.2}$$

Another useful property of PL, especially in the comparison of fluorescence and phosphorescence, is the concept of the lifetime (τ) of a given excited state. The lifetime is defined as the average time the molecule exists in the excited state before the excited population is reduced by a factor of e (exponential decay). The lifetime of an excited state is therefore the inverse of the summation of the rate constants for the potential decay pathways deactivating that state and is given by the following equations:

$$\tau(S_1) = \frac{1}{k_{ic} + k_f + k_{isc}} \tag{12.3}$$

$$\tau(T_1) = \frac{1}{k_{isc'} + k_p} \tag{12.4}$$

The *natural* lifetime (τ_n) is the intrinsic lifetime of the luminophore in the absence of nonradiative decay and can be calculated from the ratio of τ to Φ. As the processes involved in deactivation of S_1 through the T_1 excited state are all spin-forbidden, the lifetimes are typically much longer for phosphorescence (10^{-4}–10^0 s) than that observed for fluorescence (10^{-11}–10^{-7} s).

12.2.2
Fluorescence Quenching

Although many types of molecule are capable of absorbing energy to become electronically excited, relatively few exhibit the phenomenon of luminescence. This is due in part to *quenching*. Essentially, any process capable of diminishing the intensity of fluorescence or phosphorescence can be considered a quenching mechanism [18], including the rapid nonradiative de-excitation of $S_1 \to S_0$ via internal conversion. External molecules called quenchers are also capable of de-exciting a molecule; this will either diminish or completely eliminate the ability of the excited molecule to luminesce. Quenching in MOFs is important from an applications aspect as it can be both detrimental, leading to a reduction in the desired property (luminescence), and beneficial in cases where the quenching of a known luminescence signal is used as a sensor mechanism for detecting the presence of the quencher moiety. Luminescence quenching is observed in one of two forms, *dynamic* and *static*. Dynamic quenching is characterized by the collision of a quencher moiety (acceptor) with an excited-state luminophore (donor), resulting in de-excitation of the donor and concomitant excitation of the acceptor. The newly excited acceptor may or may not be luminescent in its own right, but any resulting emission will always be of lower intensity than what would be observed from the donor. In dynamic quenching, the quencher and potential luminophore must come into contact during the lifetime of the luminophore excited state, and is therefore a time-dependent process. Additionally, dynamic quenching is dependent on the concentration [Q] and diffusion of the quencher molecule and obeys the Stern–Volmer equation (Equation 12.5). In static quenching, interaction between a potential luminophore and a quencher molecule takes place in the ground state, leading to the formation of a non-luminescent complex. Here the efficiency of quenching is governed by the rate of complex formation (k_c) and [Q] (Equation 12.6).

$$\frac{\Phi}{\Phi_0} = \frac{1}{1 + k_q \tau_0 [Q]} \qquad (12.5)$$

$$\frac{\Phi}{\Phi_0} = \frac{1}{1 + k_c [Q]} \qquad (12.6)$$

12.2.3
Energy Transfer

Another important process that can affect the luminescence of MOFs is energy transfer (ET). It may be possible, during the lifetime of the excited state, to transfer the excitation energy from an excited molecule (donor) to another molecule in the ground state (acceptor). This is essentially another method of quenching the donor molecule as it will become de-excited and no longer luminescent. The acceptor absorbs the energy and is excited, which may lead to luminescence (albeit at a lower intensity), or it may deactivate via a nonradiative pathway so that as a system the luminescence will be lost. There are two mechanisms for ET distinguished by the nature of the donor–acceptor interaction. In *Förster resonance energy transfer* (FRET; Förster type; dipole mechanism; coulombic mechanism), the donor and acceptor molecules are not in contact and can be separated by as much as 10 nm [13, 18, 19]. For FRET to occur, at least some spectral overlap between the emission of the donor and the excitation of the acceptor must exist and the rate of energy transfer (probability) is dependent on both the amount of overlap and also the distance between the two molecules, with the rate decreasing in relation to $(1/r)^6$. In *excitation energy transfer* (Dexter type; exchange mechanism), the orbitals of the donor and acceptor molecules must be in direct contact. Dexter-type ET is sometimes considered as a double-electron transfer process, as the highest energy electrons from the donor (located in the LUMO) and the acceptor (located in the HOMO) are exchanged. Since the acceptor gains an electron in its LUMO while giving up an electron from its HOMO, the acceptor becomes excited while conversely the donor is deactivated. While Dexter-type ET requires spectral overlap as in FRET, this process is also diffusion controlled and depends on the viscosity of the solvent (i.e., the environment) [19].

12.3
Ligand-Based Luminescence

12.3.1
Solid-State Luminescence of Organic Molecules

The study of luminescence from organic compounds in the solid state is of increasingly high importance. In addition to the interest in fundamental physical studies, implementation of organic materials as active components in solid-state devices has led to increased attention in this area. Relevant examples include organic light-emitting diodes (OLEDs), thin-film sensors, and solid-state lasers. A major challenge for the development of such devices is suitable control over the morphology and interfacial properties of emissive organic species during device fabrication. Factors such as luminophore orientation, intermolecular distance, packing density, and details of the local molecular environment often have a significant impact on the resulting emission (and absorption) wavelength, intensity, and lifetime. Highly fluorescent organic molecules often feature an extended π-conjugated system (although this is not a requirement for emissive behavior). Such molecules share features in common with ligands often used in the field of MOFs, where a rigid, π-conjugated "linker" is often used to form the struts connecting metal ions or clusters. The overall structure of a MOF is mainly determined by geometric requirements arising from the strong coordination between the metal ions or clusters and the ligand groups, and may be varied by choosing suitable starting materials and/or growth conditions. MOFs, therefore, have a degree of structural predictability and can allow for a controllable platform for arranging luminophores in order to alter the solid-state emission properties of the material. Conversely, the structure of an entirely organic solid-state material (i.e., a molecular solid) is often dictated by a multitude of weaker crystal packing interactions, including π–π stacking and hydrogen bonding, and there is little control over the structure ultimately formed in the solid state. For MOFs, the wide variety of organic ligands and coordination geometries available to afford predictable structures, and also the ability to accommodate guests, potentially allow for many degrees of freedom and, consequently, rational modulation of the emissive properties of the material.

Several possibilities exist for the type of emission displayed by MOFs containing π-conjugated ligands. In some cases, there is little or no electronic interaction between the organic and inorganic moieties, leading to ligand-based luminescence, as will be discussed in Section 12.3.2. These ligands may be isolated from each other in the structure, resulting in emission characteristic of a dilute solution of the free ligand, or the intermolecular distance and/or orientation may be such that significant interaction occurs, leading to ligand-to-ligand charge transfer (LLCT) emission with a typically broadened and red-shifted emission profile. Sections 12.3.3 and 12.3.4 describe materials for which the luminescence properties are determined not only by the organic molecule(s), but also through interaction with the metal unit(s). Two limiting cases are possible, ligand-to-metal charge transfer (LMCT) or metal-to-ligand charge transfer (MLCT). The particular type of emission observed depends on the structure of the MOF (which dictates the spacing and orientation between the ligands and metal units), the electronic configuration of the metal, and the degree of orbital overlap between the organic and inorganic moieties. Several detailed structure–property relationships have been described for luminescent MOFs, using both steady-state and time-resolved measurements, with the number of reports in this area increasing significantly over the past few years. The following sections are intended to exemplify the important features and principles involved for correlating structure with observed luminescence in MOF materials.

12.3.2
Ligand-Based Luminescence in MOFs

Ligand-centered luminescence typically occurs for MOFs in which there is little or no electronic interaction between the organic and inorganic portions of the structure. Typically, this is the case for structures containing metal ions that have no d–d or f–f transitions possible, most commonly for MOFs with d^{10} Cd^{II}, Zn^{II}, and Ag^{I} ions. An excellent example of this is provided by the IRMOF series of Yaghi and co-workers [20], in which the bonding between anionic carboxylate linkers and Zn^{II} ions is essentially ionic and involves very little charge transfer (see the discussion below). However, the presence of a d^{10} metal ion does not necessarily dictate that the structure will display ligand-based emission, as upon light absorption the ligand may

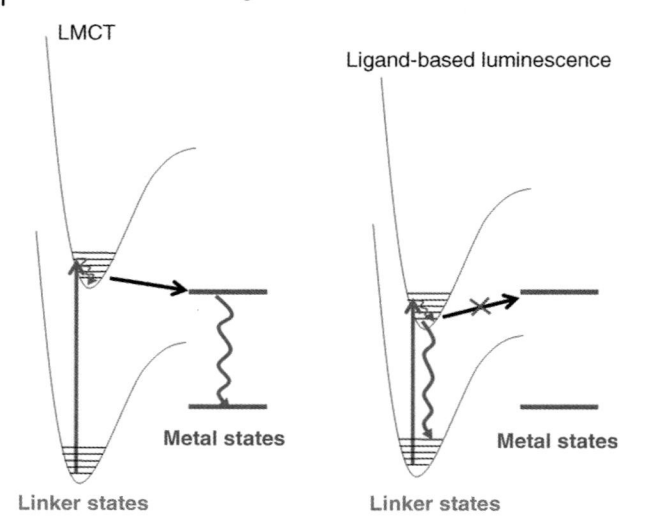

Figure 12.2 Energy level diagram depicting HOMO and LUMO states of the organic ligand, and intersystem crossing (LMCT) to emissive metal states (left). If π-conjugation in the ligand is increased, the HOMO-LUMO gap is decreased, making LMCT inefficient and, therefore, ligand-based emission is observed (right). In addition, coupling to surrounding solvent or guest molecules may also affect the efficiency of these transitions for porous structures.

facilitate charge transfer to higher energy, emissive metal states if the ligand HOMO–LUMO gap is large. Moreover, as will be seen throughout this chapter, the complexity of MOF structures can frequently produce unexpected luminescent features that defy placement within straightforward categories such as this.

The IRMOFs illustrate the effect of increasing linker conjugation. When this occurs, the $\pi-\pi^*$ energy gap decreases [13], making energy transfer to the metal less efficient. The resulting ligand-based emission resembles that of the corresponding ligand in dilute solution (Figure 12.2). A variety of IRMOFs exhibit emission similar to that of their isolated ligands, including IRMOF-11 $[Zn_4O(4,5,9,10\text{-tetrahydro-2,7-pyrenedicarboxylate})_3]_n$, IRMOF-13 $[Zn_4O(2,7\text{-pyrenedicarboxylate})_3]_n$, and $[Zn_4O(4,4'\text{-stilbenedicarboxylate})_3]_n$ [1]. Each of these structures exhibits emission similar to that of the corresponding dilute ligand in solution, indicating that ligand-centered luminescence dominates and there is little influence of the Zn_4O cluster on the electronic structure of the ligand. In addition, the absence of other bands to the red of this emission indicates there is no *significant* LLCT.

A second factor that strongly influences ligand-based MOF emission is the possibility of interpenetration within the structure. The three IRMOF structures just discussed consist of interpenetrated frameworks as a consequence of the increased length of the rigid, linear linker. MOFs with interpenetrated structures consist of two or more frameworks intertwined, and details of the structure, such as distance between the interpenetrated frameworks and dihedral angle of the organic ligands, will lead to variations observed in the emission wavelength, fine structure, and intensity. Although significant LLCT was ruled out by detailed experimental studies in the [Zn$_4$O(4,4′-stilbenedicarboxylate)]$_n$ structure [21] (as discussed in more detail below), this may not necessarily be the case for other interpenetrated MOFs, such as the pyrene- and tetrahydropyrene-based IRMOFs, for which careful studies of these two structures have not yet been carried out. However, other IRMOFs that exist as interpenetrated and non-interpenetrated versions have been probed in more detail. Meek *et al.* compared the excitation and emission spectra of the interpenetrated and non-interpenetrated versions of two IRMOFs constructed from the same ligand [22]. IRMOF-9 and -10 are constructed from 4,4′-biphenyl dicarboxylic acid (with IRMOF-9 being interpenetrated), and IRMOF-15 and -16 are constructed from 4,4′-terphenyldicarboxylic acid (IRMOF-15 is interpenetrated). In the terphenyl MOFs, the interpenetrated structure exhibits red-shifted emission compared with the non-interpenetrated variant, suggesting that interpenetration induces greater charge transfer, possibly due to smaller linker–linker and linker–metal distances. In the biphenyl structures, the interpenetrated MOF (IRMOF-9) displays a completely new band at ∼500 nm, indicating the formation of a new excited state. This emission is broad and unstructured, suggesting that it involves charge transfer between biphenyl linkers in close proximity.

One might ask why this section did not begin with a discussion of IRMOF-1, the simplest of the cubic IRMOF structures, in which the linker is 1,4-bdc (1,4-benzenedicarboxylate) [23]. Here, although the 1,4-bdc linker is fully conjugated, the HOMO–LUMO energy gap is large; the excitation maximum of the isolated linker in dilute DMSO solution occurs at 338 nm. In contrast, a green emission upon UV excitation is observed from IRMOF-1 in some cases. This has been attributed to an LMCT process, whereby energy transfer occurs from the 1,4-bdc singlet excited state to the Zn$_4$O clusters [23]. Although this mechanism is supported by excitation spectra indicating that light absorption occurs via the benzene unit and by *ab initio* studies [24], recent work has demonstrated that ligand-centered luminescence is actually the dominant process and that the green emission attributed to LMCT is due to synthetic byproducts present in the structure or degradation of the structure itself (framework collapse) [25]. This investigation emphasized that unambiguous knowledge of the structure and constituents (including contaminants) of a material is necessary for accurate interpretation of the physical causes behind observed luminescence properties. Additionally, it demonstrates that changes in framework structure, undetectable by standard characterization methods such as PXRD, can be easily identified by the significantly more sensitive luminescent methods.

The effects on the emission properties of varying the orientation and spacing between ligands have also been demonstrated in the case of Zn stilbene MOFs [21]. Two MOF structures with different topologies were synthesized from the same starting materials via changes in the growth conditions and solvent, yielding both a two-periodic net [Zn$_3$L$_3$(DMF)$_2$]$_n$ and a cubic three-periodic framework [Zn$_4$OL$_3$]$_n$ structure, where L = 4,4′-stilbenedicarboxylate (Figure 12.3). Both materials exhibit intense ligand-based luminescence, similar in shape and vibronic structure, signifying little to no influence of the different metal cluster types present in each structure. However, emission from the two-periodic structure is red shifted and

stilbene dicarboxylic acid (LH$_2$)   Zn$_3$L$_3$(DMF)$_2$ Zn$_4$OL$_3$

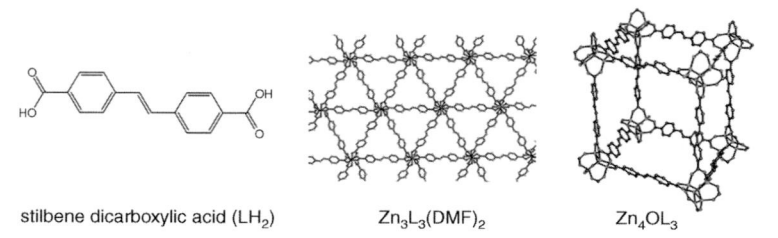

Figure 12.3 Structure of 4,4′-stilbene dicarboxylic acid, the layered 2D Zn stilbene MOF structure, and an individual cubic unit of the catenated 3D Zn stilbene MOF structure.

broadened from that of the cubic three-periodic structure, indicating a greater degree of interaction between luminophores in the two-periodic structure. This is consistent with the 6.0 Å, cofacial spacing between ligands in successive layers observed for this structure. It should be noted that the extent of this intermolecular interaction is small compared with that observed for the ligand itself in the solid state, which exhibits a significantly broader and red-shifted emission. The three-periodic cubic structure consists of anisotropically catenated cubes (edges are closer along the *b*-axis), a likely result of very moderate stilbene–stilbene interactions. The closest linkers are not cofacial in these structures, but rather tilted with a 49° dihedral angle with respect to one another. In this case, the MOF exhibits luminescence similar to a dilute solution of the ligand, a situation where LLCT is minimal.

An additional point of interest, exemplified by the stilbene MOF structures just discussed, is that the MOF materials often display increased emission intensity compared with the ligand itself. In the framework structures, the stilbene units are held tightly through strong metal coordination, and their motion is restricted. This inhibits the nonradiative *cis–trans* isomerization process that occurs readily for stilbene in solution upon absorption of light, and results in a concomitant increase in the fluorescence intensity of the rigidified ligand in the MOFs [26, 27]. An increase in the radiative lifetime typically leads to an increase in the fluorescence quantum yield by reducing the efficiency of such nonradiative pathways. Biexponential emission decay was observed for the two-periodic Zn stilbene structure, consistent with two emission pathways for the stilbene (i.e., isolated ligand-based emission and LLCT). However, the lower density cubic Zn stilbene MOF exhibited monoexponential emission decay (consistent with isolated stilbene units), with a 0.50 ns lifetime, approximately five times larger than for stilbene in solution. The increase in lifetime is indicative of the rigidity provided by incorporation into the MOF structure. As minor structural changes can lead to observable changes in luminescence properties, this allows for a natural extension of ligand-based luminescence into the field of sensing (Section 12.6). For instance, upon changing the guest solvent molecules in the cubic Zn stilbene MOF, the emission is found to shift in wavelength. This may be a result of a change in local dielectric environment for the ligand or minor changes in the overall framework structure, such as expansion or compression of the catenated units.

As luminescence can be extremely sensitive to local environment, the study of spectral properties has become a standard method for materials characterization as it is a simple, rapid indicator of structural changes and/or maintenance of structural integrity after MOFs have been subjected to a new environment. For example, recent work has shown that the ethylene bridge of stilbene in the three-periodic cubic Zn MOFs can be partially brominated without collapsing the structure [28]. However, full bromination interrupts the interpenetrated cubic units, leading to partial collapse, as was quickly determined by measurement of the luminescence spectra (revealing a broadened, red-shifted spectrum) and later verified via detailed structural studies. Another interesting aspect, supported by luminescence studies, is the change in stability to evacuation between structures upon bromination. As-prepared stilbene zinc three-periodic MOFs do not maintain their structure after full evacuation, a likely result of a change in the π–π interactions between interpenetrated cubes, which alter the stilbene–stilbene distance and disturbs the structure. This leads to lower than expected surface areas ($560\,m^2\,g^{-1}$) and a weakened, red-shifted luminescence with loss of vibronic structure upon evacuation (see Figure 12.4). However, after partial bromination, the inter-cube attractions are reduced and the structure remains intact upon evacuation, with an observed increase in surface area ($1600\,m^2\,g^{-1}$) despite the fact that the bromine atoms occupy some of the pore space. The emission spectrum of these partially brominated MOFs is nearly unperturbed from that of the as-prepared three-periodic material before evacuation or bromination, as shown in Figure 12.4 (presumably, the brominated stilbene units are substantially quenched and do not contribute significantly to the spectra).

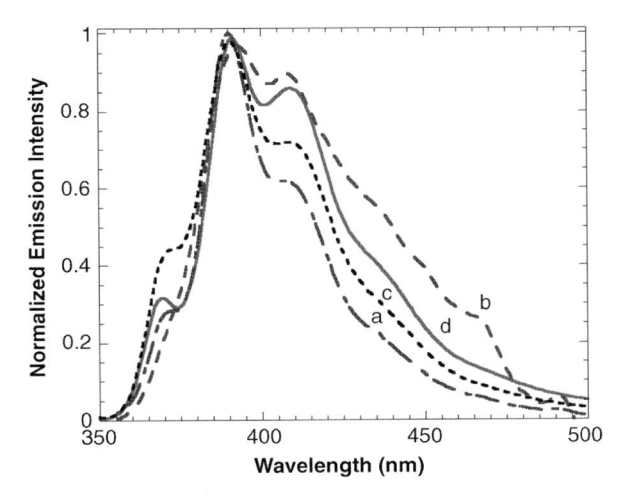

Figure 12.4 Emission spectra of 3D Zn stilbene MOFs with (a) chloroform guest molecules and (b) after full evacuation. The emission spectra of partially brominated Zn stilbene MOFs in (c) a chloroform environment and (d) *after full evacuation* reveal the stability of the structure to bromination and subsequent evacuation, as evidenced by their similarity to (a).

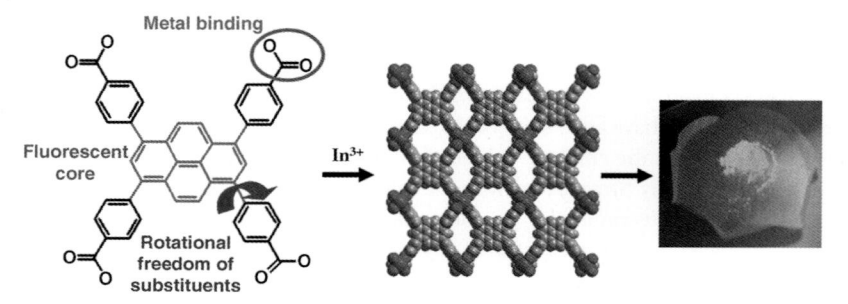

Figure 12.5 [In$_2$(OH)$_2$(TBAPy)] MOFs exhibit bright fluorescence in the extended network. The lifetime and intensity of the emission increases due to strong coordination within the framework structure. Reprinted with permission from [29]. Copyright 2010 American Chemical Society.

A recent study by Rosseinsky and co-workers also clearly demonstrates the direct influence of local ligand environment on the overall emission characteristics for MOF materials based on a luminescent pyrene core, [In$_2$(OH)$_2$(tbaPy)]$_n$ [tbaPy = 1,3,6,8-tetrakis(p-benzoic acid)pyrene] [29]. The tbaPyH$_4$ ligand itself exhibits an emission lifetime of 89 μs, which increases to 110 μs upon incorporation into the MOF, attributed to rigidification. Bright blue luminescence is observed upon coordination of the ligand in the MOF (see Figure 12.5), with a blue shift of 66 nm from that of the free ligand due a reduction of LLCT through isolation of the ligands from one another in the framework. Interestingly, upon evacuation of guest DMF molecules, the emission red shifts, the quantum yield decreases, and the fluorescence lifetime decreases. This is a direct result of the structural distortion which increases the degree of through-space coupling of the pyrene cores, elegantly demonstrated by single-crystal diffraction studies. Upon resolution with DMF, the original structure and luminescence properties are restored. Altering the identity of the guest molecules is also found to cause a change in emission, as will be discussed in Section 12.6.1.1. Other cases of luminescence enhancement due to the rigid environment in MOFs have also been reported [30–33].

12.3.3
Ligand-to-Metal Charge Transfer in MOFs

LMCT-based emission is often reported for Zn^{II} and Cd^{II} MOFs when the sufficiently absorptive ligand has a relatively large HOMO–LUMO gap (as discussed in the previous section), leading to significantly red-shifted emission compared with the ligand itself. Many MOFs displaying LMCT emission contain bdc, benzenetricarboxylate (btc), or both linkers [34–37]. In these cases, green emission is associated with UV excitation and assigned to an LMCT. However, as discussed above, this may be a consequence of an impurity (possibly ZnO). Often, ligand-based luminescence competes with LMCT, and both cases can be observed under certain circumstances. For example, {Zn$_3$(1,4-bdc)(1,3,5-btc)$_2$[NH(CH$_3$)$_2$][NH$_2$(CH$_3$)$_2$]$_2$}$_n$ emits at 430 nm due to LMCT, with a shoulder at 370 nm arising from emission from the btc ligand [34]. There are also reported examples of LMCT from MOFs containing ligands other than bdc and btc. Subtle variations in the environment can alter the efficiency of energy transfer from the ligand to the metal center, as described by Huang *et al.* for Cd^{II} 2,6-di(4-triazolyl)pyridine-based MOFs [38]. Ligand-based luminescence is observed for the hydrated structure, a result of water in the lattice preventing efficient charge transfer. However, the emission gains significant LMCT character upon dehydration as the charge-transfer process dominates.

The efficiency of LMCT is a function of the orbital overlap between ligand and metal and, therefore, the observation of ligand-based luminescence versus LMCT can also depend on the type of metal center. A recent study of 18 MOF structures containing tetrapyridyl ligands bound to either Zn^{II}, Cd^{II}, or Hg^{II} metal centers demonstrated how subtle variations in the structure can alter the emission characteristics [39]. These materials emit in a multitude of colors, from violet to yellow, depending on the structure type and metal ion. The luminescence intensity for all of the Zn^{II} and Cd^{II} structures was enhanced compared with the free ligand due to the rigidity afforded by coordination in the MOF structure. However, the Hg^{II} structures display reduced emission intensity, attributed to a heavy-metal quenching effect.

In another recent interesting study, the photoluminescence spectra of three structures containing Cd^{II}, 4,4′-bipyridine, and 2-amino-1,4-bdc were found to be characteristic of either a mixture of LMCT and ligand-centered emission, or only LMCT [40]. Upon desolation and framework distortion, the LMCT character is quenched, while ligand-centered luminescence grows in. Main-group metal complexes may also display LMCT in certain cases, for example $[Pb_4(1,3\text{-bdc})_3(\mu_4\text{-O})(H_2O)]_n$ exhibits a strong emission peak at 424 nm originating from LMCT between delocalized π-bonds of the bdc groups and p-orbitals of the Pb^{II} ions [41].

12.3.4
Metal-to-Ligand Charge Transfer in MOFs

MLCT is less commonly reported for MOFs than LMCT, and is typically observed for d^{10} Cu^I- and Ag^I-based MOFs where there is the possibility of d-electron transfer into low-lying empty ligand states (e.g., for π-acid ligands). For example, N-heterocyclic-based Cu^I MOFs displaying a multitude of geometries were shown to luminesce with varying degrees of intensity [42] (assigned to a Cu → CN MLCT process), and [Cu(1,2,4-btc)(2,2-bipyridine)]$_n$ was found to exhibit intense blue MLCT emission with a lifetime of 13.62 ns [43]. These mixed-valence complexes also exhibit intervalence charge transfer (IVCT).

Excitation of the two-periodic layered MOF {Ag[4-(2-pyrimidylthiomethyl)benzoic acid]}$_n$ at 370 nm results in an intense green emission with peak maximum at 530 nm, the origin of which was assigned to LMCT and/or MLCT modified by metal-centered (ds/dp) states mediated through Ag–Ag interactions [44]. Similarly, {[Ag(4,4′-bipyridyl)][Ag(1,2,4-Hbtc)]}$_n$ exhibits an intense emission band with a maximum at 502 nm ($\lambda_{ex} = 410$ nm) due to MLCT [43].

12.4
Metal-Based Luminescence

12.4.1
Metal Luminophores

When discussing metal ions and their salts, only the rare earth elements and some uranyl compounds are typically luminescent. For example, all of the lanthanide trivalent ions (Ln^{III}) except La^{III} and Lu^{III} are luminescent and their emissions span the spectrum from the UV (Gd^{III}) to the visible (Pr^{III}, Sm^{III}, Eu^{III}, Tb^{III}, Dy^{III}, and Tm^{III}) to the near-infrared (NIR) (Pr^{III}, Nd^{III}, Ho^{III}, Er^{III}, and Yb^{III}) region [45]. Some of these elements are fluorescent, whereas others exhibit phosphorescence, although in some instances they can exhibit both. For compounds containing uranium, only the U^{VI} uranyl ion $[UO_2]^{2+}$ is fluorescent and typically emits in the range 520–620 nm.

12.4.2
Lanthanide Luminescence and the Antenna Effect

Trivalent lanthanide ions (Ln^{III}) are attractive luminophores because they exhibit spectrally narrow emission with long lifetimes in the solid state and in solution [46]. The 4f orbital is shielded by a filled $5s^2 5p^6$ sub-shell, thus screening the inner-shell 4f–4f electronic transitions which generate well-defined energy gaps for the electronic levels allowing for the line-like emission. However, electronic transitions for Ln^{III} ions are formally forbidden by parity selection rules (Laporte's rule), typically leading to weak absorbance and, therefore, low brightness. One method of overcoming the weak absorption cross-sections of these ions has been to complex them with strongly absorbing organic molecules. Additionally, complexation often prevents quenching of luminescence by solvent molecules. When efficient vibronic coupling exists between the organic ligand and Ln^{III} moieties, direct energy transfer from the organic molecules' excited state to the metals' elevated energy levels is possible (Figure 12.6). This coupling, first identified by Weissman in 1942, leads to a large increase in luminescence intensity and is widely known as the "antenna effect" or "luminescence sensitization" [45, 47]. For Ln^{III} ions, the electronic dipole transitions are of the same order of magnitude as the magnetic dipole transitions, and therefore both are readily seen in the optical spectra.

12.4.3
Examples of Metal-Based Luminescence

12.4.3.1 Metal-Centered Luminescence
Many isostructural series of Ln^{III}-containing MOFs are known; however, typically, only the luminescence properties of the Eu^{III} and Tb^{III} analogs are reported [48–60]. The likely reason why these two metals predominate in luminescence reports is twofold: they benefit from strong coupling to most organic ligands used for MOF synthesis (antenna effect) and their red and green emission in the visible range is

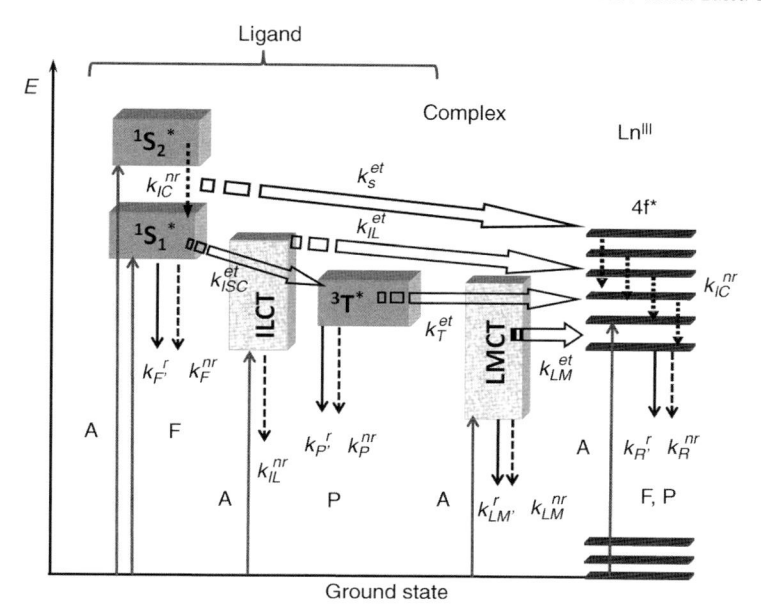

Figure 12.6 Schematic representation of energy absorption, migration, emission (plain arrows), and dissipation (dotted arrows) processes in a lanthanide complex. $^1S^*$ or S = singlet state, $^3T^*$ or T = triplet state, A = absorption, F = fluorescence, P = phosphorescence, k = rate constant, r = radiative, nr = non-radiative, IC = internal conversion, ISC = intersystem crossing, ILCT (indices IL) = intra-ligand charge transfer, LMCT (indices LM) = ligand-to-metal charge transfer. Back transfer processes are not drawn for the sake of clarity. [45] Reproduced by permission of The Royal Society of Chemistry (RSC).

easier to detect with common laboratory equipment than the longer wavelength emission of several other Ln^{III} ions. In complexes comprised of Eu^{III}, the $^5D_0 \rightarrow {}^7F_J$ transition series ($J = 0$–4) is the most readily observable, typically with the $J = 1$ and 2 transitions being both the strongest and most useful for structural determination and sensing. The dominant emission of the $^5D_0 \rightarrow {}^7F_2$ transition usually occurs around 615 nm, leading to the brilliant red color characteristic of Eu^{III} luminescence. The emissions observed for Tb^{III} come from the $^5D_4 \rightarrow {}^7F_J$ transitions ($J = 3$–6), with the strongest typically being that of $J = 5$ and 6 and giving rise to the strong green luminescence in the 540–555 nm range characteristic of Tb^{III} complexes.

An illustrative example of metal-centered luminescence was provided by Li et al., who reported a series of Ln^{III} MOFs based on $[Ln_4(OH)_4]^{8+}$ cubane-like secondary building units (SBUs) [61]. The structures, which have the unit formula $\{[Ln_4(OH)_4(3\text{-sba})_4(H_2O)_4] \cdot nH_2O\}_n$, (Ln = Eu^{III} (**1**), Gd^{III} (**2**), Tb^{III} (**3**); 3-sba = 3-sulfobenzoate), are three-periodic porous frameworks with unique (3,12) topology. The PL of **1** and **3** is characteristic of Ln^{III} metal-centered luminescence. The excitation spectra of **1**, monitored at the strongest Eu^{III} transition ($^5D_0 \rightarrow {}^7F_2$,

614 nm), displays an intense, broad band centered at 286 nm that was attributed to the ligand-centered $\pi \rightarrow \pi^*$ transition, in addition to several much weaker narrow bands attributed to direct f–f transitions within the Eu^{III} metal center ($^7F_0 \rightarrow {}^5H_4$, 318 nm; $^7F_0 \rightarrow {}^5G_6$, 362 nm; $^7F_0 \rightarrow {}^5G_{0-4}$, 377–382 nm; $^7F_0 \rightarrow {}^5L_6$, 395 nm; $^7F_0 \rightarrow {}^5D_3$, 416 nm). The emission spectra of **1** ($\lambda_{ex} = 286$ nm) exhibit several narrow spectral lines assigned as $^5D_0 \rightarrow {}^7F_J$ ($J = 0$–4) and dominated by the hypersensitive $J = 2$ peak at 614 nm, resulting in bright red luminescence. No emission corresponding to the $\pi^* \rightarrow \pi$ transition of the ligand is observed, indicating that the fluorescence is entirely metal-centered emission. The lifetime of the $^5D_0$ state was also determined as $\tau = 0.135$ ms, based on the strongest transition. In the case of **3**, the excitation spectrum was monitored at 545 nm ($^5D_4 \rightarrow {}^7F_5$) and again a strong absorption band at 288 nm was observed and assigned as the $\pi \rightarrow \pi^*$ transition. Additionally, the significantly weaker f–f transitions for Tb^{III} were also present ($^5G_2$, 351 nm; $^5D_3$, 359 nm; $^5D_2$, 369 nm; $^5D_4$, 377 nm). The emission spectra ($\lambda_{ex} = 350$ nm) of **3** is dominated by the hypersensitive $^5D_4 \rightarrow {}^7F_5$ (545 nm with a shoulder at 549 nm), but also displays the remaining transitions from the lowest excited state to the ground-state manifold ($^5D_4 \rightarrow {}^7F_{6-3}$: $J = 6$, 489 nm; $J = 4$, 528–590 nm; $J = 3$, 622 nm). The lifetime for the $^5D_4$ state was observed to be $\tau = 1.035$ ms.

In another example of metal-centered emission, Choi et al. described electron transfer and size-selective electron oxidation of organic compounds by a MOF based on Eu^{III} and 1,3,5-benzenetribenzoate [62]. When the ligand is selectively excited ($\lambda_{ex} = 285$ nm), the MOF displays characteristic Eu^{III} luminescence, emitting narrow lineshapes at wavelengths attributed to $^5D_0 \rightarrow {}^7F_J$ ($J = 1, 2, 3, 4$; 593, 615, 653, 698 nm) with no observable emission from the ligand. This suggests efficient LMCT and metal-centered emission. The authors investigated this material for the ability to facilitate electron transfer from the organic moieties to the Eu^{III} metal ions. Upon irradiation of the MOF, while in the presence of an acetonitrile solution of N,N,N',N'-tetramethyl-p-phenylenediamine (TMPD), with strong UV light (365 nm; 50 mW cm^{-1}; 10 min), the solution turned deep blue, indicative of the formation of the oxidized TMPD$^{\bullet+}$. This material also showed strong absorption at 565 and 614 nm after irradiation, both in good agreement with the presence of TMPD$^{\bullet+}$. Furthermore, the authors reported size-selective one-electron oxidation of several other organic compounds together with detailed experimental (time-resolved absorption and emission spectroscopy; confocal microscopy) and theoretical observations (Marcus theory calculations).

An interesting *solvatochromic effect* was reported by Lin et al. [63] and provides another example of metal-centered luminescence. They describes a series of porous three-periodic Ln^{III} MOFs, $\{[Ln_4(bpt)_4(dmf)_2(H_2O)_8] \cdot (dmf)_5 \cdot (H_2O)_3\}_n$ [H_3bpt = biphenyl-3,4′,5-tricarboxylic acid; Ln = Eu (**4**), Gd (**5**), Tb (**6**), Dy (**7**), Ho (**8**), or Er (**9**)]. The ligand H_3bpt was specifically targeted because it was expected to absorb UV light strongly and act as a good sensitizer of Ln^{III} ions, and because other multi-carboxylate organic molecules afford extended architectures when coordinated to Ln^{III} ions. Compound **7** displays no observable fluorescence when excited using UV radiation; however, **4** and **6** exhibit characteristic red and turquoise emissions, respectively. The emission of the $\pi^* \rightarrow \pi$ transition seen in the emission spectrum of the isolated ligand ($\lambda_{ex} = 300$ nm) was not observed for either of the MOF compounds, suggesting efficient sensitization and metal-centered luminescence. Again, the dominant peaks were attributed to the so-called hypersensitive transitions: $^5D_0 \rightarrow {}^7F_2$ (617 nm) in the case of Eu^{III} and $^5D_4 \rightarrow {}^7F_5$ (544 nm) for the Tb^{III} compound. Quantum yields were also obtained for these two compounds with $\Phi_4 = 0.1442$ and $\Phi_5 = 0.6087$ obtained for the Eu^{III} and Tb^{III} MOFs, respectively. The high value for the Tb MOF is surprising, particularly given the presence of nine-coordinated water molecules, which one would expect would quench the Ln^{III} luminescence. The compounds were annealed in air at 220 °C for 4 h, causing the quantum yields unexpectedly to *decrease* ($\Phi_{4A} = 0.052$; $\Phi_{5A} = 0.326$). The initial quantum yields were only partially recovered upon rehydration ($\Phi_{4B} = 0.12$; $\Phi_{5B} = 0.582$), suggesting a possible irreversible change in the compounds upon dehydration and subsequent rehydration. The variation in PL for the as-synthesized, dehydrated, and rehydrated forms of these two Ln MOFs suggests a possible solvatochromic effect in these materials, which is counter to the commonly understood phenomenon of water-quenched luminescence.

The efficiency of energy transfer from absorbing linkers to lanthanide ions within MOFs is not uniform, as shown by a detailed analysis conducted by Soares-Santos et al. [64]. They measured the PL of a series of Ln^{III} MOFs, showing that Eu^{III} and Tb^{III} are more efficiently sensitized than their Nd^{III} and Sm^{III} counterparts. In this investigation, 2,3-pyrazinedicarboxylate (2,3-pzdc) and oxalate (ox) combine with Ln^{III} ions to generate two-periodic bilayer structures with unit formula [Ln$_2$(2,3-pzdc)$_2$(ox)(H$_2$O)$_2$]$_n$, (Ln = Ce, Nd, Sm, Eu, Gd, Tb, or Er). In this structure, individual cationic layers interconnect via ox^{2-}. In the Nd^{III} and Sm^{III} cases, the more intense spectral peaks (Nd, $^4F_{3/2} \rightarrow {}^4I_{11/2}$; Sm, $^4G_{5/2} \rightarrow {}^6H_{9/2}$) were used to monitor the excitation; several narrow peaks corresponding to transitions from the ground states (Nd, $^4I_{9/2}$; Sm, $^6H_{5/2}$) to higher energy levels were observed. Additionally, intense, broad absorption bands from 250 to 400 nm attributed to ligand $\pi \rightarrow \pi^*$ transitions were observed for both MOFs. This ligand absorption is stronger than for the metal transitions, suggesting more efficient sensitization via ligand excited states than for direct excitation of metal transitions. The emission spectra for both the Nd^{III} and Sm^{III} compounds were characteristic of their Ln^{III} ions, with emissions in the NIR region (e.g., for the Nd^{III} compound the strongest observed peak was for $^4F_{3/2} \rightarrow {}^4I_{11/2}$ at 1064 nm). However, both of these compounds also exhibit large and broad emission bands from 380 to 540 nm, indicative of $\pi^* \rightarrow \pi$ emission from the ligands, suggesting incomplete LMCT. In the case of Gd^{III}, the excited energy levels of the metal ions are higher than that of the S_n or T_n energy levels of the ligand, precluding the possibility of energy transfer from the ligand to the metals. As such, the emission for this compound was attributed solely to the two ligands, and through detailed analysis the authors were able to determine that the emissions are due to the excited triplet states of the two ligands. Additionally, they determined the relative energies for the two triplet states and deduced that 2,3-pzdc is a better sensitizer than ox. In the case of the Tb^{III} and Eu^{III} compounds, PL similar to the other materials described in this section is observed. The excitation spectra exhibit intense, broad brands attributed to $\pi \rightarrow \pi^*$ transitions centered on the ligands and also weaker and much narrower peaks assigned to transitions from the ground state to higher energy excited states for the Ln^{III} ions, while the emission spectra are characteristic of the respective Ln^{III} ions and show no observable transitions from the ligands. This suggests that either the efficiency of ligand-to-metal energy transfer is higher in the cases of Tb^{III} and Eu^{III} than those of Nd^{III} and Sm^{III}, or that there is some nonradiative deactivation of ligand excited states in the former two cases not observed in the latter.

As a final example of metal-centered luminescence, Li et al. described a group of MOF materials in which LMCT can be essentially "switched off" and either ligand-centered or metal-centered emission is observed based on the selection of metal ion [65]. They reported a series of MOFs constructed from the same ligand, 2-(pyridine-4-yl)-1H-imidazole-4,5-dicarboxylic acid (H_3pidc), but with lanthanide or non-lanthanide metal ions (Fe^{II}, Cd^{II}, Zn^{II}, Eu^{III}, Tb^{III}, and Y^{III}). In the compounds containing Cd^{II}, Zn^{II}, and Y^{III}, the observed emission is broad and slightly blue shifted relative to the free ligand, suggesting only ligand-centered emission. However, in the compounds containing Tb^{III} and Eu^{III}, no broad ligand-centered emission is observed, while sharp, narrow emissions characteristic of the Ln^{III} ion are detected. In this study, we see that the presence of Ln^{III} ions permits efficient LMCT, preventing the $\pi^* \rightarrow \pi$ intraligand transitions while at the same time generating metal-based excited states, leading to Ln^{III} emission. When the Ln^{III} ions are replaced with d^{10} metals, no LMCT occurs and the PL is attributed solely to ligand-centered emission.

12.4.3.2 Metal-to-Metal Charge Transfer (MMCT)

The potential for energy transfer between metals within a MOF also exists. In one report, two isostructural 4,4′-bipy templated frameworks were prepared using Tb^{III} and 1:1 Eu^{III}–Tb^{III} [66]. Both of these structures exhibit efficient through-space sensitized luminescence originating from the lanthanide centers. However the compounds containing Eu^{III}–Tb^{III} show greater sensitization efficiency than the structure comprised solely of Eu^{III}. The quantum yields for the materials were reported to be $\Phi_{Eu} \approx 0.22$ and $\Phi_{Tb} \approx 0.32$. Perhaps even more interesting, the mixed-metal MOF exhibits a metal-to-metal antenna effect in which the luminescence expected from the Tb^{III} luminophore is nearly completely quenched, leaving an emission spectrum for the material dominated by the Eu^{III} luminophore. Due to the absence of the Tb^{III} emission, and also the increased quantum yield ($\Phi_{Eu/Tb} \approx 0.39$ compared with $\Phi_{Tb} \approx 0.32$), it is likely that energy transfer from the higher energy Tb^{III} $^5D_4$ state to the Eu^{III} $^5D_0$ state occurs.

Metal-to-metal charge transfer could also be useful as a structural probe. Soares-Santos et al. described a series of mixed-ligand lanthanoid frameworks constructed from 2,5-pyridinedicarboxylate and 1,4-phenylenediacetate along with Eu^{III}, Tb^{III}, and two non-equivalent mixtures of Eu^{III} and Tb^{III} [67]. Each of these MOFs demonstrates sensitized metal-centered luminescence through an antenna effect, while the mixed-metal MOFs exhibit Tb^{III}-to-Eu^{III} MMCT. In the mixed-metal frameworks, this energy transfer can serve as a structural probe by indicating the extent of metal ion dispersal throughout the material. MOFs having single-metal ion clusters would be expected to exhibit very little of this Tb^{III}-to-Eu^{III} MMCT. Conversely, when the metal ions are homogeneously dispersed throughout the structure, the Tb^{III}-to-Eu^{III} MMCT should be efficient and easily observed.

12.4.4
Lanthanide Luminescence as a Probe of the Metal-Ligand Coordination Sphere

An illustrative example of how lanthanide luminescence can be used as a structural probe of the metal-ligand coordination sphere is that of Eu^{III}. The ratio of relative intensities for the $^5D_0 \rightarrow {}^7F_J$ ($J=2$:$J=1$) transitions is very sensitive to the site symmetry of the Eu^{III} centers. Since the electric dipole transition ($J=2$) is hypersensitive to the ligand environment whereas the magnetic dipole transition ($J=1$) is practically insensitive, it is sometimes possible to elucidate some aspects of symmetry for the material even in the absence of other structural data [68]. This effect is also observed in Ln-based MOFs, particularly an Eu^{III}-doped version of MIL-78 [69], an open-framework structure comprised of one-periodic inorganic columns of eight-coordinate polyhedra connected by bridging trimesate linkers. Luminescence data depict a nearly equivalent intensity for the $^5D_0 \rightarrow {}^7F_2$ and $^5D_0 \rightarrow {}^7F_1$ transitions ($^7F_2$:$^7F_1 \approx 1$), suggesting that the Eu^{III} ions lie in an environment of high symmetry relative to the ligand, which was independently confirmed by single-crystal X-ray diffraction (XRD) data. In another example, a MOF structure exhibiting chirality shows the ability of Ln^{III} luminescence to be used as a structural probe. Yue *et al.* synthesized a group of homochiral Ln phosphonate frameworks with the enantiomerically pure ligand *s*-*N*-(phosphonomethyl)proline and several lanthanide ions (Ln = Tb, Dy, Eu, Gd) [59]. The chirality of the MOF arises from the chiral organic ligand, leading to a MOF structure comprised of a series of one-periodic triple-stranded helices. PL spectra for both the Tb^{III} and Eu^{III} analogs exhibit significant splitting in the $^5D_4 \rightarrow {}^7F_J$ series ($J=6$–4), indicative of a strong crystal field that results from the asymmetric environment of the metal center. While symmetric systems often have forbidden transitions, the asymmetry inherent in these MOF structures leads to high intensity ratios for the $^7F_2$:$^7F_1$ transitions, and also $^7F_0$ transitions that are more intense than even the $^7F_1$ transition, an unusual occurrence.

12.5
Guest-Induced Luminescence

In the previous two sections, we have discussed the phenomenon of luminescence as it arises directly from the MOF itself, from either the organic linker or the inorganic cluster. However, many MOFs have periodic scaffolds that generate well-defined, essentially permanent, nanopores and large free-volume cavities. These unfilled cavities of controllable size allow the encapsulation of a wide variety of guest molecules. In relation to MOF-based luminescence properties, the ability to encapsulate guest molecules could prove useful in a number of ways. For instance, a non-luminescent MOF could be infiltrated with a luminescent guest molecule, which lends luminescent properties to the hybrid MOF–guest material, thereby imparting desired luminescent properties into the MOF. It may also be possible that upon infiltration of the MOF, a new luminescent species is formed through MOF–guest interactions (e.g., exciplex) with different luminescence properties from that of either the pristine MOF or guest molecules by themselves. In this case, both the MOF and guest may or may not be luminescent on their own, but upon complexation they generate a new species capable of luminescence. Additionally, the inclusion of a guest molecule could lead to modulation of the inherent luminescence of the MOF host. It may then be possible to enhance (or quench) the luminescence of the MOF by incorporating specific guest molecules, thereby allowing the MOF to act as a sensor.

12.5.1
Encapsulation of Luminophores

Conceptually, the infiltration of a porous, non-luminescent MOF with a luminescent guest molecule is the simplest way of introducing desired luminescence properties in a MOF material. Müller et al. synthesized a selection of well-studied MOFs [MOF-5, MOF-177, UMCM-1, and MIL-53(Al)] and investigated the ability to incorporate strongly luminescent guest molecules into these frameworks via the vapor phase [70]. The luminophores they chose were N,N-bis(2,6-dimethylphenyl)-3,4,9,10-perylene-tetracarboxylic diimide (dxp) and (2-carboxypyridyl)bis[3,5-difluoro-2(2-pyridyl)phenyl]iridium(III) (F_2-ppy), and in each case, the infiltrated MOF exhibited PL spectra dominated by the luminophore guest. The authors attempted to remove the luminophores by washing the hybrid materials and observed that, for the F_2-ppy compounds, partial to total removal was possible, coupled with decreases in PL intensity. However, removal of dxp was not achieved, possibly indicating stronger interactions with the frameworks. The emissive properties of the dxp-containing MOFs suggest host–guest interactions such as caging effects, aggregate formation, and strong quenching of the MOF host luminescence.

In another example, this time involving infiltration of charged dye molecules, Zhang et al. reported the synthesis of nanoscale Ln^{III} MOFs (Ln = Gd, Eu, Yb) that exhibit temperature-controlled guest encapsulation and light harvesting [71]. The MOFs are constructed from highly conjugated linear linker molecules, and when the coordination spheres of the Ln^{III} ions are filled with non-bridging anions such as acetate, the lanthanide clusters are negatively charged. This allows infiltration of the MOFs with cationic dye molecules (trans-4-styryl-1-methylpyridinium iodide and methylene blue), upon which the MOF-based luminescence is quenched and FRET emission is observed. The intensity and efficiency of FRET are controllable based on the temperature of the MOF synthesis, which was shown to correlate directly with the percentage loading of the guest luminophore. Interestingly, the dye and linker together in solution (DMF) demonstrated luminescence quenching, but no energy transfer, suggesting that the highly ordered arrangement of the dye guest molecules within the MOF cavities favors light harvesting. This study is significant, not only for the sequestration of a luminophore within the cavities, but also because it demonstrates that MOF–guest interactions can have profound effects on the resultant PL properties.

It is also possible to use the presence of a guest molecule to quench the luminescence of a MOF efficiently. Park *et al.* reported the structure and luminescence spectra of a mesoporous MOF derived from a superstructure of Tb^{III} ions, triazine-1,3,5-(4,4′,4″-trisbenzoate), and dimethylacetamide [72]. The crystal structure reveals large pores (3.9 and 4.7 nm in diameter) and, upon activation, the material shows a Langmuir surface area of $3855\,m^2\,g^{-1}$. PL studies demonstrated bright green, sensitized luminescence characteristic of the Tb^{III} ions. To investigate the potential for uptake of guest molecules, vapor-phase ferrocene was introduced to the structure using elevated temperatures and reduced pressure. Figure 12.7 shows several transmission and fluorescence microscopy images, and also luminescence spectra for the as-synthesized and ferrocene-containing framework. A color change of the single crystals from colorless to dark brown, in addition to loss of Tb

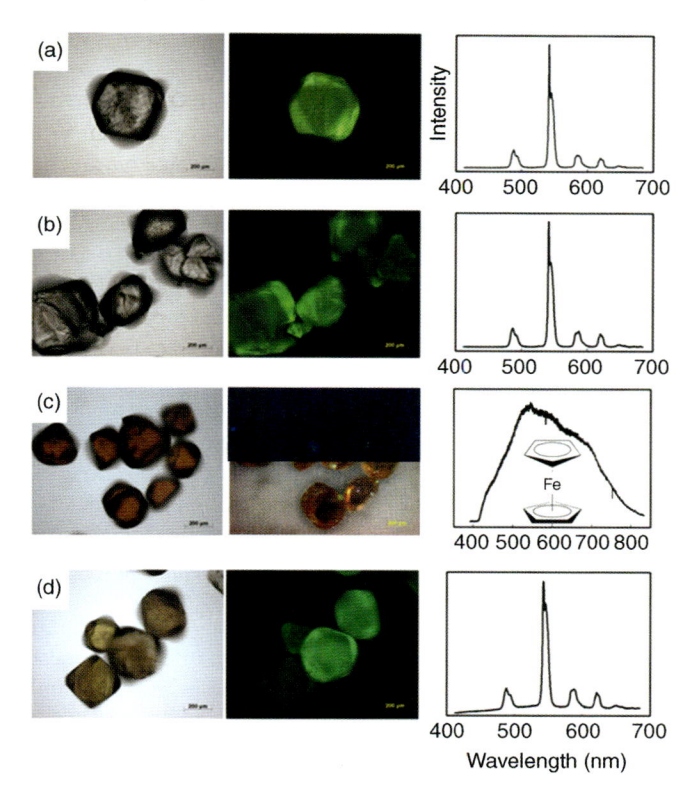

Figure 12.7 Transmission (left) and fluorescence (middle) microscope images and luminescence spectra (right) for a) a single MOF, b) multiple MOFs, c) ferrocene included MOFs, and d) regenerated initial MOFs after removal of ferrocene via evacuation at 50°C for 1 day. In the fluorescence image of c), half of the background color was subtracted to show the crystals. [72] Copyright Wiley-VCH Verlag GmbH & Co. KGaA. Reproduced with permission.

luminescence, are indicative of ferrocene inclusion. Additionally, the authors demonstrated that extraction of the ferrocene moieties is possible by using vacuum and high temperature, and that upon ferrocene removal, the framework luminescence is re-established. Elemental analysis also indicated a high uptake of ferrocene within this MOF.

12.5.2
Guest-Induced Charge Transfer: Excimers and Exciplexes

One way of introducing or modulating the luminescence of a MOF material is to generate an entirely new luminescent species in the form of an excited-state complex. When the two molecules in question are the same, they form a homodimer excited-state complex or *excimer* (excited dimer). In the case where the two molecules are different, the excited-state complex is a heterodimer called an *exciplex* (excited complex)[16]. Formation of exciplexes or excimers in MOFs is possible by two routes: (1) as a result of the framework structure itself or (2) by infiltration with a guest molecule. For the first possibility, the relative orientation of framework linkers allows the formation of an excited-state complex upon absorption of a photon. This could be achieved as a result of the framework topology itself; by "breathing" in a flexible framework that alters the structure and brings the ligands into closer contact (and in the correct orientation); or by framework interpenetration, whereby the two linker molecules are contained on different nets, but located adjacent to one another. For the second possibility, infiltration of the framework with an appropriate guest molecule, and the subsequent interaction of this molecule with the ligands, could result in an excited-state complex. In both the exciplex and the excimer, the effect on the observed luminescence is the formation of typically broad and featureless emission spectra, which are noticeably red shifted in comparison with the individual monomer(s). The exciplex or excimer emission is highly dependent on the orientation of the component molecules, so that formation of these excited-state species and observation of their PL can not only provide information on the presence of a guest molecule (sensor), but can also provide structural information regarding the interaction between the host and guest (structural probe). Cofacial arrangements of the two molecules are generally believed to be favorable to excimer and exciplex formation, and electronic structure modeling suggests that the strength of the interaction is strongly dependent on the separation distance [73].

Exciplexes have been reported in MOFs on only a few occasions. Perhaps the most extensive investigation was conducted by Zaworotko and co-workers, who synthesized several MOFs in which aromatic molecules are intercalated within the framework to allow exciplex formation. In one case, 4,4′-bipyridine (4,4′-bipy) and pyrene form a one-periodic molecular ladder coordination polymer with the unit formula $\{Zn(4,4'\text{-bipy})_{1.5}(NO_3)_2 \cdot CH_3OH \cdot 0.5\text{pyrene}\}_n$, in which the 4,4′-bipy linker assumes a cofacial arrangement with the intercalated pyrene molecule [74]. The pyrene moieties are not coordinated to the zinc ions, but rather trapped within the matrix to form a 2:1 4,4′-bipy–pyrene exciplex. The ratio of relative intensity between the third and first peaks within the vibronic structure of the weak monomer fluorescence band for pyrene can be used as a probe for the polarity of the microenvironment surrounding the pyrene [16]. The results suggest the possibility of using both excimer formation and the emission of the monomer as an internal probe of adsorbed guest (solvent) molecules. In another example [75], the same group prepared a series of Zn^{II} 4,4′-bipy-MOFs having one-periodic ladder and two-periodic square grid topologies. In these examples, the nature of the aromatic solvent molecule (benzene, chlorobenzene, o-dichlorobenzene, p-xylene, and toluene) determines the topology of the structure that results. Both λ_{max} and τ_F of the fluorescence are influenced by the structure. The emission of the one-periodic structures is red shifted ($\lambda_{max} \approx 520$ versus 460 nm) in comparison with the two-periodic emission and displays shorter lifetimes ($\tau_F \approx 43$ versus 74 ns). These features of the exciplex luminescence lead to a structural interpretation of face-to-face π–π stacking in the one-periodic structures, but for an edge-to-face C–H···π interactions in the two-periodic structure. This work is the most detailed analysis to date of the exciplex phenomenon in MOFs and demonstrates the way in which MOFs can be implemented as a platform to perturb systematically the identity and orientation of exciplex components. Conversely, the study also reveals how subtle changes in the MOF can lead to dramatic changes in the luminescence properties, demonstrating their potential use as good diagnostic handles for MOF structure.

Separately, another research group has demonstrated a case of guest-influenced luminescence in a MOF that may be related to exciplex formation. Stylianou *et al.* described the synthesis of a highly emissive MOF constructed from a strongly fluorescent linker ligand with an electronically isolated pyrene core [29] (see Section 12.3.2). For this MOF, they described solid-state PL properties in which the emission is solely the result of the ligand, which is largely isolated from other ligands or possible quenching moieties. The desolvated MOF can also be resolvated using other solvents, such as H_2O, dioxane, and p-xylene, leading to slightly altered PL. Interestingly, whereas the introduction of H_2O and dioxane produces small, but expected, red shifts in λ_{max}, the introduction of p-xylene leads to a blue-shift (~19 nm). The authors also presented PXRD data that seem to suggest that some of the p-xylene moieties are located in a face-to-face π–π stacking interaction with the pyrene core, indicating possible exciplex formation. However, the hypsochromic nature of the shift, together with the fact that the emission spectrum was observed to be no broader than the emission spectra of either the desolvated MOF or the MOFs resolvated with other solvents (DMF, H_2O, or dioxane), make classification of this luminescence emission as an example of exciplex formation in a MOF questionable and undetermined at present.

12.5.3
Encapsulation of Lanthanide Ion Luminophores

In addition to organic molecules, it is also possible to infiltrate MOFs with metal ions to act as guest luminophores. A classic strategy for the fabrication of optically active materials is to dope a compound with luminescent lanthanide ions. Recently, this strategy was shown to be a viable option for modulating MOF luminescence. Although there are numerous examples of MOFs based upon lanthanides incorporated into the framework and corresponding characterization of these materials' lanthanide-based luminescence, there are relatively few examples of MOFs infiltrated with lanthanide guests. Luo and Batten reported doping a non-lanthanide MOF with Ln^{III} cations and presented data illustrating the ability to affect the luminescence of the material [76]. In this example, a Zn^{II} MOF constructed from 1,2,4,5-benzenetetracarboxylate (H_4btec) generates permanent one-periodic channels. The overall framework structure, $\{(NH_4)_2[Zn(btec)]\cdot 6H_2O\}_n$, is anionic and the channels are filled with counterbalancing NH_4^+ cations. Subjecting the MOF to solutions of Tb^{III} [$Tb(ClO_4)_3$; 10^{-3}–10^{-6} mol L^{-1}] or Eu^{III} ($EuCl_3$; 10^{-3}–10^{-7} mol L^{-1}) results in simple cation exchange. The authors demonstrated that the luminescence of the MOF can be tuned, based on the choice of Ln^{III} dopant utilized. The as-synthesized MOF exhibits blue luminescence characterized as LMCT since the ligand itself was shown to be non-luminescent. Upon infiltration with dopants, the blue emission is no longer observed; however, strong emission characteristic of the Ln^{III} ions is generated. Additionally, it was demonstrated that the extent of infiltration plays a role in the intensity of the emission. The MOF hybrids obtained by infusing with a higher concentration of Eu^{III} are 30 times more intense than those obtained from the lower concentration. In the case of Tb^{III}, the observed intensity difference is only eight times more in the case of higher concentration.

Thirumurugan and Cheetham described a series of MOFs constructed with bismuth (Bi^{III}) in which they introduced Ln^{III} ions to modulate the luminescence of the compounds [77]. Owing to similar charge and ionic radii, bismuth-containing compounds are known to behave as good hosts for Ln^{III} ions, and this knowledge has been useful for the development of doped optical materials. Upon subjecting the MOFs to solutions of Eu^{III} and Tb^{III}, the infiltrated materials ($\sim$2 mol% Ln^{III}) exhibit PL unique from that of the pristine as-synthesized MOF. The neat MOF material has a broad emission centered at $\sim$420 nm that was characterized as ligand-centered $\pi^* \rightarrow n$ or $\pi^* \rightarrow \pi$ emission. After doping with Ln^{III} ions, the emission spectra for the hybrid material exhibit characteristic peaks for Eu^{III} ($^5D_0 \rightarrow {}^7F_J$) and Tb^{III} ($^5D_4 \rightarrow {}^7F_J$), while the main ligand-centered emission is almost completely suppressed, resulting in red and green luminescence, respectively. Additionally, the authors reported co-doping of the MOFs, using a combination of both Eu^{III} and Tb^{III} ($\sim$2 mol% each), which results in an overall orange-colored emission and clearly demonstrates the extent to which the luminescence of these materials can be fine-tuned.

Another example demonstrating the effect that encapsulating lanthanide ion luminophores within a MOF has on the PL was reported by Wang et al. [78]. They prepared a lanthanide-based MOF capable of encapsulating hydrated Ln^{III} cations by using a bent 1,2,4-trizole-bridged ligand. The encapsulated cations are confined within small volumes as they are hydrogen bonded to the framework, reducing vibrational movements originating from strong O–H oscillators. These vibrational movements normally lead to quenching of the f–f luminescence, but their reduced contribution in these materials leads to enhanced luminescence intensity, converting the practically non-luminescent $[Ln(H_2O)_8]^{3+}$ into a strongly emissive species.

12.6
Applications of Luminescent MOFs

12.6.1
Chemical Sensors

Efficient, portable sensors are increasingly required to support the development of technologies for many fields, from medicine to homeland security, and the number of reports regarding the use of MOFs for such purposes is steadily growing. The ability to tune the emissive properties of MOFs via alteration of the ligand and/or metal cluster type, along with the potential for high porosity with *accessible*, uniform pores, makes MOFs inherently multifunctional materials. In addition, the precise chemical functionality and size of the pores may be selectively tailored, allowing for specific sensing and separation capabilities. These features set MOFs apart from conventional porous materials such as porous silicas and zeolites, which offer only limited degrees of synthetic tunability. As the synthetic strategies for preparing "designer" MOFs improves, so does the ability to create MOFs with functional pores. A recent review by Chen *et al.* summarized developments in MOF materials for molecular recognition and sensing, including MOFs that can act as molecular sieves and effectively separate gases [79]. Jiang *et al.* demonstrated that MOFs can be utilized as stationary phases for liquid chromatography [40]. Additionally, the use of post-synthetic methodologies to alter MOF structure is of recent interest in this context [28, 80–84].

As was discussed in previous sections, the emission properties of solid-state materials are often highly dependent upon details of their structure, which can potentially be tailored via careful selection of the starting materials and growth conditions. Altering emission properties through "simple" changes in host–guest chemistry, such as solvent exchange, has been demonstrated for a variety of porous structures, as discussed in Section 12.3. This may lead to solvatochromic-like effects (although, of course, the guest molecule does not "dissolve" the MOF) that are caused by altering the dielectric environment of the chromophore, resulting in different emission wavelengths, intensities, or both. Changing the solvent molecule does not just alter the dielectric environment, but may also lead to small structural changes, such as lattice expansion or contraction, changes in ligand orientation, or changes in ligand–ligand distance, which can lead to observable differences in luminescence via LLCT or excimer formation. Such spectral shifts are typically relatively small compared with cases in which changing the solvent results in inhibition of LMCT versus ligand-based emission, for example. However, these early findings suggest that MOFs could be used and tailored for specific luminescent sensing capabilities [21]. Solvent incorporation may also result in major changes to the framework structure. For example, Zhu *et al.* reported a lanthanide MOF structure in which solvent incorporation or removal leads to reversible structural collapse, whereby luminescence is quenched and restored upon a dehydration–rehydration cycle [85]. A similar observation has been reported for one of the rare examples of luminescent MOFs from which emission is exclusively d-metal based, $[Cu_4I_4(dabco)_2]_n$ (dabco = 1,4-diazabicyclo[2.2.2]octane). Two different structures were prepared in water and in acetonitrile [86], and the luminescence was found to vary depending on the solvent used. The MOF prepared in water shows temperature-dependent luminescence and emits at 556 nm, whereas the acetonitrile-incorporated MOF shows little temperature dependence and emits at 580 nm. These structures can be reversibly interconverted by solvent exchange. The emission here is cluster-centered, and is a function of the Cu–Cu distance.

It should be noted that the examples in the previous paragraph were concerned with the effects of incorporated solvent on framework luminescence. There are many examples in the literature where this has been observed, and it may be considered to be an expected feature of significantly porous, luminescent structures. In the following sections, a few such effects will be highlighted, but the focus is on the design and use of MOFs for targeted sensing purposes. Although luminescence is typically considered among the more sensitive and attractive for optical sensing methods, an elegant study by Lu and Hupp demonstrated the first use of MOFs for sensing via changes in refractive index [87]. A Fabry–Pérot device fabricated from thin films of ZIF-8 (a zeolitic imidazolate framework) was found to be sensitive to various chemical vapors and gases, with analyte detection readily achievable through transmission UV–visible measurements.

12.6.1.1 Small-Molecule and Ion Sensors

Sensing of metal ions and small molecules by emission can occur by a variety of mechanisms. Ion exchange has been used to alter sensitization in several lanthanide MOFs. Chen and co-workers reported a Tb(1,3,5-btc) structure in which anionic guests solvated in methanol were incorporated, yielding anion-filled pores [88]. The luminescence from Tb^{III} was found to be enhanced, particularly in the presence of fluoride anion. This enhancement was attributed to a stronger hydrogen-bonding interaction between F^- and methanol leading to attenuation of the O–H stretching vibration, with concomitant reduction of the ability of this vibration to quench the sensitizing btc ligand. Lu and co-workers demonstrated a MOF that displays sensitivity to cation exchange [89]. Here, a Tb^{III}-based MOF exhibited a significant increase in emission intensity upon exchange of K^+ with Ca^{2+}. This was ascribed to rigidification of the imidazole-4,5-dicarboxylate ligand upon complexation with Ca^{2+}, which leads to more efficient energy transfer to the lanthanide ion. Wong et al. reported a Tb(mucicate) structure, wherein the flexible multidentate ligand is used to differentiate between several anions, with fluorescence enhancement in the case of CO_3^{2-} and CN^- [90]. Lewis basic pyridyl sites within a luminescent porous Eu(pyridine-3,5-dicarboxylate) MOF were utilized for sensing of Cu^{II} ion [91]. This material shows significant quenching upon binding to Cu^{II}. Finally, luminescent intensity modulation was also shown for two Eu^{III} 1,4,8,11-tetraazacyclotetradecane-1,4,8,11-tetrapropionate-based MOFs, whereby

Ag^I exchange significantly alters the relative transition intensities of the emissive Eu^{III} center [92].

Ion sensing is also possible using transition metal-based frameworks, typically through alteration of LMCT efficiency. For example, a net-type anionic MOF, $\{(NH_4)_2[Zn(1,2,4,5\text{-benzenetetracarboxylate})]\cdot 6H_2O\}_n$, exhibits broad luminescence centered at 440 nm, as a likely result of LMCT and/or ligand-based luminescence [93]. The open structure allows for exchange between the ammonium groups and aqueous metal cations. Titration with increasing amounts of Cu^{II} leads to quenching as a result of ion exchange, and this effect was shown for a variety of other cations in addition. Upon ion exchange with Ln^{III} ions [76], LMCT character was no longer observed, replaced by emission typical of the lanthanide ion instead. These lanthanide-incorporated structures were subsequently doped with MCl_x solutions (M = Na^+, K^+, Zn^{2+}, Ni^{2+}, Mn^{2+}, Co^{2+}, Cu^{2+}), and significant quenching was seen for Co^{II} and Cu^{II} ions.

MOFs may be prepared in which the metal center remains unsaturated, which leads to accessible, potentially luminescent, open metal sites for the sensing of small molecules. The degree of saturation of lanthanide ions in MOFs can often be manipulated via ligand exchange. For example, a porous Eu(1,3,5-btc) MOF with open metal coordination sites shows sensitivity to a variety of solvent molecules, particularly a marked decrease in emission intensity upon exposure to acetone and an increase with DMF exchange [94]. Exchange of coordinated solvents has been shown to alter emission properties in other lanthanide-based MOFs as well. A Eu^{III} MOF with 4,4′-(hexafluoroisopropylidene)bis(benzoic acid) linkers has been used to sense ethanol in air [95]. The Eu^{III} emission at 619 nm was monitored upon exposure, leading to a rapid decrease in luminescence intensity, which was recovered upon exposure to air. The initial quenching was attributed to an effect of the O–H stretching vibration upon coordination of ethanol to the Eu^{III} ions. Harbuzaru *et al.* also demonstrated similar use of an Eu^{III} MOF containing a 1,10-phenanthroline-2,9-dicarboxylic acid linker for pH sensing applications in the biologically suitable range of pH = 5–7.5 [96].

Interruption of LMCT and MLCT processes by small molecules can lead to striking differences in luminescence properties. Jiang and co-workers prepared Cd^{II} MOFs with the same building blocks, but with differing porosity and ligand–ligand separation [40, 97]. Through alteration of the solvent volume and/or temperature, the framework porosity and LMCT signal strength were found to vary reversibly. Kobayashi *et al.* prepared a Pt/Zn-containing 5,5′-dicarboxy-2,2′-bipyridine-based MOF, which exhibits thermochromism and reversible sensitivity to solvent vapors [98]. The thermochromic effect is a result of dehydration of the Pt^{II} centers and the resultant change in Pt–Pt distance, which in turn affects the triplet-sensitized MLCT process. The authors also noted solvatochromic effects with a mechanism similar to this thermochromic effect, whereby various solvents alter the intermetallic Pt distance.

Manipulation of ligand-based luminescence in MOFs has been shown in a variety of cases. As described in Section 12.3.2, Rosseinsky and co-workers prepared $[In_2(OH)_2(tbaPy)]_n$, which displays a red-shifted emission spectrum upon desolvation as a result of a reduced intermolecular ligand–ligand distance [29]. However, the spectral position and intensity nearly recover fully within 3 h upon re-exposure to DMF. The intermolecular coupling between ligands is highly distance dependent at short range, and addition of various guests leads to alteration in the position and intensity of the pyrene fluorescence. The authors were able to correlate fluorescence changes directly with structural changes, and noted that the long lifetime and quantum yield of 6.7% make these MOFs comparable to commercial lanthanide sensors. As an additional recent example, $\{[Zn_3(1,3,5\text{-btc})_2]\cdot 12H_2O\}_n$ frameworks were found to be selective for ethylamine over butylamine and propylamine, exhibiting significant emission quenching of btc-based luminescence upon increasing doses of ethylamine [99].

12.6.1.2 Oxygen Sensors

Phosphorescent MOFs functionalized with Ir^{III}-containing ligands have been synthesized as potential oxygen sensors [100]. A variety of Zn MOFs were prepared in which the ligands were $Ir(2\text{-phenylpyridine})_3$ derivatives (see Figure 12.8). Structures **11** and **12** were nonporous; however, structure **10** exhibits permanent porosity and

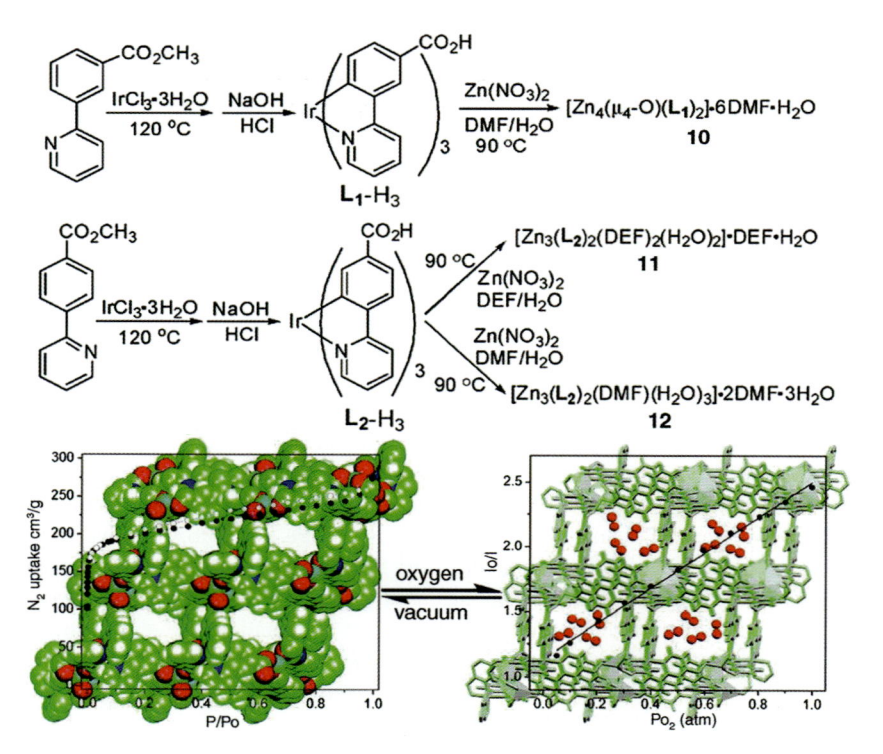

Figure 12.8 (Top) Three structures were synthesized from Ir tris(2-phenylpyridine)-based ligands. **1** was found to be porous, showing rapid and reversible luminescence quenching by oxygen, as shown schematically (bottom). Reprinted with permission from [100]. Copyright 2010 American Chemical Society.

reversible host–guest chemistry. Triplet-sensitized MLCT was observed from the Ir(2-phenylpyridine)$_3$ derivatives, at 538 nm for **10** and **12**, and at 565 nm for **11**. It is well known that dioxygen rapidly quenches triplet excited states and, as expected, a decrease in the emission intensity was observed for all three structures in the presence of O_2. Interestingly, upon exposure to O_2, only the porous structure was found to respond rapidly and reversibly. A linear Stern–Volmer plot (see Section 12.2.2; Equation 12.5) was extracted for exposure of this porous MOF to O_2, and the luminescence quenching was shown to be reversible across multiple cycles. The nonporous MOFs have limited accessibility to the interior Ir moieties, and it also is more difficult to remove the O_2 molecules from these structures. Nonetheless, **11** and **12** were also observed to display a degree of reversibility on longer timescales than **10**.

12.6.1.3 Detection of Explosives

Detection of explosives is a topic of significant contemporary relevance, and substantial work has been carried out in this area. Efficient, inexpensive detection of 2,4-dinitrotoluene (DNT) [a volatile indicator of 2,4,6-trinitrotoluene (TNT)] and 2,3-dimethyl-2,3-dinitrobutane (DMNB, an ingredient in plastic explosives), are of high interest in this regard. Typically, luminescent changes upon a redox reaction are used in current detection methods, although this is much more difficult to achieve for DMNB. The first example of a MOF that possesses the ability to detect both markers has been reported recently [101]. The structure $[Zn_2(bpdc)_2(bpee)]_n$, (bpdc = 4,4′-biphenyldicarboxylate; bpee = 1,2-bipyridylethene) was shown to absorb both molecules of interest quickly (on the order of 10s), reversibly, and efficiently, with their uptake detected through changes in the emission properties of the MOF. Thin layers of the MOFs were exposed to vapors of DMNB and DNT, upon which red-shifted, quenched emission was rapidly observed, as shown in Figure 12.9. Both analytes

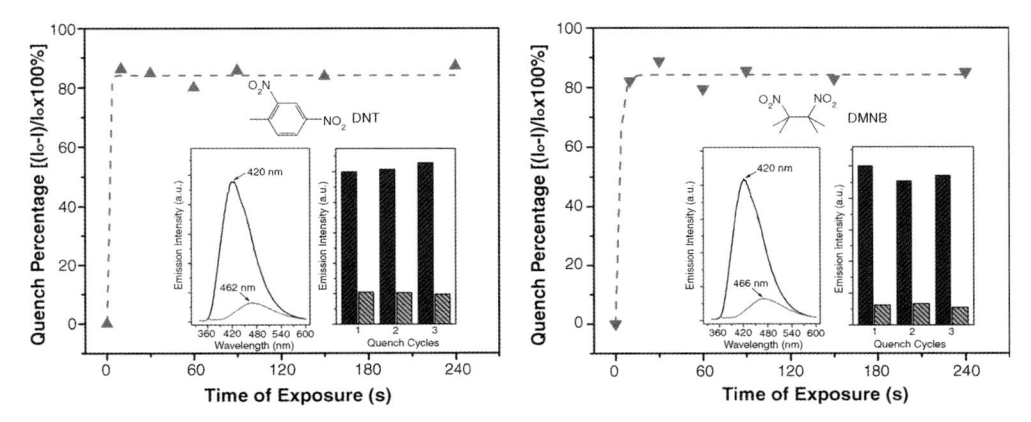

Figure 12.9 Time-dependant fluorescence quenching of MOFs by DNT and DMNB. The red-shifted fluorescence can be seen in the inset, along with three quench/fluorescence regeneration cycles. [101] Copyright Wiley-VCH Verlag GmbH & Co. KGaA. Reproduced with permission.

contain nitro groups, which are known quenchers of emission. The structure maintains a three-periodic, porous geometry in which the ligands are held at appropriate distances to minimize π–π intermolecular interactions between the ligands, and is highly emissive. It is likely that the emission is ligand-based in this MOF. DMF guests incorporated during the synthesis can be removed reversibly, although the structure is slightly distorted upon removal.

12.6.2
Radiation Detection

Radiation detection goes hand-in-hand with the aforementioned interest in explosives detection for homeland security applications. In addition, medical devices and methods in biotechnology rely on radiolabeling. Portable, inexpensive neutron detectors are currently scarce, and are often not as sensitive as required. Interestingly, the same small conjugated molecules that make for suitable ligands in MOFs can also exhibit useful scintillation properties, in particular 4,4′-stilbenedicarboxylic acid. Such molecules are known for their ability to emit light effectively upon exposure to various forms of radiation, yet their solid-state structure remains a challenge for the development of reliable, reproducible sensor devices. The structures of both two-periodic network and three-periodic cubic Zn stilbene MOFs (see Figure 12.2) provide a rigid environments for the stilbene ligands, but the relative spacing and orientation differ, leading to differences in emission wavelengths and lifetimes. The two-periodic network is denser than the porous three-periodic cubic structure. Exposure of both MOFs to high-energy (3 MeV) protons and α-particles leads to radioluminescence, although the materials have different Stokes shifts (Figure 12.10) [102]. The three-periodic stilbene MOF has a larger Stokes shift, likely another consequence of the differences in stilbene–stilbene orientation between the two structures. It was found that these MOFs have scintillation quantum yields that are comparable to those with commercial scintillators. Moreover, the stability to repeated exposure outperforms anthracene, with luminescence degrading only very slowly with integrated doses up 10 MGy (1 Gy = 1 J kg^{-1}). The versatility of MOF syntheses and the commonality between scintillating organic groups and potential as MOF ligands leaves this area of study wide open for development.

12.6.3
Solid-State Lighting

The tunable nature of MOFs and the ability to incorporate multiple luminophores, including guest molecules, ligands, and metals, and the various potential interactions between these groups, can allow for fine-scale adjustment of the emission intensity and color of hybrid MOF materials. Solid-state lighting devices, such as organic light-emitting diodes (OLEDs), represent highly attractive, energy-efficient methods for the generation of light and are the subject of much contemporary research. In particular, white light sources for ambient lighting represent particularly interesting targets [103]. In addition to the tunability of MOF structures and emissive properties

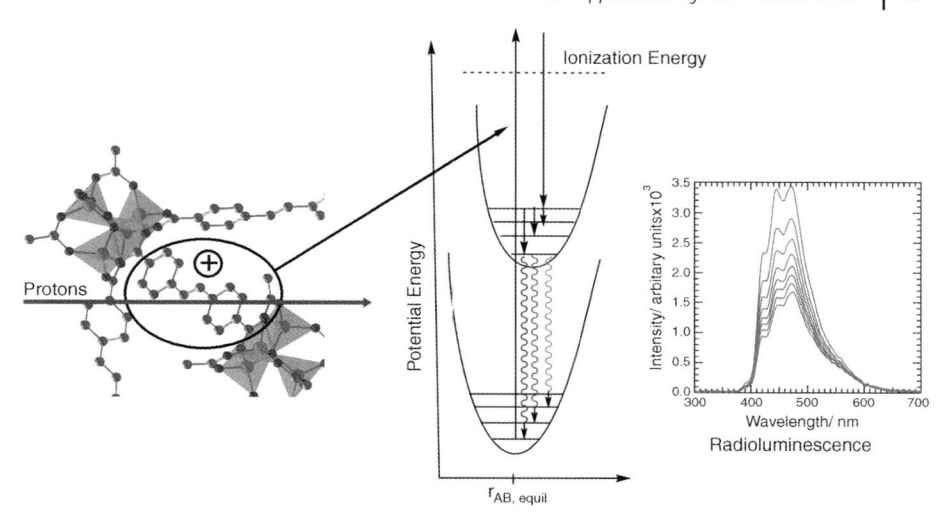

Figure 12.10 Schematic of the fast neutron detection process. Left: Interaction of high energy recoil protons with a rigid stilbene unit (a portion of the cubic 3D stilbene MOF structure is shown with stilbene dicarboxylate linkages between tetrahedral Zn_4O units), resulting in excitation, which decays (center) through radiationless pathways (solid arrows) to the lowest-lying excited state of the stilbene groups, followed by radiative decay and emission of photons (wavy arrows). The emission profile is shown on the right, revealing diminishing intensity upon repeated doses of high intensity radition.

discussed previously, their high thermal stability and often intense luminescence are desirable in this context. Current white light-emitting materials typically consist of mixtures of blue- and yellow-emitting materials [104]. Two cases of near white light-emitting MOFs have been reported to date. Ca and Sr frameworks containing 9-fluorenone-2,7-dicarboxylic acid exhibit broad, ligand-based emission that is dependent on the identity of the cation [105]. The rigidification of the ligand both by inclusion in the framework and by cooling to $-196\,°C$ leads to an enhancement of the quantum yield. The broad emission of the pure ligand is yellow, whereas the Ca structure emits green light ($\lambda_{max} = 503$ nm) and the cation-exchanged Sr structure emits orange light ($\lambda_{max} = 526$ nm). Both frameworks can be excited between 380 and 460 nm, and the FWHM of the broad emission spectra is 90 nm, comparable to Ce^{III}:YAG phosphors.

The first example of a MOF material in which variable emission processes are harnessed for the purpose of generating white light was reported recently by Wang et al. [106]. A multi-process emission profile, highly sensitive to excitation wavelength, was used to manipulate the emission color of the MOF. The authors synthesized an Ag^I(4-cyanobenzoate)-based MOF that exhibits variable emission; the color is tunable from yellow to white via variation of the excitation wavelength. The solid-state emission spectrum shows a multitude of peaks, with intensities that vary upon modulation of the excitation wavelength. The maximum emission is at $\lambda = 427$ nm when excited by 355 nm light, but shifts to 566 nm upon excitation by

330 nm light. Excitation of the MOF at 350 nm yields a white light-emitting material (to the eye) as, at this excitation wavelength, the 427 and 566 nm bands are of nearly equal intensity. There are several physical causes for the observed emission peaks, as MLCT and ligand-based emission coexist in this structure.

12.6.4
Nonlinear Optics

Nonlinear optics (NLO) properties (including second- and third-order processes) of both organic and inorganic materials are of great interest for a variety of purposes, such as communications, lighting, lasing, and frequency modulation. Second harmonic generation (SHG) is a nonlinear conversion of two photons of a particular frequency to a single photon with twice this frequency (i.e., frequency doubling). A requirement for SHG is non-centrosymmetry, which is challenging in bulk material. Donor–acceptor substituted molecules (i.e., push–pull conjugated) are often used. Current techniques involve assembly at liquid–air surfaces or electric field poling [107, 108]. As MOFs can be tailorable in function and shape, they offer potential as scaffolds amenable to the arrangement of ligands in a non-centrosymmetric manner. Several examples of MOFs exhibiting SHG are known, including Pb^{II} complexes prepared by Yang *et al.*, for which the emission can be assigned to a metal-centered transition involving the s- and p-orbitals of the metal, sensitized by the organic ligands, a relatively rare case of main-group MLCT [41]. A porous, Cd^{II} oxalate-based MOF ionic cluster exhibited among the highest SHG signals recorded for MOF materials, and this structure is also sensitive to cation exchange, illustrating the first NLO MOF with sensing potential [109]. A variety of other examples of engineered, asymmetric MOF crystals that exhibit high SHG responses have been reported [110–114]. In addition, Guo *et al.* prepared a multifunctional ferroelectric, NLO-active MOF [115].

Enhancement of standard one-photon fluorescence often leads to a concomitant increase in two-photon fluorescence, a third-order process. Although this has yet to be measured in a MOF material, the outcome may be of technological interest and warrants investigation. It has been well documented that centrosymmetric π-conjugated molecules with quadrupolar or higher degrees of charge separation have among the highest two-photon cross-sections (δ) [116]. These often have a donor–π-donor or acceptor–π-acceptor geometry and variations thereof. Such molecules are again well suited for MOF chemistry and many such MOF ligands are currently known or may be synthesized by straightforward protocols. In addition, coordination to the metal centers will likely improve the electron-withdrawing capability of the terminal carboxylate groups, and potentially improve the two-photon fluorescence signal strength.

12.6.5
Barcode Labeling

Barcoding is a method that requires reproducible, consistent materials for which readout is rapid. White *et al.* utilized the synthetic flexibility available in MOF

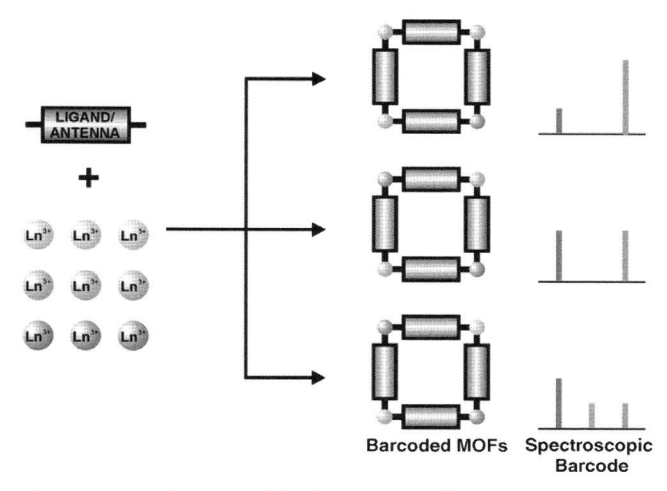

Figure 12.11 Schematic of the development of luminescent barcode materials by controlling the relative amounts of three differing emissive lanthanide ions in the MOF structures. Reprinted with permission from [117]. Copyright 2009 American Chemical Society.

syntheses to introduce multiple metal centers, each of which differs in luminescence color [117]. A mixture of lanthanide ions was used to develop MOFs with differing spectral outputs, which required excitation at a single wavelength as they are all sensitized by the same organic ligand. This allows for "encoding" of the spectral profile, as lanthanide emission is typically sharp, well resolved, and well separated (Figure 12.11). By tuning the ratio of a mixture of Ln^{III} ions (Yb, Er, Nd), a specific "code" results. The authors noted the reproducibility and wide diversity of codes possible. Upon coating in an adhesive matrix, the spectral profile was maintained, demonstrating the versatility of incorporation into other practical morphologies and composites.

12.7
Conclusion

What is so unique about MOFs, aside from the spectacular pore volumes and surface areas that they can exhibit, is that they possess the high local and long-range order characteristic of crystals, the synthetic versatility of organic polymers, the geometric predictability inherent in coordination bonding, *and* often stable nanoporosity. This amazing combination of properties is only beginning to be explored and is likely to lead to many surprises as scientists in fields beyond chemistry consider MOFs from their own perspectives and with their own specialized expertise. The examples cited in this chapter demonstrate that luminescent MOFs are not only a rapidly expanding subset with tremendous potential for a wide variety of applications, but are also an exciting venue for building a new, fundamental understanding of electronic structure, energy transfer, and light emission. It therefore seems likely that research to date represents only the beginning of a long and interesting scientific story.

Acknowledgments

This research was funded by the Defense Threat Reduction Agency under contract 0743251-0, the US Department of Energy Office of Proliferation Detection Programs, and the Sandia Laboratory Directed Research and Development Program. Sandia National Laboratories is a multi-program laboratory managed and operated by Sandia Corporation, a wholly owned subsidiary of Lockheed Martin Corporation, for the US Department of Energy's National Nuclear Security Administration under contract DE-AC04-94AL85000. C.A.B. acknowledges support from NSF Discovery Corps Fellowship Grant CHE0725176.

References

1 Allendorf, M.D., Bauer, C.A., Bhakta, R.K., and Houk, R.J.T. (2009) Luminescent metal–organic frameworks. *Chem. Soc. Rev.*, **38** (5), 1330–1352.
2 McCapra, F. (1966) Chemiluminescence of organic compounds. *Q. Rev. Chem. Soc.*, **20** (4), 485–510.
3 Dodeigne, C., Thunus, L., and Lejeune, R. (2000) Chemiluminescence as a diagnostic tool. A review. *Talanta*, **51** (3), 415–439.
4 Aslan, K. and Geddes, C.D. (2009) Metal-enhanced chemiluminescence: advanced chemiluminescence concepts for the 21st century. *Chem. Soc. Rev.*, **38** (9), 2556–2564.
5 Widder, E.A. (2010) Bioluminescence in the ocean: origins of biological, chemical, and ecological diversity. *Science*, **328** (5979), 704–708.
6 Tsien, R.Y. (1998) The green fluorescent protein. *Annu. Rev. Biochem.*, **67** (1), 509–544.
7 Meighen, E.A. (1991) Molecular biology of bacterial bioluminescence. *Microbiol. Rev.*, **55** (1), 123–142.
8 Mitschke, U. and Bauerle, P. (2000) The electroluminescence of organic materials. *J. Mater. Chem.*, **10** (7), 1471–1507.
9 Miao, W.J. (2008) Electrogenerated chemiluminescence and its biorelated applications. *Chem. Rev.*, **108** (7), 2506–2553.
10 Sweeting, L.M. (2001) Triboluminescence with and without air. *Chem. Mater.*, **13** (3), 854–870.
11 Chandra, B.P. and Rathore, A.S. (1995) Classification of mechanoluminescence. *Cryst. Res. Technol.*, **30** (7), 885–896.
12 Chandra, B.P. (1981) Mechanoluminescence and piezoelectric behavior of molecular crystals. *Phys. Status Solidi A*, **64** (1), 395–405.
13 Schulman, S.G. (1985) Luminescence spectroscopy: an overview. In *Molecular Luminescence Spectroscopy – Methods and Applications: Part 1* (ed. S.G. Schulman), John Wiley & Sons, Inc., New York, pp. 1–28.
14 Vogler, A. and Kunkely, H. (2001) Luminescence in metal complexes: diversity of excited states. In *Transition Metal and Rare Earth Compounds* (ed. H. Yersin), Springer, Berlin, pp. 183–184.
15 Ronda, C. (ed.) (2008) *Luminescence: from Theory to Applications*, Wiley-VCH Verlag GmbH, Weinheim.
16 Birks, J.B. (1970) *Photophysics of Aromatic Molecules*, Wiley-Interscience, London.
17 Valeur, B. and Brochon, J.-C. (eds) (2001) *New Trends in Fluorescence Spectroscopy:*

Applications to Chemical and Life Science, Springer, New York.

18 Lakowicz, J.R. (2006) *Principles of Fluorescence Spectroscopy*, Springer, New York.

19 Balzani, V., Bergamini, G., Campagna, S., and Puntoriero, F. (2007) Photochemistry and photophysics of coordination compounds: overview and general concepts. In *Photochemistry and Photophysics of Coordination Compounds I* (eds. V. Balzani and S. Campagna), Springer, Berlin, pp. 1–36.

20 Eddaoudi F M., Kim, J., Rosi, N., Vodak, D., Wachter, J., O'Keeffe, M., and Yaghi, O.M. (2002) Systematic design of pore size and functionality in isoreticular MOFs and their application in methane storage. *Science*, **295** (5554), 469–472.

21 Bauer, C., Timofeeva, T., Settersten, T., Patterson, B., Liu, V., Simmons, B., and Allendorf, M. (2007) Influence of connectivity and porosity on ligand-based luminescence in zinc metal–organic frameworks. *J. Am. Chem. Soc.*, **129** (22), 7136–7144.

22 Meek, S.T., Houk, R.J.T., Doty, F.P., and Allendorf, M.D. (2010) Luminescent metal–organic frameworks: a nanolaboratory for probing energy transfer via interchromophore interactions. *ECS Trans.*, **28** (3), 137–143.

23 Bordiga, S., Lamberti, C., Ricchiardi, G., Regli, L., Bonino, F., Damin, A., Lillerud, K.-P., Bjorgenb, M., and Zecchina, A. (2004) Electronic and vibrational properties of a MOF-5 metal–organic framework: ZnO quantum dot behaviour. *Chem. Commun.*, (20), 2300–2301.

24 Civalleri, B., Napoli, F., Noel, Y., Roetti, C., and Dovesi, R. (2006) *Ab-initio* prediction of materials properties with CRYSTAL: MOF-5 as a case study. *CrystEngComm*, **8** (5), 364–371.

25 Feng, P.L., Perry IV, J.J., Nikodemski, S., Meek, S.T., and Allendorf, M.D. (2010) Assessing the purity of metal–organic frameworks using photoluminescence: MOF-5, ZnO quantum dots, and framework decomposition. *J. Am. Chem. Soc.*, **132** (44), 15487–15489.

26 Saltiel, J. (1967) Perdeuteriostilbene. Role of phantom states in *cis–trans* photoisomerization of stilbenes. *J. Am. Chem. Soc.*, **89** (4), 1036–1037.

27 Waldeck, D.H. (1991) Photoisomerization dynamics of stilbenes. *Chem. Rev.*, **91** (3), 415–436.

28 Jones, S. and Bauer, C. (2009) Diastereoselective heterogeneous bromination of stilbene in a porous metal–organic framework. *J. Am. Chem. Soc.*, **131** (35), 12516–12517.

29 Stylianou, K.C., Heck, R., Chong, S.Y., Bacsa, J., Jones, J.T.A., Khimyak, Y.Z., Bradshaw, D., and Rosseinsky, M.J. (2010) A guest-responsive fluorescent 3D microporous metal–organic framework derived from a long-lifetime pyrene core. *J. Am. Chem. Soc.*, **132** (12), 4119–4130.

30 Huh, S., Jung, S., Kim, Y., Kim, S., and Park, S. (2010) Two-dimensional metal–organic frameworks with blue luminescence. *Dalton Trans.*, 1261–1265.

31 Lu, Z., Wen, L., Ni, Z., Li, Y., Zhu, H., and Meng, Q. (2007) Syntheses, structures, and photoluminescent and magnetic studies of metal–organic frameworks assembled with 5-sulfosalicylic acid and 1,4-bis(imidazol-1-ylmethyl)benzene. *Cryst. Growth Des.*, **7** (2), 268–274.

32 Park, B., Eom, G., Kim, S., Kwak, H., Yoo, S., Lee, Y., Kim, C., Kim, S., and Kim, Y. (2010) Construction of Cd(II) compounds with a chelating ligand 2,2'-dipyridiylamine (Hdpa): anion effect, catalytic activities and luminescence. *Polyhedron*, **29** (2), 773–786.

33 Zou, R., Abdel-Fattah, A., Xu, H., Burrell, A., Larson, T., McCleskey, T., Wei, J., Janicke, M., Hickmott, D., Timofeeva, T., and Zhao, Y. (2010) Porous metal–organic frameworks containing alkali-bridged two-fold interpenetration: synthesis, gas adsorption, and fluorescence properties. *Cryst. Growth Des.*, **10** (3), 1301–1306.

34 Chen, W., Wang, J.-Y., Chen, C., Yuan, Q.Y.-M., Chen, J.-S., and Wang, S.-N. (2003) Photoluminescent metal–organic polymer constructed from trimetallic clusters and mixed carboxylates. *Inorg. Chem.*, **42** (4), 944–946.

35 Dai, J.-C., Wu, X.-T., Fu, Z.-Y., Cui, C.-P., Hu, S.-M., Du, W.-X., Wu, L.-M., Zhang, H.-H., and Sun, R.-Q. (2002) Synthesis, structure, and fluorescence of the novel cadmium(II)–trimesate coordination polymers with different coordination architectures. *Inorg. Chem.*, **41** (6), 1391–1396.

36 Dai, J.-C., Wu, X.-T., Fu, Z.-Y., Hu, S.-M., Du, W.-X., Cui, C.-P., Wu, L.-M., Zhang, H.-H., and Sun, R.-Q. (2002) A novel ribbon-candy-like supramolecular architecture of cadmium (II)–terephthalate polymer with giant rhombic channels: twofold interpenetration of the 3D $8^2 10$-a net. *Chem. Commun.*, 12–13.

37 Fang, Q.-R., Zhu, G.-S., Shi, X., Wu, G., Tian, G., Wang, R.-W., and Qiu, S.-L. (2004) Synthesis, structure and fluorescence of a novel three-dimensional inorganic–organic hybrid polymer constructed from trimetallic clusters and mixed carboxylate ligands. *J. Solid State Chem.*, **177** (4–5), 1060–1066.

38 Huang, Y.-Q., Ding, B., Song, H.-B., Zhao, B., Ren, P., Cheng, P., Wang, H.-G., Liao, D.-Z., and Yan, S.-P. (2006) A novel 3D porous metal–organic framework based on trinuclear cadmium clusters as a promising luminescent material exhibiting tunable emissions between UV and visible wavelengths. *Chem. Commun.*, 4906–4908.

39 Zeng, F., Ni, J., Wang, Q., Ding, Y., Ng, S., Zhu, W., and Xie, Y. (2010) Synthesis, structures, and photoluminescence of zinc(II), cadmium(II), and mercury(II) coordination polymers constructed from two novel tetrapyridyl ligands. *Cryst. Growth Des.*, **10** (4), 1611–1622.

40 Jiang, H., Tatsu, Y., Lu, Z., and Xu, Q. (2010) Non-, micro-, and mesoporous metal–organic framework isomers: reversible transformation, fluorescence sensing, and large molecule separation. *J. Am. Chem. Soc.*, **132** (16), 5586–5587.

41 Yang, E.-C., Li, J., Ding, B., Liang, Q.-Q., Wang, X.-G., and Zhao, X.-J. (2008) An eight-connected 3D lead(II) metal–organic framework with octanuclear lead(II) as a secondary building unit: synthesis, characterization and luminescent property. *CrystEngComm*, **10** (2), 158–161.

42 He, X., Lu, C.Z., Wu, C.D., and Chen, L.J. (2006) A series of one- to three-dimensional copper coordination polymers based on N-heterocyclic ligands. *Eur. J. Inorg. Chem.*, 2491–2503.

43 Zhang, S., Wang, Z., Zhang, H.H., Cao, Y.N., Sun, Y.X., Chen, Y.P., Huang, C.C., and Yu, X.H. (2007) Self-assembly of two fluorescent supramolecular frameworks constructed from unsymmetrical benzene tricarboxylate and bipyridine. *Inorg. Chim. Acta*, **360** (8), 2704–2710.

44 Han, L., Yuan, D.Q., Wu, B.L., Liu, C.P., and Hong, M.L. (2006) Syntheses, structures and properties of three novel coordination polymers with a flexible asymmetric bridging ligand. *Inorg. Chim. Acta*, **359** (7), 2232–2240.

45 Eliseeva, S.V. and Bunzli, J.C.G. (2010) Lanthanide luminescence for functional materials and bio-sciences. *Chem. Soc. Rev.*, **39** (1), 189–227.

46 Binnemans, K. (2009) Lanthanide-based luminescent hybrid materials. *Chem. Rev.*, **109** (9), 4283–4374.

47 Sabbatini, N., Guardigli, M., and Lehn, J.M. (1993) Luminescent lanthanide complexes as photochemical supramolecular devices. *Coord. Chem. Rev.*, **123** (1–2), 201–228.

48 Chandler, B.D., Yu, J.O., Cramb, D.T., and Shimizu, G.K.H. (2007) Series of lanthanide-alkali metal–organic frameworks exhibiting luminescence and permanent microporosity. *Chem. Mater.*, **19** (18), 4467–4473.

49 Gandara, F., Garcia-Cortes, A., Cascales, C., Gomez-Lor, B., Gutierrez-Puebla, E., Iglesias, M., Monge, A., and Snejko, N. (2007) Rare earth arenedisulfonate metal–organic frameworks: an approach toward polyhedral diversity and variety of functional compounds. *Inorg. Chem.*, **46** (9), 3475–3484.

50 Huang, Y., Yan, B., Shao, M., and Chen, Z.X. (2007) A new family of dimeric lanthanide(III) complexes:

synthesis, structures and photophysical property. *J. Mol. Struct.*, **871** (1–3), 59–66.
51 Huang, Y.G., Wu, B.L., Yuan, D.Q., Xu, Y.Q., Jiang, F.L., and Hong, M.C. (2007) New lanthanide hybrid as clustered infinite nanotunnel with 3D Ln–O–Ln framework and (3,4)-connected net. *Inorg. Chem.*, **46** (4), 1171–1176.
52 Mahata, P. and Natarajan, S. (2007) A new series of three-dimensional metal–organic framework, $[M_2(H_2O)][C_5NH_3(COO)_2]_3 \cdot 2H_2O$, M=La, Pr, and Nd: synthesis, structure, and properties. *Inorg. Chem.*, **46** (4), 1250–1258.
53 Mahata, P., Ramya, K.V., and Natarajan, S. (2007) Synthesis, structure and optical properties of rare-earth benzene carboxylates. *Dalton Trans.*, 4017–4026.
54 Song, X.-Q., Liu, W.S., Dou, W., Wang, Y.W., Zheng, J.R., and Zang, Z.P. (2008) Structure variation and luminescence properties of lanthanide complexes incorporating a naphthalene-derived chromophore featuring salicylamide pendant arms. *Eur. J. Inorg. Chem.*, 1901–1912.
55 Sonnauer, A., Nather, C., Hoppe, H.A., Senker, J., and Stock, N. (2007) Systematic investigation of lanthanide phosphonatoethanesulfonate framework structures by high-throughput methods, Ln $(O_3P–C_2H_4–SO_3)(H_2O)$ (Ln=La–Dy). *Inorg. Chem.*, **46** (23), 9968–9974.
56 Sun, Y.Q. and Yang, G.Y. (2007) Organic–inorganic hybrid materials constructed from inorganic lanthanide sulfate skeletons and organic 4,5-imidazoledicarboxylic acid. *Dalton Trans.*, 3771–3781.
57 van der Horst, M.G., van Albada, G.A., Ion, R.M., Mutikainen, I., Turpeinen, U., Tanase, S., and Reedijk, J. (2008) Extended networks generated from the interaction of rare-earth (III) ions and pyridine-2-carboxamide-based ligands. *Eur. J. Inorg. Chem.*, 2170–2176.
58 Xu, H.T. and Li, Y.D. (2004) The organic ligands as template: the synthesis, structures and properties of a series of the layered structure rare-earth coordination polymers. *J. Mol. Struct.*, **690** (1–3), 137–143.
59 Yue, Q., Yang, J., Li, G.H., Li, G.D., and Chen, J.S. (2006) Homochiral porous lanthanide phosphonates with 1D triple-strand helical chains: synthesis, photoluminescence, and adsorption properties. *Inorg. Chem.*, **45** (11), 4431–4439.
60 Zhang, X.J., Xing, Y.H., Sun, Z., Han, J., Zhang, Y.H., Ge, M.F., and Niu, S.Y. (2007) A series of two-dimensional metal–organic frameworks based on the assembly of rigid and flexible carboxylate-containing mixed ligands with lanthanide metal salts. *Cryst. Growth Des.*, **7** (10), 2041–2046.
61 Li, X., Sun, H.L., Wu, X.S., Qiu, X., and Du, M. (2010) Unique (3,12)-connected porous lanthanide–organic frameworks based on Ln_4O_4 clusters: synthesis, crystal structures, luminescence, and magnetism. *Inorg. Chem.*, **49** (4), 1865–1871.
62 Choi, J.R., Tachikawa, T., Fujitsuka, M., and Majima, T. (2010) Europium-based metal–organic framework as a photocatalyst for the one-electron oxidation of organic compounds. *Langmuir*, **26** (13), 10437–10443.
63 Lin, Z.-J., Xu, B., Liu, T.-F., Cao, M.-N., Lu, J., and Cao, R. (2010) A series of lanthanide metal–organic frameworks based on biphenyl-3,4′,5-tricarboxylate: syntheses, structures, luminescence and magnetic properties. *Eur. J. Inorg. Chem.*, 3842–3849.
64 Soares-Santos, P., Cunha-Silva, L., Paz, F., Ferreira, R., Rocha, J., Carlos, L., and Nogueira, H. (2010) Photoluminescent lanthanide–organic bilayer networks with 2,3-pyrazinedicarboxylate and oxalate. *Inorg. Chem.*, **49** (7), 3428–3440.
65 Li, X., Wu, B.L., Niu, C.Y., Niu, Y.Y., and Zhang, H.Y. (2009) Syntheses of metal-2-(pyridin-4-yl)-1*H*-imidazole-4,5-dicarboxylate networks with topological diversity: gas adsorption, thermal stability and fluorescent emission properties. *Cryst. Growth Des.*, **9** (8), 3423–3431.
66 de Lill, D.T., de Bettencourt-Dias, A., and Cahill, C.L. (2007) Exploring lanthanide

luminescence in metal–organic frameworks: synthesis, structure, and guest-sensitized luminescence of a mixed europium/terbium–adipate framework and a terbium–adipate framework. *Inorg. Chem.*, **46** (10), 3960–3965.

67 Soares-Santos, P.C.R., Cunha-Silva, L., Paz, F.A.A., Ferreira, R.A.S., Rocha, J., Trindade, T., Carlos, L.D., and Nogueira, H.I.S. (2008) Photoluminescent 3D lanthanide–organic frameworks with 2,5-pyridinedicarboxylic and 1,4-phenylenediacetic acids. *Cryst. Growth Des.*, **8** (7), 2505–2516.

68 Richardson, F.S. (1982) Terbium(III) and europium(III) ions as luminescent probes and stains for biomolecular systems. *Chem. Rev.*, **82** (5), 541–552.

69 Serre, C., Millange, F., Thouvenot, C., Gardant, N., Pellé, F., and Férey, G. (2004) Synthesis, characterisation and luminescent properties of a new three-dimensional lanthanide trimesate: M $((C_6H_3)–(CO_2)_3)$ (M=Y, Ln) or MIL-78. *J. Mater. Chem.*, **14** (10), 1540–1543.

70 Muller, M., Devaux, A., Yang, C.H., De Cola, L., and Fischer, R.A. (2010) Highly emissive metal–organic framework composites by host–guest chemistry. *Photochem. Photobiol. Sci.*, **9** (6), 846–853.

71 Zhang, X., Ballem, M.A., Ahren, M., Suska, A., Bergman, P., and Uvdal, K. (2010) Nanoscale Ln(III)–carboxylate coordination polymers (Ln=Gd, Eu, Yb): temperature-controlled guest encapsulation and light harvesting. *J. Am. Chem. Soc.*, **132** (30), 10391–10397.

72 Park, Y.K., Choi, S.B., Kim, H., Kim, K., Won, B.H., Choi, K., Choi, J.S., Ahn, W.S., Won, N., Kim, S., Jung, D.H., Choi, S.H., Kim, G.H., Cha, S.S., Jhon, Y.H., Yang, J.K., and Kim, J. (2007) Crystal structure and guest uptake of a mesoporous metal–organic framework containing cages of 3.9 and 4.7 nm in diameter. *Angew. Chem. Int. Ed.*, **46** (43), 8230–8233.

73 Cornil, J., dos Santos, D.A., Crispin, X., Silbey, R., and Bredas, J.L. (1998) Influence of interchain interactions on the absorption and luminescence of conjugated oligomers and polymers: a quantum-chemical characterization. *J. Am. Chem. Soc.*, **120** (6), 1289–1299.

74 Wagner, B.D., McManus, G.J., Moulton, B., and Zaworotko, M.J. (2002) Exciplex fluorescence of $\{[Zn(bipy)_{1.5}(NO_3)_2]\cdot CH_3OH\cdot 0.5pyrene\}_n$: a coordination polymer containing intercalated pyrene molecules (bipy=4,4′-bipyridine). *Chem. Commun.*, (18), 2176–2177.

75 McManus, G.J., Perry, J.J., Perry, M., Wagner, B.D., and Zaworotko, M.J. (2007) Exciplex fluorescence as a diagnostic probe of structure in coordination polymers of Zn^{2+} and 4,4′-bipyridine containing intercalated pyrene and enclathrated aromatic solvent guests. *J. Am. Chem. Soc.*, **129** (29), 9094–9101.

76 Luo, F. and Batten, S.R. (2010) Metal–organic framework (MOF): lanthanide(III)-doped approach for luminescence modulation and luminescent sensing. *Dalton Trans.*, 4485–4488.

77 Thirumurugan, A. and Cheetham, A.K. (2010) Anionic metal–organic frameworks of bismuth benzenedicarboxylates: synthesis, structure and ligand-sensitized photoluminescence. *Eur. J. Inorg. Chem.* **2010** (24), 3823–3828.

78 Wang, P., Ma, J.P., and Dong, Y.B. (2009) Guest-driven luminescence: lanthanide-based host–guest systems with bimodal emissive properties based on a guest-driven approach. *Chem. Eur. J.*, **15** (40), 10432–10445.

79 Chen, B., Xiang, S., and Qian, G. (2010) Metal–organic frameworks with functional pores for recognition of small molecules. *Acc. Chem. Res.*, **43** (8), 1115–1124.

80 Cohen, S.M. (2009) Supramolecular chemistry: molecular crystal balls. *Nature*, **461** (7264), 602–603.

81 Ingleson, M.J., Heck, R., Gould, J.A., and Rosseinsky, M.J. (2009) Nitric oxide chemisorption in a postsynthetically modified metal–organic framework. *Inorg. Chem.*, **48** (21), 9986–9988.

82 Song, Y.-F. and Cronin, L. (2008) Postsynthetic covalent modification of metal–organic framework (MOF) materials. *Angew. Chem. Int. Ed.*, **47** (25), 4635–4637.

83 Wang, Z. and Cohen, S. (2009) Postsynthetic modification of metal–organic frameworks. *Chem. Soc. Rev.*, **38** (5), 1315–1329.

84 Wang, Z., Tanabe, K., and Cohen, S. (2010) Tuning hydrogen sorption properties of metal–organic frameworks by postsynthetic covalent modification. *Chem. Eur. J.*, **16** (1), 212–217.

85 Zhu, W.H., Wang, Z.M., and Gao, S. (2007) Two 3D porous lanthanide–fumarate–oxalate frameworks exhibiting framework dynamics and luminescent change upon reversible de- and rehydration. *Inorg. Chem.*, **46** (4), 1337–1342.

86 Braga, D., Maini, L., Mazzeo, P., and Ventura, B. (2010) Reversible interconversion between luminescent isomeric metal–organic frameworks of [$Cu_4I_4(DABCO)_2$] (DABCO=1,4-diazabicyclo[2.2.2]octane). *Chem.-Eur. J.*, **16** (5), 1553–1559.

87 Lu, G. and Hupp, J. (2010) Metal–organic frameworks as sensors: a ZIF-8 based Fabry–Pérot device as a selective sensor for chemical vapors and gases. *J. Am. Chem. Soc.*, **132** (23), 7832–7833.

88 Chen, B.L., Wang, L.B., Zapata, F., Qian, G.D., and Lobkovsky, E.B. (2008) A luminescent microporous metal–organic framework for the recognition and sensing of anions. *J. Am. Chem. Soc.*, **130** (21), 6718–6719.

89 Lu, W.G., Jiang, L., Feng, X.L., and Lu, T.B. (2009) Three-dimensional lanthanide anionic metal–organic frameworks with tunable luminescent properties induced by cation exchange. *Inorg. Chem.*, **48** (15), 6997–6999.

90 Wong, K.-L., Law, G.-L., Yang, Y.-Y., and Wong, W.-T. (2006) A highly porous luminescent terbium–organic framework for reversible anion sensing. *Adv. Mater.*, **18** (8), 1051–1054.

91 Chen, B.L., Wang, L.B., Xiao, Y.Q., Fronczek, F.R., Xue, M., Cui, Y.J., and Qian, G.D. (2009) A luminescent metal–organic framework with Lewis basic pyridyl sites for the sensing of metal ions. *Angew. Chem. Int. Ed.*, **48** (3), 500–503.

92 Liu, W., Jiao, T., Li, Y., Liu, Q., Tan, M., Wang, H., and Wang, L. (2004) Lanthanide coordination polymers and their Ag^+-modulated fluorescence. *J. Am. Chem. Soc.*, **126** (8), 2280–2281.

93 Liu, S., Li, J., and Luo, F. (2010) The first transition-metal metal–organic framework showing cation exchange for highly selectively sensing of aqueous Cu(II) ions. *Inorg. Chem. Commun.*, **13** (7), 870–872.

94 Xiao, Y., Wang, L., Cui, Y., Chen, B., Zapata, F., and Qian, G. (2009) Molecular sensing with lanthanide luminescence in a 3D porous metal–organic framework. *J. Alloys Compd.*, **484** (1–2), 601–604.

95 Harbuzaru, B.V., Corma, A., Rey, F., Atienzar, P., Jorda, J.L., Garcia, H., Ananias, D., Carlos, L.D., and Rocha, J. (2008) Metal–organic nanoporous structures with anisotropic photoluminescence and magnetic properties and their use as sensors. *Angew. Chem. Int. Ed.*, **47** (6), 1080–1083.

96 Harbuzaru, B.V., Corma, A., Rey, F., Jorda, J.L., Ananias, D., Carlos, L.D., and Rocha, J. (2009) A miniaturized linear pH sensor based on a highly photoluminescent self-assembled europium(III) metal–organic framework. *Angew. Chem. Int. Ed.*, **48** (35), 6476–6479.

97 Jiang, C., Yu, Z., Jiao, C., Wang, S., Li, J., Wang, Z., and Cui, Y. (2004) Luminescent Zn and Cd coordination polymers. *Eur. J. Inorg. Chem.*, 4669–4674.

98 Kobayashi, A., Hara, H., Noro, S., and Kato, M. (2010) Multifunctional sensing ability of a new Pt/Zn-based luminescent coordination polymer. *Dalton Trans.*, 3400–3406.

99 Qiu, L., Li, Z., Wu, Y., Wang, W., Xu, T., and Jiang, X. (2008) Facile synthesis of nanocrystals of a microporous metal–organic framework by an ultrasonic method and selective sensing of organoamines. *Chem. Commun.*, 3642–3644.

100 Xie, Z.G., Ma, L.Q., deKrafft, K.E., Jin, A., and Lin, W.B. (2010) Porous

phosphorescent coordination polymers for oxygen sensing. *J. Am. Chem. Soc.*, **132** (3), 922–923.

101 Lan, A.J., Li, K.H., Wu, H.H., Olson, D.H., Emge, T.J., Ki, W., Hong, M.C., and Li, J. (2009) A luminescent microporous metal–organic framework for the fast and reversible detection of high explosives. *Angew. Chem. Int. Ed.*, **48** (13), 2334–2338.

102 Doty, F.P., Bauer, C.A., Skulan, A.J., Grant, P.G., and Allendorf, M.D. (2009) Scintillating metal–organic frameworks: a new class of radiation detection materials. *Adv. Mater.*, **21** (1), 95–101.

103 Nakamura, H. (2009) Recent development of white LEDs and solid state lighting. *Light Eng.*, **17** (4), 13–17.

104 Chang, C., Chen, C., Wu, C., Chang, S., Hung, J., and Chi, Y. (2010) High-color-rendering pure-white phosphorescent organic light-emitting devices employing only two complementary colors. *Org. Electron.*, **11** (2), 266–272.

105 Furman, J.D., Warner, A.Y., Teat, S.J., Mikhailovsky, A.A., and Cheetham, A.K. (2010) Tunable, ligand-based emission from inorganic–organic frameworks: a new approach to phosphors for solid state lighting and other applications. *Chem. Mater.*, **22** (7), 2255–2260.

106 Wang, M.S., Guo, S.P., Li, Y., Cai, L.Z., Zou, J.P., Xu, G., Zhou, W.W., Zheng, F.K., and Guo, G.C. (2009) A direct white-light-emitting metal–organic framework with tunable yellow-to-white photoluminescence by variation of excitation light. *J. Am. Chem. Soc.*, **131** (38), 13572–13573.

107 Prasad, P.N. and Williams, D.J. (1991) *Introduction to Nonlinear Optical Effects in Molecules and Polymers*, John Wiley & Sons, Inc., New York.

108 Nalwa, H.S. and Miyata, S. (1997) *Nonlinear Optics of Organic Molecules and Polymers*, CRC Press, Boca Raton, FL.

109 Liu, Y., Li, G., Li, X., and Cui, Y. (2007) Cation-dependent non-linear optical behavior in an octupolar 3D anionic metal–organic open framework. *Angew. Chem. Int. Ed.*, **46** (33), 6301–6304.

110 Evans, O.R. and Lin, W.B. (2002) Crystal engineering of NLO materials based on metal–organic coordination networks. *Acc. Chem. Res.*, **35** (7), 511–522.

111 Liu, G., Zhu, K., Xu, H., Nishihara, S., Huang, R., and Ren, X. (2010) Five 3D metal–organic frameworks constructed from V-shaped polycarboxylate acids and flexible imidazole-based ligands. *CrystEngComm.*, **12** (4), 1175–1185.

112 Liu, T., Lu, J., Guo, Z., Proserpio, D., and Cao, R. (2010) New metal–organic framework with uninodal 4-connected topology displaying interpenetration, self-catenation, and second-order nonlinear optical response. *Cryst. Growth Des.*, **10** (4), 1489–1491.

113 Wen, L.-L., Dang, D.-B., Duan, C.-Y., Li, Y.-Z., Tian, Z.-F., and Meng, Q.-J. (2005) 1D helix, 2D brick-wall and herringbone, and 3D interpenetration d^{10} metal–organic framework structures assembled from pyridine-2,6-dicarboxylic acid N-oxide. *Inorg. Chem.*, **44** (20), 7161–7170.

114 Xiong, R.-G., Zuo, J.-L., You, X.-Z., Abrahams, B.F., Bai, Z.-P., Chec, C.-M., and Fund, H.-K. (2000) Opto-electronic multifunctional chiral diamondoid-network coordination polymer: bis{4-[2-(4-pyridyl)ethenyl]benzoato}zinc with high thermal stability. *Chem. Commun.*, 2061–2062.

115 Guo, Z., Cao, R., Wang, X., Li, H., Yuan, W., Wang, G., Wu, H., and Li, J. (2009) A multifunctional 3D ferroelectric and NLO-active porous metal–organic framework. *J. Am. Chem. Soc.*, **131** (20), 6894–6895.

116 Albota, M., Beljonne, D., Bredas, J., Ehrlich, J., Fu, J., Heikal, A., Hess, S., Kogej, T., Levin, M., Marder, S., McCord-Maughon, D., Perry, J., Rockel, H., Rumi, M., Subramaniam, C., Webb, W., Wu, X., and Xu, C. (1998) Design of organic molecules with large two-photon absorption cross sections. *Science*, **281** (5383), 1653–1656.

117 White, K.A., Chengelis, D.A., Gogick, K.A., Stehman, J., Rosi, N.L., and Petoud, S. (2009) Near-infrared luminescent lanthanide MOF barcodes. *J. Am. Chem. Soc.*, **131** (50), 18069–18071.

13
Deposition of Thin Films for Sensor Applications
Mark Allendorf, Angélique Bétard, and Roland A. Fischer

13.1
Introduction

Sensors are present everywhere: in cars, medicine, manufacturing, and homes. Although there are many transduction mechanisms by which a sensor can detect an analyte (some of which are discussed below), it is typically necessary to have some kind of "recognition layer" that provides selectivity to the desired species and may also enhances sensitivity. Organic polymer films are extensively used in this regard because of their synthetic flexibility, ease of fabrication, and low cost [1]. Nevertheless, these materials have some significant disadvantages which suggest that there is room for improvement. Polymers possess a glass transition temperature T_g, below which they become brittle. Hence the mechanical properties of such films, which are important in some transduction mechanisms, can be highly temperature dependent. Many organic polymers are unstable at relatively modest temperatures (particularly in oxidizing environments), which limits their utility.

Metal–organic frameworks (MOFs) are attractive materials for sensing because of their micro- and mesoporosity, which can be tailored with respect to both size and chemical functionality of the internal coordination space available for hosting guest molecules. This porosity leads to very high surface areas, and in some particular cases can be combined with responsive properties of the framework involving reversible structural transformations and flexibility [2–4]. These features suggest that high analyte sensitivity and selectivity with respect to interfering species could be achieved by appropriate tuning of the pore structure and chemical functionality. The organic "linker" groups in MOFs, which must be more or less rigid to provide the pores with structural integrity, permit considerable chemical versatility, as is the case for the monomers in organic polymers. However, because of the metal ions connecting the linkers, an additional tunable structural component is available for MOFs. Vacant or exchangeable coordination positions at these metal ions allow the creation of sites with much higher binding energies than are associated with physisorption. Structural flexibility, in which unit cell volumes can depend strongly on the presence or absence of guest molecules in the pores, creates a third dimension to the design

Metal-Organic Frameworks: Applications from Catalysis to Gas Storage, First Edition. Edited by David Farrusseng.
© 2011 Wiley-VCH Verlag GmbH & Co. KGaA. Published 2011 by Wiley-VCH Verlag GmbH & Co. KGaA.

possibilities. Moreover, MOFs do not undergo a glass transition like organic polymers, nor do they melt. In fact, many frameworks are stable to temperatures in excess of 300 °C, making it possible both to heat them to remove even more strongly adsorbed guests and to use them in sensing applications that involve elevated temperatures.

It would be of great interest to process the MOFs as thin films, directly on top of appropriate substrates for integration into sensing devices. In this case, detection modes such as fluorescence and infrared spectroscopy or mechanical stress can be used. In addition, analyte-dependent changes in the magnetic, conductive, or redox properties of MOFs are also possible. Relevant MOF structures include the well-known $[Cu_3(btc)_2]$ (CuBTC; btc = 1,3,5-benzenetricarboxylate) [5], which exhibits an open coordination site once dehydrated, and structurally flexible MOFs such as $[Zn_2(bdc)_2(dabco)]$ (bdc = 1,4-benzenedicarboxylate, dabco = 1,4-diazabicyclo[2.2.2]octane) [6] or the "breathing" structures belonging to the MIL family (see [2] for a review on breathing porous materials).

Since the fabrication of MOF thin films in general has recently been reviewed [7], this chapter focuses primarily on aspects of MOF film growth that are directly relevant to chemical sensing. We present a brief review of the work to date in which sensors incorporating MOF thin films have been demonstrated, then describe the various transduction mechanisms that are possible with these novel materials. Aspects that should be considered when selecting or designing a MOF (thin film) to detect a particular molecule are summarized. Finally, methods reported to date that have potential for growing sensor-quality coatings are surveyed.

13.2
Literature Survey

To the best of our knowledge, so far (by October 2010) only a few publications have described the use of MOF films for sensing applications, and they are summarized in Table 13.1. A variety of detection modes have been demonstrated (see Section 13.3 for more details); but only a few MOF structures have been used, the most popular being CuBTC because of its interesting properties and easy synthesis.

13.3
Signal Transduction Modes

For a MOF to serve as a component of a sensor it must exhibit several properties. First, an interaction with the analyte is required. Second, this interaction should (in most cases) have some specificity for the analyte of interest. Third, the response should preferably be reversible, although this is not required in all cases. Finally, and obviously, the interaction must be detectable. Molecule–MOF interactions can range from physisorption as a result of weak van der Waals interactions ($\sim$5–10 kJ mol^{-1}),

Table 13.1 Published work in which a MOF layer was studied for sensing properties. The film fabrication methods refer to the terminology used in the last section of this chapter.

MOF formula	Film fabrication method	Detected molecules	Detection mode	Ref.
CuBTC	Aged solvothermal mother solution	H_2O vapor	QCM	[8]
CuBTC	LPE	H_2O, MeOH, EtOH vapor	Stress on microcantilevers	[9]
MIL-101(Cr)	Assembly of colloids	H_2O vapor	Ellipsometry	[10]
FeBTC	Screen printing of MOF powder	H_2O, MeOH, EtOH vapor	Impedancemetry	[11]
$[Zn_3(btc)_2]$	Zinc substrate + ligand in solvothermal conditions	Diethylamine	Fluorescence	[12]
ZIF-8	Repeated immersions in fresh mother solutions	EtOH, propane, n-hexane vapors	UV–Vis spectroscopy	[13]
CuBTC	LPE	CO_2	SPR	[14]

to much stronger chemisorption, in which chemical bonds form between the adsorbing layer and the analyte. Such interactions produce a mass change upon adsorption, suggesting the use of resonator-type devices such as surface acoustic wave (SAW) sensors [15], quartz crystal microbalances (QCMs) [15], and microcantilevers (MCLs) [16].

Gas sorption in SAW devices is typically detected by measuring the frequency shift (typically a decrease) of acoustic waves traveling parallel to the surface that are generated by an oscillator (usually quartz) vibrating in the 25–500 MHz range. QCMs operate on a similar principle, but in this case the acoustic waves propagate throughout the bulk of the crystal and travel perpendicular to the surface. In these devices, the surfaces exposed to a coating process are either gold (for electrodes and transducers) or quartz. MCLs, which are microelectromechanical systems (MEMS) devices created by "machining" techniques based on complementary metal–oxide–semiconductor (CMOS) fabrication methods, can also detect mass uptake as a result of a change in their oscillation frequency. The surfaces of these devices can be tailored for the purpose of effectively binding a coating. Of these three device types, microcantilevers present the greatest challenge for MOF coatings because of their small dimensions (tens of microns wide and typically a few hundred microns long) and relative fragility.

MCLs can also operate in a static mode. In this case, adsorbates are detected as a result of a change in interfacial stress that causes the cantilever beam to bend. Two materials with different mechanical properties (e.g., silicon nitride and silicon) comprise the cantilever beam, with their thicknesses determined such that the stress between the two is balanced. Thus, adsorption causes the device to bend, which can be detected either optically or by a built-in piezoresistive sensor. Beam displacements of <1 nm can be detected using MOF-coated devices that we recently tested [9, 17]. The structural flexibility of MOFs should be an advantage here, since even small changes in unit cell dimensions can result in large tensile or compressive stresses at the interface between the cantilever and the MOF thin film.

Transduction mechanisms based on changes in the electrical properties of MOFs may also be feasible. For example, impedance-based gas detection has been demonstrated [11]. Other circuit types, however, such as transistors and resistors, will likely require the development of new MOFs that are not insulators, as are almost all known frameworks. Changes in optical properties offer additional possibilities. A Fabry–Pérot interferometer was recently demonstrated, in which changes in a MOF's refractive index occur as a result of adsorption [13]. Using films of different thicknesses, a colorimetric fingerprint can be produced that indicates the presence of an analyte. Luminescence quenching as a transduction mechanism has been demonstrated [12]; however, the advantage of a thin film in this case is unclear and would most likely require an exceptionally strong emitter to produce a sufficient signal to be detected.

13.4
Considerations in Selecting MOFs for Sensing Applications

The potential for the rational design of MOFs to achieve sensitivity and selectivity towards certain analytes naturally begins with the consideration of characteristics such as pore size and functional groups within the pore that can interact with the molecules of interest. It is critical to realize, however, that the design process cannot end there. Many factors must be taken into account before a successful integration of a MOF with a particular device can be accomplished. In this section, we discuss a number of these factors, which include film thickness, thermal stability, and mechanical properties. This discussion demonstrates that in addition to chemistry, properties that fall within the realm of materials science and mechanical engineering must become a much more important part of MOF science than they are today if these materials are to mature to the point of becoming manufactured products.

13.4.1
Pore Dimensions

MOFs with pore diameters in the range ∼3–46 Å are known, which is broad enough to allow the detection of many molecules of interest. However, when compared with biological molecules, these pore dimensions limit the utility of MOFs to the detection of relatively small species (Figure 13.1). For example, sugars such as glucose and amino acids could pass through some MOF pores, but these openings are much too small to accommodate enzymes and proteins. Note, however, that sensing of such species via surface adsorption on MOF films is not precluded, but has not been demonstrated.

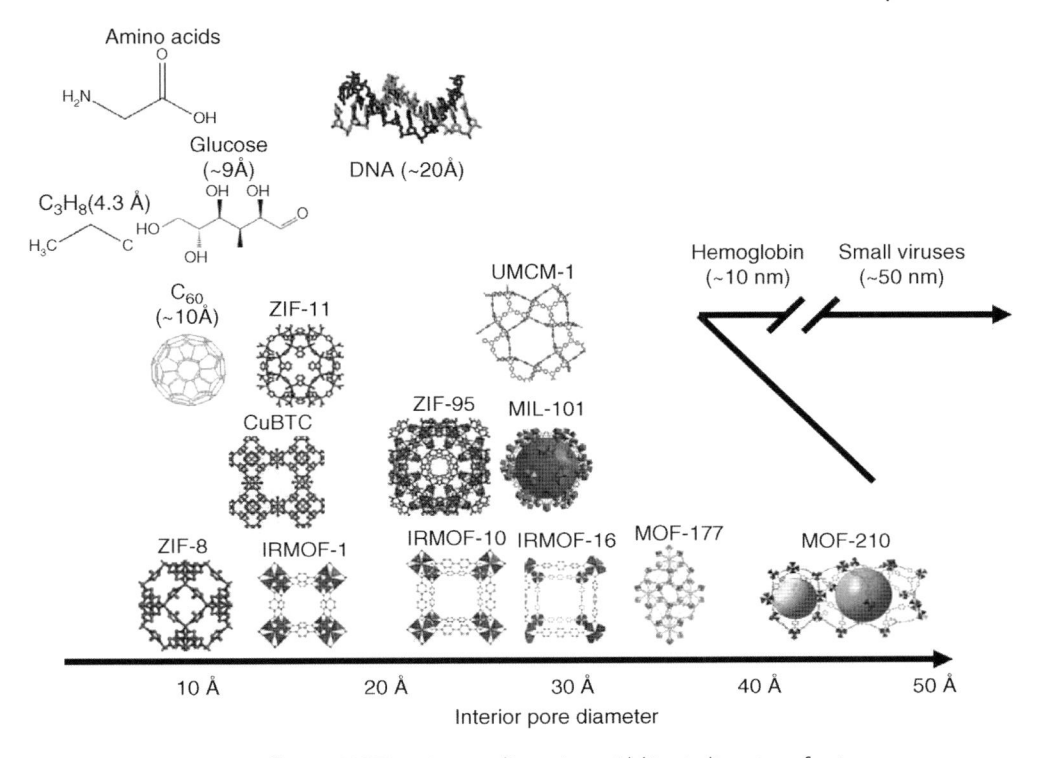

Figure 13.1 Comparison of known MOF interior pore dimensions with kinetic diameters of various molecules and biological species.

13.4.2
Adsorption Thermodynamics

The nanoporosity of MOFs indicates that chemical detection will involve adsorption on the pore surfaces rather than solvation, as in dense organic polymers. For adsorption to occur, $\Delta G°(\text{ads}) = -[Q_{st} - T\Delta S°(\text{ads})]$ must be negative. The amount adsorbed thus depends upon the isosteric heat of adsorption Q_{st} (typically reported as a positive number for an exothermic adsorption process), a property that has been measured for a number of MOFs and small molecules. Assuming that Trouton's rule for the entropy of vaporization ($\sim$85 J mol^{-1} K^{-1}) is applicable (a reasonable approximation for nonpolar organic compounds), $\Delta G°(\text{ads})$ will be negative at 298 K if $Q_{st} \geq$ 25 kJ mol^{-1} (Figure 13.2). Clearly, molecules such as methane, hydrogen, and other weakly interacting gases will be difficult to detect under ambient conditions. For example, $Q_{st}(H_2)$ is 4–7 kJ mol^{-1} [18] and values of Q_{st} measured at low coverage on IRMOF-1 for small hydrocarbons and CO_2 is no greater than 25 kJ mol^{-1}, although incorporating an amine group into the framework linker (IRMOF-3) increases values by $\sim$25% for CH_4 and CO_2 (Figure 13.3). We expect that many other small molecules will also have $Q_{st} \leq 25$ kJ mol^{-1} (MOFs having open coordination sites could be

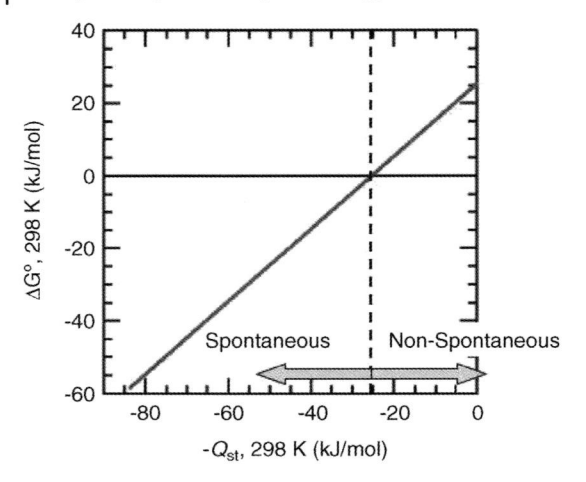

Figure 13.2 Change in Gibbs free energy of adsorption as a function of the isosteric heat of adsorption at 298 K, assuming Trouton's rule to estimate the enthalpy of adsorption.

exceptions, however), so the high surface areas found in many MOFs (10 000 m^2 g^{-1} was recently reported [19]) are advantageous when attempting to detect such species. In general, Q_{st} data for MOFs are sparse and many of the reported experimental and computational adsorption isotherms were obtained for high-pressure conditions suitable for gas storage and chemical separations, rather than the much lower

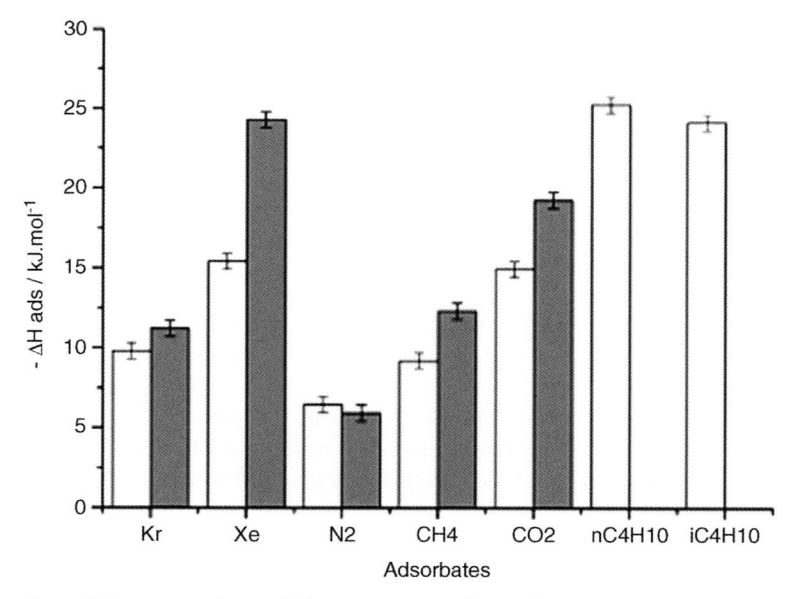

Figure 13.3 Isosteric heats of adsorption measured at very low coverages on IRMOF-1 (white) and IRMOF-3 (gray). Reproduced with permission from [20].

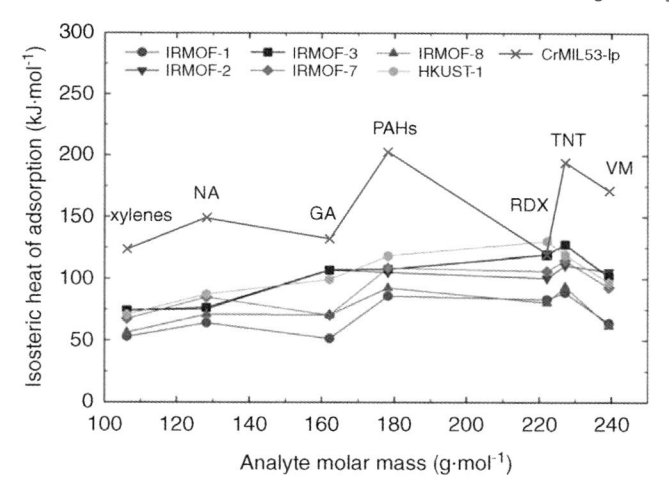

Figure 13.4 Isosteric heat of adsorption (Q_{st}) for various analyte classes by selected MOFs, as predicted by grand canonical Monte Carlo simulations. Reproduced with permission from [21].

concentrations typically of interest for chemical sensing. It is also important to note that Q_{st} typically decreases with increasing coverage as a result of repulsive forces, adding an additional impetus to maximize device sensitivity.

Fortunately, it appears that Q_{st} for molecules representing important classes of analytes, including polyaromatic hydrocarbons, volatile organics, explosives, and chemical weapons, are >50 kJ mol^{-1}. This is illustrated in Figure 13.4, where Q_{st} values obtained from grand canonical Monte Carlo simulations are displayed for selected molecules within these categories [21]. These values, some of which exceed 160 kJ mol^{-1}, allow uptake at pressures as low as 10 ppb (Figure 13.5). It is also encouraging that these simulations predict that selectivity between relatively similar molecules can be achieved (e.g., xylenes versus TNT by CrMIL53 versus CuBTC; Figure 13.5). Therefore, by using an array of sensors coated with different MOFs, it should be possible to generate a "fingerprint" for a given molecule, allowing both identification and detection to be achieved.

13.4.3
Film Attachment

If analyte adsorption involves anything more than very weak physisorption, it is likely that sensor regeneration will involve heating the device. Continuous operation at elevated temperatures may also be desirable to reduce the uptake of atmospheric gases, in particular water vapor. Under ambient conditions, adsorption of water vapor is unavoidable in such high surface area materials, so thermal stability of the coating maybe essential in some cases. The presence of adsorbed water does not necessarily preclude the detection of other species, however. For example, microcantilevers coated with films of the MOF CuBTC respond to alcohols with or without adsorbed

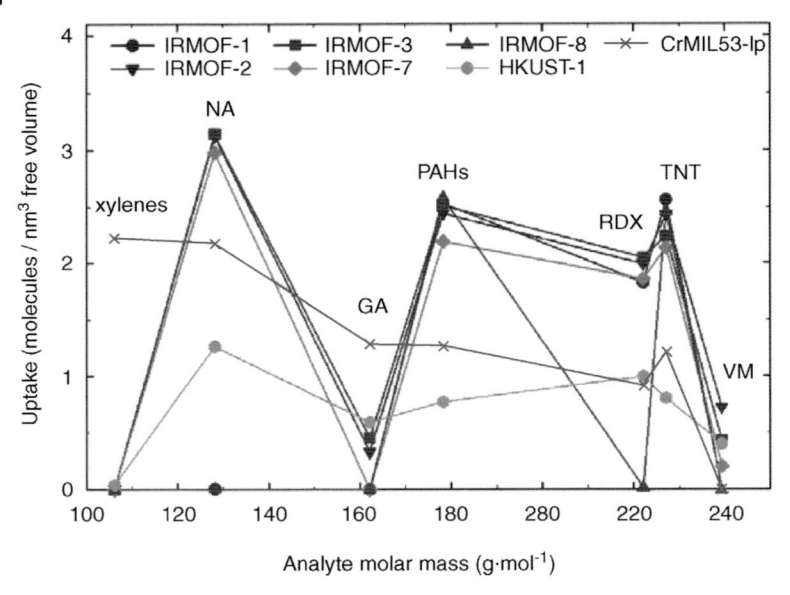

Figure 13.5 Uptake of various molecule types by selected MOFs, as predicted by grand canonical Monte Carlo simulations. Reproduced with permission from [21].

water vapor [9]. However, in cases where open metal coordination sites are incorporated within the framework to allow tight and/or selective binding of some species, the relatively high concentration of water in the atmosphere can easily cause all of these sites to become occupied, preventing uptake of molecules that bind less strongly. Under these conditions, it may be necessary to remove water vapor from the analyte stream prior to the sensor.

The effectiveness of film attachment may also influence the sensitivity of stress-based detection using microcantilevers. Adsorbate-induced structural changes in MOFs can be fairly small and thus produce modest (but readily detectable) changes in interfacial stress. For example, the maximum resistance change produced in a CuBTC-coated microcantilever by 6 vol.% H_2O vapor was only $0.016 \pm 0.002\%$ [9] The film in this instance was attached to the microcantilever using a self-assembled monolayer (SAM) formed from 11-mercaptoundecanoic acid and was highly polycrystalline. We estimate, based on the signal-to-noise ratio, that the water vapor detection limit of this device is 20 ppm. In the absence of evidence concerning the adhesion of the MOF film and the extent to which the SAM introduces slip that damps the transmission of stress from the MOF to the microcantilever, it is difficult to predict what sensitivity could ultimately be achieved by an "optimized" device. Nevertheless, it seems likely that replacing the relatively weak thiol–gold bonding to the surface with the much stronger covalent bonding available by binding the MOF to an oxide surface should substantially improve device sensitivity.

Most MOFs are sufficiently stable that they can be heated to temperatures well above 100 °C, so thermal stability is more likely limited by the chemistry attaching the film to the underlying substrate. Several methods of attaching MOFs to substrates have been demonstrated. Although casting particles directly on to device surfaces may work in some cases [11], in general this is less desirable because the mechanical stability of the MOF layer is typically rather poor. Attaching the MOF by growing it on the surface is preferable in this regard. One route is to deposit an intervening SAM with reactive groups that attach to the substrate surface and also provide nucleation sites for the MOF. Both thiol-based SAMs on gold surfaces and organosilicon SAMs bound to oxide surfaces such as silica and aluminum oxide can be used [7]. However, thiol-based SAMs will not tolerate temperatures above about 80 °C. SAMs involving covalent Si−O−M (M = surface atom) bonds are far stronger and can withstand temperatures above the decomposition temperature of many MOFs. Devices such as QCMs and SAWS have silicon oxide surfaces that can be hydroxylated to provide binding sites for organosilane SAMs. Oxide surfaces can also be created by plasma chemical vapor deposition, atomic layer deposition (ALD), sputtering, or evaporation on devices not having exposed oxide. Surface hydroxyl groups can be created by, for example, reaction with potassium hydroxide [22]. In the case of silicon oxide, a fully hydroxylated surface has 4–6 OH groups per nm^2 [23], providing many nucleation sites for MOF growth.

The choice of interface chemistry has additional implications for the sensitivity of static MCL sensors. Computational modeling of these devices shows that the mechanical properties of the dielectric (top) layer of the device, to which the MOF film could be bound, is significantly increased if silica rather than alumina is used. This is illustrated in Figure 13.6, in which the resistance change induced by a 0.1%

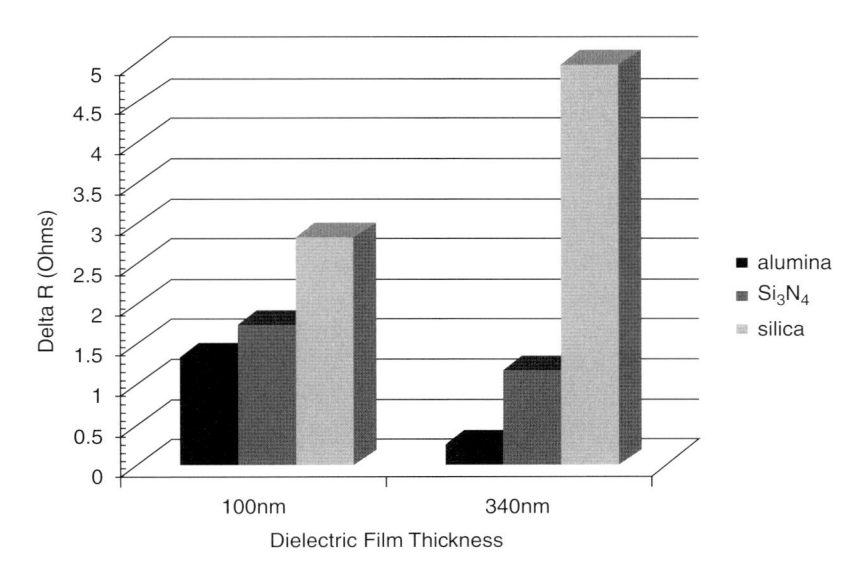

Figure 13.6 Effect of composition of dielectric layer and thickness on sensor response for equivalent MOF and dielectric layer thicknesses. Reproduced with permission from [17].

volumetric expansion of the MOF layer is shown for three different dielectric materials [17]. Assuming equivalent thicknesses for the dielectric and MOF layers, the response to unit stress for a 100 nm silica dielectric layer is more than double that of an alumina layer and more than 20 times higher for a 340 nm thick layer. This effect is a consequence of the smaller Young's modulus of silica compared with alumina (70–75 versus 345 GPa). Not all MOFs grow equally well on all oxides, however [25, 26]. This problem could be solved by depositing a second oxide layer on top of the silica dielectric (e.g., alumina could be deposited by ALD), but this increases the complexity of device fabrication and is therefore undesirable. As is clear from this discussion, adapting MOFs to sensing platforms is not a simple matter and requires consideration of a number of important factors.

13.4.4
Film Thickness and Morphology

Apart from the general consideration that thinner films lead to shorter response times (see below), little is known about the optimal MOF thickness for sensing. Almost certainly this will depend on the specifics of signal transduction in the device. The growth methods described below can produce dense films with thicknesses of 100 Å or more, corresponding to 3–10 unit cells. In our experience, these dimensions are sufficient to permit rapid analyte adsorption using QCM, SAW, and MCL devices without excessive regeneration times. QCMs and SAWs are likely relatively insensitive to the MOF thickness. In both cases, signal transduction relies on the propagation of acoustic waves through the device, so the relative softness and low density of MOFs may ultimately impose a thickness limit. With respect to static MCLs, the situation is more complicated, since the device design must take into account the thickness and elastic properties of all materials composing the beam [17]. One can only conclude at this point that film growth methods that maximize control over thickness are highly desirable. The methods described later in this chapter can in some cases produce "layer by layer" (i.e., unit cell by unit cell) growth. Since MOF unit cells are large, however, the minimum layer thickness deposited per cycle could be as much as 3 nm or more.

The effects of film morphology on sensor behavior likewise have not been characterized for MOFs. The smoothest films produced thus far have an average roughness of $\sim$5 nm, which corresponds to step heights of only two unit cells [27] and was obtained using a thiol-based SAM as the attachment layer. Binding to more stable oxide surfaces produces much rougher films (Figure 13.7). For example, we find that CuBTC films grown on the silica surfaces of SAWs yield an average roughness of $\sim$10 nm for films <100 nm; thicker films are rougher. In this case, the growth mode produces polycrystalline films with typical grain sizes of $\sim$20 nm. However, surface crystallites with sizes of $\sim$100 nm are also observed. Rougher films provide larger surface areas than smooth films, which may increase response times. However, rough polycrystalline films will scatter acoustic waves to a greater extent than amorphous films that lack grain boundaries, reducing the sensitivity of SAW and QCM devices.

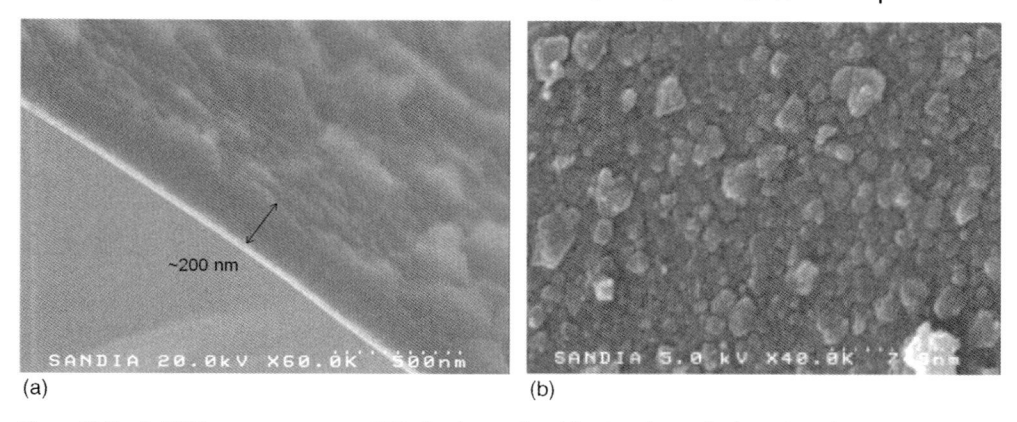

Figure 13.7 CuBTC layers grown on an ALD alumina surface (a) using the method in [28] and on the quartz surface of a SAW (b) using the method in [29].

13.4.5
Response Time

The response time of the sensor will depend on the rate of molecular transport into the film, which in turn is related to the diffusion rate. Naturally, this is a motivation to keep films as thin as possible, but predicting how quickly an analyte will diffuse into a MOF layer is difficult to do without molecular simulations. Li *et al.* reviewed the work concerning diffusion in MOFs and remark that both experimental and computational modeling studies are limited [30]. However, simulations indicate that molecular diffusion in these materials is similar to zeolites with respect to both the mechanism and magnitude of the rates. Surface diffusion, rather than Knudsen diffusion, is dominant [31]. As a result, diffusion rates in MOFs are not straightforwardly linked to the dimensions of the pore openings. Although the diffusion rate of ethane in MOF-5 is nearly two orders of magnitude higher than in zeolite NaX, most likely as a result of larger pore dimensions [32–34], correlation among adsorbed molecules can lead to Fickian diffusion constants that are influenced by both the chemical nature of the adsorbate and the environment of the MOF pores. Predicted diffusion activation barriers are low for small, weakly interacting molecules such as alkanes [34]. For example, the diffusion constant for *n*-pentane at 298 K in CuBTC obtained from NMR measurements is $4 \times 10^{10}\,\text{m}^2\,\text{s}^{-1}$. At this rate, the characteristic time for diffusion through a 100 nm film is only 25 µs, which is adequate for sensing applications. However, introduction of unsaturation (propane versus propene, for example) can increase the activation energy substantially. Gas-phase diffusion constants for larger molecules are likely to be much lower, however, considering the high isosteric heats of adsorption that have been predicted (Figure 13.4). (Note that diffusion through liquids is much slower; for example, infiltration of MOF-177 with dyes in solution requires several days [35].)

13.4.6
Mechanical Properties

The mechanical properties of MOFs must also be considered in the design of practical sensors. MOFs are relatively soft; for example, the hardness of IRMOF-1 determined from continuous stiffness measurements is 48 MPa [36]. This indicates that a MOF layer on a sensor might require protection to avoid damage. In our experience, adhesion of carboxylate-based MOFs (e.g., CuBTC) is fairly strong, suggesting that coating detachment of these materials will likely not be a problem. MOFs are also much less stiff than conventional materials used in microelectronics manufacturing. The average reduced modulus E of IRMOF-1 (MOF-5), obtained from continuous stiffness nanoindentation measurements, is 2.7 ± 1.0 GPa at room temperature [36]. For comparison, silicon and silicon dioxide have much larger Young's moduli of 179 GPa [37] and 73–80 GPa [38], respectively. Systematic investigations of the mechanical properties of MOFs as a function of morphology have not been performed. However, it would not be surprising to find values measured for polycrystalline films, which are typically produced by current MOF deposition methods, that are considerably different from the single-crystal values and display higher standard deviations. For example, average moduli of $\sim 12 \pm 4$, 0.45 ± 0.34, and $\sim 37 \pm 2$ GPa were measured for CuBTC, MOF-508, and NDC-508 (the naphthalene dicarboxylate analog of MOF-508), respectively [17].

13.5
MOF Thin Film Growth: Methods, Mechanisms, and Limitations

In this section, we focus on the different methods that can be used to grow sensor-quality MOF thin films. As discussed in the previous section, the main features desired for such films are the control of film thickness, homogeneous dense microstructure, smoothness, and crystallite orientation. Various examples of films grown directly under solvothermal conditions (temperature $>100\,^\circ$C) by immersing the substrates in the mother liquor show either poor morphology [26] or limited control over orientation and thickness [12, 39]. Furthermore, the integration of such films into devices seems difficult, and no masking or patterning techniques have been developed yet for such growth conditions. Therefore, we will largely focus our discussion here on alternative deposition methods operating at low temperatures and allowing more control over the important film parameters. Many materials can be used as substrates for MOF deposition. In order to achieve some control over the crystallite orientation, an appropriate functionalization of the substrate, for example with an organic SAM, may be necessary.

13.5.1
Growth From Aged Solvothermal Mother Solutions

A very common film growth method relies on the appropriate preparation of a mother solution so that the growth can take place at room temperature. Typically, a solution containing the MOF precursors is heated so as to induce the start of crystallization. After filtration and cooling to room temperature, the substrates are placed vertically or face down in the supersaturated solution; a time span ranging from 24 to 100 h is allowed for film growth. Such a method was successfully applied in the cases of CuBTC [40], MIL-88B [41], and MOF-5 [42]. The authors claimed that growth species are present at low concentration in the solution, and that with time they bind to the surface and form a continuous film. The most important parameters to control here are the composition of the mother solution and the conditioning (heating) program. These determine the size and concentration of nuclei in solution, and the type of surface which is used. Most experiments were carried out using SAM functionalized gold. Thanks to the mild conditions, both patterned substrates and sensitive devices can be used.

In the case of MOF-5, the use of patterned COOH/CF_3-terminated SAM-modified substrates demonstrated the principle of patterned MOF film growth, namely that MOF-5 cannot grow on CF_3- but only on COOH- terminated SAMs, allowing the formation of a 500 nm thick non-oriented film made of 100 nm sized cubic crystallites after 24 h immersion [42]. Solvent exchange, drying, and loading of the film were carried out successfully in order to prove its porosity. In the case of CuBTC, a very dilute mother solution was used for the growth, thus extending the deposition time to 100 h to obtain a continuous film [40]. Different substrates, COOH-, OH-, and CH_3-terminated SAMs on gold, were tested. It was shown that a highly oriented growth can take place: in the [100] direction on COOH-terminated SAMs and the [111] direction on OH-terminated SAMs (Figure 13.8). The authors suggested that suitable growth species attach to the surface according to their affinity with the binding groups (OH or COOH), then grow into oriented crystals. However, polycrystalline film deposition also occurred at the CH_3-terminated SAM surface. The response of CuBTC films towards water vapor was investigated using a QCM and showed a water uptake which is consistent with bulk samples [8] under identical humidity conditions.

Similar growth conditions also proved successful for MIL-88B(Fe) film formation [41]. Here again, a highly oriented growth, although slow, takes place and affords a continuous layer. What makes MIL-88B interesting for sensor applications is its large breathing effect [43], which was shown to be retained by the film [44], as the cell volume can increase up to 40% during water uptake without delamination of the film.

In some cases, the MOF structure can be obtained at room temperature without the need for a heating step. However, the crystallization may then be too fast to permit homogeneous film growth. Repeated immersion steps in freshly prepared mother solutions allow the attachment of seeds [13, 45]. The substrates can be SAM-functionalized gold in addition to silicon or glass. In a typical experiment, stock solutions of linker and metal are prepared. MOF formation starts as soon as some quantities of the two solutions are mixed, in the presence of a substrate. Typically, 30 min to a few hours are necessary for completion of the reaction. The substrate is then washed, dried, and immersed in a fresh mixture again. In the case of $[Cu_2(pzdc)_2(pyz)]_n$ (CPL-1; pzdc = pyrazine-2,3-dicarboxylate, pyz = pyrazine), the MOF crystals tend to form plate-like crystals in which the b plane is much larger than others. As a consequence, after five cycles a continuous layer of crystals forms a

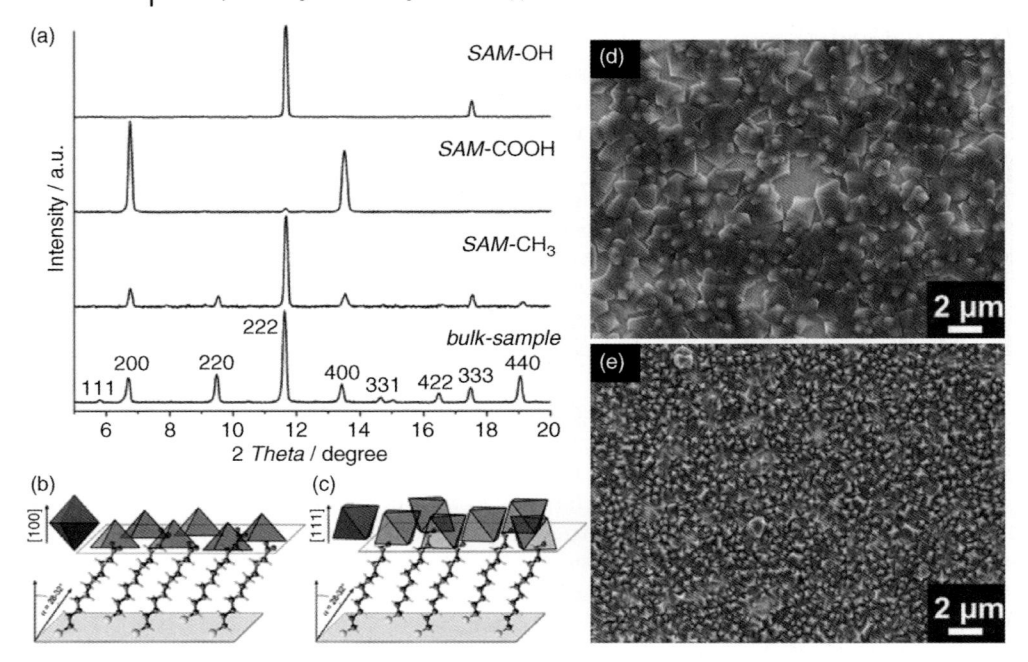

Figure 13.8 (a) X-ray diffraction patterns (background corrected) of thin films of CuBTC on functionalized gold surfaces, compared with a randomly oriented CuBTC powder sample measurement. Each pattern is normalized to the most intensive reflection. Schematic illustrations of oriented growth of CuBTC nanocrystals controlled via surface functionalization: (b) on an 11-mercaptoundecanoic acid SAM, and (c) on 11-mercaptoundecanol-modified gold surfaces. (d, e) Scanning electron micrographs of CuBTC on (a) 1-mercaptoundecan SAM and (b) mercaptohexadecanoic acid after an immersion time of 112 h. Figure adapted from [40]. Copyright 2007 American Chemical Society.

film with preferential b orientation [45]. Lu and Hupp studied the potential of such a growth method [13]. They chose ZIF-8 as a model system and controlled the thickness of the film by the number of deposition cycles. No preferred orientation was found, but very precise control of film thickness was achieved as the film thickened by 100 nm every cycle. The ZIF-8 layer is colorless, but behaves as a Fabry–Pérot interferometer and therefore, depending on the thickness of the film, the underlying silicon substrate appears variously colored (see Figure 13.9). Another solution to the fast crystallization problem was developed by Bein and co-workers [46]. It relies on a polymer gel layer which is loaded with the metal ion and laid on a surface (in the case studied, a gold substrate functionalized with a SAM), on top of which a linker solution slowly diffuses. The SAM then provides a nucleation interface and allows the formation of a continuous and sometimes oriented film. The sizes of the crystals and also the thickness of the layer are controlled by the length of the polymer chains and the concentration of metal ions in the gel.

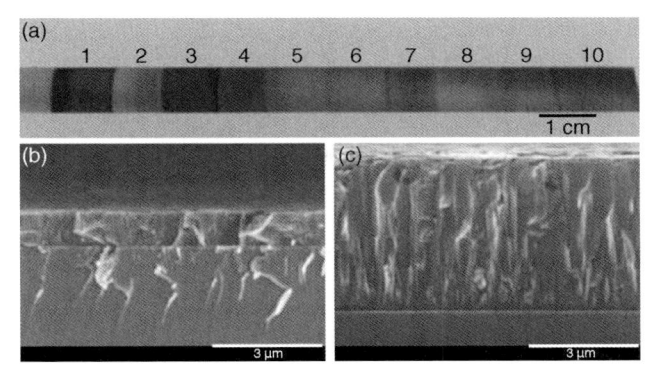

Figure 13.9 (a) Photograph of a series of ZIF-8 films of various thicknesses grown on a silicon substrate. Cross-sectional scanning electron microscopy images of ZIF-8 films grown on silicon substrates with (b) 10 and (c) 40 cycles. Adapted from [13]. Copyright 2010 American Chemical Society.

It appears from this discussion that the use of appropriately prepared mother solutions for film growth at room temperature can be a very powerful method. If sufficient time is allowed, continuous films are achieved, and in some cases they are oriented. Functionalized substrates often help in directing the orientation. Thickness can be controlled via the immersion time or number of immersion cycles, as illustrated in the ZIF-8 example. However, the fabrication of such a film requires patience, a lot of time (up to 2–3 weeks), and a very good knowledge of the MOF system concerned in order to find the optimum conditions (heating procedure, solvents, additives, etc.)

13.5.2
Assembly of Preformed MOF (Nano-) Particles or Layers

Films can also be prepared by the auto-assembly of preformed objects, typically nanocrystals and colloids. The still sparse reports on the synthesis of well-defined MOF nanocrystals have recently e been reviewed [47] and we discuss two examples of thin-film deposition using MOF colloids [10, 48]. In such a procedure, the first step consists in obtaining stable colloidal solutions. For example, careful choice of the metal precursor, temperature and time of the reaction, and dilution condition leads to a sol containing MIL-89 particles 20–40 nm in size which slowly grow into a gel [48], instead of a powder. In a second step, the sol was used to process a film by dip-coating on a bare silicon wafer (with a native SiO_2 coating). Control of the thickness was achieved through repetitive depositions followed by washing and drying; one layer has a thickness of about 40 nm. The deposited layers were of good quality and the film growth could be monitored by ellipsometry. More specifically, environmental ellipsometry measurements [49] were performed and the thickness of the film was monitored as a function of the water vapor pressure. It should be noted that a reversible swelling behavior comparable to that of the bulk material was observed. Similar work resulted in a MIL-101(Cr) film [10]. MIL-101(Cr) nanoparticles 22 nm in

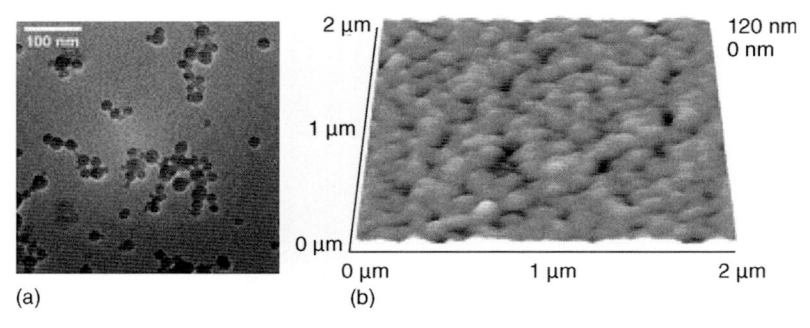

Figure 13.10 (a) Transmission electron microscopy and (b) atomic force microscopy images of nanoparticles and a thin film of MIL-101(Cr). Reproduced from [10] by permission of The Royal Society of Chemistry.

size were first synthesized and characterized (see Figure 13.10a), then redispersed in ethanol. The film was processed the same manner as described above: several cycles of dip-coating followed by washing and drying. Interestingly, the film was of low roughness (12 nm r.m.s. for a 97 nm thick film) and able to adsorb ethanol and 2-propanol reversibly.

Clearly, the main advantage of this route is the easy control over morphology and film thickness on the scale of the size of the primary MOF nanoparticles. The properties of the parent MOF bulk materials are nicely transferred to the films. In addition, the packing of the particles gives rise to intergrain boundaries, and thus mesoscopic porosity which allows easier diffusion of analytes in the film. The optical quality and low roughness of the films are also very interesting properties. The numerous advantages are counterbalanced, however, by the fact that stable colloids with appropriate particle size (ideally 20–50 nm) are needed. In addition, they should assemble in a homogeneous and dense fashion. These two conditions may require careful adjustment of the synthesis conditions, including solvent, and/or the addition of monofunctional linkers which can act as a stabilizers [50]. Nevertheless, such nanoMOFs have already been described for several structures [47], and more can be expected. No orientation has been obtained, but the use of anisotropic colloids instead of round-shaped particles may permit some progress in this direction [51]. More studies on morphology control of MOF crystallite growth in solution are needed.

An original approach was developed recently by Makiura *et al.* [52]. They grew a MOF by the successive deposition of ultra-thin two-dimensional (2D) sheets made by Langmuir–Blodgett techniques. The 2D sheets consist of square-planar cobalt-containing porphyrin tetra-acid building units linked together by copper ions. Pyridine groups are linked to the cobalt ions in the apical positions. One after another, sheets are prepared in a Langmuir–Blodgett apparatus and transferred on to the substrate. The layers interact by π–π stacking between the pyridine groups, forming an interdigitated three-dimensional (3D) structure with perfect crystalline orientation. Control of porosity may be possible by variation of the pyridine spacer ligands. The structure obtained is called NAFS-1 and, interestingly, it has not been synthesized as bulk material but only as a crystalline film. The method permits fine control of thickness by the number of deposition cycles. Of course, the method is limited to layered-based MOFs and to the precise assembly of one 2D layer at the liquid/solid interface before transfer to the substrate. It seems that the weak interactions between the transferred 2D sheets, rather than strong covalent or coordinative bonds, are preferable for this technique. Thus, the absence of stronger bindings between the layers may affect the stability of the structure. Moreover, the stability of the interface between such a film (made either of colloids or layers) and a substrate may be questioned. It is reasonable to expect some delamination when the thickness increases as no strong bonding between the MOF layer and the substrate is created.

13.5.3
Electrochemical Deposition

An electrochemical route to synthesize MOFs efficiently was initially developed at BASF for large-scale bulk MOF production. It consists in anodic dissolution of a metal electrode in a solution containing the appropriate organic linker and a conduction salt. To the best of our knowledge, only one example of electrochemical growth of a MOF thin film, $[Cu_3(btc)_2]$ (CuBTC), has been reported [53]. It was found that by varying the solvent (here a mixture of water and ethanol) and the voltage applied to the copper anode (2–25 V), densely packed films of thickness ranging from 2 to 50 μm can be obtained on top of the copper anode. The crystals are remarkably monodisperse, and the overall roughness of the film is related to the height of the crystals: 4–5 μm for a 20 μm thick film, less than 1 μm for a 2 μm thick film. It was suggested that the metal ions are released only in the vicinity of the anode, where they can react with the linker, and that the voltage controls the concentration of metal ions. Thus, a higher voltage induces a higher concentration of metal ions near the surface, leading to more nucleation events and smaller crystals compared with lower voltages.

Here, close control over the film morphology and thickness is easy. Another notable advantage is the very short time required for growth, typically less than 1 h, compared with several days for the methods described above. Furthermore, the absence of heating and aggressive chemical solutions allows deposition on integrated devices and also on patterned substrates with very good selectivity (see Figure 13.11). However, this method does not allow control over the orientation of the crystals, and so far has not been applied to other systems. Also, it is limited by the availability of suitable metals for anodic oxidation within a potential window not affecting the organic components of the MOF.

13.5.4
Liquid-Phase Epitaxy

Although well known for polymer systems, and especially polyelectrolytes [54], the layer by layer deposition technique has been applied to typical MOFs only recently [29]. Before this work, structurally defined coordination polymers were grown in a layer by layer fashion in 1993 [55, 56] using the Hofmann clathrate [Ni(bipy)Pt(CN)$_4$] (bipy = 4,4'-bipyridine). More recently, Bousseksou and co-workers used a similar

Figure 13.11 Electrochemical growth of [Cu$_3$(btc)$_2$] on patterned copper. (a) Copper tracks covered with electrochemically grown MOF. (b) Detail of a track edge in (a) showing the growth of MOF crystals strictly limited to the underlying pattern. Reproduced from [53]. Copyright 2009 American Chemical Society.

method to grow the microporous coordination polymer [Fe(pz)Pt(CN)$_4$] (pz = pyrazine) [57, 58]. The chemical identity of the film was proved by Raman spectroscopy. Magnetic susceptibility measurements were also carried out and showed that these films represent the first example of a spin-crossover phenomenon with hysteresis at room temperature in such unmixed, multilayer thin films. The deposition should be carried out at −60 °C and quickly, in order to avoid desorption of adsorbed species; furthermore, no X-ray diffraction data of the film were reported.

In typical layer by layer depositions of MOFs, the metal precursor and the organic linker are kept in separates beakers, and the substrate is immersed in each of them stepwise, after intermediate washing steps. During immersion, adsorption of the components at the surface of the growing film takes place and the MOF structure develops in the course of stepwise ligand-exchange reactions (see Figure 13.12).

The "natural" substrates for such deposition are thiol-based SAMs on gold (and also silane-based SAMs on silica and related oxides), for two reasons. First, the surface is then covered with functional groups which are able to bind metal ions (COOH, OH, and pyridine groups, for example), and second, the quality of the

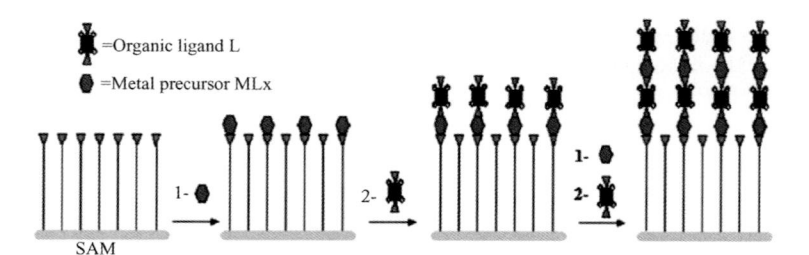

Figure 13.12 Schematic diagram for the step by step growth of the MOFs on the SAM, by repeated immersion cycles, first in solution of metal precursor and subsequently in a solution of organic ligand. Here, for simplicity, the scheme simplifies the assumed structural complexity of the carboxylic acid coordination modes. Reproduced from [29]. Copyright 2007 American Chemical Society.

surface is very high (i.e., low roughness down to the atomic level, good 2D crystallinity, long-range order and single-crystalline domains 10–20 nm in size [59]). Nevertheless, recently some layer by layer depositions have been performed on bare oxide substrates that allowed the deposition of an oriented and continuous, although very rough, [$Cu_2(ndc)_2(dabco)$] thin film [60]. In the latter case, however, the growth mechanism is different from that encountered on a SAM-modified substrate. In the following we focus on films grown on SAM on gold.

As shown by *in situ* surface plasmon resonance (SPR) studies [61], the mechanism for growth is strictly self-terminated, giving rise to a linear function between the number of deposition cycles and the deposited MOF material (i.e., every cycle results in the same increase in thickness). This leads to exceptionally smooth films in the best cases, similar to the above-mentioned Langmuir–Blodgett method of 3D assembly of preformed 2D sheets. For example, a surface roughness of 5–6 nm over areas up to 100 µm^2 was measured by atomic force microscopy for CuBTC films, corresponding to step heights of two unit cells [27] (Figure 13.13).

Such a self-terminated mechanism obviously favors the growth of oriented layers. By varying the terminating group of the SAM (COOH or OH), highly crystalline films of CuBTC grown in two different directions ([100] and [111], respectively) were found after 40 deposition cycles [61]. The orientation control here is induced by the SAM. Indeed, the [100] lattice plane contains Cu_2 dimeric units which can interact with the COOH surface groups, whereas the [111] lattice plane contains the apical position of Cu^{2+} that is usually occupied by a solvent molecule and is very likely to

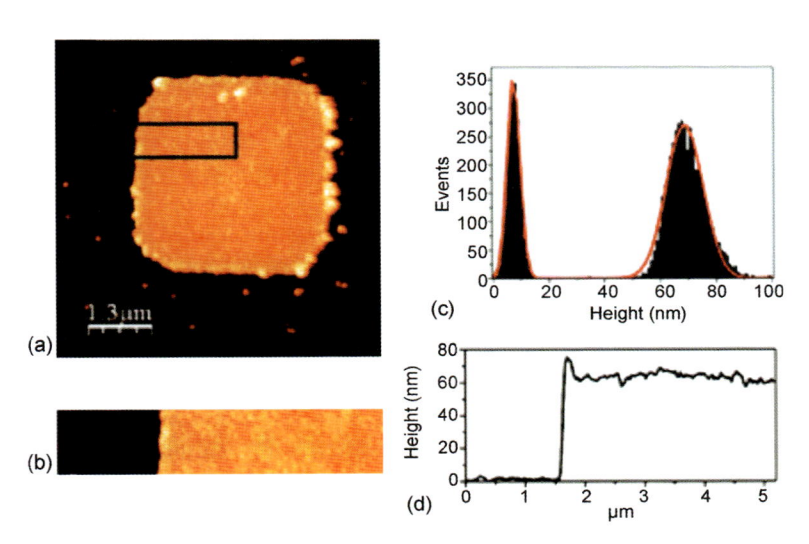

Figure 13.13 Two different ways of measuring thickness (averaged profiles and histograms). (a) Topographic image (6.5 × 6.5 mm) and (b) a selected area for accurate thickness estimation. (c) Corresponding height histogram and (d) averaged profile calculated over the whole area in (b). Red lines in the histogram are the corresponding Gaussian fits. Reproduced from [27] by permission of the PCCP Owner Societies.

interact with OH groups. Such a consideration can be used not only to explain the observed orientation, but also to predict the orientation likely to be obtained. The liquid-phase epitaxy (LPE) growth of [$Cu_2(ndc)_2(dabco)$] on SAM-modified gold substrates is a good example of orientation prediction and complying observation [62]. Deposition can also be carried out on patterned substrates because no binding and therefore no growth can occur on a CH_3-terminated SAM [42]. In addition, depositions are commonly carried out at 25 °C, sometimes up to 50 °C. The last two observations make LPE suitable for sensitive or integrated devices. The main drawbacks of the method are the length of the deposition procedure (up to a few days) and the lack of proof of wide applicability across the diversity of MOF structures. Up to now, only CuBTC, [$Cu_2(ndc)_2(dabco)$], [$Zn(bdc)(4,4\text{-bipy})_{0.5}$] (4,4-bipy = 4,4-bipyridine) and closely related structures have been grown as crystalline, smooth, and perfectly oriented multilayer films by LPE. All these MOFs share the same secondary building unit: a binuclear copper or zinc paddle-wheel unit, which is already formed in the copper or zinc acetate ethanolic solutions used for deposition. To the best of our knowledge, no other MOF system has been grown in a crystalline fashion using the layer by layer method, except maybe the MOF-5 case described below. Note that the Hofmann clathrate films mentioned above may be regarded in this context as special cases of MOFs, lacking the full range of variability of the linker functionality, which is limited to the tailoring of the dinitrogen pillar ligands.

13.5.5
Toward Heteroepitaxial Growth of Multiple MOF Layers

MOFs offer unique versatility in terms of isostructural network modifications. The best example is the IRMOF series, built with functionalized or various sized linkers but sharing the same topology [63]. In the field of sensors we can imagine multilayer sensors where the outer layers would act as filters to select the right analyte and direct it to the inner layer where it would be detected; the work of Kreno *et al.* is a first step in this direction [14]. The outer MOF layer could also be a protective layer which enhances the chemical stability of an active film. So far some detailed studies have been carried out on free-standing core–shell MOF single crystals, synthesized under solvothermal conditions. Kitagawa and co-workers were the first group to report the epitaxial growth of a [$Cu_2(ndc)_2(dabco)$] shell around a [$Zn_2(ndc)_2(dabco)$] core [64]. Related work involved the growth of MOF-5@IRMOF-3@MOF-5 and IRMOF-3@MOF-5@IRMOF-3 Matryoshka crystals [65]. The same MOF-5–IRMOF-3 system was selected to grow the first example of a heteroepitaxial hybrid MOF film [66]. Solvothermal conditions were used with glass as substrate: a substrate bearing a seeding layer of MOF-5 was immersed in an IRMOF-3 mother solution. These studies suggest the possibility of growing heterostructured MOF multilayers under mild conditions such as those described above. Deposition methods which rely on sequential immersion of the substrate in a solution are particularly well suited for such a purpose as one of the building blocks (metal, linker) can easily be replaced at some point in the synthesis. Figure 13.14 illustrates a proof of principle of this concept using LPE [62]: a perfectly oriented and highly crystalline film was obtained

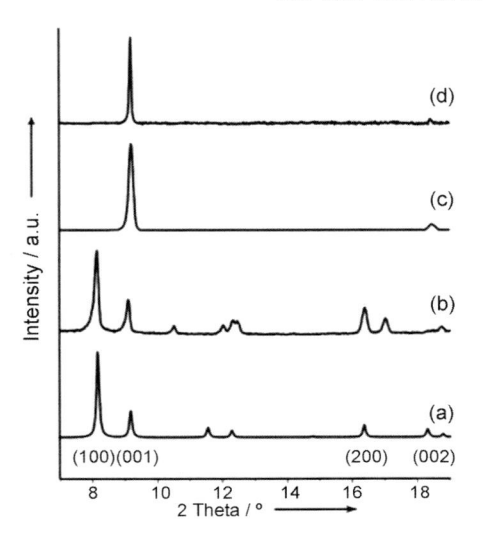

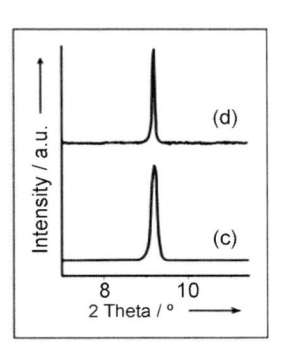

Figure 13.14 X-ray diffraction patterns (background corrected) of a 30 cycle [Zn(BME-bdc)(dabco)$_{0.5}$] deposited on top of a 30 cycle [Cu(ndc)(dabco)$_{0.5}$] MOF in [00l] orientation on pyridine-terminated SAM (d), compared with the pattern of the 30 cycle [Cu(ndc)(dabco)$_{0.5}$] MOF alone. Calculated patterns for bulk [Cu(ndc)(dabco)$_{0.5}$] (a) and experimental data of bulk [Zn(BME-bdc)(dabco)$_{0.5}$] (b). Adapted from [62] with permission. Copyright Wiley-VCH Verlag GmbH & Co. KGaA.

on growing 30 cycles of a Zn-based MOF on top of a similar Cu-based MOF. Moreover, the thinning of the [001] diffraction peak line upon addition of cycles [see curves (c) and (d)] indicates an increase in the crystallite size, and quantitative analysis of the development of the line profile as a function of the number of cycles confirmed a linear growth mode also for this heterostructured film.

13.5.6
Growth of MOF Films in Confined Spaces

Ultimately, technological sensor or analytical device fabrication may require the generation of uniform MOF layers in challenging geometries, such as capsules, capillaries, or cavities in general. In this respect, layer by layer or at least cyclic coating protocols may be beneficial because they are largely independent of concentration variations provided that sufficient reactants are offered to the growing MOF layer in each cycling step to saturate the surface with either the metal or the linker component. A demonstration of this fact was given by Mertens and co-workers, who coated a narrow fused-silica gas chromatographic capillary (length 10 m, clear diameter 0.53 mm) first with a SAM prepared from HOOC(CH$_2$)$_9$SiCl$_3$ and then grew MOF-5 on it by alternately offering basic zinc acetate in DMF (0.01 M) as precursor and terephthalic acid in DMF (0.03 M) as linker solution [67]. These components were pumped alternately through the capillaries at room temperature via an automated apparatus (see Figure 13.15a) interrupted by the removal of excess

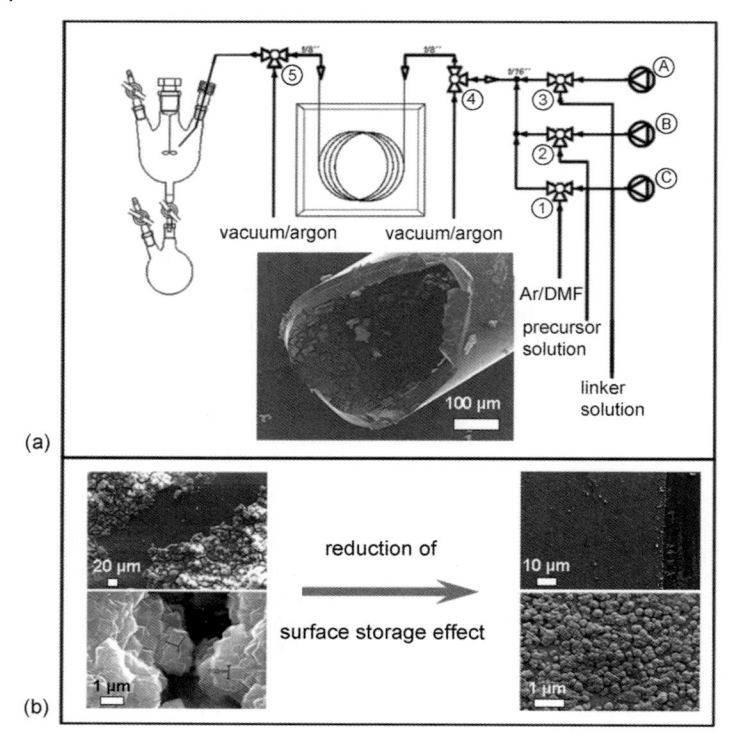

Figure 13.15 (a) Setup for the automated preparation of MOF-coated fused-silica gas chromatographic capillaries. The computer-controlled syringe pumps for the application of the precursor, the linker, and the rinsing solution are denoted by the letters A, B, C. The three-way (1–5) valves allow the refilling of the syringes, the flushing of the capillaries with argon, and the evacuation of the whole apparatus.
(b) Demonstration of the surface storage effect: MOF-5@SAM@SiO$_2$ produced by the controlled SBU approach was differently desolvated between the cyclic preparation steps, left by dropwise running 0.125 ml of DMF across the sample surface over a period of 3 s, and right by dropwise running 0.45 ml across the sample surface over a period of 38 s. Based on [68] and F.O.R.L. Mertens, personal communication.

solvent by a stream of argon, instead of rinsing with the standard solvents DEF, DMF, or dichloromethane. This coating procedure is the cyclic version of the recently developed controlled secondary building unit (SBU) approach for IRMOFs [67]. Although the procedure did not lead to a strict layer by layer growth mode, the grown MOF material uniformly covered the capillary across its entire length. The 80 cycles applied led to a coating thickness of ~1 μm, which is roughly 10 times larger than what would be expected from a layer by layer growth mode. This number is at least partly explained by the surface storage effect, that is, by the insufficient desolvation of the already formed MOF material, by which an additional amount of reactants compared with that which is already bound to the MOF layer is carried over from one reaction cycle step to the next. Figure 13.15b demonstrates that the type and the

Table 13.2 Performance of films grown by diverse methods with respect to sensing criteria.

Criteria/methods	Uniformity/density of the layers	Suitable pore orientation	Bonding strength	Response
Mother solution	Low	Good	Good	Good
Assembly of colloids/layers	Good	Low/good	?	Good
Electrochemical	Good	Low	Good	Good
LPE	Good	Good	Good	Good

intensity of the rinsing procedure between the different cycle steps can be a crucial factor determining the morphology of layers to be grown. The growth of high-quality films in confined spaces may therefore depend very strongly on the accessibility of the MOF coating to desolvation agents.

13.5.7
Comparison of the Different Methods for MOF Thin Film Growth

Many methods are available for growing MOF films; however, none of them is general. Each MOF system which is desirable as a film should first be well understood and a growth method adapted to its special case. As discussed earlier in this chapter, the most interesting films for sensors applications are those which allow layers of homogeneously sized crystals, with suitable pore orientation or easy diffusion, ideally strong interaction with the substrate, and, most importantly, which feature a noticeable response which can be detected. The properties of the films described previously are summarized in Table 13.2. Layer by layer deposition in the LPE fashion and Langmuir–Blodgett assembly of preformed MOF 2D single layers lead to highly oriented crystalline films with the potential also to grow heteroepitaxial structures. The ultimate control over thickness and surface roughness is possible only with LPE and Langmuir–Blodgett techniques.

13.6
Conclusions and Perspectives

The use of MOFs as active components in sensing devices is just emerging, but developing quickly. Proof of principle using various transduction methods has been established, often using the MOF layer as a small-molecule vapor sensor. Higher sensitivities are expected in the detection of large organic molecules, such as polyaromatics or explosives, due to the high heats of adsorption predicted for these molecules. Direct synthesis of MOF films on suitable substrates (e.g., patterned substrates and MEMS devices such as microcantilevers) is still a challenge, even though a number of low-temperature, "soft" methods have been developed. At present, the most promising methods are layer by layer deposition techniques.

Further progress in this area, particularly regarding the number of MOF structures accessible for film growth, will accelerate the development of MOF-based sensing technologies.

References

1 Adhikari, B. and Majumdar, S. (2004) Polymers in sensor applications. *Prog. Polym. Sci.*, **29** (7), 699–766.
2 Férey, G. and Serre, C. (2009) Large breathing effects in three-dimensional porous hybrid matter: facts, analyses, rules and consequences. *Chem. Soc. Rev.*, **38** (5), 1380–1399.
3 Fletcher, A.J., Thomas, K.M., and Rosseinsky, M.J. (2005) Flexibility in metal–organic framework materials: impact on sorption properties. *J. Solid State Chem.*, **178** (8), 2491–2510.
4 Kitagawa, S., Kitaura, R., and Noro, S. (2004) Functional porous coordination polymers. *Angew. Chem. Int. Ed.*, **43** (18), 2334–2375.
5 Chui, S.S.-Y., Lo, S.M.-F., Charmant, J.P.H., Orpen, A.G., and Williams, I.D. (1999) A chemically functionalizable nanoporous material [$Cu_3(TMA)_2(H_2O)_3$]$_n$. *Science*, **283** (5405), 1148–1150.
6 Uemura, K., Yamasaki, Y., Komagawa, Y., Tanaka, K., and Kita, H. (2007) Two-step adsorption/desorption on a jungle-gym-type porous coordination polymer. *Angew. Chem. Int. Ed. Engl.*, **46** (35), 6662–6665.
7 Zacher, D., Shekhah, O., Wöll, C., and Fischer, R.A. (2009) Thin films of metal–organic frameworks. *Chem. Soc. Rev.*, **38** (5), 1418–1429.
8 Biemmi, E., Darga, A., Stock, N., and Bein, T. (2008) Direct growth of $Cu_3(BTC)_2(H_2O)_3 \cdot xH_2O$ thin films on modified QCM-gold electrodes – water sorption isotherms. *Micropor. Mesopor. Mat.*, **114** (1–3), 380–386.
9 Allendorf, M.D., Houk, R.J.T., Andruszkiewicz, L., Talin, A.A., Pikarsky, J., Choudhury, A., Gall, K.A., and Hesketh, P.J. (2008) Stress-induced chemical detection using flexible meta–organic frameworks. *J. Am. Chem. Soc.*, **130** (44), 14404–14405.
10 Demessence, A., Horcajada, P., Serre, C., Boissière, C., Grosso, D., Sanchez, C., and Férey, G. (2009) Elaboration and properties of hierarchically structured optical thin films of MIL-101(Cr). *Chem. Commun.*, **101** (46), 7149–7151.
11 Achmann, S., Hagen, G., Kita, J., Malkowsky, I.M., Kiener, C., and Moos, R. (2009) Metal–organic frameworks for sensing applications in the gas phase. *Sensors*, **9** (3), 1574–1589.
12 Zou, X., Zhu, G., Hewitt, I.J., Sun, F., and Qiu, S. (2009) Synthesis of a metal–organic framework film by direct conversion technique for VOCs sensing. *Dalton Trans.*, (16), 3009–3013.
13 Lu, G. and Hupp, J.T. (2010) Metal–organic frameworks as sensors: a ZIF-8 based Fabry–Pérot device as a selective sensor for chemical vapors and gases. *J. Am. Chem. Soc.*, **132** (23), 7832–7833.
14 Kreno, L.E., Hupp, J.T., and Van Duyne, R.P. (2010) Metal–organic framework thin film for enhanced localized surface plasmon resonance gas sensing. *Anal. Chem.*, **82** (19), 8042–8046.
15 Grate, J.W. (2000) Acoustic wave microsensor arrays for vapor sensing. *Chem. Rev.*, **100** (7), 2627–2648.
16 Goeders, K.M., Colton, J.S., and Bottomley, L.A. (2008) Microcantilevers: sensing chemical interactions via mechanical motion. *Chem. Rev.*, **108** (2), 522–542.
17 Lee, J.-H., Houk, R.T.J., Robinson, A., Greathouse, J.A., Thornberg, S.M., Allendorf, M.D., and Hesketh, P.J. (2010) Investigation of microcantilever array with ordered nanoporous coatings for selective chemical detection. *Proc. SPIE*, **7679**, 7679–27.

18 Murray, L.J., Dincă, M., and Long, J.R. (2009) Hydrogen storage in metal–organic frameworks. *Chem. Soc. Rev.*, **38** (5), 1294–1314.

19 Furukawa, H., Ko, N., Go, Y.B., Aratani, N., Choi, S.B., Choi, E., Yazaydin, A.Ö., Snurr, R.Q., O'Keeffe, M., Kim, J., and Yaghi, O.M. (2010) Ultra-high porosity in metal–organic frameworks. *Science*, **329** (5990), 424–428.

20 Farrusseng, D., Daniel, C., Gaudillère, C., Ravon, U., Schuurman, Y., Mirodatos, C., Dubbeldam, D., Frost, H., and Snurr, R.Q. (2009) Heats of adsorption for seven gases in three metal–organic frameworks: systematic comparison of experiment and simulation. *Langmuir*, **25**, 7383–7388.

21 Greathouse, J.A., Ockwig, N.W., Criscenti, L.J., Guilinger, T.R., Pohl, P., and Allendorf, M.D. (2010) Computational screening of metal–organic frameworks for large-molecule chemical sensing. *Phys. Chem. Chem. Phys.*, **12** (39), 12621–12629.

22 Somasundaran, P. (ed.) (2006) *Encyclopedia of Surface and Colloid Science*, Taylor & Francis, Boca Raton, FL.

23 Zhuravlev, L.T. (1987) Concentration of hydroxyl groups on the surface of amorphous silicas. *Langmuir*, **3**, 316–318.

24 Hermes, S., Zacher, D., Baunemann, A., Wöll, C., and Fischer, R.A. (2007) Selective growth and MOCVD loading of small single crystals of MOF-5 at alumina and silica surfaces modified with organic self-assembled monolayers. *Chem. Mater.*, **19**, 2168–2173.

25 Zacher, D., Baunemann, A., Hermes, S., and Fischer, R.A. (2007) Deposition of microcrystalline [$Cu_3(btc)_2$] and [$Zn_2(bdc)_2(dabco)$] at alumina and silica surfaces modified with patterned self assembled organic monolayers: evidence of surface selective and oriented growth. *J. Mater. Chem.*, **17** (27), 2785–2792.

26 Munuera, C., Shekhah, O., Wang, H., Wöll, C., and Ocal, C. (2008) The controlled growth of oriented metal–organic frameworks on functionalized surfaces as followed by scanning force microscopy. *Phys. Chem. Chem. Phys.*, **10** (48), 7257–7261.

27 Shekhah, O. (2010) Layer-by-layer method for the synthesis and growth of surface mounted metal–organic frameworks (SURMOFs). *Materials*, **3** (2), 1302–1315.

28 Shekhah, O., Wang, H., Kowarik, S., Schreiber, F., Paulus, M., Tolan, M., Sternemann, C., Evers, F., Zacher, D., Fischer, R.A., and Wöll, C. (2007) Step-by-step route for the synthesis of metal–organic frameworks. *J. Am. Chem. Soc.*, **129** (49), 15118–15119.

29 Li, J.-R., Kuppler, R.J., and Zhou, H.-C. (2009) Selective gas adsorption and separation in metal–organic frameworks. *Chem. Soc. Rev.*, **38**, 1477–1504.

30 Skoulidas, A.I. and Sholl, D.S. (2005) Self-diffusion and transport diffusion of light gases in metal–organic framework materials assessed using molecular dynamics simulations. *J. Phys. Chem. B*, **109**, 15760–15768.

31 Mueller, U., Schubert, M., Teich, F., Puetter, H., Schierle-Arndt, K., and Pastre, J. (2006) Metal–organic frameworks – prospective industrial applications. *J. Mater. Chem.*, **16** (7), 626–636.

32 Stallmach, F., Groger, S., Kunzel, V., Karger, J., Yaghi, O.M., Hesse, M., and Muller, U. (2006) NMR studies on the diffusion of hydrocarbons on the metal–organic framework material MOF-5. *Angew. Chem. Int. Ed.*, **45** (13), 2123–2126.

33 Wehring, M., Gascon, J., Dubbeldam, D., Kapteijn, F., Snurr, R.Q., and Stallmach, F. (2010) Self-diffusion studies in CuBTC by PFG NMR and MD simulations. *J. Phys. Chem. C*, **114** (23), 10527–10534.

34 Chae, H.K., Siberio-Pérez, D.Y., Kim, J., Go, Y., Eddaoudi, M., Matzger, A.J., O'Keeffe, M., and Yaghi, O.M. (2004) A route to high surface area, porosity and inclusion of large molecules in crystals. *Nature*, **427**, 523–527.

35 Bahr, D.F., Reid, J.A., Mook, W.M., Bauer, C.A., Stumpf, R., Skulan, A.J., Moody, N.R., Simmons, B.A., Shindel, M.M., and Allendorf, M.D. (2008) Mechanical properties of cubic zinc carboxylate IRMOF-1 metal–organic framework crystals. *Phys. Rev. B*, **76** (18), 184106.

36 Bhushan, B. and Li, X. (1997) Micromechanical and tribological characterization of doped single-crystal silicon and polysilicon films for microelectromechanical systems devices. *J. Mater. Res.*, **12** (1), 54–63.

37 Munro, G. (2002) *Elastic Moduli Data for Polycrystalline Ceramics*, National Institute of Standards and Technology, Gaithersburg, MD.

38 Guo, H., Zhu, G., Hewitt, I.J., and Qiu, S. (2009) "Twin copper source" growth of metal–organic framework membrane: $Cu_3(BTC)_2$ with high permeability and selectivity for recycling H_2. *J. Am. Chem. Soc.*, **131** (5), 1646–1647.

39 Biemmi, E., Scherb, C., and Bein, T. (2007) Oriented growth of the metal–organic framework $Cu_3(BTC)_2(H_2O)_3 \cdot xH_2O$ tunable with functionalized self-assembled monolayers. *J. Am. Chem. Soc.*, **129** (26), 8054–8055.

40 Scherb, C., Schödel, A., and Bein, T. (2008) Directing the structure of metal–organic frameworks by oriented surface growth on an organic monolayer. *Angew. Chem. Int. Ed.*, **47** (31), 5777–5779.

41 Hermes, S., Schröder, F., Chelmowski, R., Wöll, C., and Fischer, R.A. (2005) Selective nucleation and growth of metal–organic open framework thin films on patterned $COOH/CF_3$-terminated self-assembled monolayers on Au(111). *J. Am. Chem. Soc.*, **127** (40), 13744–13745.

42 Serre, C., Mellot-Draznieks, C., Surblé, S., Audebrand, N., Filinchuk, Y., and Férey, G. (2010) Role of solvent–host interactions that lead to very large swelling of hybrid frameworks. *Science*, **315** (5820), 1828–1831.

43 Scherb, C., Koehn, R., and Bein, T. (2010) Sorption behavior of an oriented surface-grown MOF-film studied by *in situ* X-ray diffraction. *J. Mater. Chem.*, **20** (15), 3046–3051.

44 Kubo, M., Chaikittisilp, W., and Okubo, T. (2008) Oriented films of porous coordination polymer prepared by repeated *in situ* crystallization. *Chem. Mater.*, **20** (9), 2887–2889.

45 Schoedel, A., Scherb, C., and Bein, T. (2010) Oriented nanoscale films of metal–organic frameworks by room-temperature gel-layer synthesis. *Angew. Chem. Int. Ed.*, **49** (40), 7225–7228.

46 Spokoyny, A.M., Kim, D., Sumrein, A., and Mirkin, C.A. (2009) Infinite coordination polymer nano- and microparticle structures. *Chem. Soc. Rev.*, **38** (5), 1218–1227.

47 Horcajada, P., Serre, C., Grosso, D., Boissière, C., Perruchas, S., Sanchez, C., and Férey, G. (2009) Colloidal route for preparing optical thin films of nanoporous metal–organic frameworks. *Adv. Mater.*, **21** (19), 1931–1935.

48 Boissiere, C., Grosso, D., Lepoutre, S., Nicole, L., Bruneau, A.B., and Sanchez, C. (2005) Porosity and mechanical properties of mesoporous thin films assessed by environmental ellipsometric porosimetry. *Langmuir*, **21** (26), 12362–12371.

49 Hermes, S., Witte, T., Hikov, T., Zacher, D., Bahnmüller, S., Langstein, G., Huber, K., and Fischer, R.A. (2007) Trapping metal–organic framework nanocrystals: an *in-situ* time-resolved light scattering study on the crystal growth of MOF-5 in solution. *J. Am. Chem. Soc.*, **129** (17), 5324–5325.

50 Tsuruoka, T., Furukawa, S., Takashima, Y., Yoshida, K., Isoda, S., and Kitagawa, S. (2009) Nanoporous nanorods fabricated by coordination modulation and oriented attachment growth. *Angew. Chem. Int. Ed.*, **48** (26), 4739–4743.

51 Makiura, R., Motoyama, S., Umemura, Y., Yamanaka, H., Sakata, O., and Kitagawa, H. (2010) Surface nano-architecture of a metal–organic framework. *Nat. Mater.*, **9** (6), 1–7.

52 Ameloot, R., Stappers, L., Fransaer, J., Alaerts, L., Sels, B.F., and De Vos, D.E. (2009) Patterned growth of metal–organic framework coatings by electrochemical synthesis. *Chem. Mater.*, **21** (13), 2580–2582.

53 Quinn, J.F., Johnston, A.P.R., Such, G.K., Zelikin, A.N., and Caruso, F. (2007) Next generation, sequentially assembled ultrathin films: beyond electrostatics. *Chem. Soc. Rev.*, **36** (5), 707–718.

54 Bell, C.M., Arendt, M.F., Gomez, L., Schmehl, R.H., and Mallouk, T.E. (1994) Growth of lamellar Hofmann clathrate films by sequential ligand exchange reactions: assembling a coordination solid one layer at a time. *J. Am. Chem. Soc.*, **116** (18), 8374–8375.

55 Yang, H.C., Aoki, K., Hong, H.G., Sackett, D.D., Arendt, M.F., Yau, S.L., Bell, C.M., and Mallouk, T.E. (1993) Growth and characterization of metal(II) alkanebisphosphonate multilayer thin films on gold surfaces. *J. Am. Chem. Soc.*, **115** (25), 11855–11862.

56 Cobo, S., Molnár, G., Real, J.A., and Bousseksou, A. (2006) Multilayer sequential assembly of thin films that display room-temperature spin crossover with hysteresis. *Angew. Chem. Int. Ed.*, **45** (35), 5786–5789.

57 Molnár, G., Cobo, S., Real, J.A., Carcenac, F., Daran, E., Vieu, C., and Bousseksou, A. (2007) A combined top-down/bottom-up approach for the nanoscale patterning of spin-crossover coordination polymers. *Adv. Mater.*, **19** (16), 2163–2167.

58 Hobara, D., Sasaki, T., Imabayashi, S.-i., and Kakiuchi, T. (1999) Surface structure of binary self-assembled monolayers formed by electrochemical selective replacement of adsorbed thiols. *Langmuir*, **15** (15), 5073—2078.

59 Yusenko, K., Meilikhov, M., Zacher, D., Wieland, F., Sternemann, C., Stammer, X., Ladnorg, T., Wöll, C., and Fischer, R.A. (2010) Step-by-step growth of highly oriented and continuous seeding layers of $[Cu_2(ndc)_2(dabco)]$ on bare oxide and nitride substrates. *CrystEngComm*, **12** (7), 2086–2090.

60 Shekhah, O., Wang, H., Zacher, D., Fischer, R.A., and Wöll, C. (2009) Growth mechanism of metal–organic frameworks: insights into the nucleation by employing a step-by-step route. *Angew. Chem. Int. Ed.*, **48** (27), 5038–5041.

61 Zacher, D., Yusenko, K., Bétard, A., Henke, S., Meilikhov, M., Ladnorg, T., Shekhah, O., Wöll, C., Tertfort, A., and Fischer, R.A. (2011) Liquid phase epitaxy of multi-component layer-based metal–organic frameworks $[M(L)(P)_{0.5}]$: importance of deposition sequence on the oriented growth. *Chem. Eur. J.*, **17** (5), 1448–1455.

62 Eddaoudi, M., Kim, J., Rosi, N.L., Vodak, D., Wachter, J., O'Keeffe, M., and Yaghi, O.M. (2002) Systematic design of pore size and functionality in isoreticular MOFs and their application in methane storage. *Science*, **295** (5554), 469–472.

63 Furukawa, S., Hirai, K., Nakagawa, K., Takashima, Y., Matsuda, R., Tsuruoka, T., Kondo, M., Haruki, R., Tanaka, D., Sakamoto, H., Shimomura, S., Sakata, O., and Kitagawa, S. (2009) Heterogeneously hybridized porous coordination polymer crystals: fabrication of heterometallic core–shell single crystals with an in-plane rotational epitaxial relationship. *Angew. Chem. Int. Ed.*, **48** (10), 1766–1770.

64 Koh, K., Wong-Foy, A.G., and Matzger, A.J. (2009) MOF@MOF: microporous core–shell architectures. *Chem. Commun.* (41), 6162–6164.

65 Yoo, Y. and Jeong, H.-K. (2010) Heteroepitaxial growth of isoreticular metal–organic frameworks and their hybrid films. *Cryst. Growth Des.*, **10** (3), 1283–1288.

66 Hausdorf, S., Baitalow, F., Böhle, T., Rafaja, D., and Mertens, F.O.R.L. (2010) Main-group and transition-element IRMOF homologues. *J. Am. Chem. Soc.*, **132** (32), 10978–10981.

67 Münch, A., Seidel, J., Obst, A., Weber, E., Mertens, F. O. R. L. (2011) High Separation Performance of Chromatographic Capillaries Coated via the Controlled SBU Approach for MOF-5. *Chem Eur. J.* in press.

Part Six
Large-Scale Synthesis and Shaping of MOFs

14
Industrial MOF Synthesis
Alexander Czaja, Emi Leung, Natalia Trukhan, and Ulrich Müller

14.1
Introduction

Several hundred different metal–organic frameworks (MOFs) have been identified. The self-assembly of metal ions, which act as coordination centers, linked together with a variety of polyatomic organic bridging ligands has resulted in tailored nanoporous host materials as robust solids with high thermal and mechanical stability. Interestingly, unlike other solid matter, such as zeolites, carbons, and oxides, a number of coordination compounds are also known to exhibit high framework flexibility and shrinkage/expansion due to interactions with guest molecules [1].

In 1965, roughly three decades before the commonly assumed discovery of MOFs, Tomic mentioned materials that would nowadays be called MOFs, metal–organic polymers or supramolecular structures [2]. Bi- and trivalent aromatic carboxylic acids were used to form frameworks with zinc, nickel, iron, aluminum, thorium, and uranium, and some interesting features of MOFs, such as high thermal stability and high metal content, were reported. Interest in the field was again kindled by Yaghi and co-workers, who published the structure of zinc-based MOF-5 in late 1999 [3], and the concept of reticular design, with totally different carboxylate linkers, in 2002 [4–6].

The most striking difference from state-of-the-art materials is probably the total lack of non-accessible bulk volume in MOF structures. Although high surface areas are already known from activated carbons and zeolites, it is the absence of any "dead volume" in MOFs which principally gives them, on a weight-specific basis, the highest porosities and surface areas. It was reported, for example, that in the case of MOF-177 the surface area reaches $5640 \, m^2 \, g^{-1}$ [7] and for MIL-101 up to $5900 \, m^2 \, g^{-1}$ [8]. Of course, properties such as the drastically increased velocity of molecular traffic through these open structures are also closely related to the regularity of pores of nanometer size [9].

The combination of so unprecedented levels of porosity, surface area, pore size, and wide chemical inorganic–organic composition has recently brought these materials to the attention of many researchers in both academia and industry, with over 1000 publications on "metal–organic frameworks" and "coordination polymers" per annum [1].

Metal-Organic Frameworks: Applications from Catalysis to Gas Storage, First Edition. Edited by David Farrusseng.
© 2011 Wiley-VCH Verlag GmbH & Co. KGaA. Published 2011 by Wiley-VCH Verlag GmbH & Co. KGaA.

In this chapter, we focus on the industrial synthesis of MOF materials and explain some considerations on the economics of MOF synthesis. Additionally, two selected "real-world" examples – MOFs used for storage of natural gas in a car and MOFs embedded in polymers as ethylene adsorbents to control or prevent ripening of fruits – are presented.

14.2
Raw Materials

Every synthesis starts with the selection of the starting materials. In this section, some considerations for the raw material selection for industrial MOF synthesis processes are discussed. Since MOFs are comprised of an inorganic and an organic part, the metal and the linker will be treated separately.

14.2.1
Metal Sources

From a small-scale point of view, selection of the metal source may be one of the less critical points for synthesis planning. The only factor under consideration is the influence of the anion on the reaction pathway. Less coordinating anions, such as tetrafluoroborate and hexafluorophosphate, may be better suited to yield the desired MOF structure than anions such as chloride.

On the large scale, the availability of the metal salt plays an important role. Generally, the metal oxides, nitrates, sulfates, and chlorides are readily available. More complex anions are too expensive to be considered for large-scale applications. However, nitrates pose an inherent safety hazard, since they are oxidizing anions and could trigger an explosion when combined with organic materials such as the MOF linkers. Chlorides, on the other hand, demand more from the reactor material in terms of corrosion resistance. This means higher investment costs since higher quality materials have to be used for the set-up. The oxides and sulfates remain the most preferable choices for scale-up. Since these metal sources generally have low solubility, new challenges for the synthesis process arise. Just from these few considerations it can be seen that even the choice of the metal source is a complex problem that has to be solved for each new scale-up project.

14.2.2
Linkers

Linker molecules are an essential part of every MOF. They range from simple molecules, such as terephthalic and formic acid, to very complex structures, for example, catenane linkers [10]. For industrial applications in general, the simplest solution to a given problem is the most preferred solution. For linker selection, this means that a bulk chemical such as terephthalic acid would always be the first choice. In case of unsatisfactory performance, moderately complex linker structures would be targeted next. It has to be kept in mind that the overall production cost for a chemical is composed of the variable costs, that is, costs that scale with the amount of chemical produced, for example, raw materials and utilities, and the fixed costs, for example, labor and depreciation on the production equipment.

Although it is difficult to estimate the manufacturing costs for a large-scale chemical synthesis, some general guidelines can be given:

1) The fewer synthesis steps, the better.
2) Use simple reactions, for example, base-catalyzed are to be preferred over transition metal-catalyzed transformations.
3) Use atom-efficient reactions, for example, the Diels–Alder reaction.

As a case study, motor vehicles powered by gaseous fuels (hydrogen and hydrocarbons) will be considered. According to the Organization Internationale des Constructeurs d'Automobiles (OICA), about 47 million cars were built worldwide in 2009 [11], which is a sharp decrease from the almost 53 million cars built in 2008. However, for the purpose of this illustration, a worldwide car production of 50 million cars will be assumed. The United States Department of Energy and Department of Transportation assessed in its "hydrogen posture plan" of 2006 the viability of the production of 0.5 million hydrogen-powered automotive units per year for the US market alone [12]. Of course, not all gas-powered vehicles will include a MOF as a storage medium. However, it seems reasonable, when looking at the U.S. figures, to assume that 1% of the motor vehicles produced worldwide will contain MOF storage systems for hydrogen and natural gas. This would mean about 0.5 million cars a year. Assuming that every tank contains about 50 kg of MOF material, this would mean an annual demand for MOFs of 25 kt. If this material used terephthalate as a linker, sufficient market supply would not be an issue, as the world market for terephthalic acid was around 39 000 kt in 2008 [13].

If, for example, MOF-177, the most promising hydrogen storage material, is considered, one has to pay closer attention to linker availability and linker synthesis. 1,3,5-Benzenetribenzoic acid (H_3BTB) is not a commodity chemical and does not, to our knowledge, have any technical application. Therefore, a synthesis route for this material has to be found. The literature offers a variety of ways to prepare H_3BTB, as shown in Figure 14.1.

Looking at the syntheses in detail reveals that most use the acid-catalyzed trimerization of a methyl phenyl ketone derivative to form the central benzene ring. This reaction is very atom efficient, that is, every carbon of the starting material ends up in the product. Only water is generated as an environmentally friendly by-product. Route V takes a different approach. The last step – trimerization of a phenylacetylene – is even more atom efficient than the base-catalyzed trimerization of the methyl phenyl ketones. However, a major side product will be a polyene, which could impede isolation of the product and requires handling of the polymeric residue. Furthermore, synthesizing the alkyne from ethyl 4-bromobenzoate is a three-step reaction requiring the use of transition metal catalysts and includes a deprotection step (low atom efficiency) and a neutralization step where NaCl is formed. A large amount of salt formed during a reaction, even with a low toxicological potential such as NaCl,

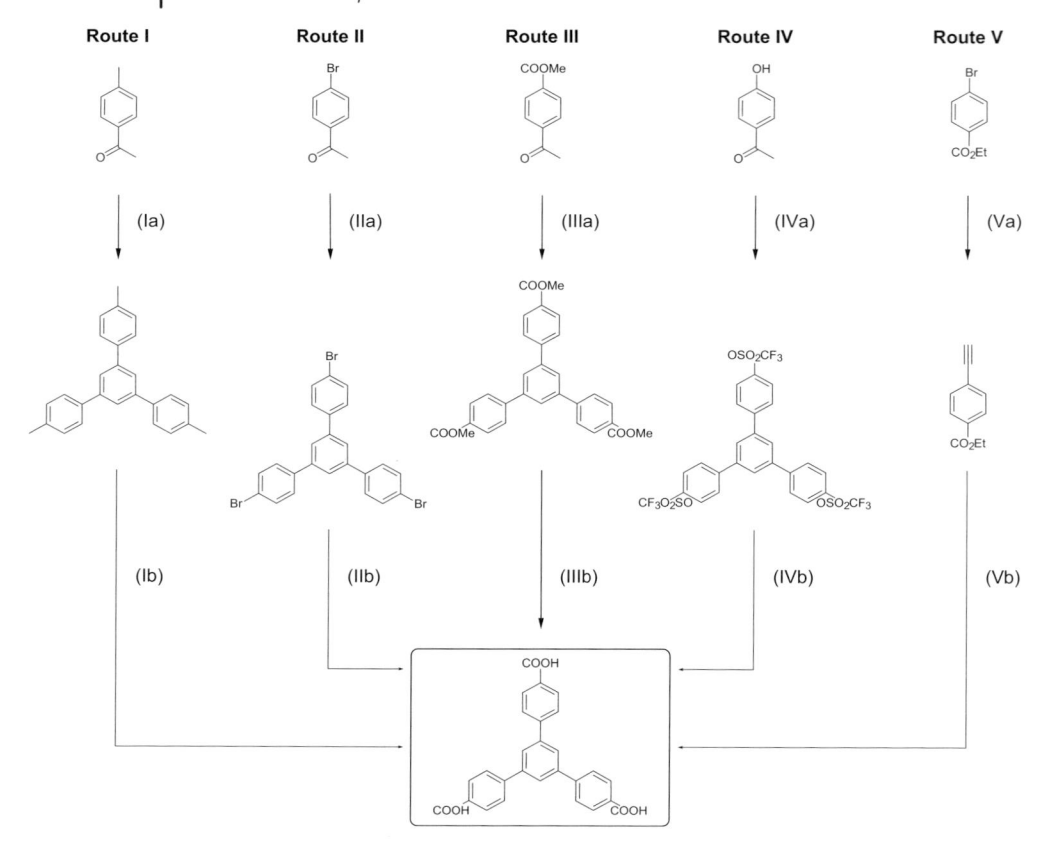

Figure 14.1 Different routes to H₃BTB. Route I: (Ia) Al$_2$O$_3$, $T=210\,°C$, yield 20% [14]; (Ib) HNO$_3$ (3 M), yield 100% [15]/HNO$_3$ (conc.), yield 52% [14]. Route II: (IIa) H$_2$SO$_4$/K$_2$S$_2$O$_7$, $T=180\,°C$, $t=14\,h$, yield 80% [16]; (IIb) n-BuLi, $T=-60\,°C$, yield 60% [17]/57% [18]. Route III [19]: (IIIa) pTosOH, yield not given; (IIIb) NaOH, yield not given. Route IV [20]: (IVa) (CF$_3$SO$_2$)$_2$O, $T=-50\,°C$, yield 97%; (IVb) CO$_2$, cat. [PdCl$_2$(PPh$_3$)$_2$], $T=90\,°C$, $E=1.6\,V$ vs. SCE, yield 75%. Route V [21]: (Va) (i) 1,1-dimethylprop-3-en-1-ol, cat. [PdCl$_2$(PPh$_3$)$_2$], CuI, PPh$_3$, (ii) NaOH, (iii) HCl, yield 44%; (Vb) $T=125\,°C$, yield 21%, not isolated.

is a matter of concern for a large-scale synthesis. There are strict legal regulations on the amount of salt in waste-water after water treatment and, since removal is costly, the formation of salt should be prevented as far as possible. Therefore, this route has low significance for a potential large-scale synthesis.

The other four routes are differentiated by the substituent on the benzene ring of the starting methyl phenyl ketone. Substituted methyl phenyl ketones are not a commodity, either. Most conveniently, these materials can be synthesized industrially by Friedel-Crafts acylation of the corresponding benzene derivatives using either acetic anhydride or acetyl chloride. Regioselectivity of this reaction starting from the monosubstituted benzenes is favorable for the required materials in that *para* substitution will be preferred. Reducing the problem to the level of the substituted benzenes leads back to commodity chemicals that are readily available on a large scale – in this case toluene, bromobenzene, methyl benzoate/benzoic acid, and phenol.

Differentiation of these routes becomes possible when looking at the second step. Route II, starting from bromobenzene, requires the use of an organolithium compound in the second step. Handling of these compounds on a large scale poses a safety hazard which should be avoided if possible. Furthermore, the yield is generally low, since the intermediate lithium aryls that are formed are highly reactive and prone to undergo side reactions. Therefore, this reaction route, although technically possible, is not considered to be promising for scale-up. The same applies to route III. The reasons for this is that the second step – saponification of the methyl ester – again demands a neutralization step, giving rise to the same issue as discussed for route V – too much salt as by-product.

This leaves routes I and IV. Both start from very easily accessible starting materials, toluene and phenol, respectively. Route IV uses a transition metal catalyst in the second step and trifluoromethyl sulfonic anhydride as an (expensive) reagent. Nevertheless, the published high yields and the use of otherwise simple and efficient to handle components, for example, electricity, make this route attractive.

Finally, route I appears the most attractive in terms of reagents and starting materials used. The use of alumina for trimerization in a heterogeneously catalyzed reaction is most favorable, since separation of the catalyst can be effected by simple filtration. Nitric acid is a cheap oxidant. Care has to be taken with the evolved nitrous oxides. However, in an integrated chemical plant, depending on the scale of the linker synthesis, NO_x may even be recycled by feeding it back to a nitric acid plant. One matter of concern with this reaction route is that the reaction time for the trimerization is given as 3–8 weeks. Definitely this has to be optimized, and surely the trimerization conditions given for the other routes also have to be tested.

The considerations given above give only a brief, qualitative glimpse into the needs and challenges of a large-scale linker synthesis. In laboratory tests, the best way to obtain H_3BTB would be determined and the procedure optimized. These tests would be accompanied by simulations of a large-scale process which also include the actual raw material costs, utility costs, and investment necessary. Based on these data, a decision on the most feasible synthesis route would be made.

Of course, these considerations should not be the major concern of an academic MOF researcher. However, when targeting a large-scale application such as gas storage, keeping these considerations in mind definitely facilitates realization. At the end of the day, the performance of the material has to justify the complexity of the synthesis.

14.3
Synthesis

Usually, laboratory-scale synthesis of MOFs is straightforward, using salts as the source of the metal component (see Section 14.2.1). The organic ingredients (see Section 14.2.2) are supplied in a polar organic solvent, typically an amine (triethyl-amine) or amide (diethylformamide, dimethylformamide). After combination of these inorganic and organic components, the metal–organic structures are formed.

When scaling up the preparation to production scale, attention has to be paid to safety issues, for example, when high nitrate concentrations are involved, especially in addition to the build-up of large surface area volumes in adiabatic reactor vessels. In Section 14.3.1, the scale-up for the classical hydrothermal approach is described. To tackle the potential safety hazards of the classical route, the salt-free electrochemical production of MOFs is described in Section 14.3.2.

14.3.1
Hydrothermal Synthesis

Metal–organic structures are formed by self-assembly after mixing of the inorganic and organic components at temperatures starting at room temperature and up to solvothermal conditions at about 200 °C during several hours. A typical scheme for a semi-technical process is shown in Figure 14.2. The scheme indicates the different steps of preparation, and also further processing of the dried powder into shaped material.

It must be noted that an issue of prime importance for production scale-up is the space–time yield (STY: kilograms of MOF product per cubic meter of reaction mixture per day), which should be as high as possible. In Table 14.1, STYs for some industrially feasible MOF syntheses in comparison with zeolite production are presented. Unoptimized small-scale laboratory syntheses as described in the literature usually have an STY between 0.1 and 1 $kg\,m^{-3}$ per day or even lower. The STY term per definition includes the mass concentration of the solid product (kilograms of MOF per cubic meter of reaction mixture). When a more expensive organic solvent is required for the reaction, this concentration becomes very important. Also, for extremely dilute processes, the largest contribution to raw materials costs comes from the organic solvent used. Furthermore, the smaller the STY, the larger the reaction vessel has to be or the more reaction vessels are needed to produce a defined amount of MOF in a given amount of time. This increases the cost of investment and makes the synthesis less economically attractive.

The same consideration should be taken into account for the filtration/washing stage. The minimization of the amount of organic solvent used, while not decreasing the performance of the final porous product, is one of the most important steps in the synthesis transfer from small to technical scale.

Mixing of reagents	Crystallization Precipitation	Filtration Washing	Drying	Shaping
• Raw materials cost • Availability of reagents	• Synthesis time • Concentration	• Duration • Solvent quantity	• Gentle conditions	• Additive selection

Figure 14.2 Simplified flow diagram of industrial MOF synthesis procedure via the solvothermal route with parameters important for scale-up indicated.

Table 14.1 Space–time yield for industrially feasible MOF syntheses.

	Material	Langmuir surface area ($m^2 g^{-1}$)	Space–time yield ($kg\,m^{-3}$ per day)
Basolite A100	Al^{III} terephthalate	1100–1500	160
Basolite A520	Al^{III} fumarate	1100–1500	200
Basolite C300	Cu^{II} benzene-1,3,5-tricarboxylate	1500–2100	225
Basolite Z1200	Zn^{II} 2-methylimidazolate	1300–1800	60–160
Basolite M050	Mg^{II} formate	400–600	~3000
Zeolites		300–800	50–150

The filtration properties of the MOF material obtained become significant in technical-scale production. Filtration can even be the most time-limiting stage in the whole production process. Filter cake resistance becomes an important factor to determine filterability. Particle size distribution, shape of the particles, as well as packing density and orientation of the particles are the major material properties that influence filter cake resistance. Usually, well filterable products are characterized by a specific filter cake resistance α_H of $10^{11}\,m^{-2}$ (based on filter cake thickness) and poorly filterable products $10^{16}\,m^{-2}$ [22, 23]. The particle size distribution can be affected by, for example, agitation during the crystallization of the product. Agitation is a very important and irreplaceable factor in technical production, providing proper dispersion of the reagents and convective heat transfer. The latter is necessary to control the temperature in huge reaction volumes, which otherwise would behave pseudo-adiabatically. Since the crystallization of MOFs is an exothermic process, agitation becomes an important safety issue [24].

Taking the synthesis of Basolite® C300 [copper(II) benzene-1,3,5-tricarboxylate] as an example, the morphology of the crystals obtained when using two different agitators and stirring conditions during crystallization are presented in Figure 14.3. As expected, the smaller particles are observed under the synthesis conditions where a larger amount of mechanical energy is introduced to the system by the stirrer.

After filtration, drying of MOFs has to be carried out very carefully. Owing to their high porosity and surface area, MOFs may easily carry 50–150 wt% of occluded solvent, which is an order of magnitude higher than in a zeolite or base metal oxide preparation. Therefore, it is necessary first to remove most of the adsorbed solvent under mild pressure and temperature conditions. Subsequently, high thermal activation conditions can be applied. Commonly, rotary evaporators or spray driers are used for drying in the industrial production of solid materials [25]. All parameters influencing large pilot-scale production of MOFs are given in Figure 14.2.

14.3.2
Electrochemical Synthesis

On a larger production scale, attention has to be paid to safety issues whenever high nitrate concentrations are involved, especially since MOF synthesis is an

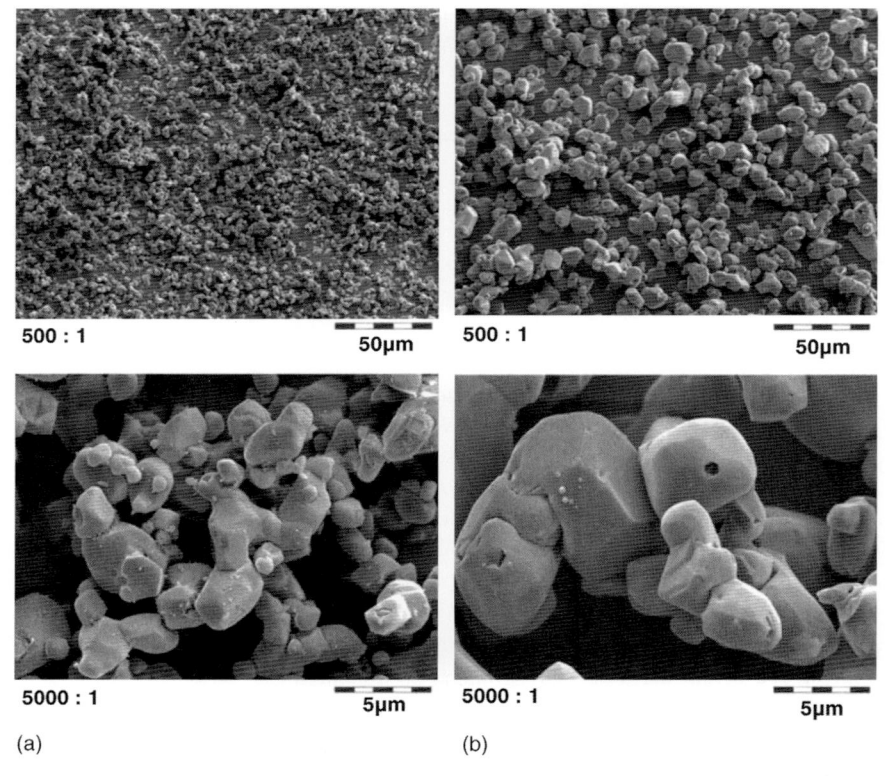

Figure 14.3 Morphology of Basolite C300 crystals analyzed by scanning electron microscopy (SEM): (a) crystallization using a multiple turbine blade agitator at 500 rpm; (b) crystallization using a blade agitator at 250 rpm.

exothermic reaction. As described in Section 14.2.1, metal chlorides could pose a severe corrosion issue and are not recommended for production. Therefore, an alternative salt-free electrochemical procedure was developed by BASF SE [26]. Bulk metal sacrificial anodes are oxidized in the presence of dissolved linker molecules (e.g., carboxylic acids or imidazoles) in an electrochemical cell (Figure 14.4). At the cathode, protons are reduced to hydrogen, thereby generating the deprotonated linker in exactly the correct stoichiometry for MOF formation. Simple recovery of the precipitated product by filtration directly yields the final MOF powder after drying. This procedure is especially beneficial for MOF structures containing open metal sites, as no anions from added salts can interact with the open metal sites and thereby block them. This method was successfully used for the syntheses of Basolite Z1200 [zinc(II) 2-methylimidazolate] and Basolite C300 [copper(II) benzene-1,3,5-tricarboxylate] [27, 28]. In the next development step, a continuous-flow pilot plant capable of producing MOFs on the kilograms per day scale was developed (Figure 14.5).

```
Mixing of  >  Electrolysis  >  Filtration  >  Drying  >  Shaping
reagents                      Washing
```

- Raw materials cost   • Synthesis time   • Duration   • Gentle conditions   • Additive selection
- Availability of   • Concentration   • Solvent quantity
  reagents

Figure 14.4 Simplified flow diagram for industrial MOF synthesis procedure via the electrochemical route with parameters important for scale-up indicated.

14.4 Shaping

In academic research, the shaping process is not an important factor and therefore is mainly neglected or treated at a very basic level. Using shaped material is necessary to ensure proper fluid dynamics in the experimental set-up. The simplest method of shaping is to press and subsequently split and sieve material for use in further studies. The desired particle size can be isolated by sieving using sieves with an appropriate mesh size. However, this is a very inefficient process and a large amount of material is lost. Upon scale-up of a process, the form in which a material is used gains importance and sometimes it is decisive for the ultimate performance of the unit. Generally, pore diffusion effects in the particles and pressure drop across the material bed govern the process of geometry optimization of shaped materials [29]. Unfortunately, pore diffusion is most favorable in small particles whereas the pressure drop over the bed is minimal with large particles. Therefore, a compromise between these two effects has to be found. In addition to these parameters, the shaped bodies also have to fulfill some secondary but nonetheless important needs. Upon

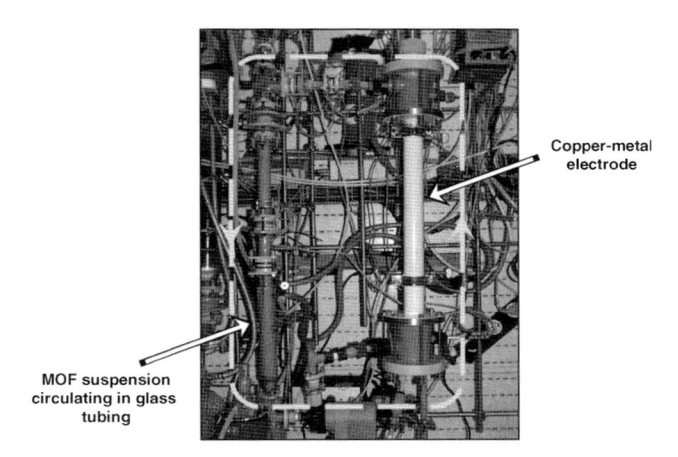

Figure 14.5 Continuous-flow pilot plant (scale ∼kg of MOF per day) for electrochemical preparation of copper-based MOFs using copper plates as electrode material. The reaction mixture is recirculated.

transportation, filling of the containment, during operation, and also when discharging the material from its containment, the material is exposed to mechanical stress. This means that the strength and the abrasion resistance have to be sufficiently high to preserve its structural integrity. Furthermore, this stability prevents the formation of fine particulates, that is, dust, which would cause problems while charging or discharging the containment.

MOFs are generally obtained as small crystals, which cannot be put to use in most applications for the aforementioned reasons. Therefore, MOFs have to be shaped into forms that allow broad use in most applications. The decrease in the outer surface area by compression during the shape-forming step is not only due to a physical reduction of the MOF surface area, but is also caused by the destruction of the porous structure within MOFs [30]. This may be explained by the very high pore volume in a highly fragile MOF structure compared with zeolite structures where the "walls" around the pores are thicker and have higher mechanical stability.

Most MOFs do not show sufficient strength after extrusion [31]. Therefore, it is important to apply binders and other additional substances that stabilize the materials to be agglomerated. Most of the binder substances may be removed from the shaped bodies by drying or calcination stages. Since in these stages substantial shrinkage of the shaped MOF occurs, crack formation or breaking can easily happen. Therefore, the calcination must be carried out under well-controlled conditions. Emphasis has to be laid on a defined, well controlled temperature, humidity, and atmosphere composition during the drying process. The final shaped body is characterized by the crush strength, which can be measured with conventional hardness grading devices.

For MOFs, the classical shaping process had to be modified to minimize the loss of surface area during shaping. At BASF SE, a process in which pressure is applied to the sample during shaping has been established [32]. The pressure may range from atmospheric pressure to several hundred bar and the temperature between room temperature and 300 °C. The composition of the atmosphere has to be well controlled during this process.

The shaping of MOF-5 will be presented as an example. Shaped bodies were formed by pressing MOF-5-containing powder (99.8 wt% of MOF and 0.2 wt% of graphite) using an eccentric press. A matrix with a hole diameter of 4.75 mm was used. The parameters of the eccentric press were chosen so that pellets with a circular base of diameter 4.75 mm and a height of 3 mm were gained.

Table 14.2 shows that the surface area of a MOF-5 shaped body (1137–1532 $m^2 g^{-1}$) is lower than that of powder (1796 $m^2 g^{-1}$). Although the weight specific surface area of shaped material decreases, the volume specific surface area increases when applying a larger crush strength. The effect of structural collapse of the framework with larger crush strength is surprisingly low.

With this process, it is very easy to achieve a high degree of compaction as expressed by the ratio between volume specific BET surface area of pellet to powder. In this example, a ratio of 1.8 was achieved by reducing the destruction of the framework. For most applications, the maximum amount of MOF used is predetermined by the volume of the containment vessel – a reactor, separation column, or

Table 14.2 Surface area of MOF-5 shaped bodies. The MOF-5 powder used had the following characteristics: weight per unit volume 220 g l^{-1}; surface area (BET) per unit volume 395 × 10^3 m^2 l^{-1}; BET surface area 1796 m^2 g^{-1}.

Crush strength (N)	Pellet weight (g)	Pellet density (g cm^{-3})	Weight per liter (g l^{-1})	BET (m^2 g^{-1})	BET/volume (1000 m^2 l^{-1})	BET/volume ratio, pellet:powder
Powder			220	1796	395	
10	0.0310	0.583	396	1532	607	1.5
28	0.0438	0.824	560	1270	711	1.8
51	0.0486	0.914	622	1137	707	1.8

tank for storage purposes. The volume specific view on the property of interest is therefore of prime importance. Shaped bodies are a way to increase these properties by simply packing the single particles more densely while still maintaining or not significantly blocking access to them.

14.5
Applications

In this section, two "real world" applications of MOFs are presented. It purposely does not deal with separations or gas storage in general, but focuses on the first field test of a MOF adsorbent for natural gas storage in a car and the use of MOFs embedded in polymers to control or prevent ripening of fruits.

14.5.1
Natural Gas Storage for Automobile Applications

The EcoFuel Asia Tour took place in 2007, in which a Volkswagen Caddy EcoFuel using natural gas was driven from Berlin to Bangkok [33]. The tour was the ideal opportunity to test MOFs as new natural gas storage materials under real conditions, such as mechanical stress on the particles in the tank during driving and different qualities of the natural gas used. The goal of this challenge was to assess this new storage material's ability to pass such an extreme practical test successfully. Basolite C300 was used for this test since it shows a high capacity for natural gas [28].

A container filled with Basolite C300 can hold up to 30% more natural gas than the empty container. This translates to an approximately 20% increase in operational radius before refueling. After 32 000 km, the tour was successfully completed at Bangkok after 10 weeks.

During the whole tour, the car showed an average consumption of 7 kg of natural gas per 100 km and emitted roughly 1.3 t less CO_2 than a comparable 1.6 l gasoline-powered car. After the tour, one of the EcoFuel's tanks was examined. Even after the harsh conditions experienced during the tour, the MOF had retained its porosity and only a small amount of sulfur contamination was found. Due to the presence of sulfur-based odorants in natural gas, a much larger sulfur build-up was expected, which would have a negative impact on the MOF's capacity for natural gas. A gradual deterioration of the capacity on repeated refueling was expected, but surprisingly not found. A significant amount of fine particles was formed in the tank by the abrasive action of the adsorbent particles on themselves due to movement of the particles relative to each other under operating conditions. This shows that the mechanical parameters of the shaped bodies have to be improved for the next generation of storage materials. Fine particulates may clog tubing and damage valves in the on-board gas infrastructure or may even cause engine damage.

Although there is still a need to improve the attrition for long-distance driving, it was proven that a MOF can be successfully used for such an extreme practical test.

14.5.2
Ethylene Adsorption for Food Storage

Ethylene accelerates the ripening of fruit and vegetables, and ethylene treatment can be used to control fruit ripening. However, the ethylene emission by the fruit itself into its environment during transport causes unwanted fast ripening and therefore shortens the shelf-life of the product.

Some attempts have been made to remove ethylene from the surrounding atmosphere of the fruit in closed packaging. Various absorbents, such as activated carbon, zeolites, and silica, have been tested in packaging [34, 35]. In addition to the desired ethylene capture property, any packaging material has to meet environmental requirements. It should be easy to dispose of or be easily recyclable. For application in the food industry, which has a high demand for packaging materials, a toxicologically harmless material is required. Due to the enhanced environmental awareness of the consumer, biodegradability is also very desirable.

Using the biodegradable polymer Ecoflex® from BASF SE and Basolite M050, a magnesium-based MOF material using formic acid as linker, a new combination of two biodegradable materials was designed. Basolite M050 is made from formic acid using magnesium salts without any toxic organic solvent [36]. Basolite M050 is practically porous magnesium(II) formate and is expected to be biodegradable, like other formates. Ecoflex is a biodegradable material – a polyester comprised from aliphatic and aromatic dicarboxylic acids and aliphatic dihydroxy compounds. The composite of Ecoflex and Basolite M050 was applied to a glass plate with a film applicator to form a thin film with a thickness of ~100 μm. These films were bonded together by heat treatment to convert the films to a plastic bag to store the food.

It was shown for Basolite M050 that the ethylene capacity is 7.5 wt% at 298 K and 1 bar. Practical longer shelf-life of fruit using Basolite M050 was demonstrated. The material was kept in the bottom of a desiccator containing five bananas (600 g). As a control experiment, another batch of five bananas (600 g) was kept in a desiccator without any adsorbent. After 12 days, the bananas with MOF still looked fresh, smelled sweet, and were edible without any mold on the surface. On the other hand, the bananas without the adsorbent smelled and looked rotten, were not edible, and

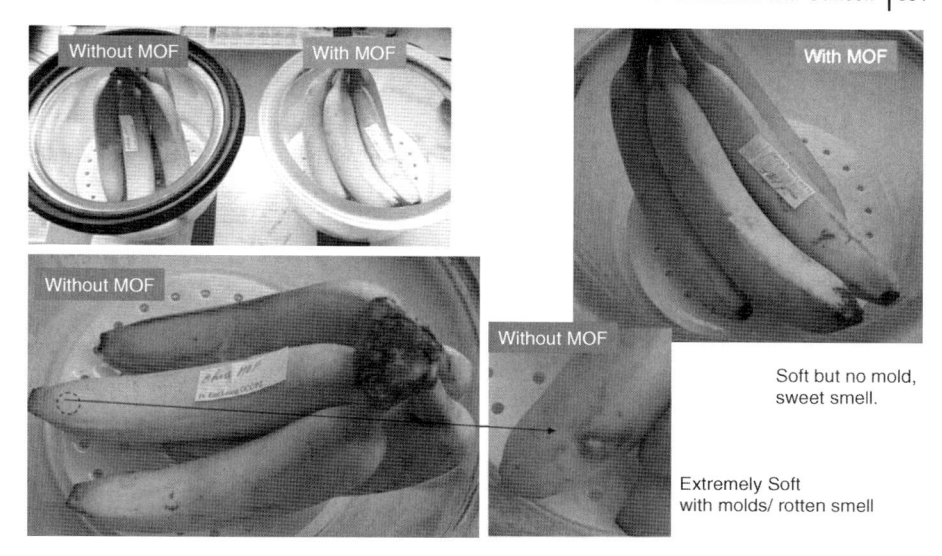

Figure 14.6 Ripening experiments with bananas – the effect of Basolite M050.

had mold on their surface (Figure 14.6). This shows that the ripening process of bananas can be controlled by applying a MOF (Basolite M050), opening up a whole new field of application for these materials.

14.6
Conclusion and Outlook

MOF materials are already beyond the status of a "laboratory curiosity." While the quest for "world record" materials in whatever respect continues in academia, the chemical industry has shown the feasibility of large-scale MOF synthesis. However, the complexity of some of the materials is a matter of concern. The availability of both the linker and the metal for a newly developed MOF are of prime importance. It always has to be kept in mind that large-scale applications demand simple, that is, inexpensive, solutions. For smaller, more specialized applications, more complex materials may also be viable. In the end, the costs for a material have to be justified by its performance, and the performance has to be significantly higher than for state-of-the art materials to make MOFs attractive as their replacement. From the virtually unlimited number of possible combinations of linker molecules and metal ions, application-oriented research should be concentrated on simple linkers and inexpensive and abundant metals. The parameter space constructed from these constraints, especially when considering multi-metal and multi-linker materials, is still huge. High-throughput methods may be the key to success here.

References

1 Kitagawa, S., Kitaura, R., and Noro, S. (2004) *Angew. Chem. Int. Ed.*, **43**, 2334.
2 Tomic, E.A. (1965) *J. Appl. Polym. Sci.*, **9**, 3745.
3 Li, H., Eddaoudi, M., O'Keeffe, M., and Yaghi, O.M. (1999) *Nature*, **402**, 276.
4 Eddaoudi, M., Kim, J., Rosi, N., Vodak, D., Wachter, J., O'Keefe, M., and Yaghi, O.M. (2002) *Science*, **295**, 469.
5 Yaghi, O.M., Eddaoudi, M., Li, H., Kim, J., and Rosi, N. (2002) WO 2002/088148.
6 Chae, H.K., Siberio-Pérez, D.Y., Kim, J., Go, Y.B., Eddaoudi, M., Matzger, A.J., O'Keeffe, M., and Yaghi, O.M. (2004) *Nature*, **427**, 523.
7 Wong-Foy, A.G., Matzger, A.J., and Yaghi, O.M. (2006) *J. Am. Chem. Soc.*, **128**, 3494.
8 Férey, G., Mellot-Draznieks, C., Serre, C., Millange, F., Dutour, J., Surblé, S., and Margiolaki, I. (2005) *Science*, **309**, 2040.
9 Stallmach, F., Gröger, S., Künzel, V., Kärger, J., Yaghi, O.M., Hesse, M., and Müller, U. (2006) *Angew. Chem.*, **118**, 2177; *Angew. Chem. Int. Ed.*, **45**, 2123.
10 Li, Q., Zhang, W., Miljanić, O.Š., Knobler, C.B., Stoddart, J.F., and Yaghi, O.M., (2010) *Chem. Commun.*, **46**, 380.
11 Organisation Internationale des Constructeurs d'Automobiles, Production Statistics, http://www.oica.net/category/production-statistics/, last accessed 11 June 2010.
12 U.S. Department of Energy, U.S. Department of Transportation, Hydrogen Posture Plan, December 2006, http://www.hydrogen.energy.gov/pdfs/hydrogen_posture_plan_dec06.pdf, last accessed 11 June 2010.
13 Terephthalic Acid (TPA): 2010 World Market Outlook and Forecast Been Recently Released by MarketPublishers.com, Business Wire, press release, 24 March 2010.
14 Kothe, G. and Zimmermann, H. (1973) *Tetrahedron*, **29**, 2305.
15 Claus, A. (1890) *J. Prakt. Chem*, **41**, 396.
16 Palomero, J., Mata, J.A., González, F., and Peris, E. (2002) *New J. Chem.*, **26**, 291.
17 Weber, E., Hecker, M., Koepp, E., Orlia, W., Czugler, M., and Csöregh, I. (1988) *J. Chem. Soc., Perkin Trans. 2*, 1251.
18 Vasylyev, M.V. and Neumann, R. (2004) *J. Am. Chem. Soc.*, **126**, 884.
19 Svirbely, W.J. and Weisberg, H.E. (1959) *J. Am. Chem. Soc*, **81**, 257.
20 Jutand, A. and Négri, S. (1998) *Eur. J. Org. Chem.*, **1998**, 1811.
21 Nijus, J.M., Sandman, D.J., Yang, L., and Foxman, B.M. (2005) *Macromolecules*, **38**, 7645.
22 Gasper, H., Oechsle, D., and Pongratz, E. (2000) *Handbuch der Industriellen Fest-Flüssig-Trennung*, Wiley-VCH Verlag GmbH, Weinheim.
23 Matteson, M.J. and Orr, C. (1987) *Filtration. Principles and Practices*, 2nd edn, Marcel Dekker, New York.
24 Dittmeyer, R., Keim, W., Kreysa, G., and Oberholz, A. (2004) *Chemische Technik. Prozesse und Produkte*, Band 1, Wiley-VCH Verlag GmbH, Weinheim.
25 Stiles, A.B. and Koch, T.A. (1995) *Catalyst Manufacture*, 2nd edn, Marcel Dekker, New York.
26 Mueller, U., Schubert, M., Teich, F., Puetter, H., Schierle-Arndt, K., and Pastré, J. (2006) *J. Mater. Chem.*, **16**, 626.
27 Mueller, U., Richter, I., and Schubert, M. (2007) WO 2007131955.
28 Mueller, U., Puetter, H., Hesse, M., and Wessel, H. (2005) WO 2005/049892.
29 Spencer, M.S. (1989) In *Catalyst Handbook*, 2nd edn (ed. M.V. Twigg), Wolfe Publishing, London, p. 17.
30 Lobree F L., Müller., U., Hesse, M., Yaghi., O.M., and Eddaoudi, M. (2003) Shaped bodies containing metal–organic frameworks, WO/2003/102000.
31 Perego, C. and Villa, P.L. (1997) *Catal. Today*, **34**, 281.
32 Hesse, M., Müller, U., and Yaghi, O.M. (2004) Shaped bodies containing metal-organic frameworks, US Patent 7 524 444.
33 Challenge4, EcoFuel Asia Tour, www.ecofuel-asia-tour.com, last accessed 9 June 2010.
34 Sextl, E., Heindl, F., Gaitzsch, T., and Full, R. (2000) Verfahren zur Adsorption von Ethylen, European Patent EP 1 106 233.
35 Frauchinger, U. and Pfenninger, A. (2004) Adsorption von Ethen zur Steuerung der Reifung von Früchten und Obst, European Patent EP 1 525 802.
36 Czaja, A., Trukhan, N., and Müller, U. (2009) *Chem. Soc. Rev.*, **38**, 1284.

15
MOF Shaping and Immobilization

Bertram Böhringer, Roland Fischer, Martin R. Lohe, Marcus Rose, Stefan Kaskel, and Pia Küsgens

15.1
Introduction

In chemical synthesis, metal–organic frameworks (MOFs) are generated as fine particle powders. As almost every process in the chemical industry requires an appropriate shaping, immobilization of powders has always been a topic of paramount importance. The great challenge is ideal shaping of the material in terms of mechanical stability without decreasing the performance of the porous framework. The latter is demanding especially for MOFs since binder systems used for zeolites are often not suitable. Moreover, in zeolites binders can be removed or condensed at high temperature (calcination) whereas MOFs would be destroyed. Owing to the large pore size in some MOFs, traditional binders may also penetrate and block the pore system.

The shaping of MOFs is always connected to a specific application, optimizing capacity, accessibility, and mass transport in the system. In this context, specific solutions were established aimed at gas filtration, textile integration, and drying, each of them requiring a specific shape.

MOF paper sheets, electrospun MOF fiber composites and cellulose or nonwoven MOF fiber composites are suitable solutions for immobilization of MOFs for filtration applications. Compared with pellets, beads, and monoliths, all approaches offer high flexibility of the substrate and the use of such composites, for example in protective clothing, is thereby simplified.

Nevertheless, the use of additives is also necessary in most cases. MOF paper sheets, which are manufactured through the classical papermaking process, do not offer a homogeneous distribution of powders and adequate adhesion of particles and substrate. An alternative route that is not based on the use of additives is the direct crystal growth of MOF particles on to cellulose fibers. Such fibers can be processed to paper sheets and yield a homogeneous distribution of particles in addition to satisfying adhesion of particles and fiber substrate. In the case of electrospun fibers, the polymeric fiber constitutes both the carrier material and the adhesive. Other

Metal-Organic Frameworks: Applications from Catalysis to Gas Storage, First Edition. Edited by David Farrusseng.
© 2011 Wiley-VCH Verlag GmbH & Co. KGaA. Published 2011 by Wiley-VCH Verlag GmbH & Co. KGaA.

alternatives for processing nonwoven MOF fiber composites include the use of bicomponent fibers. Moreover, classical coating processes such as spray drying, foam coating, and dip coating are useful immobilization techniques. Other methods that are not in need of binding agents, among others the Fibroline® process, are addressed in this chapter.

In adsorption applications such as gas separation and gas storage, beads, pellets, or monolithic bodies are the ideal choice. They provide a low pressure drop even at high flow rates. Mobile applications, for example automotive devices, demand high mechanical stability. Formed bodies are therefore supported with binding agents, which provide both low abrasion and the required hardness. Suitable binders are of either polymeric or mineral nature. This chapter describes the extrusion of MOF monolithic structures and also the manufacture of MOF polymeric beads. In the case of extrusion processes, plasticizers have to be employed to attain a suitable molding batch with the required flow properties.

MOF gels offer a secondary highly porous structure, which is highly efficient in adsorption processes. As mechanical stability is not given in this case, gels have to be incorporated in suitable carrier materials, for example, in metallic foam. Requirements for adsorbents in protective applications are also discussed in this chapter.

15.2
MOF@Fiber Composite Materials

15.2.1
MOF-Containing Paper Sheets

The immobilization of MOF powders in paper sheets provides a useful and convenient method; since the paper industry is well advanced and can permit inexpensive and reliable production even on a large scale. Other approaches to immobilizing MOFs include the use of α-alumina or copper mesh as substrates [1, 2].

Paper represents a more suitable substrate, as it is easily pliable. Paper sheets allow different possibilities of filter geometry, for example, star-shaped, folded, or rolled solutions. Moreover, choosing different pulp qualities influences the properties of the filter element and enables the device to be engineered according to the desired application. Another possibility for tailoring the product is the use of additives. Properties that can be influenced include thickness, permeability, hydrophilicity, and flexural rigidity. Thereby, the respective paper sheet or filter element can be designed for almost any application, for example, for operation in liquid or air filtration.

These possibilities account for the extensive and variable potential of paper sheets as substrates for MOF immobilization.

The introduction of MOF particles into the sheet can be accomplished by utilizing processes for introducing fillers or pigments, once again a well-known process in paper technology [3]. A useful method for investigating the effects of pulp modifications is to manufacture paper hand sheets. These are usually made in a Rapid Köthen sheet mold [4]. A Rapid Köthen sheet mold was used to fabricate

15.2 MOF@Fiber Composite Materials

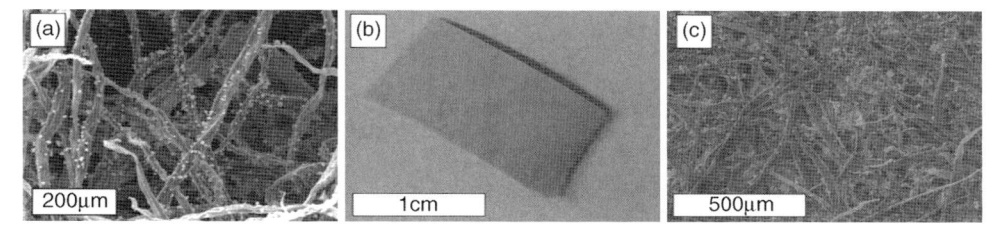

Figure 15.1 (a) MOF@CTMP pulp fibers, (b) MOF-containing paper sheet, and (c) SEM image of the sheet.

MOF-containing paper sheets by adding MOF powder to pulp slurry. $Cu_3(btc)_2$ (btc = 1,3,5-benzenetricarboxylate) was chosen as a representative MOF, since it was among the first MOFs to become commercially available [5]. The resulting paper sheets showed an inhomogeneous distribution of crystals in the fibrous network (Figure 15.1).

Determination of the MOF content in the sheet, using thermogravimetric analysis, confirmed these results. The average content of $Cu_3(btc)_2$ ranged from 6.5 to 14.6 wt%. The inhomogeneity was reflected also in the specific surface area (S_a) of the samples, which was determined by nitrogen physisorption at 77 K. The average S_a of the samples was about 170 $m^2 \, g_{filter}^{-1}$. Thus, incorporating MOFs in paper sheets is a useful and practical approach for immobilization of these materials.

However, inhomogeneities required more advanced techniques, as presented in the following.

15.2.2
MOF@Pulp Fibers

An alternative solution to produce paper sheets is the direct growth of MOF crystals on the surface of pulp fibers. The resulting composites can be fabricated as sheets or filter elements. The direct growth allows for both a regular distribution of powder within the paper sheet and enhanced adhesion between substrate and crystals. First attempts to deposit other porous materials such as zeolites on vegetable fibers have been made by Valtchev and co-workers [6, 7].

In order to obtain MOF@pulp fibers, $Cu_3(btc)_2$ was synthesized in the presence of pulp fibers of different qualities [8]. Two pulp samples were obtained with the Kraft process, and another sample was processed according to a chemothermomechanical pulping (CTMP) process. These samples differed mainly in their residual lignin content. Whereas CTMP is a fairly mild method, Kraft pulping leads to more pronounced delignification of the fibrous matter. Owing to the different compositions of the chosen pulp qualities, significant differences regarding the coverage of crystals were expected. Table 15.1 summarizes the results of the experiments.

The residual lignin content is indirectly specified with the κ-number, indicating the lignin content, usually by determining the consumption of a sulfuric permanganate solution of the selected pulp sample [9].

Table 15.1 Pulp sample characteristics.

Pulp sample	$Cu_3(btc)_2$ content (wt%)[a]	k[b]	S_a (m^2 g^{-1})[c]
CTMP	19.9	114.5	314
Kraft unbleached	10.7	27.6	165
Kraft bleached	0	0.3	10

a) Determined by thermogravimetric analysis.
b) Determined according to ISO 302.
c) Single-point BET surface area determined by nitrogen gas sorption at 77 K.

The CTMP sample provides the largest S_a with 315 m^2 g^{-1} and is also the sample with the highest lignin residue. As shown in the scanning electron microscopy (SEM) image in Figure 15.1c, the crystals are regularly distributed on the fiber surface. Fixation at the surface is significantly more abundant compared with the paper sheet obtained using the slurry approach, where the crystals only loosely fill the spaces between the fiber bundles. The unbleached pulp sample showed a slightly lower content of MOF crystals and S_a. Almost no MOF adhered to the bleached Kraft sample, which was also free of any lignin.

Lignin is a high molecular weight structure consisting mainly of phenylpropane units linked with several radicals [10]. During the pulping process, both lignin and hemicellulose are removed mostly chemically from the wood material to obtain cellulose fibers as remaining material. The definitive structure of lignin is not fully understood, but the occurrence of functional groups such as carbonyl and carboxylic acids is well known [11]. The facts that lignin contains these functions may explain why pulp samples with increasing κ-number show a better affinity towards $Cu_3(btc)_2$ crystals. A similar observation was made by Zacher et al., who studied the deposition of $Cu_3(btc)_2$ and other selected MOFs on self-assembled organic monolayers (SAMs) [12], which were terminated by either CF_3 or COOH groups. The results showed that $Cu_3(btc)_2$ adheres selectively to the COOH-terminated SAMs whereas CF_3-terminated organic surfaces were completely inert. In contrast to lignin, cellulose can be considered as a polyvalent alcohol without any carboxylic functions. The nucleation of $Cu_3(btc)_2$ crystals on highly delignified fibers therefore scarcely occurs.

15.2.3
Electrospinning of MOF@Polymer Composite Fibers

Electrospinning is a versatile method for the production of ultrathin fibers of nearly every polymer, but also for the immobilization of small particles in and on polymeric fibers [13, 14]. In recent decades it has been, and still is, in the focus of research for the production of high-performance nonwovens. Several thousand publications are available just from the last few years. This strong interest can be explained by the versatility and simplicity of the method. The general principle is the polarization of polymers in a solution or melt in a strong electric field. The polymer is drawn from

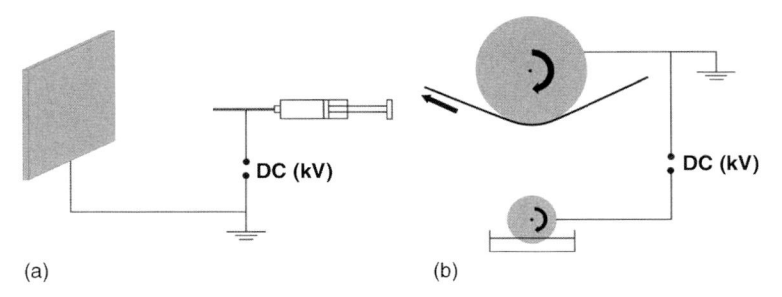

Figure 15.2 Schematic diagrams of a laboratory-scale (a) and a continuously working industrial (b) electrospinning device.

the droplet on one electrode to a counter-electrode. During this process, the solvent evaporates or, if a polymer melt is used, it solidifies and fibers are deposited on the counter-electrode. Applying optimized conditions, thin nonwovens can be obtained instead of single fibers. Even continuous nonwoven layers can be deposited on common textiles used as substrates. Electrospun fibers offer unique properties in comparison with other production methods for nonwovens such as the melt-blown process. They combine fiber diameters in the lower nanometer range with a very narrow fiber size distribution, resulting in excellent filter properties with regard to the particle separation rates.

Laboratory-scale electrospinning is carried out with a syringe containing the spinning solution, optionally connected to a syringe pump (Figure 15.2a). The conducting needle is connected to the high-voltage power supply. The counter-electrode is a grounded metal plate that can be covered with substrates such as textiles or aluminum foil. The electrospinning process can be carried out horizontally, but also vertically from top to bottom and the other way around. On the industrial scale, the electrospinning setup consists of two horizontal, parallel aligned and rotating cylinders made of a conducting metal (Figure 15.2b). The lower cylinder is rotating and wetted continuously with the spinning solution, while the upper cylinder is continuously transporting the textile substrate. By this method, the production of homogeneous electrospun layers on yard ware is possible. Larger laboratory scale-ups to industrial electrospinning devices with a substrate width of several meters are available from Elmarco (Liberec, Czech Republic).

Electrospinning can also be useful in the functionalization of textiles with adsorptive properties. Electrospun fibers with intrinsic porosity can be produced from polymeric precursors by transformation to activated carbon [15] or carbide-derived carbon fibers [16]. Other attempts at the production of porous fibers from inorganic materials such as zeolites were successful. However, these fibers are much too brittle for applications on flexible textile substrates. Therefore, the alternative means of functionalization with good adsorptive properties is the incorporation of porous materials into polymeric fibers used as binders, providing sufficient flexibility for application on textile substrates. Composite fibers made by electrospinning have been the subject of intense research in recent years, since this method provides the know-how for the integration of particles in fibers for advanced functionality. Hence the integration of MOF particles in electrospun fibers seems feasible. So far, it has been shown that salts [17], organic and inorganic particles [18], and even carbon nanotubes [19] can be processed to composite fibers. The purpose of such solid additives is the improvement of properties such as mechanical strength, electrical conductivity, and even catalytic activity. Up to now only a few results have been reported on the incorporation of porous materials in composite fibers for applications related to adsorptive processes. Mainly zeolites and the mesoporous silica SBA-15 were used [20–23].

Within the European Project NanoMOF, the immobilization of MOF particles by electrospinning was investigated for application of MOFs as adsorptive compounds in protective clothing. First results were obtained recently by Rose et al. [24] and in parallel by Smarsly et al. [25]. Preliminary tests were conducted using a horizontal laboratory-scale set-up. A syringe was connected to a syringe pump. As counter-electrode, a metal plate with a diameter of about 12 cm covered with the textile substrates was used. This metal plate was connected to a rotor to obtain homogeneous layers on the substrate. HKUST-1 and MIL-100(Fe) were chosen as model substances for the first tests, since they provide sufficient stability under the processing conditions. Furthermore, polystyrene (PS) in tetrahydrofurane (THF), polyvinylpyrrolidone (PVP) in ethanol, and polyacrylonitrile (PAN) in dimethylformamide (DMF) were used as solutions for spinning. Homogeneous layers could be obtained in all cases. Using PS as binder, loadings of up to 40 wt% could be achieved. Due to the large fiber diameter in comparison with the particle diameter, a significant ratio of the binder is unnecessary. The MOF particles were arranged like a pearl necklace (Figure 15.3a). The use of PVP and PAN as binder resulted in much higher loadings of up to 80 wt%. In this case, the fiber diameter is on the nanometer scale while the particles are several microns in size. A necklace-like morphology is not possible due to the significant differences in particle and fiber size. In fact, the morphology is much more closely comparable to a spider's web in which the MOF particles are comparable to trapped flies (Figure 15.3b and c). However, this morphology seems to be the most promising, since the fixation of the particles with a mass of binder as low as possible due to the small fiber diameter is allowed. These parameters were used on a medium-sized industrial scale electrospinning device

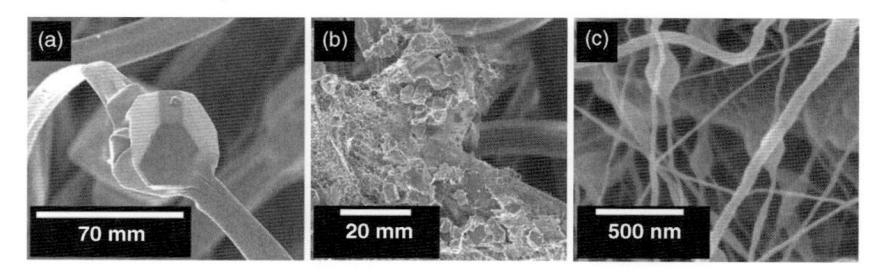

Figure 15.3 SEM images of electrospun composite fibers: (a) HKUST-1/PS spun from a solution in THF resulting in a pearl necklace-like morphology; (b, c) HKUST-1/PAN spun from a solution in DMF resulting in a spider web-like morphology.

(from Elmarco) at Norafin (Mildenau, Germany). Using this set-up, it was also possible to obtain homogeneous coated textiles with composite fiber layers in a continuous process.

The integrity of the MOFs after the electrospinning step was characterized using X-ray powder diffraction (XRPD). No change in structure could be observed. Also, the porosity of the composite fibers was investigated using nitrogen physisorption measurements at $-196\,°C$. The results showed that in the case of the high loadings, the specific surface area and the pore volume decreased by only 20%, which corresponds to the 20 wt% binder in the form of the electrospun fibers. Hence no pore blocking by the polymer was observed and the full adsorption capacity was available after the processing step.

The selection of the best suited polymer for the immobilization of MOF particles on flexible textile substrates is crucial since it has to fulfill certain requirements. The polymer should be processable from a solution containing a nontoxic solvent, in the ideal case water. However, if it is soluble in water, another problem is the solubility after the electrospinning process. The electrospun layer has to be stable against water during the use of the protective clothing containing it. In the ideal case it should even be washable. Hence a post-electrospinning treatment is necessary for the cross-linking of the polymer and the fixation of the fiber shape. Chemical cross-linking is not possible in most cases, since the stability of most MOF materials is not sufficient. A convenient method could be cross-linking with high-energy radiation such as UV light or an electron beam. Alternatively, a polymer such as PAN could be used, which is not soluble in water and displays the above-mentioned properties.

However, the immobilization of MOF particles by electrospinning with a polymer as binder is a promising method for the production of functionalized textiles for protective clothing. Especially the advantages of this versatile method, and the simplicity of the process resulting in the production of ultrathin composite fibers, might prove useful.

15.2.4
MOF Fixation in Textile Structures

A shift towards highly functional and added-value textiles is now recognized as being essential to the sustainable growth of the textile and clothing industry in developed countries. The demand for tailored surface modifications for water repellence, long-term hydrophilicity, enhanced adhesion, and antibacterial properties, is therefore increasing.

Nonwoven materials consisting of natural or synthetic polymer fibers provide an excellent platform for the integration of functional structures to improve the performance of the material for a variety of applications. There are many possibilities to fix particles on fibers (e.g., sputter coating with copper to deposit functional nanostructures on the surface of polypropylene spunbonded nonwoven material [26], plasma technology [27–29], multifunctional textile finishing produced according to principles of green nanotechnology [30], and functional particles fixed by binders [31–33]).

A nonwoven is a manufactured sheet, web, or batt of directionally or randomly oriented fibers, bonded by friction, and/or cohesion, and/or adhesion, excluding paper and products that are woven, knitted, tufted, stitchbonded incorporating binding yarns or filaments, or felted by wet milling, whether or not additionally needled. The fibers may be of natural or man-made origin. They may be staple or continuous filaments or be formed *in situ* [34].

When particles such as coal or MOFs are used in filtration applications, it is advantageous to deposit them regularly on the surface (two-dimensional) of the nonwoven or even better spatially (three-dimensional) into nonwoven structures. If it is the aim to obtain a maximum particle load into textiles, for example, for adsorption (gas filtration) or storage processes (protective clothing), the three-dimensional particle deposition in the textile is a necessity. Owing to the relatively high surface area of the fibers, nonwoven materials are a convenient substrate for spatial particle fixation.

The choice of the material is dependent on the application. Important selection criteria are aspects such as skin contact, heat resistance, strength, mass area, or other defined requirements, for example, electrical and/or heat conductibility. Hence the material has to be carefully chosen.

Another important task is the implementation of the particles into the nonwoven structure, aiming at a spatially consistent particle arrangement with maximum possible loading capacity. In most cases, high air and/or water permeability is required.

15.2.4.1 Pretreatment

After choosing the appropriate basic material (e.g., nonwoven material) for the corresponding application, a suitable adhesive for the particle fixation on the fibers must be found. The adhesive should guarantee strong interconnection between particles and fibers. In most cases, suitable adhesives or their mixtures are water-based dispersions, based on polyvinyl acetate (PVA), polyacrylate, and polyurethane, which form a homogeneous, soft, and sticky film at the fiber surface.

The requirements for the adhesive itself are multifarious, for example, easy to handle, nontoxic, and biologically degradable. If the nonwoven material has skin contact, it is necessary to use eudermic adhesive, and application in protective clothes, such as uniforms for fire-fighters, requires a flame- and/or temperature-resistant system, which is non-melting in a defined temperature range. In many cases, another demand on protective clothing is the washing stability. If the clothes are re-usable, the particle–fiber bonding must be stable, even after many washing processes. It is advantageous if the coating process can be carried out in an aqueous medium, for example, in an aqueous dispersion or solution. In order to obtain a homogeneous film on the fiber surface, the polymer has to be cross-linked by temperature or pressure.

The most important and widely used impregnation technique for adhesives is padding (Foulard technique). Padding can be understood as the dipping of a product (e.g., nonwoven) into an impregnating bath, which is followed by a squeezing process by which the excess liquid is removed (see Figures 15.4 and 15.5). During the coating

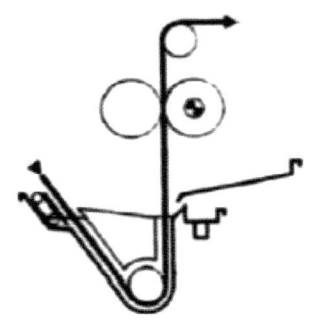

Figure 15.4 Principle of padding.

procedure, the coating components (viscose polymer solutions, dispersions, melts, or coating powders) are applied to the textile. The application is achieved by calendering, spraying, pressure, or printing processes.

The coating- and lamination systems consist of the following segments:

- roll-off device for the basic material
- application unit for the coating substance
- drying and gelation channel for the coated material
- roll-on device for the finished material.

With the Foulard machine, it possible to create a homogeneous adhesive film on the fibers for particle fixation.

Figure 15.5 Typical padding apparatus [67].

15.2.4.2 Wet Particle Insertion

When choosing a liquid as solvent and transport medium for the particles, the nature of the liquid has great importance for the particle insertion. Concerning environmental compatibility, water is in most cases the best choice as it is sustainable, cheap, abundant, and biologically degradable. A problem that might occur when water is the solvent of choice is the generation of hydrogen bridges between both particles and fibers of the nonwoven, which can hinder the mass transport. The use of small particles can promote agglomeration due to high surface energy. To avoid agglomeration, a dispersant or a medium with a suitable pH value (acidic, neutral, or alkaline) can be employed.

Another way to apply particles on the surface is to use, for example, glue dots [35]. The disadvantage of this procedure is that most of the particles are coated with adhesive, which leads to a reduced capacity of the active material in adsorption processes.

Further methods for particle insertion include the use of suspensions or foams. These techniques are standard methods in the field of textile finishing. For the process of foam coating, the particles are first dispersed. Subsequently, the emulsion/suspension is applied to the material by using a knife as a roll [36]. When using the foam coating technique, it should be noted that the penetration is low and the particles are only applied to the surface rather than penetrating the nonwoven material.

Typical foam in most cases consists of a polymer emulsion (such as polyurethanes, acrylics, or rubbers), a foaming agent, a thickening agent, and filler. In some cases, it is also required to use a cross-linking agent. Air is blown into the mixture to generate the foam structure. By varying the volume of air used, the density of the foam can be tailored [36]. Foam application is followed by a drying step making use of a drying oven with a temperature gradient to prevent cracking that can arise when fast heating is applied.

At this point in the process, a distinction between the so-called crushed foam coating and the foam finishing has to be made. In the process of crushed foam coating, input of mechanical energy by stainless-steel rolls is necessary. The energy input occurs when the water content in the foam is in the range of 5–10% during the drying. The applied pressure is fitted to the required texture of the finished film. If the objective lies in a porous coating film, the pressure has to be chosen at a lower level. If a compact film is desired, then the pressure has to be higher. After the crushing, the coated material is cured to 160–170 °C, fitted to the type of polymer.

In the case of foam finishing, the emulsion used contains a lower solid content. As a result, the so-produced foam collapses during the drying and leaves an invisible film behind.

Another wet method to impregnate textile substrates with MOFs could be the process of dip coating. For the treatment procedure, the MOF particles have to be dispersed in a liquid, and the resulting suspension is filled into a tank (see Figure 15.6).

In the next step, the textile is immersed in the filled tank for a defined time (dwell time). At the end of the dwell time, the textile is transported through the squeezing

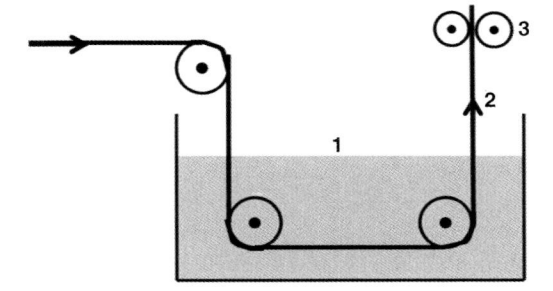

Figure 15.6 Principle of dip coating: (1) tank, (2) textile, (3) rolls for squeezing [36].

rolls to extract the excess suspension. The amount of absorbed suspension is dependent on the solid content in the suspension and the adsorption capacity of the textile [36]. After the impregnation of the textile, a last step is needed to finish the product, namely drying. This drying process can also be combined with thermofixation. Generally, the thermal drying steps are very energy intensive whereas mechanical dewatering of the textile is very energy efficient. The decrease in humidity of the product allows reduced drying times for the thermal treatment.

The common mechanical dewatering methods are

- foulards
- section bars
- centrifugation.

According to the heat supply, there is a distinction within the thermal drying processes:

- convection drying (tenter, hot flue, screening drum dryer, tumbler)
- contact drying (calender, cylinder dryer)
- infrared dryer
- diathermy dryer.

The heating could be performed indirectly by thermal oils, vapor, hot water, or directly via natural gas, propane, or butane.

15.2.4.3 Dry Particle Insertion

After pretreatment of the nonwoven with an adhesive binder system, it is also possible to insert the particles on the dry path, preferably in a three-dimensional structure. The main advantage of dry particle insertion is the solvent-free particle bonding into the nonwoven material. By use of, for example, water- or foam-based technology, the micropores of the particles are blocked irreversibly, resulting in a significant loss of adsorption capacity.

Using a dry method of powder insertion, the particles have to possess high mechanical stability in order to move through the nonwoven.

Important influence factors for the particle insertion are the nature of the particles (e.g., size, shape, hydrophobicity, hydrophilicity), the nature of the basic nonwoven material (e.g., mass, density, water and/or air permeability) and the particle energy during the penetration process into the nonwoven.

Particle acceleration before distribution in the nonwoven can occur in a strong electric field, for example, in the Fibroline® process (see below). To benefit from this treatment, the particles need to be electrostatically chargeable. In the case of activated carbon or MOFs, this requirement is fulfilled.

In the Fibroline process, two face-to-face electrodes are connected to an alternative high-tension generator ($U = 10–50\,kV$). The electrodes are protected by a suitable dielectric material. The distance between the two electrodes varies according to the thickness of the fibrous support.

The fibrous support is placed between the two dielectrics, and a strong alternating electric field is created by the high voltage applied to the electrodes. The particles that are distributed on the substrate through a scattering unit will be moved inside the porous substrate when it passes through the electrodes (Figure 15.7). Typically, a 10 s resistance time is sufficient and leads to homogeneous particle insertion through the cross-sectional area of the nonwoven. Metallic particles with high electrical conductivity cause problems.

An alternative method for particle fixation on the fibers is the use of bicomponent (BiCo) fibers with a melting shell. BiCo fibers are comprised of two polymers of different chemical and/or physical properties extruded from the same spinneret with both polymers within the same filament. This type of fiber consists of a core with a higher and a coat with a lower melting point. If the fibers are heated in a temperature range that lies between the melting points of the shell and core, the shell melts first and the particles are fixed on the fiber surface. On cooling of the textile, the particles adhere to the fibers. The polymers listed in Table 15.2 can be used as components in the cross-sections [37]. In practice, a defined amount of BiCo fibers is mixed with conventional fibers and used for particle bonding and thermofixation (see Figure 15.8).

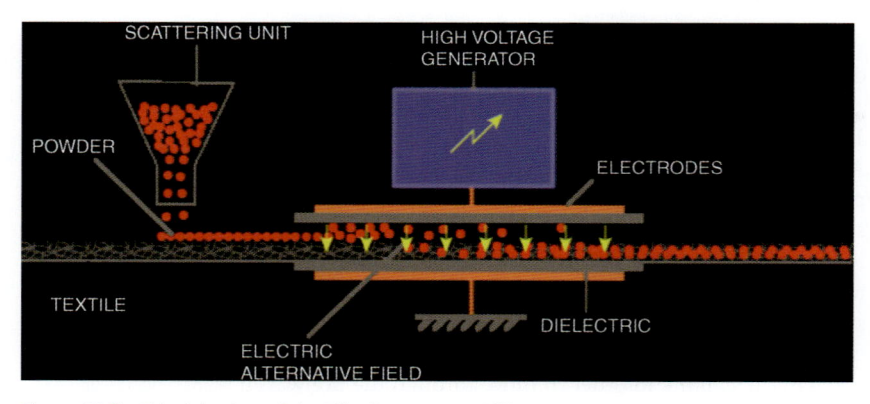

Figure 15.7 Principle view of the Fibroline process [68].

Table 15.2 Polymers used as either of the components in the cross section.

PET (polyester)	PEN-polyester
Nylon 6.6	PCT-polyester
Polypropylene	PBT-polyester
Nylon 6	Co-polyamides
Polylactic acid	Polystyrene
Acetal	Polyurethane
Soluble co-polyester	HDPE, LLDPE

BiCo thermal binder provides a uniform distribution of adhesive and thereby a homogeneous distribution of particles on the nonwoven sheet. The core–shell principle preserves the structure of the fibers and thereby adds integrity. A wide range of bonding temperatures and the individual choice of BiCo fibers promote the process. Additionally, it is environmentally friendly and recyclable.

The air flow method represents another common dry particle insertion approach that operates under low-pressure conditions. A precondition is that the nonwoven material has a relatively low density ("open structure"), enabling the particles to penetrate into the textile. The strongest influence factors of this method are the particle shape, size, mass, the affinity to the fibers, and of course the nature of the fibers and particles. To obtain a homogeneous and high particle loading of the substrate, it is fundamental that both the pore size of the substrate and the particle size are in an optimal relationship. Admittedly, it is possible that the particles acquire such a high kinetic energy that they slightly open the pores. Therefore, it is more favorable to use particle sizes smaller then the middle pore size. The principle is shown in Figures 15.9 and 15.10. For generating the air flow and to give the particles the necessary kinetic energy, a section bar is generally used. In most cases, an important condition for particle infiltration into a porous nonwoven material is homogeneous application on the substrate. This can be achieved by a scattering unit. Another option is particle-spraying (see Section 15.4) on to the substrate surface.

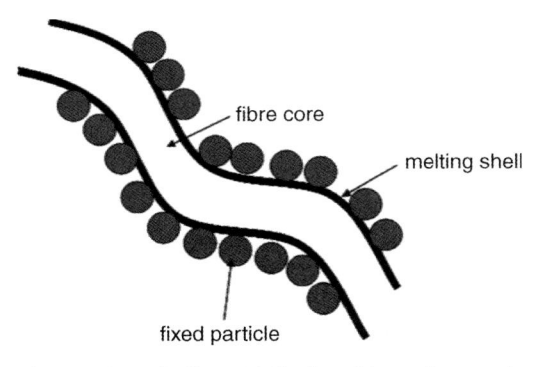

Figure 15.8 BiCo fiber with fixed particles and conventional fiber.

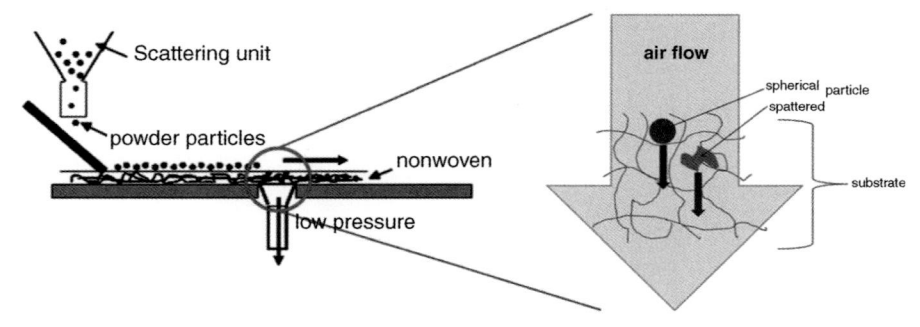

Figure 15.9 Principle of the air flow method.

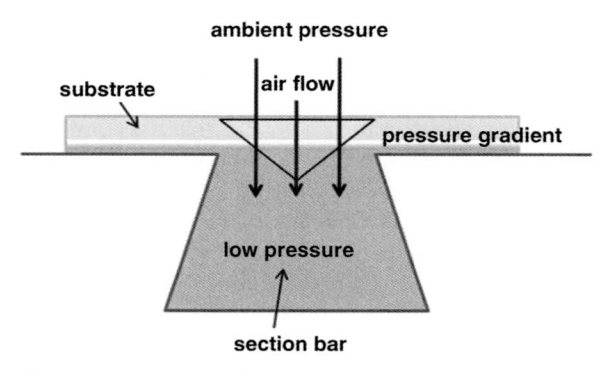

Figure 15.10 Process of air flow.

Figure 15.11 describes the process in detail. A low pressure on the bottom side of the substrate generates a pressure gradient and in this way an air flow. The air molecules could be accelerated up to sonic velocity and acquire a very high kinetic energy, which facilitates particle movement. The generated air–particle flow is influenced by the properties of the substrate such as its air permeability and the particles. If a nonwoven is taken as substrate, these properties can be controlled by its thickness, density, and connectivity. These parameters could be adjusted in these loading processes.

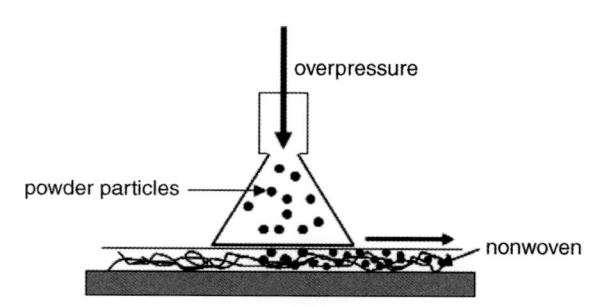

Figure 15.11 Principle of particle spraying.

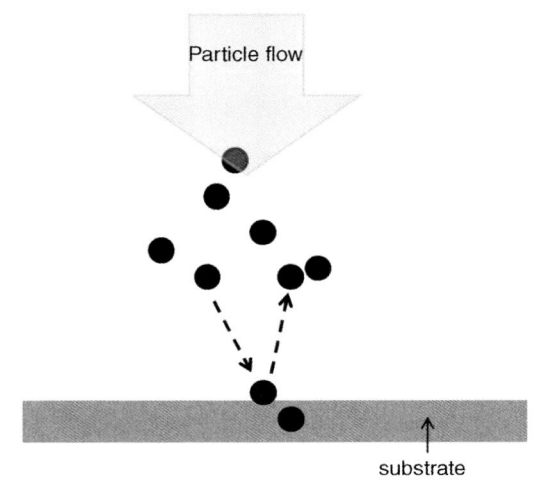

Figure 15.12 Principle of the energy loss in particle spraying.

Another method for particle insertion into the substrate is particle spraying on to or into the nonwoven. If the particles are spattered they will be fixed in areas close to the surface, whereby they will block adjacent particles and the insertion will only occur at the surface, not in the depth of the nonwoven. On the other hand, the spattered particles can adhere more uniformly to the fibers and less energy is necessary for this process. Therefore, the particles have to be accelerated and directed to the surface by air flow. A schematic illustration of the process is shown in Figure 15.12.

The difference between the pressure method and the spraying method is the method of energy creation and leading. In the case of the pressure method, the air flow is generated by a pressure gradient. In the spraying technique, the particles are accelerated in the spray unit and deposited on the substrate surface. The inserting effect of the spray method is lower in comparison with the pressure method, because in the spray technology the air–particle stream is reflected and brakes the particle flow in the direction of the nonwoven. These reflected particles collide with the input particles with a decelerating effect. As a consequence, the particles lose part of their energy.

15.3
Requirements of Adsorbents for Individual Protection

15.3.1
Relevant Protective Clothing Applications

Whereas air-permeable protective clothing for chemically contaminated environments has been based on activated carbon types for decades, the application of MOFs in protective clothing is an entirely new approach [38].

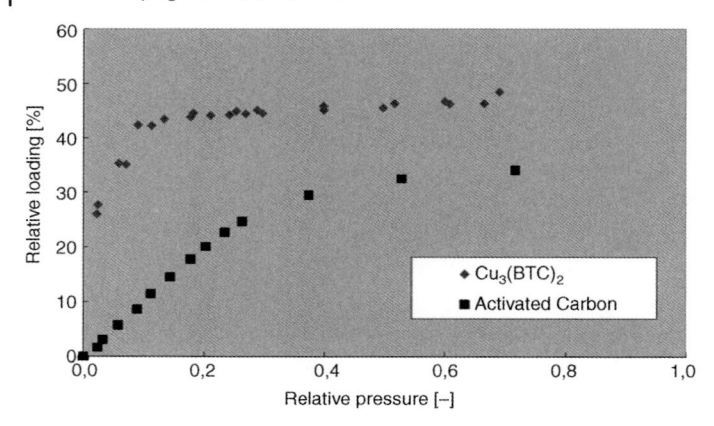

Figure 15.13 Adsorption isotherms for ammonia at room temperature for $Cu_3(btc)_2$ and activated carbon (spherical activated carbon).

The activated carbons used in this application are spherical activated carbon (e.g., Saratoga®), activated carbon fibers, and sometimes granular activated carbon. These products show excellent adsorptive properties for organic molecules; therefore, they are used for protection against skin-permeating chemical warfare agents (CWAs) [39].

In an industrial environment, the presence of such chemicals is rare. The chemicals in an industrial environment are less toxic but include also common inorganic chemicals, such as acids and ammonia. Activated carbons adsorb such chemicals to only a small extent, limiting the application of activated carbons in industrial protective clothing.

MOFs with their typical architecture of inorganic clusters and organic linkers are likewise adsorbents for inorganic as well as organic molecules. Figure 15.13 shows the difference in the adsorption of ammonia between activated carbon and $Cu_3(btc)_2$ [40].

Although industrial protection focuses on liquid protection, the most common inorganic molecules are gases. Hence contamination with gases and vapors is even more likely than with liquids as vapors are always present when chemicals are used. Most classes of chemical protective clothing are not fully encapsulated and therefore will exhibit, after a certain time, a certain concentration of gases also on the inside of such suits, where it will contaminate the skin. The use of adsorptive material will prevent this [39].

15.3.2
Filter Performance

When applying MOFs as an adsorptive layer, especially classes 4, 5, and 6, which are described in EN 14605 (Type 4), EN 13982-1 (Type 5), and EN 13034 (Type 6) [41–43], respectively, seem applicable for chemical protective clothing.

Table 15.3 Type 4 equipment (EN 14605).

Abrasion resistance + waterproof evaluation	EN 530
	EN 14325
Flex cracking at standard temperatures + waterproof evaluation	EN ISO 7854-B
Flex cracking at low temperatures + waterproof evaluation	EN ISO 7854-B
Test methods for nonwovens – determination of tear resistance	EN ISO 9073-4
Puncture resistance	EN 863
Tensile strength	ISO 13934-1
Permeation calibration	EN 374-3
Measurement of permeation resistance	EN ISO 6529/EN 374-3
Measurement of gas permeation resistance	EN ISO 6529/EN 374-3
Resistance to ignition	EN 13274-4 method 3
Seam tensile properties of fabrics and made-up textile articles – Part 2: determination of maximum force to seam rupture using the grab method	EN ISO 13935-2
Determination of resistance to penetration by spray (spray test)	EN 17491-4
Determination of resistance to penetration by a jet of liquid (jet test)	EN 17491-3

Tables 15.3–15.5 describe the requirements and possible performances for chemical protective clothing according to the official standards (4.5.4) for Type 4, 5, and 6 protective garments.

Protective garments against chemicals are used in different environments. There are fields of application where the protective clothes are exposed to significant, mainly mechanical, stress. For instance, during rescue operations after disasters, regardless of whether the rescue takes place outside or inside a building, the protective clothes may be exposed to heavy mechanical loads.

The protective clothes must show an appropriate strength, so that mechanical stress cannot cause damage to the material of the clothing (e.g., perforation). In the case of mechanical damage, the protecting ability of the protective clothes will be significantly reduced or cease entirely.

Table 15.4 Type 5 equipment (EN 13982-1).

Abrasion resistance + waterproof evaluation	EN 530
	EN 14325
Flex cracking at standard temperatures + waterproof evaluation	EN ISO 7854-B
Test methods for nonwovens – determination of tear resistance	EN ISO 9073-4
Puncture resistance	EN 863
Resistance to ignition	EN 13274-4 method 3
Seam tensile properties of fabrics and made-up textile articles – Part 2: determination of maximum force to seam rupture using the grab method	EN ISO 13935-2
Protective clothing for use against solid particulates – Part 2: test method of determination of inward leakage of aerosols of fine particles into suits	EN ISO 13982-2

Table 15.5 Type 6 equipment (EN 13034).

Determination of abrasion resistance (for 50 000 cycles)	ISO 12947-2
Abrasion resistance + waterproof evaluation	EN 530
Test methods for nonwovens – determination of tear resistance	EN ISO 9073-4
Protection against liquid chemicals – test method for resistance of materials to penetration by liquids	EN 368/EN 469/EN ISO 6530/NFPA 1994
Resistance to ignition	EN 13274-4 method 3
Seam tensile properties of fabrics and made up textile articles – Part 2: determination of maximum force to seam rupture using the grab method	EN ISO 13935-2
Determination of resistance to penetration by spray (spray test)	EN 468
	EN 13034
Tensile properties: Determination of maximum force and elongation at maximum force using the strip method	EN ISO 13934-1
Puncture resistance	EN 863

There are fields of application when the protecting clothes are used in normal (not extreme) circumstances, for example, for laboratory work or manufacturing processes when hazardous chemicals are used. In these cases, the mechanical effects affecting the clothes are generally considerably smaller, hence protective clothes with lower strength may be applied, giving the worker a higher level of comfort.

For the protection of the wearer, it is required that personal protection equipment (PPE) also does not release harmful substances, which might damage or even penetrate the human skin. To ensure this, the requirements of Oeko-Tex® Standard 100, especially the requirements for the extractability of metals, should be taken into account [44].

The application of MOFs in protective clothing also requires hydrolytic stability of the adsorbent, as the application requires that the MOF is in contact with ambient humidity, sweat of the wearer, and water (during laundering), and this even at elevated temperatures (tumble drying). This limits the number of MOFs that can be selected for the application, as some are not stable towards hydrolysis.

Figure 15.14 shows the water vapor isotherms of spherical activated carbon as a reference and $Zn(bdc)(dabco)_{0.5}$. The spherical activated carbon shows the typical adsorption/desorption behavior of a hydrophobic adsorbent. $Zn(bdc)(dabco)_{0.5}$ shows a small adsorption of water, which for its BET surface area of about 2000 $m^2 g^{-1}$ is remarkably low. During desorption, only a small portion of the water is released. This means that the water is irreversibly bound, and therefore $Zn(bdc)(dabco)_{0.5}$ is not applicable as an adsorbent in water contact.

Figure 15.15 shows the nitrogen adsorption isotherms of $Zn(bdc)(dabco)_{0.5}$ and $Cu_3(btc)_2$ before and after storing in an atmosphere of $65 \pm 5\%$ relative humidity at $20 \pm 2\,°C$. (4.5.3)There, $Cu_3(btc)_2$ shows a limited reduction in nitrogen adsorption capacity. The capacity of $Zn(bdc)(dabco)_{0.5}$ after storage is very low, confirming that this type of MOF is not usable for the intended application.

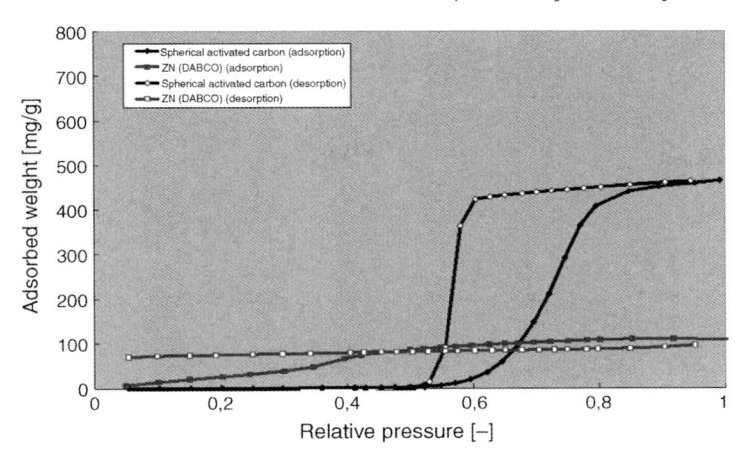

Figure 15.14 Water vapor isotherm of $Zn(bdc)(dabco)_{0.5}$ and spherical activated carbon as a reference.

15.3.3
Testing the Chemical Protection Performance of Filters

The actual test methods available to test the penetration of chemicals through protective materials/clothing are focused on high levels of contamination with liquid or gas. This does not take into account that for high concentrations only fully encapsulated protective suits (class 1) are required.

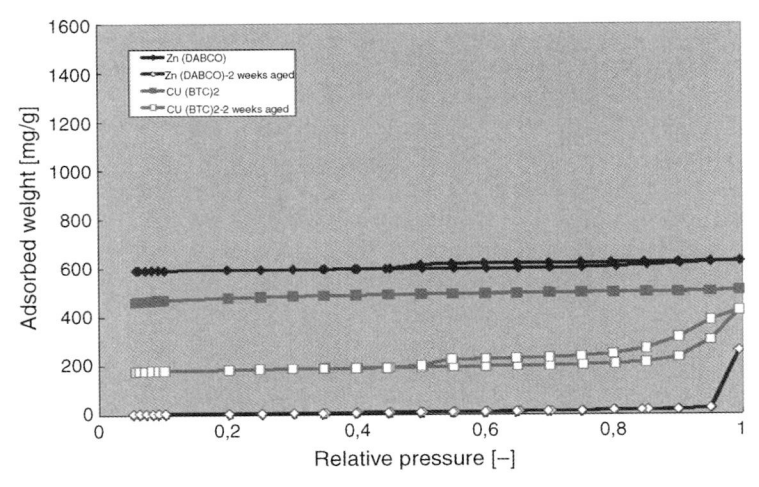

Figure 15.15 Nitrogen adsorption isotherms (77 K) of $Cu_3(btc)_2$ and $Zn(bdc)(dabco)_{0.5}$ original and aged in an environment of $65 \pm 5\%$ relative humidity and a temperature of $20\,°C$ for 2 weeks.

Breakthrough Test

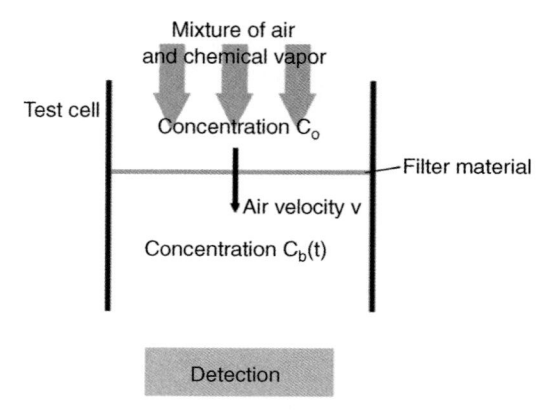

Figure 15.16 Sketch of a vapor test set-up.

Class 2, 3, 4, or 5 protective garments are defined to be used in situations where predominantly vapor or small remaining amounts of liquids are present after an incident/accident. This is because in real applications, the contamination is diluted fairly quickly, ending up in medium vapor concentrations, which may last for longer times. However, the testing for permeation of chemicals usually ignores this and testing is done with large amounts of neat liquid per unit area.

We suggest a more realistic permeation test with vapors to ensure that class 2, 3, 4, and 5 air-permeable equipment is testing realistically.

The test set-up is depicted in Figure 15.16. A mixture of air and chemical with a defined constant initial concentration c_0 is drawn through a filter sample with a fixed air speed. The breakthrough concentration c_b is either measured in real time using an online detection method such as flame ionization detection, or quantified via accumulation over defined time intervals. Temperature and relative humidity have to be defined and kept constant during testing as they affect the adsorption behavior. If the filter is intended to be fully saturated during testing within a reasonable timeframe, an appropriately high initial concentration c_0 has to be chosen. This will be significantly higher than in a realistic threat scenario. Therefore, the test results from these test methods are used for quality control purposes or to compare different filters among each other, and only indirectly to draw conclusions about the protection performance. Figure 15.17 shows the typical outcome of such a breakthrough experiment, an S-shaped breakthrough curve. Depending on both the filter material and the chemical tested, there is an initial efficiency, correlating with the porosity of a filter material, but also the kinetic properties of the adsorptive material in the thin layer. The slope of the breakthrough curve reveals information on the kinetic properties of the adsorptive material and finally the capacity of the filter material for a specific chemical under the test conditions (gas velocity, concentration, temperature) chosen.

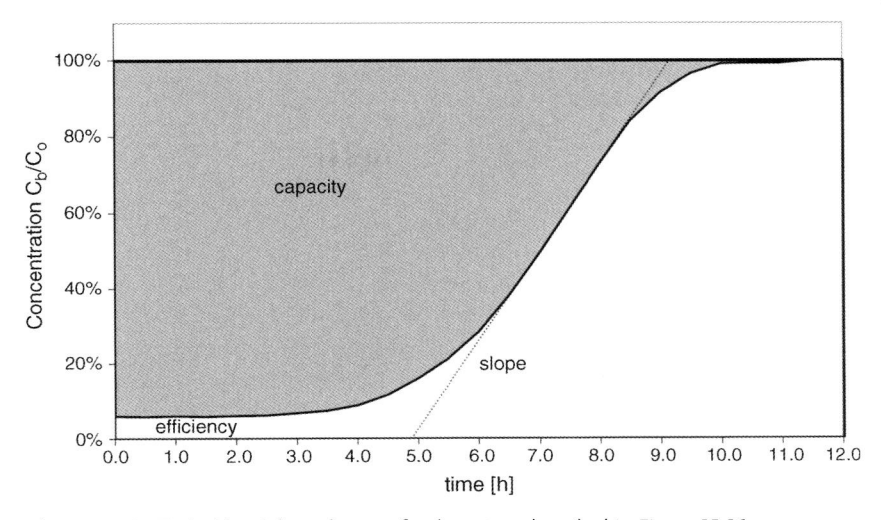

Figure 15.17 Typical breakthrough curve for the set-up described in Figure 15.16.

A test set-up such as this would allow testing of chemical protective materials with MOFs in a more realistic way.

15.3.4
Concepts for Application

In addition to the long list of performance requirements, the technical challenge is how to fix and homogeneously distribute MOFs on a textile surface that is to be applied in a chemical protective garment. Promising concepts have already been demonstrated (Section 15.2.3):

- use of electrospinning to distribute a binder and the MOFs homogeneously on a carrier
- fixing of MOFs with a layer of adhesive on the carrier
- agglomerating the MOFs to larger particles (for example, by using a binder) and fixing the agglomerated particles on a carrier.

Future work is required to identify the optimum method and combination with selected MOFs and textiles to fulfill the requirements for protective clothing.

15.4
MOFs in Monolithic Structures

Porous materials synthesized solvothermally are mostly generated as fine particle powders. As any industrial process requires an appropriate shape or form of the active material owing to the necessity for a low pressure drop, some strategies for integrating MOFs in matrices are given in this section.

15.4.1
MOF@Polymeric Beads

Recently, several solutions for the processing of MOFs into shaped bodies were developed using polymeric binders [45, 46].

A versatile method for the production of composite beads containing MOF powders and a polymeric binder has been patented [47]. Poly(ethylene terephthalate) (PET) is dissolved in the solvent 1,1,1,3,3,3-hexafluoroisopropanol (HFIP), and a defined amount of the MOF (e.g., HKUST-1) is dispersed in the solution. The resulting dispersion is dropped into another solvent acting as coagulant, in which the polymer is not soluble. In the case of PET as binder polymer, acetone was used. The polymer–MOF composite precipitates and immediately solidifies in the form of small spheres with a small aggregate size distribution. The spheres exhibit high mechanical stability. Despite a binder content of about 25 wt%, no pore blocking was observed. Determination of the specific surface area from nitrogen physisorption measurements resulted in a decrease due to the mass of the nonporous binder polymer, but it exhibits the full available inner surface of the MOF particles.

15.4.2
Extruded MOF Bodies

Monolithic structures are already well-known substrates, which are used for example as catalysts supports in three-way catalytic converters.

In order to fabricate MOF-containing monoliths, techniques that made use of zeolitic materials were studied first. There are two main techniques to produce zeolitic monoliths known from industrial applications. The first technique makes use of a washcoating process. A porous material is mixed with a binder and coated on a monolith substrate. In the second technique, the material is applied on long bands of fabric, which are then wound into monolith-like structures. Both techniques are relatively complex compared with extrusion. Moreover, they reduce the solid fraction of the active material extensively. Monolithic bodies are advantageous compared with beaded or pelleted packed beads, since here abrasion is reduced to minimum without a significant decrease in pressure drop [48]. Additionally, high mechanical stability is also ensured when employing monoliths in industrial processes.

The fabrication of $Cu_3(btc)_2$ was based on the work of Trefzger [49], who examined the process technology for the manufacture of zeolitic monoliths. In the first step, $Cu_3(btc)_2$ was mixed in a laboratory-scale kneader with binding agent and plasticization agent, yielding a molding batch of homogeneous appearance. In the second step, the batch was extruded in a ram extruder (Figure 15.18). The monolith was cut into 200 mm pieces, which were treated by microwave drying for 20 min.

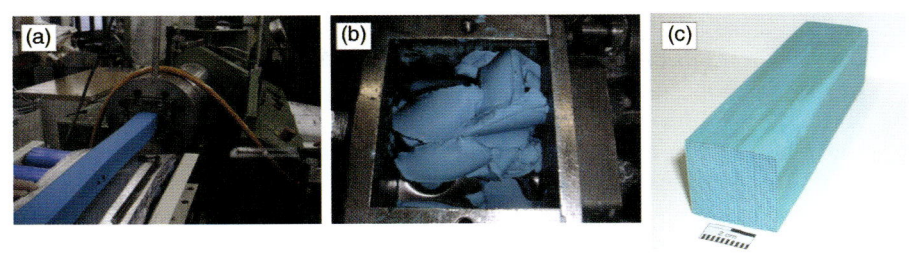

Figure 15.18 (a) Extrusion, (b) kneading, and (c) monolith.

Methyl siloxane ether was employed as binding agent. Other possible materials include attapulgite, montmorillonite, kaolin, and bentonite [50]. Silicone resins can also be used as permanent binders [51]. These possess partly cross-linked properties and are available as liquids. Therefore, they are suitable as binding agents in extrusion processes, as the binder is added in its liquid state to the molding batch. After shaping, complete cross-linking is induced at elevated temperatures.

Since $Cu_3(btc)_2$ shows no plastic behavior after immersion in water, an appropriate plasticization additive has to be used. Methylcellulose is a suitable material, because of its high solubility and nontoxicity. Additionally, it is commercially available in large quantities, owing to its extensive use as a thickening agent and plastic flow additive in the food industry [52]. In the case of zeolites, the plastic behavior of methylcellulose is based on the interaction of hydrated layers between the particles, which may also apply to $Cu_3(btc)_2$ [53].

The extrusion experiments yielded $Cu_3(btc)_2$ monoliths of reasonable mechanical stability (see Figure 15.18). The crushing strength was determined on a Zwick material testing equipment. Both the as-made and activated samples (150 °C) were tested, whereas the as-made samples showed a slightly lower stability, 281 N. Activated samples were harder, 320 N.

Several experiments were performed to examine the specific surface area (S_a) of the monoliths. In a first run, S_a amounted to 480 m² g⁻¹, which is an excellent result. After a certain time, S_a decreased to 287 m² g⁻¹. Apparently, storage under ambient conditions promoted the decomposition of the porous framework, which shows insufficient stability towards water [54].

Nevertheless, the two-step extrusion of $Cu_3(btc)_2$ mold yielded mechanically stable MOF monoliths [55].

15.4.3
Monolithic MOF Gels

Another novel field of research on MOFs is the production of monolithic structures without the use of any binders or the pelletizing of dry powders. Use of the sol–gel approach opens up new possibilities in the preparation of amorphous MOF structures.

Sol–gel chemistry is well developed for various inorganic compounds such as silica, alumina, and titanium oxide. The different stages in such a sol–gel process are well classified [56]. Starting with a sol, which is basically a colloidal suspension of particles in an arbitrary solvent, different processes can lead to aggregation of the sol particles to form a gel. The gel differs from a simple precipitate in that the sol particles aggregate to form a continuous network and the pores are filled with solvent. To dry such gels, that is, to replace the continuous liquid phase with a gas, several processes can be used. Standard drying uses atmospheric or low pressures and/or elevated temperatures, resulting in so-called xerogels in most cases. These xerogels undergo considerable shrinkage during the drying process due to the capillary forces of the solvent, which is called syneresis. Other techniques that allow conservation of the original gel morphology are freeze- and supercritical drying processes. Cryogels are prepared if the solvent is frozen and subsequently sublimed from the pores at low pressure. In the case of aerogels, the solvent is heated above its critical point pressure and subsequently removed while maintaining the high temperature.

In recent years, it was shown that all these processes and techniques from classical sol–gel chemistry can also be applied to coordination polymer-based materials such as MOFs. MOF gels are promising candidates, since the amorphous network allows for the presence of larger transport pores in close vicinity to the micropores used for storage. This leads to high storage capacities but also high adsorption rates, whereas crystalline MOFs with a monomodal size distribution suffer in terms of adsorption rates as soon as larger crystals are present in the product. The same applies for applications in catalysis, where the larger transport pores can significantly improve mass transport, thus minimizing the kinetic limitations often observed in crystalline microporous systems.

A good review on recent developments in various supramolecular metal- and anion-binding gels was given by Piepenbrock *et al.* [57]. However, many of the known structures are formed through hydrogen bonding, solvophobic effects, van der Waals forces, and π–π stacking. The fibrillar morphologies observed are comparable to those of purely macroporous, polymeric organo-gels which do not exhibit any microporosity. MOF-based gels therefore emerge as a separate subgroup of coordination polymers or metallogels in general. The well-known concept of secondary building units (SBUs) is adopted and combined with classical sol–gel chemistry to retrieve materials of unique hierarchical architecture. On the molecular scale, the versatile modular building block concept is used to generate clusters analogous to the SBUs that can be found in crystalline MOFs (Figure 15.19a). These clusters then condense to microporous nanoparticulate structures (Figure 15.19b), which then connect to form networks similar to those found in colloidal silica gels [58].

So far, little is known about the detailed structural motifs and the building mechanisms that lead to the formation of coordination polymer gels. In detail, the question of the conditions under which gelation is favored over crystallization still remains a significant challenge. Westcott *et al.*, who studied metallogels with tripodal cyclotriveratrylene-type and 1,3,5-substituted benzene-type ligands, mentioned that gel formation is only observed under very specific conditions for each metal–ligand pair investigated [59]. On the other hand, our own experiments have shown that MOF gels, based on typical linkers such as trimesic acid, can be produced at almost any concentration if an appropriate solvent is found. Moulton and co-workers, for

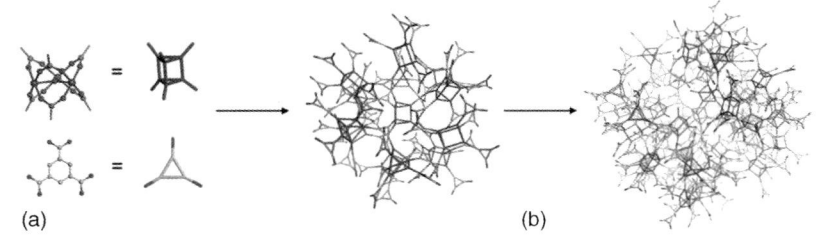

Figure 15.19 Scheme of the formation and growth of amorphous MOF particles based on trimesic acid as the linker and a trigonal prismatic $M^3O(COO^-)_6^+$ cluster as connector as can be found in iron trimesate gels.

instance, prepared a whole family of coordination polymer gels based on a $Mn_{12}O_{12}(CH_3COO)_{16}$ cluster with 13 different linkers in dimethyl sulfoxide and examined the structures and mechanical properties [60]. Another example is iron(III) nitrate and benzene-1,3,5-tricarboxylic acid (btc) in ethanol, which also seems to be the first known true MOF-like metallogel exhibiting significant microporosity even in the xerogel state. Iron trimesate also gelates if significant amounts of non-coordinating polymer precursors like methyl methacrylate are present in the reaction medium. Therefore, Wei and James used it as a soft template for the preparation of macroporous poly(methyl methacrylate) polymer powders. Since it easily dissolves in 1 M hydrochloric acid, this preparation method is really fast and straightforward compared with conventional templating methods such as emulsion techniques and suspended silica particles [61]. Later, this templating approach was further developed to produce monolithic polymer molds with good performance in the separation of proteins and a significantly reduced back-pressure compared with conventional particle-filled columns [62]. Macroporous polymers prepared with such a soft templating method thus demonstrate themselves to be promising materials for high-throughput separations of proteins and enzymes.

Of course, the possible applications of MOF gels are not limited to such short-lived, destructive methods as the templating approach where the gel is simply dissolved after polymerization of the matrix. Recently, it was shown that iron carboxylate-based gels can be easily produced with linkers other than trimesic acid, such as 5-*tert*-butylisophthalic acid, 5-diphenylphosphanylisophthalic acid and 5-1*H*-benzo[*d*]imidazoleisophthalic acid (see Figure 15.20a–c, respectively). The last two acids can easily immobilize Pd^{II} species due to strong HSAB (hard and soft acids and bases) matching and were successfully tested in Suzuki–Miyaura cross-coupling reactions [63, 64].

It is noteworthy that although they show a similar, colloidal morphology, the gels prepared from these dicarboxylic acids exhibit a significantly reduced surface area in the aero- and xerogel state compared with pure trimesic acid-based gels. Co-condensation of different linkers may consequently enhance microporosity and could lead to compounds with even higher catalytic activity. Basically, all of the contemplated possible applications concentrate on wet gels and so far only a few

Figure 15.20 Alternative linkers for the production of gels with iron(III) nitrate: (a) 5-*tert*-butylisophthalic acid, (b) 5-diphenylphosphanylisophthalic acid, and (c) 5-1*H*-benzo[*d*]imidazoleisophthalic acid.

groups have utilized supercritical drying techniques to characterize the synthesized gels in the aerogel state. Research in our group is concentrated on dry gels [65]. The combination of micro- and macroporosity in a monolithic structure is promising for applications in storage and separation, due to both good adsorption rates and capacities. The monolithic nature of the aero- and also the xerogels makes them suitable for applications where fine powders are unfavorable. MOFs that are produced on an industrial scale are always of relatively small size. For larger monolithic molds, the risk of continuous elution of the catalyst particles from a reaction vessel under high gas or liquid flows is reduced significantly.

A detailed investigation of the iron trimesate system is leading to the conclusion that it is basically an amorphous MIL-100(Fe) [65]. This is an interesting finding, given that the crystalline phase has to be prepared under very harsh conditions [66]. As for MOFs, the washing of the samples prior to drying plays a key role in the preparation of highly porous compounds. To extract reagents from even large monolithic samples of several cm^3 in volume, a classical Soxhlet extraction set-up has been found useful. Monolithic aerogels of relatively large volume (Figure 15.21a) can be prepared using supercritical drying techniques and also specifically molded xerogels such as two millimeter-sized compact spheres (Figure 15.16c) if the gel is prepared in a metal foam template (Figure 15.21b) and is easily removed after drying and shrinkage under elevated temperature and atmospheric pressure.

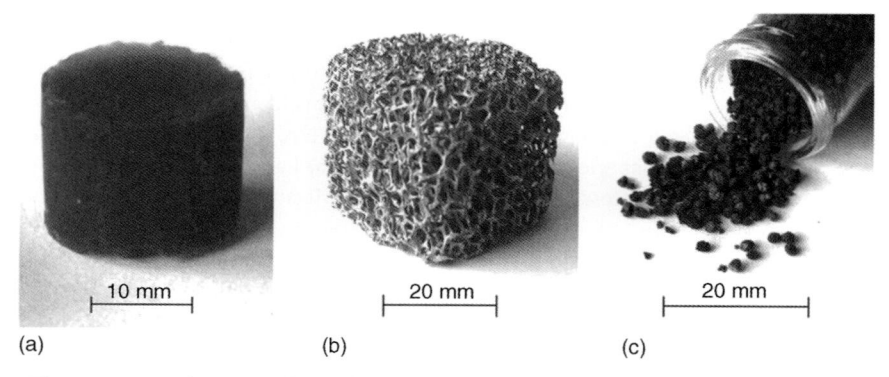

Figure 15.21 Fe(btc) aerogel monolith (a) and aluminum foam (b) suitable for production of uniform xerogel particles (c).

References

1 Gascon, J., Aguado, S., and Kapteijn, F. (2009) Manufactue of dense coatings of $Cu_3(BTC)_2$ (HKUST-1) on α-alumina. *Micropor. Mesopor. Mater.*, **113**, 132–138.
2 Guo, H. *et al.* (2009) Twin copper source: growth of metal–organic framework membrane: $Cu_3(BTC)_2$ with high permeability and selectivity for recycling H_2. *J. Am. Chem. Soc.*, **113**, 646–647.
3 Krogerus, B. (1999) In *Papermaking Science and Technology* (eds. J. Gullichsen and H. Paulapuru), Fapet Oy, Helsinki, pp. 117–149.
4 Töppel, O. (1993) In *Prüfung von Pappe, Papier, Zellstoff und Holzstoff* (ed. W. Franke), Springer, Heidelberg, p. 43.
5 Mueller, U. *et al.* (2006) Metal–organic frameworks – prospective industrial applications. *J. Mater. Chem.*, **16**, 626–636.
6 Mintova, S. and Valtchev, V. (1996) Deposition of zeolite A on vegetal fibers. *Zeolites*, **16**, 31.
7 Valtchev, S., Vulchev, I., and Lazarova, V. (1994) Influence of reactive radicals in cellulose fibres on the formation of zeolite coatings. *Chem. Commun.*, 2087–2088.
8 Küsgens, P., Siegle, S., and Kaskel, S. (2009) Crystal growth of the metal–organic framework $Cu_3(BTC)_2$ on the surface of pulp fibers. *Adv. Eng. Mater.*, **11**, 93–95.
9 Kappa Number of Pulp. TAPPI, TAPPI Standard T214, TAPPI, Norcross, GA.
10 Sarkanen, K. and Ludwig, C. (1971) *Lignins*, John Wiley $ Sons, Inc., New York.
11 Pearl, I.A. (1967) *The Chemistry of Lignin*, Marcel Dekker, New York, p. 339.
12 Zacher, D. *et al.* (2007) Deposition of microcrystalline $[Cu_3(btc)_2]$ and $[Zn_2(bdc)_2(dabco)]$ at alumina and silica surfaces modified with patterned self assembled organic monolayers: evidence of surface selective and oriented growth. *J. Mater. Chem.*, **17**, 2785.
13 Agarwal, S., Wendorff, J.H., and Greiner, A. (2009) Progress in the field of electrospinning for tissue engineering applications. *Adv. Mater.*, **21**, 1–9.
14 Greiner, A. and Wendorff, J.H. (2007) Electrospinning: a fascinating method for the production of ultrathin fibers. *Angew. Chem. Int. Ed.*, **46**, 5670–5703.
15 Bui, N.-N. *et al.* (2009) Activated carbon fibers from electrospinning of polyacrylonitrile/pitch blends. *Carbon*, **47**, 2538–2539.
16 Rose, M. *et al.* (2010) High surface area carbide-derived carbon fibers produced by electrospinning of polycarbosilane precursors. *Carbon*, **48**, 403–407.
17 Bognitzki, M. *et al.* (2006) Preparation of sub-micrometer copper fibers via electrospinning. *Adv. Mater.*, **18**, 2384–2386.
18 Wang, M. *et al.* (2004) Field-responsive superparamagnetic composite nanofibers by electrospinning. *Polymer*, **45** (16), 5504–5514.
19 Almecija, D. *et al.* (2009) Mechanical properties of individual electrospun polymer-nanotube composite nanofiber. *Carbon*, **47**, 2253–2258.
20 Di, J. *et al.* (2008) Fabrication of zeolite hollow fibers by coaxial electrospinning. *Chem. Mater.*, **20** (11), 3543–3545.
21 Madhugiri, S. *et al.* (2003) Electrospun MEH-PPV/SBA-15 composite nanofibers using a dual syringe method. *J. Am. Chem. Soc.*, **125** (47), 14531–14538.
22 Srinivasan, D., Rao, R., and Zribi, A. (2006) Synthesis of novel micro- and mesoporous zeolite nanostructures using electrospinning techniques. *J. Electron. Mater.*, **35** (3), 504–509.
23 Cucchi, I. *et al.* (2007) Fluorescent electrospun nanofibers embedding dye-loaded zeolite crystals. *Small*, **3** (2), 305–309.
24 Rose, M. *et al.* (2011) MOF processing by electrospinning for functional textiles. *Adv. Eng. Mater.*, **13** (4), 356–360.
25 Bernd M. Smarsly *et al.* (2011) Metal–organic framework nanofibers via electrospinning. *Chem. Commun.*, 442–444.
26 Wie, Q. *et al.* (2008) Characterization of nonwoven material functionalized by

sputter coating of copper. *Surf. Coating Technol.*, **202** (12), 2535–2539.

27 Guimond, S. *et al.* (2010) Plasma functionalization of textiles: specifics and possibilities. *Pure Appl. Chem.*, **82** (6), 1239–1245.

28 Hegemann, D. (2006). Plasma polymerization and its applications in textiles. *Ind. J. Fibre Text.*, **31**, 99–115.

29 Shishoo, R. (2007) *Plasma Technologies for Textiles*, Woodhead Publishing, Cambridge.

30 Rodie, J.B. (2008) GreenShield™ Nanoscale multitasking Textile World, Quality fabrics of the month, Aug, 2008, p.26.

31 Hansen, M.R., Young, R. H. (1997), Particle binding to fibers, US-Patent No. 5641561 Date of patent: Jun 24/1997.

32 Hansen, M.R., Young, R. H. (1994), Method for binding particles to fibers using reactivatable binders, US-Patent No. 5352480 Date of patent: Oct. 4/1994.

33 Soane, D.S. *et al.* Nanoparticle-based permanent treatments for textiles US-Patent application publication, publ.-No. US 2003/0013369 A1 publ. date Jan 16/2003.

34 Nonwoven training course Edana European disposables and nonwovens Association Definition of nonwovens, ISO 9092, EN 29092 .–EDANA-definition Module 1.2, Jan 2005.

35 Kaskel, S. Sorptionsfiltermaterial und seine Verwendung DE-Patent DE10200 8005218 A1 Blücher GmbH, Erkrath Anmeldetag: 18.01.2008.

36 Kumarsen, A. (2008) *Coated Textiles: Principles and Applications*, Taylor & Francis, Boca Raton, FL.

37 Lewin, M. and Preston J. (1985) High technology fibers, IN: Handbook of fiber science and technology, part A, Vol III, New York/Basel, Marcel Dekker Inc., Ed. Menachem Lewin and Jack Preston, p. 397, ISBN 0-8247-7279-2.

38 Truong, Q. and Wilusz, E. (2008) Chemical and biological protection. In *Military Textiles* (ed. E. Wilusz), Woodhead Publishing, Cambridge, pp. 242–280.

39 Böhringer, B. (2009) Lightweight CBRN protective materials with enhanced protection and wear comfort. Presented at the 4th European Conference on Protective Clothing (ECPC).

40 Giebelausen, J.M., Fichtner, S., and A.-T. GmbH (2008).

41 CEN (2005) EN 13034:2005. *Protective Clothing Against Liquid chemicals – Performance Requirements for Chemical Protective Clothing Offering Limited Protective Performance Against Liquid Chemicals (Type 6 and Type PB [6] Equipment)*. European Commission for Standardization, Brussels.

42 CEN (2005) EN 14605. *Protective Clothing Against Liquid Chemicals – Performance Requirement for Clothing with Liquid-Tight (Type 3) or Spray-Tight (Type 4) Connections, Including Items Providing Protection to Parts of the Body*. European Commission for Standardization, Brussels.

43 CEN (2004) EN ISO 13982-1. *Protective Clothing for Use Against Solid Particulates – Part 1: Performance Requirement for Chemical Protective Clothing Providing Protection to the Full Body Against Airborne Solid Particulates (Type 5 Clothing)*. European Commission for Standardization, Brussels.

44 Oeko-Tex (2010) www.oeko-tex.com (last accessed 15 April 2011), Oeko-Tex® Standard 100, Limit Values and Fastness, Ed. 01.01.2010.

45 Mueller, U. *et al.* (2003) Shaped bodies containing metal–organic frameworks, BASF, Ludwigshafen. US Patent Application, Pub. Number US/2003 0222023 A1.

46 Finsy, V. *et al.* (2009) Separation of CO_2/CH_4 mixtures with the MIL-53(Al) metal–organic framework. *Micropor. Mesopor. Mater.*, **120**, 221–227.

47 Kaskel, S. (2008) Sorption filter material and use thereof. WIPO Patent Application WO/2009/056184.

48 Fritz, H.-G. and Hammer, J. (2005) Aufbereitung zeolithischer Formmassen und ihre Ausformung zu Adsorptionsformteilen. *Chem.-Ing.-Tech.*, **77**, 1588–1600.

49 Trefzger, C. (2002) Herstellung zeolithischer Wabenkörper, Institut für Kunststofftechnik, University of Stuttgart, Stuttgart.

50 Mitchell, W. and Moore, W. (1961) Bonded Molecular Sieves, US Patent 2.973.327, Union Carbide Corporation, New York.
51 Wusirika, R. and Morris, I. (1995) Low Expansion Molecular Sieve and Method of Making the Same, European Patent EP 0 706 824 A1, Corning Inc., Corning, NY.
52 Lapasin, R. and Pricl, S. (1995) *Rheology of Industrial Polysaccharides, Theory and Applications*, Blackie, Glasgow.
53 Schuetz, J. (1986) *Ceram. Bull.*, **65**, 12.
54 Küsgens, P. *et al.* (2009) Characterization of metal–organic frameworks by water adsorption. Methylcellulose polymers as binders for extrusion of ceramics, *Micropor. Mesopor. Mater.*, **120**, 325.
55 Küsgens, P. *et al.* (2009) Metal–organic frameworks in monolithic structures. *J. Am. Ceram. Soc.*, 93 (9), 2476–2479.
56 Aleman, J. *et al.* (2007) Definitions of terms relating to the structure and processing of sols, gels, networks, and inorganic–organic hybrid materials. *Pure Appl. Chem.*, **79**, 1801–1827.
57 Piepenbrock, M.O.M. *et al.* (2010) Metal- and anion-binding supramolecular gels. *Chem. Rev.*, **110**, 1960–2004.
58 Hüsing, N. and Schubert, U. (1998) Aerogels airy materials: chemistry, structure, and properties. *Angew. Chem. Int. Ed.*, **37**, 23–45.
59 Westcott, A. *et al.* (2009) Metallo-gels and organo-gels with tripodal cyclotriveratrylene-type and 1,3,5-substituted benzene-type ligands. *New J. Chem.*, **33**, 902–912.
60 Luisi, B.S., Rowland, K.D., and Mouton, B. (2007) Coordination polymer gels: synthesis, structure and mechanical properties of amorphous coordination polymers. *Chem. Commun.*, 2802–2804.
61 Wei, G. and James, S.L. (2005) A metal–organic gel used as a template for a porous organic polymer. *Chem. Commun.*, 1555–1556.
62 Yin, J.F. *et al.* (2007) Macroporous polymer monoliths fabricated by using a metal–organic coordination gel template. *Chem. Commun.*, 4614–4616.
63 Huang, J. *et al.* (2010) Dynamic functionalised metallogel: an approach to immobilised catalysis with improved activity. *J. Mol. Catal. A Chem.*, **317**, 97–103.
64 Zhang, J.Y. *et al.* (2009) Metal–organic gels as functionalisable supports for catalysis. *New J. Chem.*, **33**, 1070–1075.
65 Lohe, M.R., Rose, M., and Kaskel, S. (2009) Metal–organic framework (MOF) aerogels with high micro- and macroporosity. *Chem. Commun.*, 6056–6058.
66 Horcajada, P. *et al.* (2007) Synthesis and catalytic properties of MIL-100(Fe), an iron(III) carboxylate with large pores. *Chem. Commun.*, 2820–2822.
67 2-WALZEN-FOULARD VERTIKAL TYP (2011) MathisAG www.mathisag.com/de/dyn_ output.html?content.void=197&b52e162625f4e1864c4b5663d63da530, last accessed 20 April 2011.
68 The Fibroline Process (2008) Fibroline www.fibroline.com/general-description.htm, last accessed 20 April 2011.

Index

a

absorbents   350
– CO_2 adsorbents in H_2-PSAs   103
– concepts, for application   373
– filter performance   368–371
– relevant protective clothing applications   367, 368
– testing chemical protection performance, of filters   371–373
acid gas removal (AGR)   103
– process   103
activity tests   231
– drug-containing MOFs   231–233
– NO-loaded samples   233, 234
– silver coordination polymers   234, 235
adsorbate–adsorbent interactions   73, 184
adsorbents
– for gas separation and purification   74
– hydrophobic   107
– selection criteria   111, 114
adsorption
– analytical methods   62, 63
– based system   158
– enthalpy   87
– in flexible porous solids   58
– – direct molecular simulation   58–60
– gate-opening pressure for   65
– mechanism   164
– of normal C_4 over higher alkanes and alkenes   89
– potential energy for carbon dioxide   85
– process, interpretation   162
– purification by   69
– thermodynamics, for selecting MOFs for sensing applications   313–315
adsorption isotherms   61, 81, 83, 145
– for ammonia   368
– of C_8-alkylaromatics and n-octane   180

– with gate-opening behavior   9
– nitrogen   370, 371
aerogels   376
air flow method, for dry particle insertion   365
air purification   69
– CO production from oxidation of carbon   71
– cryogenic separation   71
– evolution of Claude Linde process   69
– and gas separation, general principles   72, 73
– separation of components   69
– and separation process on industrial plant (SIAD)   70
alkanes   73, 89
alkene oxidation   202
amide coupling   32
amine-grafted MIL-101
– catalytic properties   197
2-aminobenzenedicarboxylic acid (ABDC)   13, 259
O-(2-aminoethyl)-O'-(2-azidoethyl) nonaethylene glycol   35
3-aminopropyltriethoxysilane   29
2-aminoterepththalate   32, 38
ammonia borane (AB)   18
– technical challenges   18
analyte-dependent changes   310
analytical methods, for adsorption   62, 63
anisotropic colloids   324
anodic aluminum oxide (AAO)   132
anodic oxidation   325
antenna effect. See metal-based luminescence
anthropogenic CO_2 emissions   105
argon   69
aryl chlorides   28

b

barium sulfate   252
Basolite M050   350
benzaldehyde
– acetalization of   193
– catalytic sites are not prone to reduction by   193
– cyanosilylation of   207, 208
– Knoevenagel condensation   196, 197, 201
– oxidation of benzyl alcohol   206
1,3,5-benzenetricarboxylate ligands   28
5-1H-benzo[d]imidazoleisophthalic acid   378
BET surface area   34, 41
BiCo thermal binder   365
bioluminescence   270
biomineralization   131
BioMOF formation, from bioactive linker   227, 228
breathing phenomenon   54, 83
Brønsted acidity   201
Brunauer–Emmett–Teller (BET) specific surface areas   34, 41, 154
buoyancy effects   157

c

calcination   353
C_8-alkylaromatic isomers   88
calorimetry   61
carbonation–decarbonation cycles   106
carbon capture and storage (CCS)   105
carbon-free fuel. See hydrogen
carboxylate-based MOFs   90
carboxylimidazolates   34
C_8-aromatics loop   173
catalytic hydrogenation   205
catalytic reactions   194
cathodoluminescence   270
cellulose acetate   104
Chahine's rule   159
chemical vapor deposition (CVD)   17
chemical warfare agents (CWAs)   368
chemiluminescence   270
chemothermomechanical pulping (CTMP) process   355
chiral L-proline derivatives   29
chiral organic catalytic functions   199, 200
chlorobenzimidazole ligands   112
cisplatin   259
Claude Linde process   69
Claus process   104
click chemistry   34
– DMOF-1-NH$_2$ for click functionalization   37
– generic post-functionalization route, from amino-derived MOFs   38
cluster models   109
CMIL materials, for asymmetric aldols reactions   30
coadsorption, applications   63–66
– ideal adsorbed solution theory   64
– osmotic framework adsorbed solution theory   63, 64
– predicted coadsorption properties, of coordination polymer   64, 65
– predictions of mixture, of CO_2 and CH_4 in   66
co-adsorption experiments, for CO_2/CH_4 and CO_2/CO selectivities   109
CO_2 capture agents   99
CO_2 capture process
– by absorption with chemical solvent   106
– H_2 from syngas, production   101–103
– metal–organic framework (MOF)   108–115
– MOFs, opportunities for   99
– post-combustion capture   105–108
– pressure swing adsorption (PSA)   99–101
– removal from natural gas   103–105
– schematic presentation   106
CO_2–CH_4 mixtures   83
– tetrahydrothiophene removal from   90
CO_2 in flue gases   105
– requirements for adsorbents for   107, 108
combinatorial synthesis   26
combustion processes   99
complementary metal–oxide–semiconductor (CMOS) fabrication   311
complex hydrides, alanate   154
composite beads, containing MOF powders   374
computed tomography (CT)   252, 261
– contrast agents   252
condensation
– of aldehyde to amino moiety of UMCM-1-NH$_2$ 33
– of amine to formyl moiety of ZIF-90   34
CO_2/N_2 selectivity   108
continuous chromatographic countercurrent process   174
continuous fixed-bed reactor   205
controlled drug release technology   221–227
– antitumoral molecules, encapsulated from nanoparticles of   226
– cancer/AIDS drugs, encapsulated in different porous MOFs   225
– drug capacities on iron carboxylate nanoparticles   226
– ibuprofen as model drug   222, 223

– – delivery kinetics   224, 225
– – encapsulation kinetics   224
– metal-organic framework (MOF)
– – flexible, excellent candidates for   224
– – use of porous   222
– ordered mesoporous silica materials   221
– ordered porous materials for   221
– pore openings of MIL-53 solid: water   223
– pore size and pore volume, of different porous solids   223
– stability of solids   224
coordination polymers   3, 4, 251
copper(II) coordination polyhedra   84
copper net-supported HKUST-1 membrane   138
CPO-27 materials, heat of adsorption   115
cryogels   376
crystallographic density   158
Cu^{2+} paddle-wheel cluster   6
Cu–S coordinative bonds   86
cyclic hydrocarbons   73
cyclohexane catalytic oxidation   203
cyclohexene
– allylic oxidation   194
– epoxidation   193, 202
– oxidation with H_2O_2 202
cytotoxicity assays   259

d

dehydration of HKUST-1   29
dehydration–rehydration cycle   294
densification process   166
depressurization   103
5,5′-dicarboxy-substituted binol ligands   198
dichlorogold(III) complex   40
diethylenetriamine   29
N,N-diethylformamide (DEF)   35, 132
diethylformamide molecules   15
diffusion coefficients   127
2,3-dimethyl-2,3-dinitrobutane   297
N,N-dimethylformamide (DMF)   133, 135
– solvent for synthesis of MOFs   108
dimethyl terephthalate (DMT)   174
2,4-dinitrotoluene (DNT)   297
1,4-dioxane   132
dip-coating   363
– technique   133
5-diphenylphosphanylisophthalic acid   378
dipole–induced dipole interactions   85
2015 DOE system   166
doping, of metal sites   15
drug delivery   251
drug–matrix interactions   224

dry gel method
– steam-assisted crystallization (SAC)   128
– vapor-phase transport (VPT)   128
Dubinin–Astakhov (DA) micropore filling model   162
dynamic quenching   273

e

elastic modulus
– $vs.$ hardness materials   146
electric dipole moment   86
electrochemical route, to synthesize MOFs   325
electroluminescence   270
π-electron stacking   3
electrospinning   356–358
ellipsometry   323
enantioselectivity   24, 198, 200
energy transfer   273, 274
equilibrium shift principle   145
ethidium bromide monoazide   35
ethylbenzene (EB)   88, 173, 175, 181, 183, 184
– adsorption isotherms   180, 182
– chromatograms   179
– liquid mixture separation   186
– separation curves   185
ethylene adsorption, for food storage   350, 351
ethylenediamine   29
ethylene emission, by fruit   350
EXAFS analysis   41
excitation energy transfer   274
expansion coefficients   146
extruded MOF bodies   374, 375

f

Fabry–Pérot interferometer   312
faujasite-type zeolites   177
Fe(btc) aerogel monolith   378
fibroline process   364
film attachment, for MOFs   315–318
film thickness
– measurement   327
– and morphology, for selecting MOFs   318
flexible MOFs
– adsorption isotherms of   82
– for enhanced adsorption selectivity   81–86
– possible applications of   54, 55
– terephthalates   83
– zero-order kinetics   178
flexible porous solids   56
– adsorption in   58
flue gases   107
fluorescence quenching   273

food storage 350, 351
formate route 132
Förster resonance energy transfer 274
fossil fuels 101, 105
– combustion 105
framework
– aluminosilicate framework 71
– concept of 4
– dynamics in hybrid organic–inorganic materials 53
– with high surface area 5, 6
– for mixed-ligand systems 14
– soft-type PCP 14
– theoretical methods, describing adsorption and flexibility 55, 56
– – GCMC method 56
– – ideal adsorbed solution theory 55
– of [$Zn_4O(bdc)_3$] 15
Friedel–Crafts alkylation 206

g

gadolinium carboxylate NMOFs 253–257
gas dehydration 69
gasification 102
gas impurities 69
gas purification 69
gas separation 51–52
– design of MOFs for 73, 75, 77
– performances 144
– physicochemical parameters 76
gate-opening 54, 178
– materials 55
– pressure 82
gate-type adsorption profile 82
glass transition temperature 309
global warming 102
gmelinite (GME) topology 112
grafting, of amide functional groups 196–198
grand canonical Monte Carlo (GCMC) simulation 56, 61, 160
gravimetric hydrogen adsorptions, schematic representations 157
guest encapsulation
– within NMOFs 263
– peptide–polyoxometalate (POM) spheres for 263
– temperature-controlled 288
guest–guest interactions 58
guest-induced flexibility, taxonomy 62, 63
guest-induced luminescence 287
– charge transfer 290–291
– encapsulation

– – of lanthanide ion luminophores 291–293
– – of luminophores 288–290
guest-induced swelling 53

h

H-bonds 3
^{1}H NMR spectroscopy 208
heat of absorption 105
Henry constants 112, 180
heterogeneous catalysis 28, 71
heterogeneous porphyrinic catalysts 196
1,1,1,3,3,3-hexafluoroisopropanol (HFIP) 374
hexatopic carboxylate ligands 6
high-pressure compressed gas system 167
high-temperature carbonation–decarbonation cycles 106
Hofmann clathrate films 328
homochiral MOFs 198
– chiral organic catalytic functions 199, 200
– CMIL-1, catalytically active 29
– with intrinsic chirality 198
– metalloligands 200, 201
host–guest interactions 58, 89
Huisgen cycloaddition 34, 36, 37
hybrid organic–inorganic frameworks 54
hydrocarbon separation 88
hydrodesulfurization reaction 91
hydrogen
– adsorption enthalpy, comparison 165
– adsorption isotherms 163
– bonds 85
– fuel cell electric vehicles (FCEVs) 151
– Gemini, modified PSA process 102
– pressure vessels 152
– production of H_2 from syngas 101–103
– steam reforming plant 101
– storage capacity 160
– storage technologies 158
– – challenges 151, 152
– two-bed PSA system 102
hydrogen sulfide 69
hydrophilic polymers 255, 263
hydrophobic adsorbents 107
hysteresis 85

i

ideal adsorbed solution theory (IAST) 55, 56, 64
imine condensation 33
– hydrophobization of SIM-1 with C_{12}-alkylamine 35
immobilization of MOFs

– for filtration applications 353
– MOF@fiber composite materials 354
– – electrospinning of MOF@polymer composite fibers 356–359
– – MOF-containing paper sheets 354, 355
– – MOF@pulp fibers 355, 356
– MOF fixation in textile structures 359, 360
– – dry particle insertion 363–367
– – pretreatment 360, 361
– – wet particle insertion 362, 363
industrial MOFs synthesis 339
– electrochemical synthesis 345, 346
– hydrothermal approach 344, 345
– laboratory-scale synthesis 343
– linker molecules 340–343
– metal sources 340
– shaping process 347–349
inorganic adsorbents, for gas separation 74
inorganic hydroxides 69
inorganic membranes, classification 121
inorganic porous materials 71
interdigitated framework $Cu_2(dhbc)_2(bpy)$, 3D crystal structure 83
intrinsic biodegradability 252
iodinated NMOFs 260–262
ion sensing 295
IRMOF-3 catalyst 197
iron carboxylate NMOFs 258–260
IR spectroscopy 193
isoalkanes 73
isophthalic acid (IPA) 174

k

kinetic energy 365
kinetic selectivity 73
kinetic trapping mechanism 165
Knoevenagel condensation 29, 196, 197
Knudsen diffusion 125
Knudsen separation factor 141
kraft pulping 355

l

Langmuir–Blodgett assembly of preformed MOF 2D single layers 331
Langmuir model 110, 157
lanthanide-based nanocages 203
lanthanide luminescence 287
– and antenna effect 282
– as probe of metal-ligand coordination sphere 287
lanthanide nucleotide NMOFs 262, 263
– formation 262
Le Chatelier principle 143
Leonard-Jones potential 125

Lewis acid 192, 207
Lewis acidic frameworks 6–8
ligand-based luminescence 274
– ligand-to-metal charge transfer in MOFs 280, 281
– metal-to-ligand charge transfer in MOFs 281
– in MOFs 274–281
– solid-state luminescence of organic molecules 274, 275
ligand-base solid solution 14
ligand ratio 14
lignin 355, 356
liquefied natural gas (LNG) 103
– production 103
liquid-phase epitaxy (LPE) growth 328
liquid/solid interface 325
luminescence 27, 254, 269, 270, 274, 278, 282, 287, 291
– quenching 296, 312
luminescent MOFs, applications 293
– barcode labeling 300
– chemical sensors 293–298
– – detection of explosives 297, 298
– – ion sensors 294–296
– – oxygen sensors 296, 297
– – small-molecule 294–296
– nonlinear optics 300
– radiation detection 298
– solid-state lighting 298–300

m

magnetic behavior 15
magnetic resonance imaging (MRI) 235, 252, 262
magnetic susceptibility measurements 326
manganese carboxylate NMOFs 257, 258
manganese ions 252
mass transport 125
$MCM-41-NH_2$ catalyst 198
mechanical dewatering methods 363
mechanical properties, for selecting MOFs for sensing applications 320
12-membered ring (MR) straight channels 134
membrane permeation processes 69
membrane separation and distillation (MDI) 122
membrane separation processes 121, 124
– ideal separation factor 124
mercaptans 104, 105, 113
mesoporous silicate (MS) materials 23
mesoscopic porosity 324
metal-based luminescence 282

– lanthanide luminescence 282
– metal luminophores 282
metal carboxylate-based membranes 132, 137, 138
metal-centered luminescence 282–286
metal ions 3, 191
metal–ligand bonds 155, 159
metal–ligand interactions 28
metalloligands 200, 201
metalloporphyrins 25, 203, 204
metal nanoparticles 204–206
metal-organic frameworks (MOFs) 3, 99, 108–115
– as adsorptive hydrogen storage options 154–156
– advantages 154
– applications 113
– for bioapplications 216–221
– building by post-synthetic modification 23–26
– challenges 116
– characterization 157
– classification 82
– containing organometallics 39
– with coordination unsaturated metal centers 86–88
– design 147
– diagnostics, applications
– – magnetic resonance imaging 235, 236
– – optical imaging 236, 237
– dynamic materials in 52–54
– features, and physicochemical properties 41
– films for sensing applications 311
– functionalization 113
– – rate, and material efficacy 43, 44
– functionalized materials, characterization of 44, 45
– gas-phase loading with perylene derivative 27
– for H_2-PSA 109–113
– large synthesis and stability 108, 109
– membranes, application 137
– – gas separation 137–141, 144
– – limitations 143, 145, 146
– – shaped structured reactors 141, 142
– mesoporous 206–208
– MOF-177 5, 28, 109, 165, 206, 319, 339, 341
– MOF-5 thin films, SEM images 131
– molecule interactions 310
– monolithic, gels 375–378
– opportunities for separation processes 73
– Pd-containing MOF 41
– piano stool arene–chromium complex 31
– post-functionalization

– – by covalent bonds 31–39
– – by host–guest interactions 26–28
– post-modification, of IRMOF-3 with salicylaldehyde 39
– properties 99
– stability 108
– supported chiral titanium–BINOL complex 30
– supported Cu/Fe catalysts, post-synthetic modification 40
– symmetric anhydrides, differing from IRMOF-3 family 33
– synthesis of UMCM-1-supported Pd^{II} complexes 42
– synthetic restrictions 43
– tubular support 122
metal–organic polymers 339
metal-to-metal charge transfer (MMCT) 286, 287
methanol 135
methylaromatics 89
methyl benzoylformate 24
methylcellulose 375
5-methylisophthalate ligands 193
methyl siloxane 375
microcantilevers (MCLs) 311
microelectromechanical systems (MEMS) 311
microporous adsorbents 167
microporous metal–organic frameworks (MMOFs) films 136
microwave-assisted solvothermal synthesis 134
midazolate-2-carboxaldehyde 33
MIL-53 179
– pore structure 182
MIL-101 6, 7, 28, 29, 112, 124, 201, 202, 222, 240, 323
– encapsulating nanoparticles 206
– post-synthetic modification 199
MIL-88B crystals, growth 131
MIL-47 pore system 181
– X-ray powder diffraction patterns 181
mixed ligands 13–16
– structure of $[Zn_4O(bdc)_x(abdc)_{3-x}]$ 14, 15
mixed matrix membranes (MMMs) 143
mixed metals 13–16
molecular metal–organic hybrid 4
molecular sieve membranes synthesis 127–136
– MOF membranes and films, preparation 129–136
– zeolite membranes synthesis 128, 129
molecular sieves 69, 73

– MOF materials as   79–81
molecular simulation methods   58–60
– in flexible porous solids   58
monoethanolamine (MEA) process   105
– drawbacks   106
– in post-combustion CO_2 capture   105
Monte Carlo simulation   57, 59
multifunctional frameworks   10
– preparation of   13

n

nanomaterials   251–253
nanoparticles   256
– encapsulation   27, 28
– MOF-supported Pd nanoparticles   28
– Ru nanoparticles   206
– synthesis   237, 238
nanoscale metal–organic frameworks (NOMEs)   252
2,6-naphthalenedicarboxylate   35
natural gas   103
– CO_2–CH_4 separation, requirements for adsorbents   104, 105
– CO_2 removal from   103
– MOFs for CO_2 removal from   113
– storage for automobile applications   349, 350
negative thermal expansion   52, 53
nephrogenic systemic fibrosis (NSF)   252
nicotinic acid   228
nitric oxide   228
– adsorption/desorption isotherm   230
– insufficient/excess in human body, diseases related   229
– interacting with available CUSs   230
– release profiles for Ni and Co CPO-27   230, 231
– storage   230
nitric oxide synthase (NOS)   228
4,4′,4″-nitrilotris(benzene-4,1-diyl)tris(ethyne-2,1-diyl)triisophthalate (ntei)   6
nitrogen adsorption isotherms   370, 371
N–O ligand   39
nucleation–growth mechanisms   143

o

on-board cryo-adsorptive hydrogen storage. See also storage systems, for hydrogen fuel
– hydrogen storage options   152
– material research results   159–165
– metal–organic frameworks, research status   151
– research problem, and significance   151–154

open metal sites (OMSs)
– coupling, and polarity gradients in   86
– high interaction with adsorbate   86
– useful for enhancing efficiency in   86
optical imaging   252
organic amines   69
organic functional groups   73
organic ligands   177, 195
organic lightemitting diodes (OLEDs)   298
organic linkers   156
organic polymer films   309
organic reactions catalysts
– organic framework linkers, catalytic functionalization   195–198
organic reactions catalysts, MOF   191
– with catalytically active metal nodes   191–195
– encapsulated catalytically active guests   201–206
– homochiral   198–201
– mesoporous   206–208
osmotic ensemble   56
– isothermal–isobaric ensemble   57
– pseudoensembles   57
– restricted, use of   60–62
– semi-grand ensemble   57
– uses in molecular simulation   57, 58
osmotic framework adsorbed solution theory   63, 64
oxides   13, 30, 69, 131, 216, 252, 318, 326, 340, 343

p

padding
– apparatus   361
– principle   361
palladium atom   41
palladium-loaded catalysts
– catalytic activities   207
paper sheets   354, 355
particle spraying   366
– energy loss   367
particle-spraying   365
Pd-based catalyst   28
Pd-supported MOF-5   28
permeation. See also air purification; immobilization of MOFs
– in H_2–n-C_4H_{10} binary mixture   126
– mixed gas permeation   125
– profiles   126
– realistic permeation test with vapors   372
– vs. pressure and temperature   126
– Wicke–Kallenbach permeation cell with 1:1 mixture of H_2 and N_2 135

phenylacetylene 37
photoluminescence 270–274
piezoluminescence 270
polydimethylsiloxane (PDMS) membranes 143
poly(ethylene terephthalate) (PET) 88, 374
polyethylenimine (PEI) solution 133
polyimide 104
polymer components 365
polymeric binder 374
polymeric organo-gels 376
polymer–MOF composite 374
polyoxometalates 17, 201–203
polyoxotungstates immobilization 202
polystyrene 88
polyvinylpyrrolidone (PVP) 254
pore dimensions, for selecting MOFs for sensing applications 312
pore size 92, 365. See also permeation
– Knudsen diffusion 125, 126
– mass transport and separation mechanism as function of 125
– surface diffusion mechanism 125
– viscous flow and 125
porosity 4–5
– and conductivity/dielectricity 12
– and magnetism 10–12
porous adsorbents characterization, methods
– adsorption isotherms, measurements 77
– Brunauer–Emmett–Teller (BET) equation 78
– Clausius–Clapeyron equation 78
– Dubinin–Radushkevich method 78
– establishment of permanent porosity 77
– gas chromatographic methods 78
– monocomponent isotherms 78
– separation efficiency 78
– variable-temperature spectroscopic methods 78
porous coordination polymers (PCPs) 3
– [Al(bdc)(OH)] (MIL-53) 15
– control of structural flexibility 14
– core–shell system of 16, 17
– [Cu_3(btc)$_2$] (HKUST-1) 15, 17
– design of frameworks, perspectives 18, 19
– and nanoparticles 17, 18
porous flexibility. See also flexible porous solids
– and catalysis 12, 13
porous MOFs, in controlled delivery of drugs 222
porphyrin functional groups 195, 196
post-combustion CO_2 capture
– MEA process, drawbacks 106
– MOFs for 113, 115

– PSA and VSA processes in 106, 107
POST-1 synthesis, ligand employed in 199
pressure swing adsorption (PSA) systems 99–101, 103, 121, 123. See also adsorption
– basic steps 100
– characteristics 104
– and membrane technology 123
– pure component isotherms and working capacities 100
– steps 100
– two-bed system 102
proton exchange membrane fuel cells (PEMFCs) 121
Prussian Blue compounds 4
pure component isotherms, representation 100
purified terephthalic acid (PTA) 174
2-pyridinecarboxaldehyde 33, 41
pyridine ligands 29

q

quadrupole moment, of CO_2 85
quartz crystal microbalances (QCMs) 311
quenching 237, 259, 263, 273, 288, 291, 294, 295. See also dynamic quenching; fluorescence quenching; luminescence quenching

r

radioluminescence 270
raffinate 100
Raman spectroscopy 326
Rapid Köthen sheet mold 354
real gas law 157
reference catalyst 205
resins 174
response time, of sensor 319
rho-zeolite-like metal–organic framework (*rho*-ZMOF) 203
Rietveld refinement 183
Ru nanoparticles 206

s

scanning electron microscopy (SEM) 254, 355, 358
– Al_2O_3 support 123
– crystals grow on surface of bead 142
– iodinated NMOFs 261
– lanthanide nucleotide NMOFs 262
– MMOF membrane 136
– of MOF-5 thin films 130
– sheet 355
– SIM-1 membrane 135

– ZIF-7 membrane   134
– ZIF-69 membrane   134
secondary building units (SBUs)   129, 154
seeding processes   129
seeding-supported crystallization   127
self-assembled organic monolayers (SAMs)   130
– carboxylate-terminated areas   131
– deposition of HKUST-1 silane-based SAMs on   131
– deposition of MOF-5 crystals   130
– heterogeneous nucleation   130
– organosilanes   130
– oriented growth of MIL-88B   131
separation factor   124
separation mechanism   125
separation processes   100
– role in   123
shaping, for MOFs   239
– approaches   240, 241
– chemiluminescence measurements, NO release   241
– homogeneous powders, pellets, or tablets   240
– NO-loaded hydrocolloid   241, 242
– thin films   240
Sieverts' method, to determines amount of hydrogen uptake   157
silicone resins   375
silylation, of OH bridging groups in MIL-53(Al)   39
SIM-1 membrane   135, 142
simulated moving bed (SMB) technology   174
single-crystal X-ray crystallography   4
sodalite (SOD) topology   112
soft porous coordination polymers   14
soft porous crystals   8–10, 54
sol–gel chemistry   375, 376
sol–gel technique   128
solid oxide fuel cells (SOFCs)   121
solvent-free aza-Michael condensation   198
solvent-resistant nanofiltration (SRNF)   143
solvothermal synthesis   130
– metal carboxylate-based membranes   131, 132
– microwave-assisted   134, 238
– shaping   239
sorbate–solid interactions   51
spin-crossover phenomenon   326
π–π stacking interactions   207
steam-assisted crystallization   128

steam reforming furnace   101
steric constraints   181
steric effect   73
steric restrictions   183
Stern–Volmer equation   273
storage systems, for hydrogen fuel   101, 151, 152
– adsorbed hydrogen phase, nature of   162–165
– development   101
– from laboratory-scale materials to engineering   165–167
– MOFs as adsorptive hydrogen storage options   154–156
– options for on-board applications   152
– requirements   153
– structure–hydrogen storage properties correlations   159–162
– thermodynamic assessment of porous MOFs for   156–158
structure-directing agent (SDA)
– for zeolites   127
styrene   88
– hydrogenation   204
substrate surfaces, chemical modifications   136, 137
superparamagnetic iron oxides (SPIOs)   252
supramolecular structures   339
surface acoustic wave (SAW) sensors   311
surface diffusion mechanism   125
surface engineering   239
surface plasmon resonance (SPR)   327
syneresis   376

t

terephthalic acid   329
5-tertbutylisophthalic acid   378
tetrachloroethene   61
tetraethyl orthosilicate (TEOS)   254
2,3,5,6-tetraiodo-1,4-benzenedicarboxylic acid   260
thermal drying processes   363
thermal expansion coefficients   53
thermal stability   6, 15, 73, 178, 316
thermogravimetric analysis (TGA)   32
thin film growth   320
– from aged solvothermal mother solutions   320–323
– assembly   323–325
– in confined spaces   329–331
– different methods, comparison of   331
– electrochemical deposition   325
– heteroepitaxial growth   328, 329

– liquid-phase epitaxy 325–328
thin films metal–organic framework (MOF) manufacture
– application 137–143
– limitations 143–146
– mass transport and separation mechanism 123–127
– membrane technologies, advantages and limitations 121–123
– molecular sieve membranes synthesis 127–136
toluene
– alkylation 195
– in condensation of an aldehyde to 33
– conversion 195
– disproportionation 174
transesterification reaction, catalytic properties 199
transition metals 6, 73
– nodes 192–194
transmission electron microscopy (TEM) 256, 258, 324
triboluminescence 270
3-[(trimethylsilyl)ethynyl]-4-[2-(4-pyridinyl)ethenyl]pyridine 35
2,4,6-trinitrotoluene (TNT) 297
1,3,5-tris(3-ethynylbenzonitrile) benzene ligand 26, 27

u

UMCM-1-supported Fe^{III} and Cu^{II} complexes 40, 41
UMCM-1-supported Pd^{II} complexes 42

v

vacuum swing adsorption (VSA) 100
van der Waals interactions 3, 153, 178, 310
vapor-phase transport 128
vapor pressure 54, 85
vapor test set-up 372
Vegard's law 13
vehicular hydrogen storage system 152
volatile organic compounds (VOCs) 88
– capture 89, 90
Volkswagen Caddy EcoFuel 349

w

waste gas 101
water vapor isotherm of Zn(bdc)(dabco)$_{0.5}$ 371

Wicke–Kallenbach permeation cell 135
Wicke–Kallenbach technique 141

x

xerogels 376
X-ray crystal structure, of Mn (HCOO)$_2$ 81
X-ray powder diffraction (XRPD) 32, 109, 133, 253, 270
– of 30 cycle [Zn(BME-bdc)(dabco)$_{0.5}$] deposited on 329
– one-dimensional structures MOFs 260
– of thin films of HKUST- 1 322
xylene isomers separation 88, 173–176
– limitations 176
– MOFs properties *vs.* zeolites 176–178
– *o*-, *m*- and *p*-xylene 174
– using MIL-47 and MIL-53 178–185
– – low-coverage gas-phase adsorption properties 179, 180
– – molecular packing 180–184
– – xylene-mixtures, separation of 184–185

z

zeolite adsorbents 104
zeolite films preparation 129
zeolite imidazolate frameworks (ZIFs) 133
zeolite membrane
– direct nucleation–growth 128
– synthesis 127, 128
– – limitations 123
– – pore plugging 129
– – secondary growth 129
– – *in situ* synthesis 129
– variety 128
zeolites 23, 24, 52, 73, 221, 355
– expansion coefficients 145
zeolitic adsorbents 180
zeolitic imidazolate frameworks (ZIFs) 155
zero-order kinetics 178
ZIF-8 films, of various thicknesses 323
ZIF-69 framework 134
ZIF-7 membrane
– cross-section SEM images of 134
ZIF-8 [Zn (2-methylimidazole)$_2$] 89
zinc imidazolate-based membranes 133–135, 138–141
– gases, permeances 138
zirconia membranes 121
ZMS-5 membrane 127
Zn-loaded porphyrin ligands 196
Zn$_4$O clusters 6
ZSM-5 type membranes 123